MACHINE LEARNING
IN BIOINFORMATICS

Wiley Series on
Bioinformatics: Computational Techniques and Engineering

Bioinformatics and computational biology involve the comprehensive application of mathematics, statistics, science, and computer science to the understanding of living systems. Research and development in these areas require cooperation among specialists from the fields of biology, computer science, mathematics, statistics, physics, and related sciences. The objective of this book series is to provide timely treatments of the different aspects of bioinformatics spanning theory, new and established techniques, technologies and tools, and application domains. This series emphasizes algorithmic, mathematical, statistical, and computational methods that are central in bioinformatics and computational biology.

Series Editors: **Professor Yi Pan** and **Professor Albert Y. Zomaya**
pan@cs.gsu.edu   zomaya@it.usyd.edu.au

Knowledge Discovery in Bioinformatics: Techniques, Methods, and Applications
Xiaohua Hu and Yi Pan

Grid Computing for Bioinformatics and Computational Biology
Edited by El-Ghazali Talbi and Albert Y. Zomaya

Bioinformatics Algorithms: Techniques and Applications
Ion Mandiou and Alexander Zelikovsky

Analysis of Biological Networks
Edited by Björn H. Junker and Falk Schreiber

Machine Learning in Bioinformatics
Edited by Yan-Qing Zhang and Jagath C. Rajapakse

MACHINE LEARNING IN BIOINFORMATICS

Edited by

Yan-Qing Zhang

Jagath C. Rajapakse

A JOHN WILEY & SONS, INC., PUBLICATION

Library of Congress Cataloging-in-Publication Data:
Zhang, Yan-Qing.
    Machine learning in bioinformatics / edited by Yan-Qing Zhang, Jagath C. Rajapakse.
        p. cm.
    Includes bibliographical references.
    ISBN 978-0-470-11662-3 (cloth)
1.   Bioinformatics.   2.   Machine learning.   I. Rajapakse, Jagath Chandana. II. Title.
    QH324.2.Z46 2008
    572.80285'61–dc22
                                                                        2008017095

10 9 8 7 6 5 4 3 2 1

CONTENTS

Foreword   ix

Preface   xi

Contributors   xvii

1   **Feature Selection for Genomic and Proteomic Data Mining**   1
   Sun-Yuan Kung and Man-Wai Mak

2   **Comparing and Visualizing Gene Selection
   and Classification Methods for Microarray Data**   47
   Rajiv S. Menjoge and Roy E. Welsch

3   **Adaptive Kernel Classifiers Via Matrix Decomposition
   Updating for Biological Data Analysis**   69
   Hyunsoo Kim and Haesun Park

4   **Bootstrapping Consistency Method for Optimal Gene
   Selection from Microarray Gene Expression Data for
   Classification Problems**   89
   Shaoning Pang, Ilkka Havukkala, Yingjie Hu, and Nikola Kasabov

5   **Fuzzy Gene Mining: A Fuzzy-Based Framework for Cancer
   Microarray Data Analysis**   111
   Zhenyu Wang and Vasile Palade

6   **Feature Selection for Ensemble Learning and Its Application**   135
   Guo-Zheng Li and Jack Y. Yang

7 **Sequence-Based Prediction of Residue-Level Properties in Proteins**   157
   Shandar Ahmad, Yemlembam Hemjit Singh, Marcos J. Araúzo-Bravo, and Akinori Sarai

8 **Consensus Approaches to Protein Structure Prediction**   189
   Dongbo Bu, ShuaiCheng Li, Xin Gao, Libo Yu, Jinbo Xu, and Ming Li

9 **Kernel Methods in Protein Structure Prediction**   209
   Jayavardhana Gubbi, Alistair Shilton, and Marimuthu Palaniswami

10 **Evolutionary Granular Kernel Trees for Protein Subcellular Location Prediction**   229
   Bo Jin and Yan-Qing Zhang

11 **Probabilistic Models for Long-Range Features in Biosequences**   241
   Li Liao

12 **Neighborhood Profile Search for Motif Refinement**   263
   Chandan K. Reddy, Yao-Chung Weng, and Hsiao-Dong Chiang

13 **Markov/Neural Model for Eukaryotic Promoter Recognition**   283
   Jagath C. Rajapakse and Sy Loi Ho

14 **Eukaryotic Promoter Detection Based on Word and Sequence Feature Selection and Combination**   301
   Xudong Xie, Shuanhu Wu, and Hong Yan

15 **Feature Characterization and Testing of Bidirectional Promoters in the Human Genome—Significance and Applications in Human Genome Research**   321
   Mary Q. Yang, David C. King, and Laura L. Elnitski

16 **Supervised Learning Methods for MicroRNA Studies**   339
   Byoung-Tak Zhang and Jin-Wu Nam

17 **Machine Learning for Computational Haplotype Analysis**   367
   Phil H. Lee and Hagit Shatkay

18 Machine Learning Applications in SNP–Disease Association Study **389**

Pritam Chanda, Aidong Zhang, and Murali Ramanathan

19 Nanopore Cheminformatics-Based Studies of Individual Molecular Interactions **413**

Stephen Winters-Hilt

20 An Information Fusion Framework for Biomedical Informatics **431**

Srivatsava R. Ganta, Anand Narasimhamurthy, Jyotsna Kasturi, and Raj Acharya

Index **453**

FOREWORD

Machine learning is a subfield of artificial intelligence and is concerned with the development of algorithms and techniques that allow computers to learn. It has a wide spectrum of applications such as natural language processing, search engines, medical diagnosis, bioinformatics and cheminformatics, stock market analysis, computer vision, and game playing. Recently, the amount of biological data requiring analysis has exploded and many machine learning methods have been developed to deal with this explosion of data. Hence, machine learning in bioinformatics has become an important research area for both computer scientists and biologists.

The aim of this book is to provide applications of machine learning to problems in the biological sciences, with particular emphasis on problems in bioinformatics. The book consists of a number of stand-alone chapters that explain and apply machine learning methods to central bioinformatics problems such as feature selection, sequence-based prediction of residue-level properties, promoter recognition, protein structure prediction, gene selection, and SNPS selection, classification, and data mining. This book represents the unification of two important fields in sciences— biology and computer science—with machine learning as a common theme. The chapters are written by well-known researchers in these interdisciplinary areas, and applications of various machine learning methods to different bioinformatics problems are presented. Students and scientists in biology and computer science will find this book valuable and accessible.

Several books in the similar areas have been published. However, this book is unique in that it presents cutting-edge research topics and methodologies in the area of machine learning methods when applied to bioinformatics. Many results presented in this book have never been published in the literature and represent the most advanced technologies in this exciting area. It also provides a comprehensive and balanced blend of topics, implementations, and case studies. I firmly believe that this book will further facilitate collaboration between machine learning researchers and bioinformaticians.

Both editors, Dr. Yan-Qing Zhang and Dr. Jagath C. Rajapakse, are rising stars in the areas of machine learning and bioinformatics. They have achieved a lot of research results in these areas. Their vision of creating such a book in a timely manner deserves our loud applause. This book is ideally suited both as a reference and as a text for a graduate course on machine learning or bioinformatics. This book can also serve as a repository of significant reference materials because the references cited in each chapter serve as useful sources for further study in this area.

I highly recommend this timely and valuable book. I believe that it will benefit many readers and contribute to the further development of machine learning in bioinformatics.

Atlanta, Georgia
August 2008

DR. YI PAN
Chair and Professor
Georgia State University

PREFACE

In recent decades, machine learning techniques have been widely applied to bioinformatics. Many positive results have indicated that machine learning methods are useful for solving complex biomedical problems too difficult to solve by experts. Traditionally, researchers do biomedical research by using their knowledge and intelligence, performing experiments by hands and eyes, and processing data by basic statistical and mathematical tools. Due to huge amounts of biological data and a very large number of possible combinations and permutations of various biological sequences, the conventional human intelligence-based methods cannot work effectively and efficiently. So artificial intelligence techniques such as machine learning can play a critical role in complex biomedical applications.

Experts from different domains have contributed chapters to this book, which feature novel machine learning methods and their applications in bioinformatics. Relevant machine learning methods include support vector machines, kernel machines, feature selection, neural networks, evolutionary computation, statistical learning, fuzzy logic, supervised learning, clustering, ensemble learning, Bayesian networks, linear regression, principal components analysis, hidden Markov models, entropy-based information methods, and many others. The 20 chapters of the book are organized in a convenient order, based on their contents, so as to enable the readers to easily gather information in a progressive manner. A concise summary of each chapter follows.

In Chapter 1, Kung and Mak present feature selection methods such as the support vector machine recursive feature elimination (SVM-RFE), filter methods, and wrapper methods, in application to microarray data. Filter methods are based on input and output correlation statistics between input and predictions, or signal-to-noise (SNR) statistics, independent of the classifier or predictor. The development of microarray technology has brought with it problems that are interesting, both from statistical and biological perspectives. One important problem is to identify important genes that are relevant to distinguish cancerous samples from benign samples, or different cancer types. In the SVM-RFE, the magnitude of the weight connected to a particular feature is used as the ranking criteria for selection. The methods are illustrated in selection of important genes and in prediction of protein subcellular localization. In the protein subcellular localization, whether a protein lies in the cytoplasm, nuclear, extracellular, mitochondrial, or nuclear location is predicted from its amino acid sequence.

In Chapter 2, Menjoge and Welsch give a new feature selection method using 1-norm SVM and 2-norm SVM techniques, where the weights are used as regularization terms of 1-norm and 2-norm forms. Results show that these methods perform well as compared with other methods. The elastic net, in particular, demonstrates excellent classification accuracy. However, none of the methods dominate the other methods in both selecting a small number of variables and classifying data sets.

In Chapter 3, Kim and Park discuss adaptive supervised machine learning algorithms since the adaptive classifiers avoid expensive recomputation of the solution from scratch. Both an adaptive KDA/RMSE (aKDA/RMSE) based on updating the QR decomposition and an adaptive KDA/MSE based on updating the UTV decomposition KDA/MSE-UTV is proposed. These new kernel classifiers can be applied to compute leave-one-out cross-validation efficiently for bioinformatics applications.

In Chapter 4, Pang, Havukkala, Hu, and Kasabov propose a new gene selection method with better bootstrapping consistency for reliable microarray data analysis. The method ensures the reliability and generalizability of microarray data analysis, which thereby leads to an improvement of disease classification performance. Compared with the traditional gene selection methods without using consistency measurement, bootstrapping consistency method provides more accurate classification results. More importantly, results demonstrate that gene selection with the consistency measurement is able to enhance the reproducibility and consistency in microarray data analysis and proteomics-based diagnostics systems.

In Chapter 5, Wang and Palade introduce a series of fuzzy-based techniques, including the fuzzy gene selection method, the fuzzy C-mean clustering-based enhanced gene selection method, and the neuro-fuzzy ensemble approach for building a microarray cancer classification system. Three benchmark microarray cancer data sets, namely, the leukemia cancer data set, colon cancer data set, and lymphoma cancer data set, are used for simulations. The experimental results show that fuzzy-based systems can be efficient tools for microarray data analysis.

In Chapter 6, Li and Yang provide an ensemble learning method with feature selection to improve generalization performance of single classifiers from three aspects. Experiments on benchmark data show that genetic algorithm-based multitask learning (GA-MTL) is more effective than the earlier heuristic algorithms. The algorithms are demonstrated on a brain glioma data set to show the use of the algorithm as an alternative tool for bioinformatics applications.

In Chapter 7, Ahmad, Singh, Araúzo-Bravo, and Sarai study machine learning methods such as neural networks and support vector machines to predict one-dimensional features of protein structures, such as secondary structure, solvent accessibility, and coordination number, and more recently one-dimensional functional properties such as binding sites. The prediction techniques have been shown to have good performance even in the absence of known homology to other proteins. The computational similarities of the methods are highlighted. Common standards for making such sequence-based predictions are also developed.

In Chapter 8, Bu, Li, Gao, Yu, Xu, and Li give a new protein structure prediction method. Despite significant progresses made recently, every protein structure prediction method still possesses limitations. To overcome such shortcomings, a natural idea is integrating the strengths of different methods to obtain more accurate structures by boosting some weaker predictors into a stronger one. As suggested by recent CASP competitions, the consensus-based prediction strategies usually outperform others by generating better results.

In Chapter 9, Gubbi, Shilton, and Palaniswami investigate different kernel machines in relation to protein structure prediction. Amino acids arrange themselves in 3D space in stable thermodynamic conformations, referred to as native conformation, and the protein becomes active in this state. Thermodynamic interactions include formation of hydrogen bonding, hydrophobic interactions, electrostatic interactions, and complex formation between metal ions. Protein molecules are quite complex in nature and often made up of repetitive subunits.

In Chapter 10, Jin and Zhang give a new method to predict protein subcellular locations based on SVM with evolutionary granular kernel trees (EGKT) and the one-versus-one voting approach. The new method can effectively incorporate amino acid composition information and combine binary SVM models for protein subcellular location prediction.

In Chapter 11, Liao discusses three applications, where the long-range correlations are believed to be essential, by using specific classification and prediction schemes: hidden Markov models for transmembrane protein topology, stochastic context-free grammars for RNA folding, and global structural profiling for antisense oligonucleotide efficacy. By first examining the limitations of present models, some expansions to capture and incorporate long-range features from the aspects of model architecture, learning algorithms, hybrid models, and model equivalence are made. The performance has been improved consequently.

In Chapter 12, Reddy, Weng, and Chiang give a novel optimization framework that searches the neighborhood regions of the initial alignment in a systematic manner to explore the multiple local optimal solutions. This effective search is achieved by transforming the original optimization problem into its corresponding dynamical system and estimating the practical stability boundary of the local maximum. Results show that the popularly used EM algorithm often converges to suboptimal solutions, which can be significantly improved by the proposed neighborhood profile search.

In Chapter 13, Rajapakse and Ho give a novel approach to encode inputs to neural networks for the recognition of transcription start sites in RNA polymerase II promoter regions. The Markovian parameters are used as inputs to three neural networks, which learn potential distant relationships between the nucleotides at promoter regions. Such an approach allows for incorporating biological contextual information at the promoter sites into neural networks and in general implementing higher-order Markov models of the promoters. Experiments on a human promoter data set show an increased correlation coefficient rate of 0.69 on average, which is better than the earlier reported by the NNPP 2.1 method.

In Chapter 14, Xie, Wu, and Yan propose three eukaryotic promoter prediction algorithms, PromoterExplorer I, II, and III. PromoterExplorer I is developed based on relative entropy and information content. PromoterExplorer II takes different kinds of features as the input and adopts a cascade AdaBoost-based learning procedure to select features and perform classification. The outputs of these two methods are combined to build a more reliable system, PromoterExplorer III. Consistent and promising results have been obtained, indicating the robustness of the method. The new promoter prediction technique compares favorably with the existing ones, including Promoter-Inspector, Dragon Promoter Finder (DPF), and First Exon Finder (FirstEF).

In Chapter 15, Yang, King, and Elnitski introduce a bidirectional promoter—a region along a strand of DNA that regulates the expression of genes that flank the region on either side. An algorithm is developed for the purpose of finding uncharacterized bidirectional promoters. Results of the analysis have identified thousands of new candidate head-to-head gene pairs, corroborated the 5′ ends of many known human genes, revealed new 5′ exons of previously characterized genes, and in some cases identified novel genes. More effective machine learning approaches to classifying these features will be useful for future computational analyses of promoter sequences.

In Chapter 16, Zhang and Nam review computational methods used for miRNA research with a special emphasis on machine learning algorithms. In particular, detailed descriptions of the case studies based on the kernel methods (support vector machines), probabilistic graphical models (Bayesian networks and hidden Markov models), and evolutionary algorithms (genetic programming) are given. The effectiveness of these methods was validated by various approaches including wet experiments and their contributions were successful in the domain of miRNA. A well-defined generative model, such as Bayesian networks or hidden Markov models, constructed from a known data set in the prediction of miRNAs, can be used for the rational design of artificial pre-/shRNAs.

In Chapter 17, Lee and Shatkay present several works on tag SNP selection and mapping disease locus based on association study using SNPs. Tag SNP selection uses redundancy in the genotype/haplotype data to select the most informative SNPs that predict the remaining markers as accurately as possible. In general, machine learning methods tend to do better than purely combinatorial methods and also are applicable to bigger data sets with hundreds of SNPs. Identifying SNPs in disease association study is more difficult, largely depends on the population under study, and often faces the problem of replication.

In Chapter 18, Chanda, Zhang, and Ramanathan elaborate the application of some well-known machine learning techniques such as support vector machines, neural networks, linear regression, principal components analysis, hidden Markov models, and entropy-based information theoretic methods to locate genetic factors for complex diseases such as cystic fibrosis and multiple myeloma. They focus on two aspects, namely, tag SNP selection or selectively choosing some SNPs from a given set of possibly thousands of markers as representatives of the remaining markers (that are not chosen) and machine learning models for detecting markers that have potential high association with given disease phenotypes.

In Chapter 19, Winters-Hilt presents a new channel current-based nanopore cheminformatics to provide an incredibly versatile method for transducing single molecule events into discernable channel current blockade levels. The DNA–DNA, DNA–protein, and protein–protein binding experiments that were described were novel in that they made critical use of indirect sensing, where one of the molecules in the binding experiment is either a natural channel blockade modulator or is attached to a blockade modulator.

In Chapter 20, Ganta, Narasimhamurthy, Kasturi, and Acharya propose an information fusion model-based analytical and exploratory framework for biomedical informatics. The framework presents a suite of tools and a workflow-based approach to analyze and explore multiple biomedical information sources through information fusion. The goal is to discover hidden trends and patterns that could lead to better disease diagnosis, prognosis, treatment, and drug discovery. However, there is a limit to the extent of knowledge that can be extracted from individual data sets. Recent focus on techniques analyzing genomic data sources in an integrated manner through information fusion could alleviate problems with individual techniques or data sets.

We sincerely thank all the authors for their important contributions and timely cooperation for publication of this book. We also thank Jung-Hsien Chiang, Arpad Kelemen, Rui Kuang, Ying Liu, Xinghua Lu, Lakshmi K. Matukumalli, Tuan D. Pham, and Changhui C. Yan for their valuable comments. We thank editors Paul Petralia and Anastasia Wasko from Wiley and Sanchari Sil of Thomson Digital for their guidance and help. We would like to thank Nguyen N. Minh for formatting the book. Finally, we would like to thank Dr. Yi Pan for his constant guidance.

Atlanta, Georgia   YAN-QING ZHANG
Nanyang, Singapore   JAGATH C. RAJAPAKSE
August 2008

CONTRIBUTORS

Raj Acharya, Pennsylvania State University, University Park, Pennsylvania.

Shandar Ahmad, Kyushu Institute of Technology, Kyushu, Japan, and Jamia Millia Islamia, New Delhi, India.

Marcos J. Araúzo-Bravo, Kyushu Institute of Technology, Kyushu, Japan.

Dongbo Bu, University of Waterloo, Waterloo, Ontario, Canada, and Institute of Computing Technology, China

Pritam Chanda, The State University of New York, Buffalo, New York.

Hsiao-Dong Chiang, Cornell University, Ithaca, New York.

Laura L. Elnitski, National Institutes of Health, Bethesda, Maryland.

Xin Gao, University of Waterloo, Waterloo, Ontario, Canada.

Srivatsava R. Ganta, Pennsylvania State University, University Park, Pennsylvania.

Jayavardhana Gubbi, The University of Melbourne, Melbourne, Australia.

Ilkka Havukkala, Auckland University of Technology, Auckland, New Zealand.

Sy Loi Ho, Nanyang Technological University, Nanyang, Singapore.

Yingjie Hu, Auckland University of Technology, Auckland, New Zealand.

Bo Jin, Georgia State University, Atlanta, Georgia.

Nikola Kasabov, Auckland University of Technology, Auckland, New Zealand.

Jyotsna Kasturi, Pennsylvania State University, University Park, Pennsylvania.

Hyunsoo Kim, Georgia Institute of Technology, Atlanta, Georgia.

David C. King, Pennsylvania State University, University Park, Pennsylvania.

Sun-Yuan Kung, Princeton University, Princeton, New Jersey.

Phil H. Lee, Queen's University, Kingston, Ontario, Canada.

Guo-Zheng Li, Shanghai University, Shanghai, China.

Ming Li, University of Waterloo, Waterloo, Ontario, Canada.

ShuaiCheng Li, University of Waterloo, Waterloo, Ontario, Canada.

Li Liao, University of Delaware, Newark, Delaware.

Man-Wai Mak, The Hong Kong Polytechnic University, Hong Kong, China.

Rajiv S. Menjoge, Massachusetts Institute of Technology, Cambridge, Massachusetts.

Jin-Wu Nam, Seoul National University, Seoul, Korea.

Anand Narasimhamurthy, Pennsylvania State University, University Park, Pennsylvania.

Vasile Palade, Oxford University, Oxford, United Kingdom.

Marimuthu Palaniswami, University of Melbourne, Melbourne, Victoria, Australia.

Shaoning Pang, Auckland University of Technology, Auckland, New Zealand.

Haesun Park, Georgia Institute of Technology, Atlanta, Georgia.

Jagath C. Rajapakse, School of Computer Engineering, and The Bioinformatics Research Center, Nanyang Technological University, Nanyang, Singapore.

Murali Ramanathan, The State University of New York, Buffalo, New York.

Chandan K. Reddy, Wayne State University, Detroit, Michigan.

Akinori Sarai, Kyushu Institute of Technology, Kyushu, Japan.

Hagit Shatkay, Queen's University, Kingston, Ontario, Canada.

Alistair Shilton, University of Melbourne, Melbourne, Australia.

Yemlembam Hemjit Singh, Jamia Millia Islamia, New Delhi, India

Zhenyu Wang, Oxford University, Oxford, United Kingdom

Roy E. Welsch, Massachusetts Institute of Technology, Cambridge, Massachusetts.

Yao-Chung Weng, Cornell University, Ithaca, New York.

Stephen Winters-Hilt, University of New Orleans, New Orleans, Louisiana.

Shuanhu Wu, City University of Hong Kong, Hong Kong, China.

Xudong Xie, City University of Hong Kong, Hong Kong, China.

Jinbo Xu, Toyota Technological Institute, Chicago, Illinois.

Hong Yan, City University of Hong Kong, Hong Kong, China.

Jack Y. Yang, Harvard University, Cambridge, Massachusetts.

Mary Q. Yang, National Institutes of Health, Bethesda, Maryland.

Libo Yu, University of Waterloo, Waterloo, Ontario, Canada.

Aidong Zhang, The State University of New York, Buffalo, New York.

Byoung-Tak Zhang, Seoul National University, Seoul, Korea.

Yan-Qing Zhang, Georgia State University, Atlanta, Georgia.

1

FEATURE SELECTION FOR GENOMIC AND PROTEOMIC DATA MINING

Sun-Yuan Kung and Man-Wai Mak

1.1 INTRODUCTION

The extreme dimensionality (also known as the curse of dimensionality) in genomic data has been traditionally a serious concern in many applications. This has motivated a lot of research in feature representation and selection, both aiming at reducing dimensionality of features to facilitate training and prediction of genomic data.

In this chapter, N denotes the number of training data samples, M the original feature dimension, and the full feature is expressed as an M-dimensional vector process

$$\mathbf{x}(t) = [x_1(t), x_2(t), \ldots, x_M(t)]^T, \quad t = 1, \ldots, N.$$

The subset of features is denoted as an m-dimensional vector process

$$\mathbf{y}(t) = [y_1(t), y_2(t), \cdots, y_m(t)]^T \qquad (1.1)$$

$$= [x_{s_1}(t), x_{s_2}(t), \cdots, x_{s_m}(t)]^T, \qquad (1.2)$$

where $m \leq M$ and s_i stands for index of a selected feature.

From the machine learning's perspective, one metric of special interest is the sample–feature ratio N/M. For many multimedia applications, the sample–feature

Machine Learning in Bioinformatics. Edited by Yan-Qing Zhang and Jagath C. Rajapakse
Copyright © 2009 by John Wiley & Sons, Inc.

ratios lie in a desirable range. For example, for speech data, the ratio can be as high as 100 : 1 or 1000 : 1 in favor of training data size. For machine learning, such a favorable ratio plays a vital role in ensuring the statistical significance of training and validation.

Unfortunately, for genomic data, this is often not the case. It is common that the number of samples is barely compatible with, and sometimes severely outnumbered by, the dimension of features. In such situation, it becomes imperative to remove the less relevant features, that is, features with low signal-to-noise ratio (SNR) [1].

It is commonly acknowledged that more features means more information available for disposal, that is,

$$I[A] \leq I[A \cup B)] \leq \cdots, \tag{1.3}$$

where A and B represent two features, say x_i and x_j, respectively, and $I(X)$ denotes information of X. However, the redundant and noisy nature of genomic data makes it not always advantageous but sometimes imperative to work with properly selected features.

1.1.1 Reduction of Dimensionality (Biological Perspectives)

In genomic applications, each gene (or protein sequence) corresponds to a feature in gene profiling (or protein sequencing) applications. Feature selection/representation has its own special appeal from the genomic data mining perspective. For example, it is a vital preprocessing stage critical for processing microarray data. For gene expression profiles, the following factors necessitate an efficient gene selection strategy.

1. *Unproportionate Feature Dimension w.r.t. Number of Training Samples.* For most genomic applications, the feature dimension is excessively higher than the size of the training data set. Some examples of the sample–feature ratios N/M are

$$
\begin{array}{lll}
\text{protein sequences} & \rightarrow & 1 : 1 \\
\text{microarray data} & \rightarrow & 1 : 10 \text{ or } 1 : 100
\end{array}
$$

Such an extremely high dimensionality has a serious and adverse effect on the performance. First, high dimensionality in feature spaces increases the computational cost in both (1) the learning phase and (2) the prediction phase. In the prediction phase, the more the features used, the more the computation required and the lower the retrieval speed. Fortunately, the prediction time is often linearly proportional to the number of features selected. Unfortunately, in the learning phase, the computational demand may grow exponentially with the number of features. To effectively hold down the cost of computing, the features are usually quantified on either *individual* or *pairwise* basis. Nevertheless, the quantification cost is in the order of $O(M)$ and $O(M^2)$ for individual and pairwise quantification, respectively (see Section 1.2).

3. *Plenty of Irrelevant Genes.* From the biological viewpoint, only a small portion of genes are strongly indicative of a targeted disease. The remaining "housekeeping" genes would not contribute relevant information. Moreover, their participation in the training and prediction phases could adversely affect the classification performance.

4. *Presence of Coexpressed Genes.* The presence of coexpressed genes implies that there exists abundant redundancy among the genes. Such redundancy plays a vital role and has a great influence on how to select features as well as how many to select.

5. *Insight into Biological Networks.* A good feature selection is also essential for us to study the underlying biological process that lead to the type of genomic phenomenon observed. Feature selection can be instrumental for interpretation/tracking as well as visualization of a selective few of most critical genes for *in vitro* and *in vivo* gene profiling experiments. The selective genes closely relevant to a targeted disease are called biomarkers. Concentrating on such a compact subset of biomarkers would facilitate a better interpretation and understanding of the role of the relevant genes. For example, for *in vivo* microarray data, the size of the subset must be carefully controlled in order to facilitate an effective tracking/interpretation of the underlying regulation behavior and intergene networking.

1.1.2   Reduction of Dimensionality (Computational Perspectives)

High dimensionality in feature spaces also increases uncertainty in classification. An excessive dimensionality could severely jeopardize the generalization capability due to overfitting and unpredictability of the numerical behavior. Thus, feature selection must consider a joint optimization and sometimes a delicate trade-off of the computational cost and prediction performance. Its success lies in a systematic approach to an effective dimension reduction while conceding minimum sacrifice of accuracy.

Recall from Equation 1.3 that the more the features the higher the achievable performance. This results in a monotonically increasing property: the more the features selected, the more the information is made available, as shown in the lower curve in Fig. 1.1a.

However, there are a lot of not-so-informative genomic features that are noisy and unreliable. Their inclusion is actually much more detrimental (than beneficial), especially in terms of numeric computation. Two major and serious adverse effects are elaborated below:

- *Data Overfitting.* Note that overoptimizing the training accuracy as the exclusive performance measure often results in overfitting the data set, which in turn degrades generalization and prediction ability.

  It is well known that data overfitting may happen in two situations: one is when the feature dimension is reasonable but too few training data are available; the other is when the feature dimension is too high even though there is a

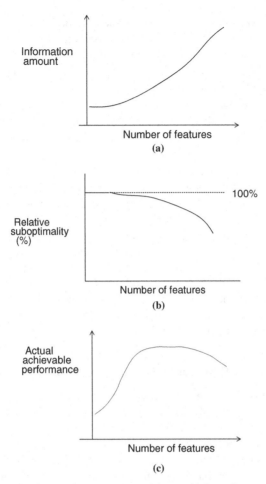

Figure 1.1   (a) Monotonic increasing property of the total information available. (b) Relative performance versus the feature size taking into consideration data overfitting and limited computational resources. (c) Nonmonotonic increasing property of the actual classification performance achievable. The best performance is often achieved by selecting an optimal size instead of the full set of available features.

reasonable amount of training data. What matters most is the ratio between the feature dimension and the size of the training data set. In short, classification/ generalization depends on the sample–feature ratio.

Unfortunately, for many genomic applications, the feature dimension can be as high or much higher than the size of the training data set. For these applications, overtraining could significantly harm generalization and feature reduction is an effective way to alleviate the overtraining problem.

- *Suboptimal Search.* Practically, the computational resources available for most researchers are deemed to be inadequate, given the astronomical amounts of genomic data to be processed. High dimensionality in feature spaces increases uncertainty in the numerical behaviors. As a result, a computational process often converges to a solution far inferior to the true optimum, which may compromise the prediction accuracy.

In conclusion, when the feature size is too large, the degree of suboptimality must reflect the performance degradation caused by data overfitting and limiting computational resource (see Fig. 1.1b). This implies a nonmonotonic property on achievable performance w.r.t. feature size, as shown in Fig. 1.1c. Accordingly, but not surprisingly, the best performance is often achieved by selecting an optimal subset of features. The use of any oversized feature subsets will be harmful to the performance. Such a nonmonotonic performance curve, together with the concern on the processing speed and cost, prompts the search for an optimal feature selection and dimension reduction.

Before we proceed, let us use a subcellular localization example to highlight the importance of feature selection.

Example 1 (Subcellular localization). *Profile alignment support vector machines (SVMs) [2] are applied to predict the subcellular location of proteins in an eukaryotic protein data set provided by Reinhardt and Hubbard [3]. The data set comprises 2427 annotated sequences extracted from SWISSPROT 33.0, which amounts to 684 cytoplasm, 325 extracellular, 321 mitochondrial, and 1097 nuclear proteins. Fivefold cross-validation was used to obtain the prediction accuracy. The accuracy and testing time for different number of features selected by a Fisher-based method [4] are shown in Fig. 1.2. This example offers an evidence of the nonmonotonic performance property based on real genomic data.* ☐

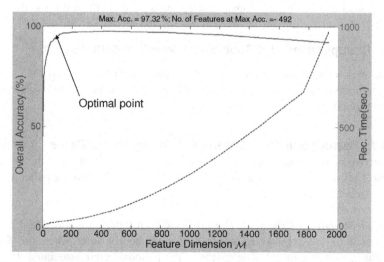

Figure 1.2   Real data supporting the monotonic increasing property. *Upper curve*: performance reaches a peak by selecting an optimal size instead of the full set of the features available. *Lower curve*: the computational time goes up (more than linear rate) as the number of features increases.

1.1.3 How Many Features to Select or Eliminate?

The question now is how many features should be retained, or equivalently how many should be eliminated? There are two ways to determine this number.

1. *Predetermined Feature Size.* A common practice is to have a user-defined threshold, but it is hard to determine the most appropriate threshold. For some applications, we occasionally may have a good empirical knowledge of the desirable size of the subset. For example, how many genes should be selected from, say, the 7129 genes in the leukemia data set [5]? Some plausible feature dimensions are as follows:

   (a) From classification/generalization performance perspective, a sufficient sample–feature ratio would be very desirable. For this case, empirically, an order of 100 genes seems to be a good compromise.

   (b) If the study concerns a regulation network, then a few extremely selective genes would allow the tracking and interpretation of cause–effect between them. For such an application, 10 genes would be the right order of magnitude.

   (c) For visualization, two to three genes are often selected for simultaneous display.

2. *Prespecified Performance Threshold.* For most applications, one usually does not know *a priori* the right size of the subset. Thus, it is useful to have a preliminary indication (formulated in a simple and closed-form mathematical criterion) on the final performance corresponding to a given size. Thereafter, it takes a straightforward practice to select/eliminate the features whose corresponding criterion functions are above/below a predefined threshold.

1.1.4 Unsupervised and Supervised Selection Criteria

The features selected serve very different objectives for unsupervised versus supervised learning scenarios (see Fig 1.3). Therefore, each scenario induces its own type of criterion functions.

1.1.4.1 Feature Selection Criteria for Unsupervised Cases

In terms of unsupervised cases, there are two very different ways of designing the selection criteria. They depend closely on the performance metric, which can be either fidelity-driven or classification-driven.

1. *Fidelity-Driven Criterion.* The fidelity-driven criterion is motivated by how much of the original information is retained (or lost) when the feature dimension is reduced. The extent of the pattern spread associated with that feature is evidently reflected in the second-order moment for each feature x_i, $i = 1, \cdots, M$. The larger the second-order moment, the wider the spread, thus the more likely the feature x_i contains useful information.

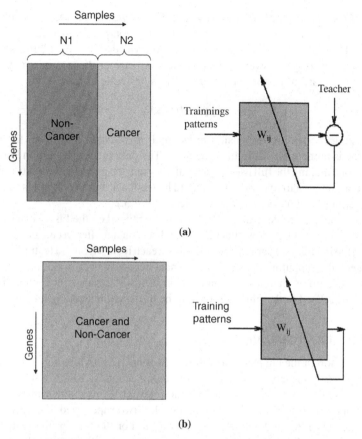

Figure 1.3 Difference between (a) supervised and (b) unsupervised feature selection.

There are two major types of fidelity-driven metrics:

- A performance metric could be based on the so-called mutual information: $I(\mathbf{x}|\mathbf{y})$.
- An alternative measure could be one which minimizes the reconstruction error:

$$\epsilon(\mathbf{x}|\mathbf{y}) \equiv \min_{\mathbf{y}\in\Re^m} ||\mathbf{x} - \hat{\mathbf{x}}_\mathbf{y}||,$$

where $\hat{\mathbf{x}}_\mathbf{y}$ denotes the estimate of $\mathbf{x}$ based on $\mathbf{y}$.

2. *Classification-Driven Criterion.* From the classification perspective, separability of data subclusters plays an important role. Thus, the corresponding criterion depends on how well can the selected features reveal the subcluster structure. The higher-order statistics, known as independent component analysis (ICA), has been adopted as a popular metric. For more discussion on this subject, see Ref. [6].

1.1.4.2 Feature Selection Criteria for Supervised Cases The ultimate objective for supervised cases lies in a high classification/predition accuracy. Ideally speaking, if the classification information is known, denoted by C, the simplest criterion will be $I(C|\mathbf{y})$. However, the comparison between $I(C|\mathbf{x})$ and $I(C|\mathbf{y})$ often provides a more useful metric. For example, it is desirable to have

$$I(C|\mathbf{y}) \rightarrow I(C|\mathbf{x}),$$

while keeping the feature dimension m as small as possible. However, the above formulation is numerically difficult to achieve. The only practical solution known to exist is the one making the full use of the feedback from the actual classification result, which is computationally very demanding. (The feedback-based method is related to the wrapper approach to be discussed in Section 1.4.5.)

To overcome this problem, an SNR-type criterion based on the Fisher discriminant analysis is very appealing. (Note that the Fisher discriminant offers a convenient metric to measure the interclass separability embedded in each feature.) Such a feature selection approach entails computing Fisher's discriminant denoted as FD_i, $i = 1, \ldots, M$, which represents the ratio of intercluster distance to intracluster variance for each individual feature. (This related to the filter approach to be discussed in Section 1.4.1.)

1.1.5 Chapter Organization

The organization of the chapter is as follows. Section 1.2 provides a systematic way to quantify the information/redundancy of/among features, which is followed by discussions on the approaches to ranking the relevant features and eliminating the irrelevant ones in Section 1.3. Then, in Section 1.4, two supervised feature selection methods, namely filter and warper, are introduced. For the former, the features are selected without explicit information on classifiers nor classification results, whereas for the latter, the select requires such information explicitly. Section 1.5 introduces a new scenario called self-supervised learning in which prior known group labels are assigned to the features, instead of the vectors. A novel SVM-based feature selection method called Vector-Index-Adaptive SVM, or simply VIA-SVM, is proposed for this new scenario. The chapter finishes with experimental procedures showing how self-supervised learning and VIA-SVM can be applied to (protein-sequence-based) subcellular localization analysis.

1.2 QUANTIFYING INFORMATION/REDUNDANCY OF/AMONG FEATURES

Quantification of information and redundancy depends on how the information is represented. A representative feature is the one that can represent a group of similar features. Denote S as a feature subset, that is, $S \equiv \{y_i\}$, $i = 1, \ldots, m$. In addition to the general case, what of most interest is either a single individual feature $m = 1$ or a pair of features $m = 2$. A generic term $I(S)$ will be used temporarily to denote the information pertaining to S, as the exact form of it has to depend on the application scenarios.

Recall that there are often a large number of features in genomic data sets. To effectively hold down the cost of computing, we have to limit the number of features simultaneously considered in dealing with the interfeature relationship. More exactly, such computational consideration restricts us to three types of quantitative measurements of the feature information:

1. *Individual Information*: The quantification cost is in the order of $O(M)$.
2. *Pairwise Information*: The quantification cost becomes now $O(M^2)$.
3. *Groupwise Information:* (with three or more features).

The details can be found in the following text.

1.2.1   Individual Feature Information

Given a single feature x_i, its information is denoted as $I(x_i)$. Such a measure is often the most effective when the features are statistically independent. This leads to the individual ranking scheme in which only the information and/or discriminative ability of individual features are considered. This scheme is the most straightforward, since each individual feature is independently (and simultaneously) evaluated. Let us use a hypothetical example to illustrate the individual ranking scheme.

Example 2 (Three-party problem—without interfeature redundancy). *The individual ranking method works the best when the redundancy plays no or minimal role in affecting the final ranking. In this example, each area in Fig. 1.4 represents one feature. The size of the area indicates the information or discriminativeness pertaining to a feature. In the figure, no "overlapping" between elements symbolizes the fact that there exists no mutual redundancy between the features. In this case, the combined information of any two features is simply the sum of two individual amounts. For example, $I(A \cup B) = I(A) + I(B) = 35 + 30 = 65$.*

When all the features are statistically independent, it corresponds to the fact that there is no overlap pictorially. All methods lead to the same and correct result. It is, however, a totally different story with the statistically dependent cases.   □

Unfortunately, the downside of considering the feature individually is that it does not fully account for the redundancy among the features. For example, it is very possible that two highest-rank individual features share a great degree of similarity. As a result, the inclusion of both features would amount to a waste of resource. In fact, one needs to take the interfeature relationship (such as mutual similarity/redundancy) into account. This problem can be alleviated by adopting either pairwise or groupwise information to be discussed next.

1.2.2   Pairwise Feature Information

Given a pair of features x_i and x_j, its information is denoted as $I(x_i \cup x_j)$. The main advantage of studying the pairwise relationship is to provide a means to identify the *similariy/redundancy* of the pair. A fundamental and popular criterion is based on

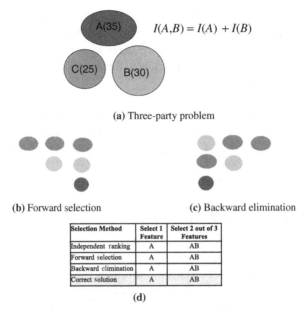

$$I(A,B) = I(A) + I(B)$$

(a) Three-party problem

(b) Forward selection   (c) Backward elimination

Selection Method	Select 1 Feature	Select 2 out of 3 Features
Independent ranking	A	AB
Forward selection	A	AB
Backward elimination	A	AB
Correct solution	A	AB

(d)

Figure 1.4   (a) Three-party problem without redundancy. No "overlapping" between elements symbolizes the fact that no mutual redundancy exists between the features. (b) Consecutive result of the step-by-step forward selection. (c) Consecutive result of the step-by-step backward elimination. (d) Table illustrating search results of different strategies.

correlation. For example, the Pearson correlation coefficient is defined as

$$r_{x_i x_j} = \frac{E[x_i x_j]}{\text{var}(x_i)\text{var}(x_j)} = E[x_i x_j].$$

Without loss of generality, here we shall simply assume that both the features x_i and x_j are zero mean with unit variance $\text{var}(x_i)\text{var}(x_j) = 1$.

From the practical decision perspective, there are again two pairwise criteria: (1) mutual predictability and (2) mutual information.

1. *Mutual Predictability.* The mutual predictability represents the ability of estimating one feature from another feature. Such a metric is also closely tied with the (Pearson) correlation coefficients, that is,

$$\hat{x}_j = E[x_j | x_i] = r_{x_i x_j} x_i. \tag{1.4}$$

When there is no correlation, that is, $r_{x_i x_j} = 0$, then $\hat{x}_j = 0$ regardless of whatever the value of x_i is. In other words, the information of x_i offers no information about x_j. In general, the predictability is a function of $r_{x_i x_j}$; the higher the correlation, the more predictable is x_j given x_i.

2. *Mutual Information.* Suppose that there exists pairwise redundancy, then $I(x_i \cup x_j) \le I(x_i) + I(x_j)$. The mutual information $I(x_i, x_j)$ is also a function of $r_{x_i x_j}$, the higher the correlation, the greater the mutual information.

Individual information and the pairwise information complement each other very well. In fact,

- Individual information reveals the separability of subclusters (termed as SNR in supervised case, see Section 1.4.1), while the pairwise information does not.
- Conversely, pairwise information reveals the redundancy between features, while the individual information does not.

1.2.3 Groupwise Information (with Three or More Features)

In order to optimally evaluate total information contained in a subset of multiple features, the most prudent approach is to have the entire group's information content, $I(\mathbf{y})$, evaluated collectively as an undivided entity.

Sometimes, it is more meaningful or direct to evaluate the information loss in terms of the difference between $I(\mathbf{x})$ and $I(\mathbf{y})$ due to the features missing from the subset. A common objective is to drop as many irrelevant features as possible but suffering from a minimum loss.

The importance of *groupwise information* is evident:

- When compared with individual ranking, the group evaluation has a clear advantage in harnessing information about interfeature redundancy. When compared with pairwise information, it can reveal the SNR as well as a fuller picture of interfeature redundancy.
- When compared with the consecutive ranking (to be discussed next), it delivers a relatively fair solution, because no feature has a better or poorer chance of being selected/eliminated just because it is being evaluated earlier than others.

The downside of group ranking is that it is computationally demanding to find the best combination of features, since it involves exhaustive search to find such an optimal subset of features. If every possible combination has to be considered, it would involve $2^M - 1$ possible combinations.[1] This represents a very formidable computation cost, rendering the group evaluation approach impractical.

1.3 CONSECUTIVE RANKING OF FEATURES

Based on the different types of information discussed previously, this section will look into computationally effective selection strategies.

In seeking a reasonable compromise between accuracy and cost, consecutive search approach arises as a very viable option. The consecutive ranking evaluates features on a one-by-one basis. Because of the interfeature redundancy, the order of

---

[1]Even if only those subsets of size m or lower are searched, it still amounts to $C_m^M \times 2^m - 1$ possible combinations. The search space can be substantially reduced to only C_m^M groups, if the size of subsets is predetermined to be exactly m.

feature selection can significantly affect the outcome of the selection. There are two ways to conduct a consecutive search:

1. *Forward Selection (Augmenting)*.   Search usually begins at an empty feature set.
2. *Backward Elimination*.   Search can start at either the full set (i.e., backward elimination) or, more generally, any subset. This can be considered as a special case of the overselect-then-prune strategy.

1.3.1   Forward Search: Most Innovative First Admitted

Forward search is a recursive search scheme, which begins at an empty feature set. It adopts a most innovative first admitted (MIFA) strategy. In this scheme, one feature is added to the existing chosen subset at each step. (Once a feature is admitted, it will not be dropped anymore, except in an iterative procedure.) The *importance* of the candidate features is not judged by its individual merit but rather by how much extra value it can add to the existing subset. In other words, the usefulness of a feature depends on how well does it complement the current subset. The search continues until either a preset number of features is reached or a prespecified performance is achieved.

Example 3 (Three-Party problem—with interfeature redundancy. *The scenario is illustrated in Fig. 1.5. If only one feature is to be selected, the optimal and correct solution is A. This is because*

$$I(A) > I(B) > I(C).$$

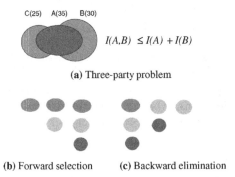

C(25)   A(35)   B(30)

$$I(A,B) \leq I(A) + I(B)$$

(a) Three-party problem

(b) Forward selection   **(c)** Backward elimination

Selection Method	Select 1 Feature	Select 2 out of 3 Features
Independent ranking	A	AB
Forward selection	A	AB
Backward elimination	B	BC
Correct solution	A	BC

(d)

Figure 1.5   (a) Three-party problem with redundancy, where *A*(35), *B*(30), and *C*(25) exhibit a peculiar overlapping pattern. "Overlapping" between elements symbolizes the fact that mutual redundancy exists between the features. (b) Consecutive result of the step-by-step forward selection. (c) Consecutive result of the step-by-step backward elimination. (d) Table illustrating search results of different strategies.

If more than one feature are to be selected, then the redundancy among the features selected plays a role. For example, it is often the case that a gene A is very discriminative on its own, and so is gene B. But the information revealed by gene A overlaps significantly with that of gene B. The interfeature redundancy implies that

$$I(A \cup B) \leq I(A) + I(B). \tag{1.5}$$

In Fig. 1.5, such interfeature redundancy is pictorially displayed as "overlapping" of two corresponding elliptic areas. The size of overlap indicates the amount of "information discount" in Equation 1.5.

In this example, there is a large overlap between A and B, which is, visually speaking, about half of the size of B($\approx 30/2 = 15$). Therefore,

$$I(A \cup B) \approx 50 \leq 55 = I(A) + I(B).$$

In contrast, there is virtually no overlap between B and C.

$$I(B \cup C) \approx 55 = I(A) + I(B).$$

Therefore, $I(A \cup B) \leq I(B \cup C)$. If a two-feature subset is to be selected, the optimal and correct solution is B and C. Note that the individual ranking would select A and B, because they have the two highest scores, which is an incorrect selection.

In the forward ranking, A will be chosen in the first round. In the second round of selection, B will be a better choice to be teamed up with A. So the best two features would be A and B, again an incorrect selection. See Fig. 1.5b.   □

1.3.2 Backward Elimination: Least Useful First Eliminated

This is a recursive search scheme, usually starting at the full set. In this scheme, at each step, one feature is removed from the current subset based on strategy in which the feature with least useful (or indispensable) is first eliminated.[2] The *impact* of the candidate features is not judged by its individual merit but rather by how much loss of information is caused by its removal from the current subset. The one that incurs the least damage is deemed to have the least importance (or impact), which is also a natural candidate for elimination at that step. In this sense, the elimination rule is that the feature causing least loss gets first eliminated.

To minimize the incurred loss of information, we compute

$$\text{Loss} = I(S) - I(S^-),$$

where S denotes the current subset and S^- denoted the subset after the elimination.

Example 4 (Backward elimination: three-party problem). *Let us continue the example illustrated in Fig. 1.5. In backward elimination, A will be eliminated in the first round. This is a surprising result, because A would be considered to be the "most important"*

---

[2]Once a feature is eliminated, it will not be reselected anymore.

according to independent ranking. However, according to the backward elimination rule, A is considered to be the "least important" because its elimination will incur least loss. Therefore, A is eliminated and the best two-feature subset is B and C (see Fig. 1.5c). This matches with the correct answer derived via the group evaluation approach. In this case, only the backward elimination reaches the correct solution; cf. Fig. 1.5d.

Unfortunately, this also means that the backward approach would result in an incorrect selection if only one feature is to be selected, because the correct solution, that is, feature A, would be eliminated in the very first round, since it is regarded as having the least impact (and thus most dispensable) in the first iteration of the backward elimination. Among the remaining two features, feature B would beat feature C in the second round and get selected if only one feature is wanted, as shown in Fig. 1.5d. □

1.4 SUPERVISED FEATURE SELECTION AND EXTRACTION

For supervised selection, there are two types of selection criteria: closed-loop type and open-loop type.

- *Filter (Open-Loop) Approach.* In order to facilitate a quick search of candidate features (they may or may not be the final choice), some simple open-loop criterion functions may be desirable for guiding or conducting an efficient preliminary search. In this case, the selection method/criterion is self-determined, that is, it is independent of the classifier.
- *Wrapper (Closed-Loop) Approach.* The classification method is predetermined and it has substantial influence on the choice of the feature subset. For supervised learning cases, the ultimate goal is to yield the highest possible classification accuracy while using only a subset of features within the given size constraint. In order to estimate the exact accuracy, the only known approach is via a closed-loop solution, that is, it involves the complete training and testing phase of the adaptive classification. This unfortunately demands an enormous computational burden, which may prove too costly.

While the filter method is simple to compute, it fails to consider the critical inter-feature redundancy. The wrapper approach, however, can fully rectify this problem, although its closed-loop process is very computational demanding. It is natural to combine or consider both methods in order to reach an optimal strategy.

1.4.1 Filter Method—An Open-Loop Approach

In the filter method, feature selection and classifier design are separated in that a subset of features is first selected and then classifiers are trained based on the selected features. From the structural dependence perspective, while the classifier has to depend on the selection strategy, the features are selected with no regard of what classifier will be adopted.

All the filter approaches are open-loop approaches, that is, no information on the postprocessing classifier is needed. In addition, they are all based an individual ranking scheme, mostly using an SNR-type criterion. An SNR-type criterion, which is intuitively a normalized distance, is defined as follows:

$$\text{SNR} = \frac{\text{signal}}{\text{noise}} = \frac{\text{distance}}{\text{variance}},$$

where the signal is the numerator representing the distance (separation) between the two centroids (one for positive class and the other negative class) and the noise is the denominator representing the variance of each class of data.[3]

Example 5. *Golub's SNR:*

$$\text{SNR}(j) = \text{signed-FDR}(j) = \frac{\mu_j^+ - \mu_j^-}{\sigma_j^+ + \sigma_j^-}, \tag{1.6}$$

where μ_j^+, μ_j^-, σ_j^+, and σ_j^- represent the class-conditional means and standard derivations of the jth feature, respectively. Note that this criterion follows the idea of Fisher discriminant, and therefore, will be referred to as signed Fisher discriminant ratio (signed-FDR) in the sequel.   □

If our goal is to identify a single feature, then the best choice can be obtained as

$$\arg\max_j \text{SNR}(j). \tag{1.7}$$

Such formulation can be easily extended to multiple-feature selection, where we retain the *m* features with the highest scores.

In filter approaches, because feature selection acts as a preprocessor and the classifier does not play any role in the selection process, the feature selector can be considered as a "master" governing the final set of feature to be used and the classifier can be considered as a "slave" that uses whatever features provided by the feature selector.

Several SNR-type ranking criteria have been proposed in the literature. They are summarized in Table 1.1.

1.4.1.1 First Order SNR-Type Criteria

- *Signed-FDR.* The discriminative power of features is evaluated independently according to [5]

$$\text{signed} - \text{FDR}(j) = \frac{\mu_j^+ - \mu_j^-}{\sigma_j^+ + \sigma_j^-}, \tag{1.8}$$

---

[3]Here, we assume that the positive and negative classes have the same variance. When the variances are different, then the average of the two variances will be adopted instead.

Table1.1   SNR-based filter methods

Name	Order	Criterion	Reference
Signed-FDR	First	$\dfrac{\mu_j^+ - \mu_j^-}{\sigma_j^+ + \sigma_j^-}$	[5]
t-Statistics	First	$\dfrac{\mu_j^+ - \mu_j^-}{\sqrt{\dfrac{(s_j^+)^2}{N^+} + \dfrac{(s_j^-)^2}{N^-}}}$	[7]
FDR	Second	$\dfrac{(\mu_j^+ - \mu_j^-)^2}{(\sigma_j^+)^2 + (\sigma_j^-)^2}$	[8]
SD	Second	$\dfrac{1}{2}\left(\dfrac{(\sigma_j^+)^2}{(\sigma_j^-)^2} + \dfrac{(\sigma_j^-)^2}{(\sigma_j^+)^2}\right) - 1 + \dfrac{1}{2}\left(\dfrac{(\mu_j^+ - \mu_j^-)^2}{(\sigma_j^+)^2 + (\sigma_j^-)^2}\right)$	[4]

FDR: Fisher discriminant ratio; SD: symmetric divergence. "+" and "−" represent positive and negative classes, respectively. Refer to Equation 1.10 for s_j^+ and s_j^-.

where μ_j^+, μ_j^-, σ_j^+, and σ_j^- represent the class-conditional means and standard derivations of the jth feature, respectively. Furey et al. [9] proposed to use the absolute value of signed-FDR(j) as ranking criterion. Features with high signed-FDR(j) or |signed-FDR(j)| are selected for classification. The method is intuitively appealing because high signed-FDR(j) or |signed-FDR(j)| means that the corresponding features produce maximum separation between the positive and negative classes.

- *T-Statistics.* A similar ranking criterion is based on the t-statistics [7]:

$$z_j = \frac{\mu_j^+ - \mu_j^-}{\sqrt{\dfrac{(s_j^+)^2}{N^+} + \dfrac{(s_j^-)^2}{N^-}}}, \qquad (1.9)$$

where

$$(s_j^+)^2 = \frac{\sum_{k \in C^+} (x_{kj} - \mu_j^+)^2}{N^+ - 1} \quad \text{and} \quad (s_j^-)^2 = \frac{\sum_{k \in C^-} (x_{kj} - m_j^-)^2}{N^- - 1}, \qquad (1.10)$$

where N^+ and N^- are the numbers of training samples in the positive class C^+ and negative class C^-, respectively.

1.4.1.2 *Second Order SNR-Type Criteria*

- *Fisher Discriminant Ratio (FDR).* A slight variant of signed-SNR is the FDR [8]:

$$FDR(j) = \frac{(\mu_j^+ - \mu_j^-)^2}{(\sigma_j^+)^2 + (\sigma_j^-)^2}. \qquad (1.11)$$

As an illustrative example, Fig. 1.6 shows the FDR of 7129 genes in an acute leukemia data set [5] that contains two types of acute leukemia: ALL and AML.

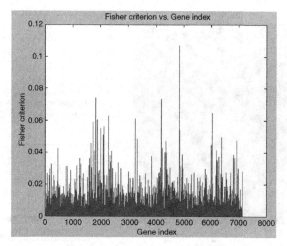

Figure 1.6 The Fisher discriminant ratio of 7192 genes in the acute leukemia data set 5. The FDR cutoff point (threshold) and the corresponding number of remaining genes are shown in Table 1.2.

Evidently, only a small number of genes have large FDR, meaning that only a few genes are useful for differentiating the two types of acute leukemia. Table 1.2 shows the numbers of selected genes at different cutoff points.

- *Symmetric Divergence (SD).* By assuming the distributions of the positive and negative classes as Gaussians, Mak and Kung [4] proposed ranking the features according to the symmetric divergence between the positive and negative class distributions:

$$
D(p(x_j^+) \| p(x_j^-)) = E\left\{ \log \frac{p(x_j^+)}{p(x_j^-)} \middle| C^+ \right\} + E\left\{ \log \frac{p(x_j^-)}{p(x_j^+)} \middle| C^- \right\}
$$

$$
= \frac{1}{2}\left(\frac{(\sigma_j^+)^2}{(\sigma_j^-)^2} + \frac{(\sigma_j^-)^2}{(\sigma_j^+)^2} \right) - 1 + \frac{1}{2}\left(\frac{(\mu_j^+ - \mu_j^-)^2}{(\sigma_j^+)^2 + (\sigma_j^-)^2} \right),
$$

$$(1.12)$$

where $p(x_j^+)$ and $p(x_j^-)$ represent the density functions of the positive and negative classes, respectively. Figure 1.7 illustrates the procedure of computing the symmetric divergence of feature j and Fig. 1.8 shows the histograms of the symmetric divergences. Apparently, only a small fraction of the features have

Table 1.2 The FDR cutoff point (threshold) and the corresponding number of remaining genes

Threshold	0.01	0.02	0.04	0.06
No. of remaining genes	596	142	19	8

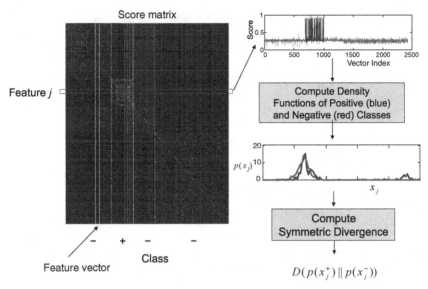

Figure 1.7   Computation of symmetric divergence (Eq. 1.12) in a subcellular localization task.

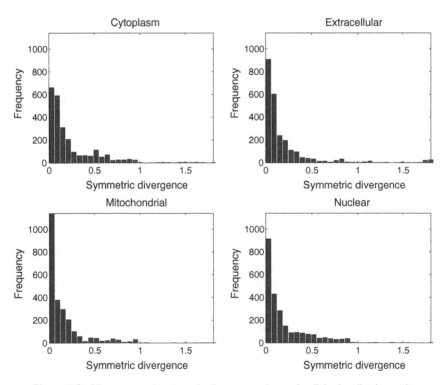

Figure 1.8   Histograms of symmetric divergences in a subcellular localization task.

large symmetric divergences, which means that only a few feature dimensions are relevant for classification. This hypothesis is confirmed by our experimental results below.

1.4.1.3 *Comparing Different Filters*
The ranking methods based on signed-FDR, t-statistics, FDR, and symmetric divergence share a common property: Features with small variances but large difference in class means will be ranked high. However, there are also important differences. For example, Equations 1.11 and 1.12 differ in the additional term that depends only on the ratio of variances. Therefore, features with high variance ratio between positive and negative classes will be ranked high under the symmetric divergence criterion. t-Statistic is used to assess the statistical significance of the difference between the sample means of two normal distributed population when the sample size is small. However, Kullback–Leibler divergence (from which symmetric divergence is derived) is a distance measure between two probability distributions. In most cases, one distribution represents the observations and the other represents a model that attempts to approximate the former one.

To compare the capability of different filter methods, we applied signed-FDR (Eq. 1.6), absolute value of signed-FDR, FDR (Eq. 1.11), and symmetric divergence (Eq. 1.12) to select features for classifying subcellular location of proteins. The data set [3] that we used contains 2427 annotated amino acid sequences (684 cytoplasm, 325 extracellular, 321 mitochondrial, and 1097 nuclear proteins) extracted from SWISSPROT 33.0. We applied pairwise profile alignment to create a score matrix for classification by RBF-SVMs [2].

Figure 1.9 shows the accuracy of subcellular localization when the number of selected features is progressively increased from 1 to the full size. Evidently, the accuracy increases rapidly at small feature size and becomes saturated at around 200

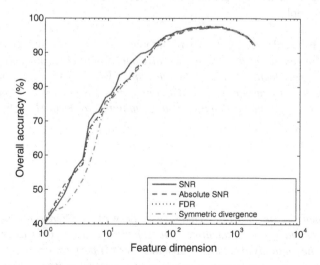

Figure 1.9 Accuracy based on features selected by signed-FDR, absolute value of signed-FDR (|Signed-FDR|), FDR, and symmetric divergence.

(the accuracy even drops slightly because of the curse of dimensionality), suggesting only one-tenth of the features are useful for classification. The results show that the performance of signed-FDR is better than that of |signed-FDR| and FDR, suggesting that the sign is important for feature selection.

1.4.1.4 Limitations of Filter Methods

Selecting features based on the filter methods suffers from several serious drawbacks. First, these methods require users to set a cutoff point in the ranking scores under which features are deemed to be irrelevant for classification. However, the optimal number of features (i.e., the best cutoff point) is usually unknown. Second, the features selected may be highly correlated because the feature set may contain many highly discriminative features but with almost identical characteristics. This means that some of these *redundant* features can be removed without affecting the classification accuracy. Third, the ranking criteria do not take the combined effect of features into account. For example, a low-ranked feature may become very useful for classification when combined with another low-ranked feature. To overcome these limitations, researchers have proposed using the performance of classifiers to guide the feature selection progress, which is to be discussed in Sections 1.4.3 and 1.4.5.

1.4.1.5 Cope with Redundancy

To develop a more appropriate open-loop criterion, one wants not only to maximize the discriminativeness of individual features, but also to minimize the redundancy between the features selected. Because features may be mutually redundant, that is, two highly ranked features may carry similar discriminative information to the target class, redundancy removal can be done by (1) forward selection based on the "most innovative first admitted" approach; and (2) backward elimination based on the "least useful first eliminated" approach. To illustrate the idea, let us consider the first strategy in the following example.

Example 6 (Forward selection of two out of three—with versus without dependence). *Without loss of generality, assume that we have a total of three features: $x_1, x_2,$ and x_3, and that x_1 is preselected a priori. The question is which of x_2 and x_3 could be the preferred choice to team up with x_1. The answer hinges upon which one can best complement x_1 rather than which is more discriminative by itself. Mathematically, let us denote*

$$\hat{x}_2 = x_2/x_1,$$
$$\hat{x}_3 = x_3/x_1,$$

where x_i/x_1 denotes the innovation component of x_i w.r.t. x_1. There is no unique way to define the innovation component for the supervised cases. One approach is to (1) identify the mutually dependent subspace and (2) perform coordinate transformation so that the innovative component is represented by its own axis, which is perpendicular to the dependent subspace. For example, the features x_1 and x_2 are

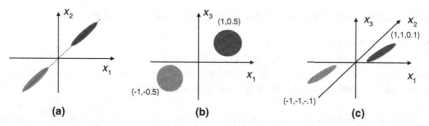

Figure 1.10 Selecting two out of three features in supervised feature selection. x_1 and x_2 are mutually dependent. x_1 and x_3 are independent. x_3 is independent of both x_1 and x_2.

mutually dependent as shown in Fig. 1.10a . Following the above-mentioned procedure, let us use a coordinate transformation:

$$\begin{bmatrix} 1 & 0 \\ -1 & 1 \end{bmatrix}.$$

After the transformation, the vertical axis would represent the innovative component $\hat{x}_2$. Thus, the (innovative) discriminative power is represented by the normalized distance (FD) of $\hat{x}_2$. In contrast, the features x_1 and x_3 are mutually independent as shown in Fig. 1.10b. Therefore, one can say, given that we already have the knowledge on x_1, the innovative component of x_3 is greater than that of x_2.

Apply the same procedures to all the features. In this example, features x_2 and x_3 can be transformed to $\hat{x}_2$ and $\hat{x}_3$ via a three-dimensional coordinate transformation:

$$\begin{bmatrix} 1 & 0 & 0 \\ -1 & 1 & 0 \\ 0 & 0 & 1 \end{bmatrix}.$$

After the transformation, the best choice (to team up with x_1) can then be identified as

$$\arg\max_{i \neq 1} \mathrm{FD}(\hat{x}_i).$$

The interfeature dependence is mostly due to the mutual information or redundancy between two features. If two features have too much redundancy (i.e., they share very similar common information), then selection of one feature would immediately obviate the selection of the other, and vice versa.

In this example, features x_1 and x_2 have by themselves high discriminative power (in FD sense), but the two features have a great statistical similarity. Therefore, selection of x_1 would naturally obviate any chance for x_2. In this case, the second best candidate is x_3. In conclusion, the best feature pair is x_1 and x_3. ☐

This formulation can be easily extended to the backward elimination strategy.

1.4.2   Weighted Voting and Linear Combination of Features

After the features are selected, say, by a filter method, the next task is classification. For this, a popular approach was proposed by Golub et al. [5]. In this method, an SNR filter-based selection is combined with a weighted voting scheme. Each selected feature independently contributes a vote for a class. More exactly,

- The vote from the jth feature is represented by $v_j(x_j) = x_j - b_j$. Here the decision threshold b_j is set halfway between the class means, that is,

$$b_j = \frac{\mu_j^+ + \mu_j^-}{2}.$$

  Each feature j casts a vote to one of the two classes. A positive value is for the positive class and negative value for the negative class.

- Furthermore, every vote will be given a different weighting w_j. The weighting factor w_j is introduced to reflect how trustworthy is the feature—it depends on how well the feature is correlated with the class distinction. In Ref. [5], the weighting factor w_j is based on the SNR of the individual feature:

$$w_i = \frac{\mu_i^+ - \mu_i^-}{\sigma_i^+ + \sigma_i^-},$$

  that is, it is a function of the mean and standard deviation of the feature values.

- This leads to

$$\text{weighted total score} = \sum_j w_j v_j(x_j).$$

If the weighted total score is positive, then the sample is identified, otherwise it is rejected.

To illustrate the role of weighting, let us take a look at the data distribution shown in Fig. 1.12b. It is obvious that a higher weighting must be placed on x_2. Such a conclusion may also be reached either via the weighted voting approach (due to the higher SNR associated with x_2) or by almost any optimal linear classifier, such as LSE, Bayesian classifier, or linear SVM.

1.4.3   Linear Feature Representation

Note that

$$\text{weighted total score} = \sum_j w_j v_j(x_j) = \sum_j w_j(x_j - b_j) = \mathbf{w}^T \mathbf{x} - b,$$

where $b = \sum_j w_j b_j$. This shows that a weighted voting is basically equivalent to the use of a linear decision boundary. Then, it is natural and appealing to pursue a more direct solution for optimal linear decision boundary. This is in fact equivalent to the pursuit of an optimal linear representation of the features:

$$y = \sum_j w_j x_j$$

and an associated threshold b for optimal linear classification.

The linear feature representation is formulated as follows. Given a set of (say N) M-dimensional vectors, $X = \{\mathbf{x}(1), \mathbf{x}(2), \ldots, \mathbf{x}(N)\}$, each of which is known to belong to either the positive class or the negative class, find an optimal linear combination of features that best separates two classes of data samples. More exactly, we are to find $\{w_1, w_2, \ldots, w_M\}$ to best classify the test vectors into two classes, depending on the sign of the discriminant function:

$$f(x) = w^T x - b.$$

For M-dimensional training vectors, the decision boundary represents an $(M-1)$-dimensional hyperplane.

In fact, linear classifiers via supervised training have a long history.

1. Rosenblatt's perceptron [10] was the first neural classifier proposed in 1958.
2. Least squares or MMSE formulation is to best approximate the teacher values.
3. Fisher's discriminant [11] laid the groundwork for statistical pattern recognition in 1936.
4. SVM developed by Vapnik [12] in 1995 and by Boser et al. [13] in 1992.

One would be naturally concerned that the restriction to linear classification could render the proposed methods in this chapter to have very limited application domains. Indeed, in general, the best prediction result can only be achieved via using a more sophisticated nonlinear decision boundary.

Fortunately, linear classifiers often work well for most genomic data because they usually posses a very low sample-to-feature ratio. For example, the sample-to-feature ratio for gene expression data is around 1:100 and that for vectorized data produced by pairwise sequence (e.g., protein or DNA sequences) comparison is 1:1. It is well known that linear classifiers work well under such situations. Therefore, it should be of no surprise that the two inherently linear schemes, that is, the weighted voting and wrapper approach, work well for genomic data. For example, Golub et al. [5] show that the weighted votes of a set of informative genes yield a good prediction of a new sample. More exactly, a 50-gene predictor derived in cross-validation tests correctly assigned 36 of the 38 samples as either AML or ALL, while the remaining two were labeled as uncertain (PS = 0.3).

1.4.4   Comparison of Weighted Voting and Linear Classification

The advantage of weighted voting is that it is numerically stable; thus, it is preferable especially when the number of samples is too small for conducting a robust statistical estimation. The disadvantage of weighted voting, however, is that it fails to consider redundancy among the features. Consequently, it could lead to a highly undesirable scenario in which two highly redundant features get to vote twice (one vote from each feature). This is very different from the optimal weighting strategy according to the Bayesian classification. This can be best explained in Example 7.

Example 7. *Without loss of generality, assume that we have two classes of data represented by three features: x_1, x_2, and x_3, and that x_1 is preselected a priori. The centroids for the positive and negative classes are [+1, +1, +1] and [−1, −1, −1], respectively. Assume that the two classes have the same intraclass covariance matrix:*

$$\begin{bmatrix} 1 & 0 & 0 \\ 0 & 1 & \epsilon \\ 0 & \epsilon & 1 \end{bmatrix}.$$

When $\epsilon \to 0$, then all the features are statistically independent (in intraclass sense); therefore, the optimal weighting will be one with the uniform weighting, $w_1 = 1$, $w_2 = 1$, and $w_3 = 1$, a result derivable via either the weighted voting or the optimal Bayesian classification. However, when $\epsilon \to 1$, then the features x_2 and x_3 are highly correlated or they have a high mutual redundancy. Notwithstanding, the weighted voting approach would still reach a uniform weighting result, $w_1 = 1$, $w_2 = 1$, and $w_3 = 1$, because it fails to take such redundancy into account. This can be effectively handled by wrapper approaches based on linear classifier.

For example, features x_2 and x_3 are mutually dependent as shown in Fig. 1.11a. It can be derived (omitted here) that the optimal Bayesian classification is [1 1 0] or [1 0 1].   □

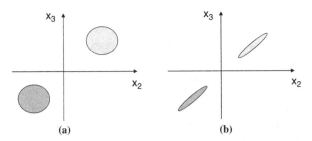

(a)   (b)

Figure 1.11   (a) Equal weights are placed on x_2 and x_3. (b) When $\epsilon \to 1$, x_2 and x_3 are highly correlated or they have a high mutual redundancy. The weighted voting approach would fail to take such redundancy into account. This can be effectively handled by wrapper approach based on linear classifier, as discussed in Example 7.

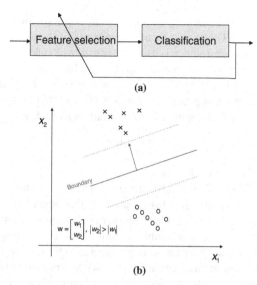

Figure 1.12 (a) The architecture of wrappers. The classifier acts as a master guiding the feature selection process. (b) Linear separable case in which x_2 will be chosen over x_1.

This example highlights the fact that the weakness of the weighted voting scheme lies in its failure to consider the important role of redundancy. There are two possible solutions:

1. *Direct Solution.* As shown in the above example, the weighted voting is very effective when the features are relatively independent. So, it would be wise that we weed out those highly redundant features during feature selection, using for example the notion of the innovation components. See Example 6.

2. *Use a Wrapper Approach.* In this case, an optimal linear classifier would have taken into account the underlying redundancy. See Fig. 1.12.

1.4.5 SVM-Based Wrapper Methods—A Closed-Loop Approach

All of the linear classifiers mentioned earlier can be considered as candidates in the wrapper approach. Nevertheless, the literature in genomic applications, especially microarray data, seem to strongly favor the SVM approach.

It is nearly impossible to have an open-loop solution (in a closed form) to faithfully reflect the exact optimal classification. Let us highlight this point via an example. Suppose we have M individual FDs, each representing the discriminative power of a feature. A closed-form criterion would mean that there exists a function of the FDs: $f(\text{FD}_1, \text{FD}_2, \ldots, \text{FD}_M)$, for example, $\sum_i \text{FD}_i$, which can be used as a bona fide measure of the classification performance. Unfortunately, it is a naive and impractical goal. In practice, there exists no such function $f(\text{FD}_1, \text{FD}_2, \ldots, \text{FD}_M)$ meeting the full spectrum of applications. Even with the benefit of the statistically independent assumption, the pessimistic assessment will remain true.

The fact that there will be no easy-to-apply open-loop criterion necessitates the use of closed-loop solution. This would incur an enormous amount of computation costs, because classification performance is used as critical feedbacks to assess whether the final feature subset can truly deliver an optimal performance (see Fig. 1.12). This feedback strategy is adopted by most wrapper approaches to be discussed in this subsection. Briefly, assuming that the full set (FS) of features delivers the best accuracy, our goal is to find a subset (SS) of features such that

$$|A(FS) - A(SS)|$$

is minimal, where $A(\cdot)$ stands for accuracy.

Because the ultimate goal of feature selection is to increase classification accuracy, it is intuitive to choose a particular classification method and use its parameters or its performance on training data to guide the feature selection process (see Fig. 1.12a). Typically, this is done by selecting a subset of features and evaluating its performance on the chosen classifier, and the process is repeated until the best performing subset is obtained. Methods based on this approach are known as wrappers in the literature [14].

From the structural dependence perspective, this approach is very much unlike the filter approach, the feature selection strategy depends on the classifier adopted. In fact, the classifiers in wrappers act as the master guiding the feature selection process.

A number of wrapper approaches have been proposed for bioinformatics data mining:

- *SVM Approach.* This kind of approach includes the recursive feature elimination (RFE) [15] and recursive SVM [16, 17].
- *Nearest Neighbor Approach.* A typical example of this approach is the ReliefF [18].

This subsection will focus on the SVM-RFE approach.

SVM-RFE. Guyon et al. [15] proposed a backward elimination algorithm, namely, SVM-RFE, that ranks features based on the weights of a linear SVM. The algorithm begins with using the full-feature training vectors $\mathbf{x}_i \in \mathfrak{R}^{D_0}$ to train a linear SVM. Features are then ranked by sorting the square of the SVM's weights $\{w_j^2\}_{j=1}^{D_0}$ in descending order, where the weight vector is given by

$$\mathbf{w} = \sum_{i \in \text{SV}} \alpha_i y_i x_i, \quad \mathbf{w} \in \mathfrak{R}^{D_0}. \tag{1.13}$$

A subset of features corresponding to the end of the sorted list (i.e., those with small w_j^2) is then removed.[4] The remaining features are used to construct a new set of training vectors $x_i \in \mathfrak{R}^{D_1}$, where $D_1 < D_0$. These vectors are then used to train another linear SVM and the process is repeated until all features have been eliminated. At the end of

---

[4]Intuitively, a small weight w_j means that the jth axis in the feature space is irrelevant to classification and can be removed without affecting performances.

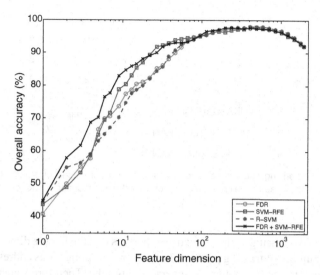

Figure 1.13 Performance of FDR, SVM-RFE, R-SVM, and the fusion of FDR and SVM-RFE in the subcellular localization task.

the iterative process, a ranked list of features is produced. Figure 1.13 compares the performance of SVM-RFE, FDR, and the fusion of SVM-RFE and FDR.

1.5 FEATURE SELECTION VIA VECTOR-INDEX-ADAPTIVE SVM FOR FAST SUBCELLULAR LOCALIZATION

The comparison of two temporal sequences is often hampered by the fact that two sequences often have different lengths whether or not they belong to the same family. To overcome this problem, pairwise comparison between a sequence and a set of known sequences has been a popular scheme for creating fixed-size feature vectors from variable-length sequences [2, 19, 20]. This process is referred to as *vectorization*. Suppose that we have M sequences, then each sequence is converted into a (column) vector of dimension M with entries representing the pairwise similarity between that sequence and all of the M sequences. In total, there will be M such M-dimensional (column) vectors. For example, in Fig. 1.14, three sequences $S^{(1)}$, $S^{(2)}$, and $S^{(3)}$ are converted to three three-dimensional column vectors. Together these vectors form an $M \times M$ matrix, named the *kernel matrix*.

In the previous sections, we have constantly used an axis to represent a feature, and therefore, feature selection/elimination means axis selection/elimination. Note that under the pairwise kernel representation, we have as many features as vectors. Then, does a feature correspond to an axis or a data point? The somewhat surprising answer is: *both*, because the symmetrical kernel matrix exhibits a useful reflexive property, which is best illustrated by an example shown in Fig. 1.15. It can be advantageous to harness such a symmetry property, which motivates the discussion of this section.

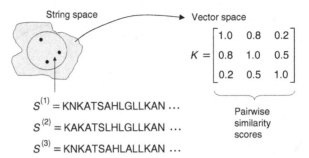

Figure 1.14 Vectorizing sequences. Because strings are composed of alphabets and have different lengths, they need to be converted to vectors of the same dimension for classification via kernel methods.

The downside of using such symmetry property is that the feature dimension is the same as the number of training patterns. In fact, for the applications addressed in this chapter, they are in the range of several thousands. Therefore, such curse of dimensionality could hurt both training and recognition speed. Because a large number of sequences are being added to sequence databases on a daily basis, it is imperative to design feature selection methods that can weed out irrelevant features to reduce training and recognition time.

In this section, we propose a method that makes use of the symmetric property of pairwise scoring matrices to select relevant features. The method considers the columns of a pairwise scoring matrix as high-dimensional vectors and uses the column vectors to train a linear SVM. Because of the symmetric property of the score matrix, the row vectors with row indexes equal to the support vector indexes are identical to the support vectors. Also, because the support vectors define the decision boundary and margins of the SVM, they are critical for classification performance. Therefore, the support vector indexes are good candidates for selecting features for the column vectors, that is, only the rows corresponding to the support vectors are retained. The column vectors with reduced dimensions are then used to train another SVM for classification. Because the indexes of support vectors are used for selecting features, we referred this method to as vector-index-adaptive SVM, or simply VIA-SVM [26].

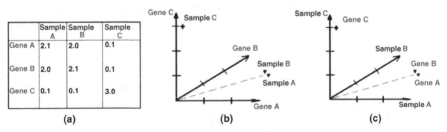

Figure 1.15 Symmetry (reflexive) property of pairwise data. (a) Table showing the feature/sample data. (b) Features are displayed as axes and samples as data vectors. (c) Features are displayed as data vectors and samples as axes.

Later in the section, we shall compare the VIA-SVM with Guyon et al.'s SVM-RFE [15] in a subcellular localization benchmark and show that the VIA-SVM not only avoids setting a cutoff point but also is insensitive to the penalty factor in SVM training.

1.5.1 Pairwise Scoring Kernels

Denote $\mathcal{D} = \{S^{(1)}, \ldots, S^{(T)}\}$ as a training set containing T protein sequences. Let us further denote the operation of PSI-BLAST[5] search given the query sequence $S^{(i)}$ as

$$\phi^{(i)} \equiv \phi(S^{(i)}) : S^{(i)} \rightarrow \{\mathbf{P}^{(i)}, \mathbf{Q}^{(i)}\},$$

where $\mathbf{P}^{(i)}$ and $\mathbf{Q}^{(i)}$ are the PSSM and PSFM of $S^{(i)}$, respectively.[6] Because these matrices are based on the information of a large number of sequences that are similar to the query sequence, they contain rich information about the remote homolog of the query sequence, which may help improve the prediction of subcellular locations and protein functions. Given the profiles of two sequences $S^{(i)}$ and $S^{(j)}$, we can apply the Smith–Waterman algorithm [23] and its affine gap extension [24] to align $\mathbf{P}^{(i)}$, $\mathbf{Q}^{(i)}$, $\mathbf{P}^{(j)}$, and $\mathbf{Q}^{(j)}$ to obtain the normalized profile alignment score $\zeta(\phi^{(i)}, \phi^{(j)})$.[7]

The scores $\{\zeta(\phi^{(i)}, \phi^{(j)})\}_{i,j=1}^{T}$ constitute a symmetric matrix $\mathbf{Z}$ whose columns can be considered as T-dimensional vectors:

$$\boldsymbol{\zeta}^{(j)} = [\zeta(\phi^{(1)}, \phi^{(j)}) \cdots \zeta(\phi^{(T)}, \phi^{(j)})]^{\mathrm{T}}, \quad j = 1, \ldots, T. \tag{1.14}$$

An M-class protein prediction problem can now be solved by M one-versus-rest SVMs:

$$f_m(S) = \sum_{j \in S_m} y_{m,j} \alpha_{m,j} K(\phi(S), \phi(S^{(j)})) + b_m, \tag{1.15}$$

where S is an unknown sequence, $m = 1, \ldots, M$, $y_{m,j} \in \{+1, -1\}$, S_m contains the indexes of support vectors, $\alpha_{m,j}$ are Lagrange multipliers, and

$$K(\phi(S), \phi(S^{(j)})) = g(\boldsymbol{\zeta}, \boldsymbol{\zeta}^{(j)})$$

is a kernel function.

Now we may consider the columns of a pairwise scoring matrix as high-dimensional vectors. This means that there are T feature vectors with dimension equal to the training

---

[5]To efficiently produce the profile of a protein sequence (called query sequence), the sequence is used as a seed to search and align homologous sequences from protein databases such as SWISSPROT [21] using the PSI-BLAST program [22].

[6]The homolog information pertaining to the aligned sequences is represented by two matrices (profiles): position-specific scoring matrix (PSSM) and position-specific frequency matrix (PSFM). Both PSSM and PSFM have 20 rows and L columns, where L is the number of amino acids in the query sequence.

[7]See http://www.eie.polyu.edu.hk/~mwmak/BSIG/PairProSVM.htm.

set size. The T T-dimensional column vectors can be used to train M SVMs. Because of the high dimensionality, linear SVM is a preferred choice, that is, $g(\zeta, \zeta^{(j)}) = \langle \zeta, \zeta^{(j)} \rangle$. The class of S can then be obtained by $y(S) = \arg \max_{m=1}^{M} f_m(S)$, where M is the number of classes.

1.5.2 VIA-SVM Approach to Pairwise Scoring Kernels

The pairwise approach always results in feature vectors with extremely high dimensions. This creates a problem known as the curse of dimensionality. An obvious solution is to reduce the feature size and yet retaining the most important information critical for classification. The challenge thus lies in how to effectively determine those relevant features. The approaches mentioned in Section 1.4 do not make use of the symmetric property of the pairwise scoring matrices in the selection process (the symmetric property is illustrated in Fig. 1.16), because they are designed for general cases. In fact, they are primarily designed for gene selections in microarray data where expression matrices are neither square nor symmetric. However, it can be advantageous to adopt a feature selection method that is tailor designed for the pairwise scoring vectors.

To design a feature selection algorithm for pairwise scoring vectors, we need to exploit the reflexive property of pairwise scoring matrices. The idea is based on the notion that support vectors are important for classification and pairwise scoring matrices are symmetric. (Namely, the elements of the ith column of $\mathbf{Z}$ are identical to those in the ith row.) This suggests a possible hypothesis:

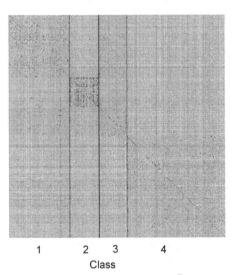

Class

Figure 1.16 Profile alignment score matrix $\mathbf{Z} = \{\zeta(\phi^{(i)}, \phi^{(j)})\}_{i,j=1}^{T}$. The training vectors have been prearranged such that the vectors belonging to the same class are all grouped together, that is, they are consecutively indexed. The three vertical lines were artificially added to divide the column vectors into four classes.

Hypothesis. *The support vector indexes are good candidates for selecting features for the column vectors, that is, only the rows corresponding to the support vectors are retained.*

Heuristically, due to the symmetry property, if the jth vector is a critical (supporting) vector for the decision boundary, then the jth feature would also be a critical feature and therefore should be selected. We refer to this selection scheme as vector-index-adaptive SVM, or simply VIA-SAM. In fact, our simulation results also strongly support this hypothesis.

1.5.2.1 Why Consider Only Support Vectors?
In VIA-SVM, the support vector indexes are reused as feature selection indexes. The use of support vectors to select relevant features is intuitively appealing because they are "critical" for establishing the decision boundary of SVM classifiers. Because of the symmetrical property of kernel matrices, the elements of the ith column of $\mathbf{Z}$ are identical to those in the ith row. If the ith column of $\mathbf{Z}$ happens to be a support vector, the corresponding feature dimension (the ith row of $\mathbf{Z}$) will also be critical for classification. However, nonsupport vectors are irrelevant for classification, so are their corresponding feature dimensions.

The diagonal dominance implies that a large value of α_i in Equation 1.34 is likely to lead to a large value of w_i. By the same token, the nonsupport vectors are those that correspond to $\alpha_i = 0$, and therefore their corresponding weight values w_i are more likely to be smaller. Therefore, only those features corresponding to $\alpha_i > 0$ are considered for selection. Those corresponding to $\alpha_i = 0$ will be eliminated automatically. This concept is further elaborated in Fig. 1.17.

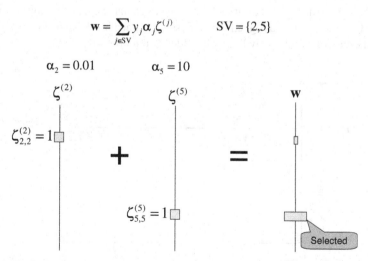

Figure 1.17 Diagram illustrating how VIA-SVM uses the values of Lagrange multipliers α_j to select features. The vertical bars represent support vectors $\zeta^{(j)}$ and the weight vector $\mathbf{w}$, and the magnitude of the vector components are proportional to the thickness along the bars. It is assumed that the matrix $\mathbf{Z}$ has been normalized such that all diagonal elements are equal to 1.

The above interpretation of VIA-SVM is consistent with how SVM-RFE selects features in that, in both methods, indexes with large weight will be chosen first. Moreover, they both prune the vectors/features corresponding to zero α_i. However, there is also an important difference, which lies in the treatment of the vectors/features corresponding to nonzero α_i. More exactly, in VIA-SVM, different types of support vectors receive different levels of preferences.

1.5.2.2 Differential Treatments of Support Vectors

Because the SVM-RFE takes the overall weight vector **w** into account, it only considers the Lagrange multipliers α_i but not the slack variables ξ_i. In contrast, the VIA-SVM considers both α_i (related to SV) and ξ_i (indicates safety margin or, sometimes, outlier). In this sense, the VIA-SVM offers a more comprehensive coverage of all the critical factors made available by the SVM classifier.

It is important to recognize the fact that *not all SVs are created equal*. Therefore, in the VIA-SVM, support vectors are differentially treated. In fact, they are divided into four levels of preferences as specified by the four regions in Fig. 1.18:

Level 1   *Most-Preferred.* The SV is on the margin, that is, $0 < \alpha_i < C$ and $\xi_i = 0$, where C is the penalty factor in SVM training.

Level 2   *Preferred.* The SV is in the fuzzy region and on the correct side of the decision boundary, that is, $\alpha_i = C$ and $0 < \xi_i < 1$.

Level 3   *Marginally Preferred.* The SV is in the fuzzy region but on the wrong side of the decision boundary, that is, $\alpha_i = C$ and $1 \leq \xi_i < 2$.

Level 4   *Nonpreferred.* The SV is regarded as an outlier, that is, $\alpha_i = C$ and $\xi_i \geq 2$.

The reason of ruling out the outlier SVs is self-explanatory. The decision to have the marginal support vectors assigned the highest preference level can be justified on the basis that they offer relatively higher confidence than the fuzzy SVs.

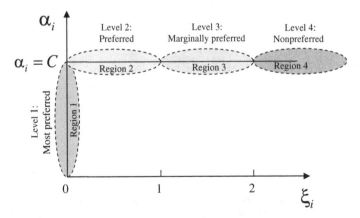

Figure 1.18 The four levels of preferences for the support vectors in VIA-SVM. Regions 1–4 correspond to preference levels 1–4, respectively.

Although it appears that two parameters (α_i and ξ_i) are required to define the preferences, ξ_i alone can already provide sufficient information to determine the preference level of an SV.[8] In fact, the preference decreases with increasing ξ_i. More exactly, $\xi_i = 0$, $0 < \xi_i < 1$, $1 \leq \xi_i < 2$, and $2 \leq \xi_i < \infty$ define Level 1 to Level 4, respectively (cf. Fig. 1.18).

1.5.3 The VIA-SVM Algorithm

Feature selection in VIA-SVM is divided into two steps:

Step 1   The score matrix $\mathbf{Z} = \{\boldsymbol{\zeta}(\phi^{(i)}, \phi^{(j)})\}$ is used to train M SVMs (Eq. 1.15) from which M sets of support vector indexes S_m are determined. This results in a set of support vectors $\zeta^{(j)} = [\zeta(\phi^{(1)}, \phi^{(j)}) \cdots \zeta(\phi^{(T)}, \phi^{(j)})]^{\mathrm{T}}$ for each class, where $j \in S_m$.

Step 2   For the mth class, the indexes in S_m are used to select the feature dimensions (rows of $\mathbf{Z}$) of the column vectors to obtain vectors $\zeta'^{(j)}$ of reduced dimensions, where $j = 1, \ldots, T$. These vectors are then used to train another SVM for classification. This process is repeated for all classes.

These two steps are iterated N times (five times in this work). Specifically, the features selected at the nth iteration are used to train a new SVM in the $(n+1)$th iteration, whose support vectors are subsequently used for determining the feature set in the $(n+2)$th iteration, and so on. The classification accuracy of the training data at each iteration is recorded. At the end of the Nth iteration, the support vectors of the SVM with the highest training accuracy are used for selecting the final set of features. The column vectors with reduced dimensions are then used to train another SVM for classification. Figure 1.19 shows the pseudocode of VIA-SVM.

1.5.3.1 Level-Dependent VIA-SVM Selection Strategies   For the actual implementation, it is worth noting that *not* all SVs are created equal. Here, we propose four level-dependent VIA-SVM selection strategies:

Strategy 1   (*Level 1 only*) Select the most-preferred SVs, that is, select only the "pure" marginal SVs ($\alpha_i < C$) while excluding those fuzzy and outlier SVs ($\alpha_i = C$).

Strategy 2   (*Levels 1 and 2*) Select the correctly classified SVs only, that is, select the "pure" marginal SVs ($\alpha_i < C$) and the correctly classified SVs that are falling on the fuzzy region ($\alpha_i = C$ and $0 < \xi_i < 1$).

Strategy 3   (*Levels 1–3*) Remove the nonpreferred outlier SVs, that is, only keep those SVs with $\xi_i < 2$.

---

[8]Under some rare situations in which extremely small numbers of features are desirable, we may use α_i to further divide Level 1 into sublevels. However, this subdivision is unlikely to be useful in bioinformatic applications.

Algorithm VIA-SVM

Input: $\mathbf{X} = [\mathbf{x}_1 \ \ \mathbf{x}_2 \ \ \cdots \ \ \mathbf{x}_T], \mathbf{y} = [y_1 \ \ y_2 \ \ \cdots \ \ y_T],$

and $C \in \mathfrak{R}$, where $\mathbf{x}_t \in \mathfrak{R}^T$, $y \in \{+1,-1\}$

Initialization: $\mathbf{X}' = \mathbf{X}$;

for k = 1 to N
do

> // *Train an SVM to obtain the indexes to the support vectors in* **i**
> $[\boldsymbol{\alpha}, b, \mathbf{i}] = \text{SVM\_train}(\mathbf{X}', \mathbf{y}, C)$;

> // *Find the support vector indexes* **j** *such that* $0 < \alpha_j \le C$ *and* $0 \le \xi_j < 2$
> // *where* $\xi_j = 1 - \alpha_j(\mathbf{x}_j \cdot \mathbf{w} + b)$ *and* $\mathbf{w} = \sum_{i \in i} \alpha_i y_i \mathbf{x}_i$
> $\mathbf{j}\{k\} = \text{find\_sv\_index}(\mathbf{X}', \mathbf{y}, C, \boldsymbol{\alpha}, b, \mathbf{i})$;

> // *Select a feature subset based on the selected support vector indexes*
> $\mathbf{X}' = \mathbf{X}(\mathbf{j}\{k\}, :)$;

> // *Train another SVM based on the selected features*
> $[\boldsymbol{\alpha}, b, \mathbf{i}] = \text{SVM\_train}(\mathbf{X}', \mathbf{y}, C)$;

> // *Computing classification accuracy on training data*
> $a(k) = \text{SVM\_test}(\mathbf{X}', \mathbf{y}, \boldsymbol{\alpha}, b, \mathbf{i})$;

end
Output: $\mathbf{j}\{k'\}$ where $k' = \arg\max_k a(k)$

Figure 1.19   Pseudocode of VIA-SVM.

Strategy 4   (*Levels 1–4*) Select ALL SVs, that is, select all marginal and fuzzy SVs with $0 < \alpha_i \le C$.

Experiments on subcellular localization support that Strategies 2 and 3 appear to produce the best performance.

1.5.3.2 Comparing VIA-SVM and SVM-RFE

Although both VIA-SVM and SVM-RFE are based on SVMs, they do have an important difference in terms of information used. Figure 1.20 illustrates the information used by VIA-SVM and SVM-RFE in selecting and ranking features. Clearly, SVM-RFE uses the weight

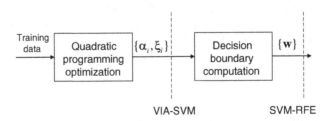

Figure 1.20   Algorithmic difference between VIA-SVM and SVM-RFE. The former uses the Lagrange multipliers α_i and slack variables ξ_i, whereas the latter uses the weight vector **w**.

vector **w**, whereas VIA-SVM uses the Lagrange multipliers α_i and slack variables ξ_i. In terms of the usefulness of the respective parameters, there are key similarity and difference:

- *Similarity.* Because of the symmetric property and diagonal dominance of the scoring matrix **Z**, a large α_i will lead to a large w_i, as exemplified in Fig. 1.17. Therefore, when all training samples are linearly separable, that is, $\xi_i = 0 \ \forall i$, VIA-SVM and SVM-RFE share a similar feature selection strategy. Such a similarity will play a role when α_i have different values (cf. Region 1 in Fig. 1.18). However, such situation can only occur in the ideal, linearly separable case.

- *Difference.* In reality, most training samples are not linearly separable even in the kernel space. In such cases, many support vectors satisfy $\alpha_i = C$ and $\xi_i > 0$, that is, they belong to Regions 2–4 in Fig. 1.18. The problem is that SVM-RFE offers no means to directly differentiate the three regions; particularly, it cannot discriminate whether the support vectors are outliers or not, because they all share the same constant $\alpha_i = C$. In contrast, VIA-SVM is sensitive to the value of $\xi_i > 0$. In particular, the VIA scheme uses ξ_i to divide support vectors into three levels of preferences (Levels 2–4 in Fig. 1.18) when $\alpha_i = C$. The uses of ξ_i to select support vectors (features) is intuitive appealing because when $\xi_i > 0$, α_i—which is a constant—can no longer provide information regarding the degree of relevance of a feature. This additional information provided by ξ_i plays an important role in selecting more vital and "relevant" features and disregarding the "misleading" (i.e., outlier) ones.

1.5.4 Combined with Other Selection Schemes

1.5.4.1 *Redundancy Removal for VIA-SVM* A common weakness of both VIA and RFE approaches is that they do not explicitly consider the redundancy factor. This will result in wasteful selection because some of the selected features can be either repetition of each other or highly redundant. To minimize feature redundancy, we propose two pruning methods for VIA-SVM.

1. *Euclidean Distance.* For those support vectors in Region 2 or 3, their pairwise Euclidean distances (eDist) in the reduced kernel space (defined by the selected feature dimensions) are examined. If two features (with similar ξ_i) are close to each other, one of them may be removed without affecting the decision boundary. The degree of closeness can be defined by multiplying the average distance of all support vectors by a user-defined constant η. More specifically, we remove feature j if $d(\zeta_r^{(i)}, \zeta_r^{(j)}) < \eta \bar{d}$, where $\bar{d}$ represents the average distance and the subscript r signifies that the distance is evaluated in the reduced kernel space.

2. *K-Means.* Support vectors that have similar ξ_i in Region 2 or 3 are clustered by *K*-means in the reduced kernel space. The SVs closest to the centers are retained and all remaining SVs in Regions 2 and 3 are removed.

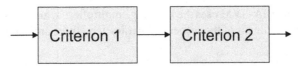

Figure 1.21   In the cascade fusion architecture, Criterion 1 is meant for coarse and fast over-selection and Criterion 2 represents pruning. The threshold for Criterion 1 can be more relaxed (i.e., its value does not have to be very close to the optimal number of features) to include more candidate features for the second stage.

If the allowable number of features is very small, the redundance-reduced feature set can be further pruned by a filter method such as symmetric divergence. Results of cascading VIA-SVM, redundancy removal, and feature pruning will be shown in Section 1.5.5.

Note that while SVM-RFE can also apply a postprocessing redundancy removal selection process (such as applying K-means or double checking the similarity score in the kernel matrix), it does not enjoy the numeric information provided by ξ_i.

1.5.4.2 Fusion of Selection Criteria   It is natural to combine/consider both the filter and wrapper methods in order to reach an optimal strategy. Specifically, the features and/or criterion functions obtained from the filter and wrapper methods can be combined via a cascaded fusion scheme shown in Fig. 1.21.

It is desirable to seek an optimal compromise between accuracy and cost. A possibility is via a comprehensive selection procedure comprising two consecutive phases. Here, we refer to these procedures as overselect-and-prune strategy. The strategy has two steps.

Step 1   *Overselection Phase.* This is a stage in which only those obviously irrelevant features are quickly weeded out. Preferably, it involves a quick and coarse (suboptimal) evaluation, for example, individual ranking. The goal is to have most unwanted features filtered out while guarantee a retention rate of the eligible features. This phase can be implemented via Criterion 1 of a cascade architecture shown in Fig. 1.21.

Step 2   *Pruning Phase.* The second stage may serve as a fine-tuning process. The goal is to remove redundant features with minimum information loss. If major redundancy exists between A and B, then one of the two may be pruned without incurring much loss of information.[9]

1.5.5 Experiments on Subcellular Localization

Two data sets were used for evaluating the performance of VIA-SVM and for comparing it against other feature selection algorithms. The first data set is provided by Reinhardt and Hubbard [3]. It comprises 2427 amino acid sequences extracted from

---

[9]The second stage may also serve as a second opinion. This can be implemented by a cascaded architecture (Fig. 1.21) with two complementary criteria applied in series. More precisely, a feature can be retained only if it has good scores in both criteria.

SWISSPROT 3.3, with each protein annotated with one of the four subcellular locations: cytoplasm, extracellular, mitochondrial, and nuclear. The second data set was provided by Huang and Li [25]. It was created by selecting all eukaryotic proteins with annotated subcellular locations from SWISSPROT 41.0 and by setting the identity cutoff to 50%. The data set comprises 3572 proteins (622 cytoplasm, 1188 nuclear, 424 mitochondrial, 915 extracellular, 26 golgi apparatus, 225 chloroplast, 45 endoplasmic reticulum, 7 cytoskeleton, 29 vacuole, 47 peroxisome, and 44 lysosome). We used fivefold cross-validation for performance evaluation so that every sequence in the data sets will be tested.

1.5.5.1 *Performance of VIA-SVM*

Now let us discuss the case study results based on the four selection strategies mentioned in Section 1.5.3.1. Because Strategy 1 includes only very few SVs, the features are extremely underselected. In contrast, because Strategy 4 includes all SVs regardless of their types, it is likely to cause overselection, particularly when the penalty factor C is small. In Strategy 2, all the SVs that are incorrectly misclassified will be excluded. However, this may lead to underselection as there are some useful SVs falling on the fuzzy regions. Strategy 3 is a compromise between the overselection in Strategy 2 and the underselection in Strategy 4. More exactly, Strategy 3 excludes all the outlier SVs. (The SVs lying beyond the margin of the opposite class are deemed to be outliers.) In this strategy misclassified SVs that lie within the margin of separation will still be selected, leading to overselection, especially when the penalty factor C is very small.

Figure 1.22 shows the performance of Strategies 1–4 when the penalty factor C varies from 0.004 to 4096. Note that the number of selected features (feature dimension) is automatically determined by the SVMs. For each strategy, a smaller penalty factor C will generally lead to a larger number of features, and vice versa for a larger C. Therefore, markers on the right region of the figure correspond mainly to small C's. The results show that the optimal number of features found by Strategy 4 is considerably higher than those found by the other strategies. Notwithstanding the larger number of features, the maximum accuracy attained by Strategy 4 is still lower than those achieved by the other strategies. This confirms our earlier hypothesis that including all SVs will lead to overselection. Results also show that Strategy 1 will lead to underselection when the penalty factor C becomes large. These case studies suggest that Strategies 2 and 3, which exclude either the nonpreferred SVs or both the marginally preferred and nonpreferred SVs (cf. Fig. 1.18), seem to be the least sensitive to the penalty factor, because they can keep the number of features within a small range and maintain the accuracy at a constant level for a wide range of C.

1.5.5.2 *Comparison Between VIA-SVM and SVM-RFE*

We compared the proposed VIA-SVM (Strategy 2) with SVM-RFE [15] in the subcellular localization benchmarks mentioned earlier.[10] Note that SVM-RFE does not make use of the

---

[10]Because the performances of SVM-RFE, R-SVM, and symmetric divergence are comparable in the two benchmarks, we only report the results of SVM-RFE for clarity of presentation.

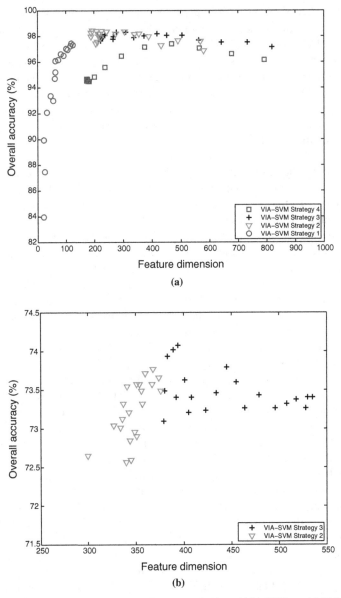

Figure 1.22 Prediction performance of different strategies of VIA-SVM on (a) Reinhardt and Hubbard's data set and (b) Huang and Li's data set when the penalty factor C varies from 0.004 to 4096. See Section 1.5.5 for details of the strategies.

symmetric property of the pairwise scoring matrices in the selection process, because it is primarily designed for gene selections in microarray data where expression matrices are neither square nor symmetric.

Figures 1.23 and 1.24 show the performance of SVM-RFE (–□– and VIA-SVM (○). Evidently, VIA-SVM is superior to SVM-RFE in two aspects: (1) It outperforms SVM-RFE at almost all feature dimensions, particularly at low feature dimensions, and (2) it automatically bounds the number of selected features within a small range. A drawback of SVM-RFE is that it requires a cutoff point for stopping the selection. However, VIA-SVM is insensitive to the penalty factor in SVM training and can avoid the need to set a cutoff point for stopping the feature selection process.

1.5.5.3 Fusion of VIA-SVM and SD

Given a particular axis, which corresponds to one feature, two factors can be considered: VIA-SVM parameters (C and ξ_i) and symmetric divergence of feature i (SD_i). We adopt the overselect-and-prune cascaded fusion architecture, as proposed in Section 1.5.4.2, to combine VIA-SVM and SD. Based on this cascade fusion strategy, the selection process is divided into two stages.

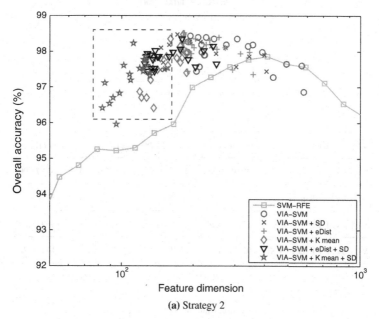

(a) Strategy 2

Figure 1.23 Prediction performance of SVM-RFE, VIA-SVM, and VIA-SVM cascaded with various pruning methods on Reinhardt and Hubbard's data set. (a) The VIA-SVM uses Strategy 2 for feature selection. (b) Strategy 3 was used. (c) The means and standard derivations (in parentheses) of the classification accuracies and feature dimensions for 21 penalty factors ranging from 0.004 to 4096; for SVM-RFE in (c), the means and standard derivations are based on 11 points in (a) and (b) whose feature dimensions range from 95 to 591. Pruning was applied according to the order indicated in the legend; for example, "VIA-SVM + eDist + SD" means that the features selected by VIA-SVM were pruned by eDis-based method followed by the symmetric divergence-based method.

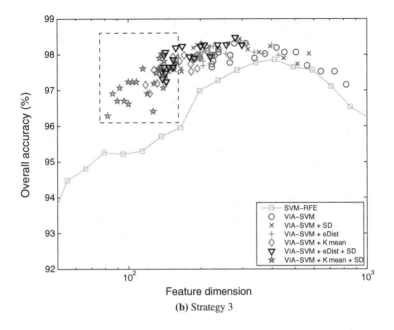

(**b**) Strategy 3

	Accuracy (%)		Feature Dimension	
Method	Strategy 2	Strategy 3	Strategy 2	Strategy 3
VIA-SVM	97.96 (0.42)	97.90 (0.28)	299 (125.20)	370 (183.18)
VIA-SVM + SD	97.84 (0.32)	97.96 (0.28)	210 (87.60)	259 (128.18)
VIA-SVM + eDist	98.05 (0.32)	98.03 (0.19)	230 (56.02)	261 (71.00)
VIA-SVM + Kmean	97.65 (0.58)	97.68 (0.34)	161 (23.47)	166 (28.81)
VIA-SVM + eDist+SD	97.82 (0.24)	97.92 (0.35)	161 (39.16)	183 (49.69)
VIA-SVM + Kmean + SD	97.30 (0.55)	97.19 (0.48)	112 (16.49)	116 (20.24)
SVM-RFE	96.81 (1.04)		279 (163.23)	

(**c**) Accuracy and feature dimension

Figure 1.23   (*Continued*).

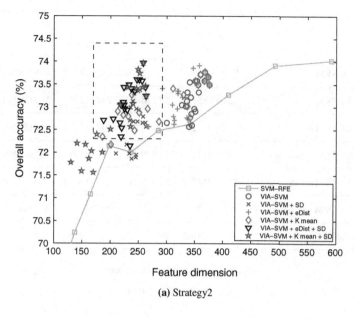

(a) Strategy2

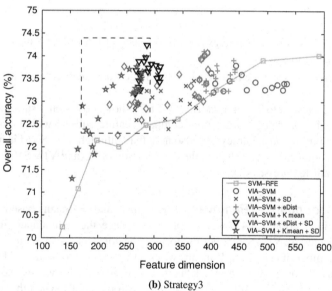

(b) Strategy3

Figure 1.24   Same as Fig. 1.23 but based on Huang and Li's data set. For SVM-RFE in (c), the means and standard derivations are based on nine points in (a) and (b) whose feature dimensions range from 114 to 492.

Method	Accuracy (%)		Feature Dimension	
	Strategy 2	Strategy 3	Strategy 2	Strategy 3
VIA-SVM	73.23 (0.38)	73.48 (0.27)	348 (17.22)	445 (55.77)
VIA-SVM + SD	72.66 (0.51)	73.12 (0.41)	244 (12.07)	312 (39.01)
VIA-SVM + eDist	73.23 (0.42)	73.56 (0.29)	332 (27.77)	408 (22.69)
VIA-SVM + Kmean	73.09 (0.47)	73.26 (0.48)	296 (65.29)	327 (61.84)
VIA-SVM + eDis + SD	73.11 (0.47)	73.63 (0.27)	232 (19.35)	286 (15.83)
VIA-SVM + Kmean + SD	72.73 (0.78)	72.93 (0.71)	207 (45.76)	229 (43.39)
SVM-RFE	71.90 (1.45)		264 (129.10)	

(c) Accuracy and feature dimension

Figure 1.24   (*Continued*).

Stage 1   Use VIA-SVM (Strategy 2 or 3) to select all-but-outlier SVs, that is, only keep those with $\xi_i < 2$.

Stage 2   Use SD to sort the features found in Stage 1 and keep the most relevant $x\%$.

In this work, we set x to 70. Figures 1.23 and 1.24 show the fusion results ($\times$), which suggest that fusion can produce more compact feature subsets without significant reduction in prediction accuracy. We also note that although VIA-SVM is inferior to SVM-RFE for large feature-set size, the combination of SD and VIA-SVM performs better at small feature-set size.

1.5.5.4 Redundance Removal   The eDist- and K-means-based methods mentioned in Section 1.5.4.1 were applied to reduce the redundance among the features selected by VIA-SVM. For the former, the constant η was set to 0.3; for the latter, the number of centers in K-means was set to 100 for Reinhardt and Hubbard's data set and 300 for Huang and Li's data set. The setting of these values was based on the observation from Fig. 1.22 that the accuracy drops rapidly when the number of features is smaller than these values. The results ($+$ and $\diamond$) shown in Figs 1.23 and 1.24 suggest that both methods can reduce the feature size without scarifying accuracy. This once again demonstrates the merit of using the slack variables ξ_i in selecting features.

To further reduce the feature size, we applied SD to prune the features after redundance removal. Figures 1.23 and 1.24 (∇ and $\star$) show that the resulting feature sets achieve a significantly higher accuracy when compared with SVM-RFE.

The tables in Figs 1.23c and 1.24c compare Strategy 2 and Strategy 3 in terms of the mean accuracy and mean feature dimension obtained by VIA-SVM and VIA-SVM cascaded with various pruning methods. Three observations can be obtained from these tables:

1. For each pruning method, there is no significant difference between the accuracy obtained by Strategies 2 and 3.
2. Strategy 2 generally leads to a smaller number of features than Strategy 3 with almost the same performance statistically.
3. SVM-RFE not only gives lower average accuracies but also leads to a larger variation in both accuracy and feature dimension.

These observations suggest that Strategy 2 is a winner because it can keep the number of features to a minimum without scarifying accuracy. Strategy 2 is also a better choice for applications where feature dimension (and hence recognition time) should be kept to a minimum. However, for applications where accuracy is of primary importance, Strategy 3 is a better option.

1.5.5.5 *Range of Desirable Feature Dimension*   In most pattern recognition problems, a smaller feature size can result in faster recognition speed, but at the expense of lower classification accuracy. The opposite situation occurs for large (but not excessively large) feature size. We advocate that there is a range of desirable feature dimension for a particular problem and that whether the lower or upper limit of the range should be used depends on the applications. For example, in biometric applications where real-time recognition is essential, we may opt for minimum feature size that produces the best compromise between recognition speed and accuracy. However, in bioinformatic applications where accuracy is far more important than speed, we may prefer using the upper limit that gives high accuracy but not to the point that causes the curse of dimensionality. This notion of desirable range of feature dimension is highlighted by the dashed rectangles in Figs 1.23 and 1.24. Evidently, the redundance removal and the cascade fusion of VIA-SVM and SD enable us to find the feature sizes falling on the desirable range. In particular, when recognition speed is a concern, we can overselect features by using VIA-SVM (Strategy 2 or 3) and then remove feature redundance by using K-means, and finally, we can further prune the features by using SD ("VIA-SVM + Kmean + SD"). On the other hand, if accuracy is more important, we may skip the final pruning stage, that is, using "VIA-SVM + Kmean", or do not use pruning at all.

1.6 CONCLUSION

This chapter discusses the feature selection methods from many different aspects, including selection criteria, the order of feature selection, and open-loop (filter-type) and closed-loop (wrapper-type) approaches.

A special method (VIA-SVM) designed exclusively for pairwise scoring kernels is introduced. This is the first method that fully utilizes the reflexive property uniquely available in this scenario. Based on several subcellular localization experiments, the VIA-SVM when combined with some filter-type metrics appears to reach a significantly dimension reduction (one-order of magnitude) while conceding minimum sacrifice of accuracy.

It is well known that fusion of complementary criteria or methods can substantially enhance classification performance. While this chapter has offered several exemplar cases of successful fusion, there are still more variants of fusion techniques that need to be pursued. This appears to be a promising direction worth further investigation.

ACKNOWLEDGMENTS

This work was in part supported by The Research Grant Council of the Hong Kong SAR (Project Nos. PolyU5241/07E and PolyU5251/08E.)

REFERENCES

1. Guyon, I. and Elisseeff, A. An introduction to variable and feature selection. *Journal of Machine Learning Research*, 3: 1157–1182, 2003.
2. Mak, M. W. Guo, J. and Kung, S. Y. PairProSVM: Protein subcellular localization based on local pairwise profile alignment and SVM. *IEEE/ACM Transaction on Computational Biology and Bioinformatics*, 5 (3): 416–422, 2008.
3. Reinhardt, A. and Hubbard, T. Using neural networks for prediction of the subcellular location of proteins. *Nucleic Acids Research*, 26: 2230–2236, 1998.
4. Mak, M. W. and Kung, S. Y. A solution to the curse of dimensionality problem in pairwise scoring techniques. *International Conference on Neural Information Processing*, pp. 314–323, 2006.
5. Golub, T. R., Slonim, D. K., Tamayo, P., Huard, C., Gaasenbeek, M., Mesirov, J. P., Coller, H., Loh, M., Downing, J. R., Caligiuri, M. A., Bloomfield, C. D., and Lander, E. S. Molecular classification of cancer: class discovery and class prediction by gene expression monitoring. *Science*, 286: 531–537, 1999.
6. Hyvarinen, A. and Oja, E. Independent component analysis: algorithms and applications. *Neural Networks*, 13: 411–430, 2000.
7. Pan, W. A comparative review of statistical methods for discovering differentially expressed genes in replicated microarray experiments. *Bioinformatics*, 18: 546–554, 2002.
8. Pavlidis, P., Weston, J., Cai, J., and Grundy, W. N. Gene functional classification from heterogeneous data. *International Conference on Computational Biology, Pittsburgh, PA*, pp. 249–255, 2001.
9. Furey, T. S., Cristianini, N., Duffy, N., Bednarski, D. W., Schummer, M., and Haussler, D. Support vector machine classification and validation of cancer tissue samples using microarray expression data. *Bioinformatics*, 16: 906–914, 2000.

10. Rosenblatt, F. The perceptron: a probabilistic model for information storage and organization of the brain. *Psychology Review*, 65: 42–99, 1958.

11. Fisher, R. A. The use of multiple measurements in taxonomic problems. *Annals of Eugenics*, 7: 179–188, 1936.

12. Vapnik, V. N. *The Nature of Statistical Learning Theory*, Springer-Verlag, New York, 1995.

13. Boser, B. E., Guyon, I. M., and Vapnik, V. N. A training algorithm for optimal margin classifiers. In: D. Haussler, Ed., *Proceedings of the 5th Annual ACM Workshop on Computational Learning Theory*, pp. 144–152, 1992.

14. Kohavi, R. and John, G. H. Wrappers for feature selection. *Artificial Intelligence*, 97(1–2): 273–324, 1997.

15. Guyon, I., Weston, J., Barnhill, S., and Vapnik, V. Gene selection for cancer classification using support vector machines. *Machine Learning*, 46: 389422, 2002.

16. Duan, K.-B., Rajapakse, J. C., Wang, H., and Azuaje, F. Multiple SVM-RFE for gene selection in cancer classification witn expression data. *IEEE Transactions on Nanobioscience*, 4(3): 228–234, 2005.

17. Zhang, X. G., Lu, X., Shi, Q., Xu, X. Q., Leung, H. C. E., Harris, L. N., Iglehart, J. D., Miron, A., Liu, J. S., and Wong, W. H. Recursive SVM feature selection and sample classification for mass-spectrometry and microarray data. *BMC Bioinformatics*, 7(197): 2006.

18. Kira, K. and Rendell, L. A practical approach to feature selection. *International Conference on Machine Learning, Aberdeen*, pp. 368–377, July 1992.

19. Liao, L. and Noble, W. S. Combining pairwise sequence similarity and support vector machines for detecting remote protein evolutionary and structural relationships. *Journal of Computational Biology*, 10(6): 857–868, 2003.

20. Kim, J. K., Raghava, G. P. S., Bang, S. Y., and Choi, S. Prediction of subcellular localization of proteins using pairwise sequence alignment and support vector machine. *Pattern Recognition Letters*, 27(9): 996–1001, 2006.

21. http://www.expasy.org/sprot.

22. Altschul, S. F., Madden, T. L., Schaffer, A. A., Zhang, J., Zhang, Z., Miller, W., and Lipman, D. J. Gapped BLAST and PSI-BLAST: a new generation of protein database search programs. *Nucleic Acids Research*, 25: 3389–3402, 1997.

23. Smith, T. F. and Waterman, M. S. Comparison of biosequences. *Advances in Applied Mathematics*, 2: 482–489, 1981.

24. Gotoh, O. An improved algorithm for matching biological sequences. *Journal of Molecular Biology*, 162: 705–708, 1982.

25. Huang, Y. and Li, Y. D. Prediction of protein subcellular locations using fuzzy K-NN method. *Bioinformatics*, 20(1): 21–28, 2004.

26. Kung, S. Y., and Mak, M. W. Feature selection for self-supervised classification with applications to microarray and sequence Data. *IEEE Journal of Selected Topics in Signal Processing, Special Issue on Genomic and Proteomic Signal Processing*, 2 (3): 297–309, 2008.

2

COMPARING AND VISUALIZING GENE SELECTION AND CLASSIFICATION METHODS FOR MICROARRAY DATA

Rajiv S. Menjoge and Roy E. Welsch

2.1 INTRODUCTION

The development of microarray technology has brought with it problems that are interesting, both from a statistical perspective and from a biological perspective. The data sets produced in this application contain the expressions of several patients' genes and indicate whether or not each patient has cancer. Two problems of interest can be solved using models from statistics and data mining. The first problem is to learn the relationship between the patient's gene expressions and his condition (cancer or no cancer), so that if a new patient's gene expressions are given, he can be diagnosed with reasonable accuracy. This problem, if executed efficiently enough, could be used for medical diagnosis. Statistically, it is a classification problem, where the predictor variables are the patient's genes and the response is the indication of whether or not the patient has cancer. The second problem is to find a relatively small subset of genes that demonstrate the differences between cancer patients and noncancer patients. This problem could help identify interesting genes for subsequent research and could

Machine Learning in Bioinformatics. Edited by Yan-Qing Zhang and Jagath C. Rajapakse
Copyright © 2009 by John Wiley & Sons, Inc.

reduce the amount of input data required for diagnosis. Statistically, this is a problem in feature selection. This chapter focuses on methods that do both to some degree. The two objectives to some extent conflict, but many methods exist that do a relatively good job at both tasks.

The major statistical challenge these problems present is that the number of predictor variables, p, is often in the thousands, while the number of observations, n, is only in the hundreds. This makes model fitting more difficult because most standard statistical methods are constructed for the case when n is greater than p and in some cases even fail to work if this is not the case. Nevertheless, there is extensive literature on this topic. Diaz-Uriarte and Alvarez de Andres [1] give a thorough comparison and review of popular classification methods and feature selection methods and analyze them on microarray data sets publicly available at ligarto.org/rdiaz/Papers/rfVS/randomForestVarsel.html [2]. This chapter displays some of the paper's results and discusses the methods used. In addition, this chapter describes other methods, which are also known to work well, and compares them with the methods described in Ref. [1] on some of the data sets. The qualitative behavior of some of the feature subsets is then explored through plots. Specific details about the data sets used in this chapter are given in Fig. 2.1.

In the descriptions of the methods below, $\mathbf{x}$ denotes the 1 by p vector of predictors and y denotes the response. Meanwhile, $\mathbf{X}$ denotes the n by p matrix that contains the n realizations of the variable $\mathbf{x}$, and $\mathbf{y}$ denotes the n by 1 vector that contains the n realizations of the variable y. Specifically, the ith row of $\mathbf{X}$, denoted $\mathbf{x}_i^T$, contains the expressions of the ith patient's genes and the ith entry of $\mathbf{y}$, denoted y_i, equals the ith patient's status (-1 if he does not have cancer and 1 if he does). In this notation, the goal of classification is to find a relationship between the $\mathbf{x}^T$ and y, given the realized values, $\mathbf{X}$ and $\mathbf{y}$. With this knowledge, an analyst, when given the expressions of a new patient's genes, would be able to predict whether or not the patient has cancer with reasonable accuracy. The goal of feature selection is to select a subset of columns of $\mathbf{x}^T$ such that if one were to use only those columns (genes) to learn the relationship between $\mathbf{x}^T$ and y, one could still predict new cases with similar accuracy.

Data Set	Genes	Patients
Leukemia	3051	38
Breast	4869	96
Colon	2000	62
Prostate	6033	102

Figure 2.1 The data sets.

2.2 METHODS

2.2.1 General Approaches

There are several general approaches to selecting variables. One is to attempt to rank variables in terms of importance. This is useful in that it can allow the analyst to choose the size of the subset of variables he wishes to study, and an analyst after studying a given number of variables can get an immediate idea of where to look next. The difficulty with these approaches is that the best subset of a certain number of variables may contain entirely different variables than the best subset of a different number of variables.

Another approach to selecting variables is to use a classification algorithm that has feature selection as a side effect. The first two methods discussed in this section are of the first type, the second two are of the second type, and the last method can, in some sense, be viewed in either way.

2.2.2 Nearest Neighbor + Variable Selection

Perhaps the most intuitive way of ranking variables in terms of importance is by considering each variable individually and computing how the variable values for patients with cancer differ from the variable values for patients without cancer. Variables, on whose values cancer patients and noncancer patients differ more, are ranked as more important. Of course, the units on which these differences are measured have to be uniform among the variables, so this difference is divided by a quantity that approximates the standard deviation. This normalized difference is called the t-statistic and its square can be used to rank variables. For column j, let X_{j0} be the column including only those patients without cancer and let X_{j1} be the column including only those patients with cancer. Let n_0 be the number of patients without cancer and let n_1 be the number of patients with cancer. The t-statistic is given by the following formula:

$$t = \frac{\bar{X}_{j0} - \bar{X}_{j1}}{\sqrt{\frac{\text{var}(\mathbf{X}_{j0}) * (n_0 - 1) + \text{var}(\mathbf{X}_{j1}) * (n_1 - 1)}{n_0 + n_1 - 2} * \left(\frac{1}{n_0} + \frac{1}{n_1}\right)}},$$

assuming the variances are approximately equal.

The F-ratio generalizes this idea to multiple classes and is a common tool in statistics for hypothesis testing. Basically, the F-ratio is the ratio between the amount that the class means (in the two-class case, the cancer patient mean and noncancer patient mean) differ from the global mean and the amount that observations in general deviate from their class means. For column j, let $\mathbf{X}_{jk}$ be the column including only those patients in class k. Let n_k be the number of patients in class k. Let $\bar{\bar{X}}_j$ denote the global mean, n denote the number of patients, and K denote the number of classes. Then the F-ratio is given by the following formula:

$$F_j = \frac{\dfrac{\sum\limits_{k=1}^{K} n_k (\bar{\mathbf{X}}_{jk} - \bar{\bar{\mathbf{X}}}_j)^2}{K - 1}}{\dfrac{\sum\limits_{k=1}^{K} \mathrm{var}(\mathbf{X}_{jk}) * (n_k - 1)}{n - K}}.$$

A high F-ratio for a column would indicate that patients in different classes differ significantly in their expressions of the corresponding gene, and hence the column is important. Variables are ranked according to their F-ratios. The F-ratio is a simple and intuitive technique for obtaining variable importance and is often used as a quick way to filter out variables that are clearly not important but may suffer from multiple comparison problems. With this simplicity, however, comes the cost that the F-ratio only provides univariate rankings and does not incorporate the interactions of variables. Also, if two variables are identical, it will select both of them, which is not optimal behavior if a very small set of variables is required.

Nevertheless, when paired with an intuitive classifier called nearest neighbor classification, the F-ratio still performs well. Nearest neighbor classification classifies in a way that is somewhat consistent with a human's heuristic for classification. Given a new sequence, $\mathbf{x}_{\text{new}}$, the algorithm compares the sequence to each row and predicts the class, y_{new}, to be the class corresponding to the most similar row, where similarity is generally measured by the Euclidean distance. This works well with the F-ratio rankings because the F-ratio selects features in a way that attempts to separate classes well when comparing individual variables.

In nearest neighbor + variable selection, these two methods are combined. The genes are ranked based on their F-ratio. The parameter to select is then the size of the subset of variables. The following procedure is used. In an iteration, the data set is split into 10 approximately equal parts. At each step, one of these parts is left-out. Using the nine remaining parts, the genes are ranked based on their F-ratio. Then, the responses for the left-out part are predicted and an error is computed. Following this, 20% of the genes with the lowest F-ratio are eliminated and the procedure is repeated until there are very few variables left. The errors are averaged over the 10 iterations and the final number of genes is the one that produces the smallest error rate [1].

2.2.3 Random Forests

Another way to rank variables is to first pick a classifier, fit the classifier, and then figure out which variable is least useful for the particular fit. That variable can then be eliminated and the classifier can be rerun. By ranking the eliminated variable last at each step, one can obtain a full ranking of the variables.

The classifier we consider in this case is one of the members of a popular family of classification models known as classification trees. The idea behind classification trees is that the relationship between x and y is a series of questions. For instance, for a given x_{new}, a response of yes to the three questions—is the first entry greater than five, is the fifth entry less than nine, and is the fourth entry greater than two—asked in that order may indicate that y_{new} should equal 1, whereas a response of no to the last question may indicate that y_{new} should equal 0, and a response of no to the first question may trigger an entirely different set of questions to be asked. Classification tree methods attempt to build these trees (question structures) in a way that the sequences of questions stay reasonably short and when a training example (given value of x_i) is fed into the tree, the predicted y_i value should equal the actual y_i value most of the time. Building such a tree perfectly is a difficult optimization problem, so the existing algorithms for classification trees try to approximate the optimal tree by adding questions in a stepwise manner, where the next question is the one that improves the model's description of the existing data the most.

Random forests are described in full detail in Ref. [3]. In short, random forests fit several classification trees on the data set. To predict the response for a new observation, x_{new}, the observation is fed into each of the classification trees and a majority vote determines the class to which the observation is thought to belong. The algorithm used to fit these classification trees is, however, different from the algorithm used to fit general classification trees. In this algorithm, a random subset of columns is chosen at each step and only questions pertaining to those columns are candidates for questions that can be asked next, whereas for general classification trees, any question is a candidate at any step. The random forest algorithm, once run, outputs an estimate of the importance of each column, based on how much worse the fit would be if the columns were not used.

Random forests have several advantages. One major advantage is that they can very easily be extended to cases where there are more than two classes as well as to cases where the response is continuous. In addition, random forests are not sensitive to monotone transformations in the columns, the need for which is sometimes difficult to detect. Random forests also only require the user to specify two parameters: the number of variables selected at each step and the number of trees. The algorithm is not very sensitive to either of these, which helps ease the computational burden. Random forests also work reasonably well when $p \gg n$ and can be used independently as a classifier. Nevertheless, more parsimonious solutions can be obtained by the method varSelRF [1].

The method varSelRF uses the method described above to rank features, in that it eliminates a variable at each step and ranks this variable the lowest among the variables it is considering before the step. The variable to be eliminated at each step is given by the random forest algorithm itself. Once features are ranked, the number of features in the final model can be chosen in a similar way to nearest neighbors + variable selection. Here, however, random forests are fit to estimate the prediction error instead of nearest neighbors, since random forests correspond to the way variables were selected. In the varSelRF s.e. $= 0$ case, the method stops eliminating features when the cross-validation error is minimum. In the varSelRF s.e. $= 1$ case, it stops when the cross-validation error is no longer within one standard error of the minimum.

This method suffers the same disadvantages that ranking methods in general face, but has been shown to select extremely small subsets of variables, while still classifying well. In addition, varSelRF will not necessarily select redundant variables. Depending on the goals of the analyst, this could potentially be a strength or weakness. The method varSelRF is also quite useful in that it selects variables that are well suited to tree models. This is particularly useful because many software packages that fit trees return a picture of the actual tree. Because of the intuitive question-like structure of these pictures, they are particularly interpretable and could be useful in providing further insight.

2.2.4 Support Vector Machines

One method that classifies and selects features at the same time, as a side effect of the classification, is the 1-norm support vector machine (SVM). To motivate the idea behind this method, we first describe the more popular 2-norm SVM.

The 2-norm SVM is a common classification tool. The idea behind the SVM is geometric. Consider a p-dimensional scatter plot of the columns of $\mathbf{X}$, where each axis of the scatter plot represents values for a particular gene. Each patient (row) can then be represented by a point in such a plot (Figs. 2.4–2.6 show examples of this for $p = 2$). The SVM attempts to separate the points $\mathbf{x}_i$ for which $y_i = 1$ from those for which $y_i = -1$ with a certain kind of boundary. In the linear case, which is the case that is generally considered for the classification of microarray data because of its simplicity, the SVM attempts to separate the points with a hyperplane. The hyperplane is chosen so that the distance to the closest point on either side is maximized. In two dimensions, one could visualize this as separating the two classes with the widest possible tube. This width of the separating tube is called the margin and the line that is at the center of the tube is the separating hyperplane. Of course, it is possible that no hyperplane can separate the two groups for a particular data set. In this case, a penalty is added for the extent to which the hyperplane fails to separate the two groups. Mathematically, the linear SVM assumes that the relationship between y and $\mathbf{x}$ is of the form $y = \text{sign}(\beta_0 + \mathbf{x}^T\boldsymbol{\beta})$. Using some mathematics, one can show that the $\boldsymbol{\beta}$ and β_0 that are closest to achieving the goal of the SVM given the realized data are those that minimize

$$\sum_{i=1}^{n} \max(0, 1 - y_i(\beta_0 + \mathbf{x}_i^T\boldsymbol{\beta})) + p_1||\boldsymbol{\beta}||^2.$$

The equation, as written in the above form, also has a different interpretation. The first term is a loss function, sometimes used for classification called the hinge loss, and the second term is the 2-norm of the coefficient vector. Hence, the first term ensures that the model describes the existing data well and penalizes the extent to which it fails to do so, and the second term can be interpreted both as an attempt to keep the margin large and as an attempt to keep the coefficient vector small, which often helps classifiers generalize to other data sets more effectively. This version of the SVM, referred to as the 2-norm SVM, works well even in the cases where $p \gg n$. However, it does not have a feature selection capability built in like the 1-norm SVM described below.

The 1-norm SVM is introduced and discussed in detail in Ref. [4]. The modification is that the 2-norm in the optimization problem is replaced by a 1-norm. This keeps the goal somewhat similar and turns the problem into a linear programming problem. Linear programming problems have the characteristic that optimal solutions, when they exist, occur at vertices of the constraint set and it turns out that when the 1-norm SVM is formulated as a linear programming problem, the vertices of the constraint set occur when many elements of the $\boldsymbol{\beta}$ vector are zero. Hence, feature selection occurs as a side effect of searching for the optimal $\boldsymbol{\beta}$ and β_0 for the SVM. Formally, where $|\boldsymbol{\beta}|$ denotes the 1-norm of $\boldsymbol{\beta}$, the model for the 1-norm SVM is $y = \text{sign}(\beta_0 + \mathbf{x}\boldsymbol{\beta})$, where β_0 and $\boldsymbol{\beta}$ are chosen to minimize

$$p_1|\boldsymbol{\beta}| + \sum_{i=1}^{n} \max(0, 1 - y_i(\beta_0 + \mathbf{x}_i^T\boldsymbol{\beta})).$$

Several efficient algorithms exist for solving the 1-norm SVM. Zhu et al. [4] propose a forward selection algorithm for solving the optimization problem. This algorithm has the advantage that it can find solutions for models, in which various different numbers of variables are desired in an efficient manner. The 1-norm SVM can also be solved using standard linear programming software, such as MATLAB. The advantage of this is that the analyst can be flexible in the way he formulates the linear programming problem. The exact formulation for the problem as it is stated above can be found by letting $\boldsymbol{\beta}^+ = \max(\boldsymbol{\beta}, \mathbf{0})$ and $\boldsymbol{\beta}^- = \max(-\boldsymbol{\beta}, \mathbf{0})$, so that $\boldsymbol{\beta} = \boldsymbol{\beta}^+ - \boldsymbol{\beta}^-$ and $|\boldsymbol{\beta}| = \boldsymbol{\beta}^+ + \boldsymbol{\beta}^-$. The formulation is

$$\min \sum_{i=1}^{n} h_i + p_1 * (\boldsymbol{\beta}^+ + \boldsymbol{\beta}^-)$$
$$\text{s.t. } \forall i, h_i \geq 0, h_i \geq 1 - y_i(\beta_0 + \mathbf{x}_i^T(\boldsymbol{\beta}^+ - \boldsymbol{\beta}^-))$$
$$\boldsymbol{\beta}^+, \boldsymbol{\beta}^- \geq 0.$$

In this case, the parameter p_1 can also be used to specify the balance between the number of variables selected and the accuracy of the model. However, the formulation can vary depending on how the analyst wants to interpret the problem. For instance, the following formulation is equivalent to the original formulation for a certain value of p_1:

$$\min|\boldsymbol{\beta}|$$
$$\text{s.t. } \sum_{i=1}^{n} \max(0, 1 - y_i(\beta_0 + \mathbf{x}_i^T\boldsymbol{\beta})) \leq g*Z.$$

Here, Z represents the minimum value of the hinge loss using $\boldsymbol{\beta}$ and β_0, and g is some constant. The value $g = 1$ would give the model the interpretation of choosing the smallest set of $\boldsymbol{\beta}$ values while the training error is as low as possible.

For this chapter, a highly efficient algorithm called the NLPSVM proposed by Fung and Mangasarian [5] is used to solve the SVM 1-norm. They provide a full description of the algorithm as well as the code in the paper itself.

The generalization of this algorithm to cases in which there are more than two classes is not easy. One way of handling this is by choosing k sets of $\boldsymbol{\beta}$ and β_0, where the kth set separates the kth class from the other classes in an optimal way with respect to the SVM. Denote the kth set as $\boldsymbol{\beta}_k$ and β_{0k}. Then the class k of new observation $\mathbf{x}_{new}$ can be chosen as the k that maximizes $\beta_{0k} + \mathbf{x}_{new}\boldsymbol{\beta}_k$.

2.2.5 Shrunken Centroids

The method of shrunken centroids, which is introduced in Ref. [6], is another method that selects features as a side effect of fitting a classification model. The idea behind shrunken centroids is as follows. The data set can be partitioned into two different sets (in the case of multiple classes, k different sets): one set for patients who have cancer and one set for patients who do not have cancer. A centroid can then be computed for each of these data sets by averaging the rows. This centroid can be thought of as the average location of the class in the p-dimensional scatter plot described above. In the same way, a centroid can be computed for the pooled data set. Let the ith coordinate (gene) of the centroid of the kth class be denoted as c_{ik} and let the ith coordinate of the centroid of the overall dataset be denoted as c_i. Let s_d^2 be the pooled within-class standard deviation for gene d, which is the denominator of the F-ratio, described above. Let s_0 be the median of the values of the standard deviations, s_i, over the set of genes and define the following terms by the expressions below:

$$m_k = \sqrt{\frac{1}{n_k} + \frac{1}{n}},$$

$$d_{ik} = \frac{c_{ik} - c_i}{m_k * (s_i + s_0)}. \tag{2.1}$$

Using these equations and assuming all these values have been calculated, we can also write c_{ik} as

$$c_{ik} = c_i + m_k * (s_i + s_0) d_{ik}. \tag{2.2}$$

If we let d_{ik} be defined by Equation 2.1, c_{ik} equals the ith coordinate of the centroid of the kth class. The method of shrunken centroids decreases the absolute value of each d_{ik} by a certain amount and then uses Equation 2.2 for c_{ik}. These recomputed class centroids are called shrunken centroids. The amount of shrinkage for the values of d_{ik} is a parameter selected by the analyst. A higher value of the parameter will produce more parsimonious solution because a column, i, is no longer in the model once the absolute value of d_{ik} is less than or equal to zero for each k. Here, the amount of shrinkage is determined as the amount that leads to the selection of the minimum number of variables among all solutions with the same minimum cross-validation error. Given a new gene expression sequence, $\mathbf{x}_{new}$, it is assigned to the class whose shrunken centroid is closest.

2.2.6 Elastic Net

The elastic net is a modification of the ordinary least squares (OLS) regression model and is introduced and described in detail in Ref. [7]. Assume here that y, $\mathbf{x}$, $\mathbf{y}$, and $\mathbf{X}$ are all standardized. Then, like the OLS model, the relationship between $\mathbf{x}$ and y is assumed to be of the form $y = \mathbf{x}\boldsymbol{\beta} + \boldsymbol{\varepsilon}$ for some $\boldsymbol{\beta}$, where $\boldsymbol{\varepsilon}$ is a normally distributed random variable with zero mean. In the elastic net, the value of the parameter $\boldsymbol{\beta}$ is estimated as a rescaling factor multiplied by

$$\text{argmin}_\beta (\mathbf{y} - \mathbf{X}\boldsymbol{\beta})^\mathrm{T}(\mathbf{y} - \mathbf{X}\boldsymbol{\beta}) + p_1|\boldsymbol{\beta}| + p_2||\boldsymbol{\beta}||^2.$$

Here, the single vertical line denotes the L1 norm and the double vertical line denotes the L2 norm. The first term ensures that the chosen model describes the existing data well. The second term pushes the coefficients of the columns to zero so that some of the columns can be deemed unnecessary, in the same way as they were deemed unnecessary in the SVM 1-norm. Finally, the third term ensures that the solution is unique and seeks a fit that has lower variance. It plays a similar role to the 2-norm in the SVM 2-norm. The lasso [8] and ridge regressions are special cases of the optimization problem above, where the penalties of the L2 and L1 norms are zero. In ridge regression, the $\boldsymbol{\beta}$ vector that solves the optimization problem is unique, but some elements of the vector typically are not shrunk all the way down to zero. In the lasso regression, many of the elements of the $\boldsymbol{\beta}$ vector do shrink to zero, but other issues arise when $p \gg n$. The elastic net, by combining these two penalties, produces parsimonious solutions, which also work well when $p \gg n$. The rescaling factor is $(1 + p_2)$. This decreases the bias caused by the double shrinkage. The penalties p_1 and p_2 are parameters that need to be specified by the analyst. Typically they are chosen to minimize an estimate of the test error, such as cross-validation error.

The variable ranking characteristic occurs because it turns out that there is a forward selection algorithm called LARS [9], which produces the same solutions as the lasso problem. Since LARS is a very efficient algorithm, it is generally used to solve the lasso optimization problem. An efficient way to solve the optimization problem in the elastic net is to augment the data matrix and the response vector to achieve the same effect as the penalty on the L2 norm and then use LARS on the modified data set. The LARS procedure operates in steps in such a way that a specific number of steps corresponds to a specific value of p_1. For the sake of simplicity, the analyst typically specifies the number of steps executed in the LARS algorithm rather than the penalty on the 1-norm.

The elastic net has several advantages. One major advantage is that its gene selection facility is implemented within the model itself. Therefore, the results have a very clear meaning. The coefficients that are zero correspond to those columns that were not needed in the model. The model itself is only fit once and eliminates columns in a continuous manner. Hence, it should not have as high a variance as a procedure that eliminates columns in stages. Also, the parameters have an intuitive meaning. The number of steps controls how many variables the analyst wants in the model and the L2 penalty controls the variance of the model. The elastic net also operates successfully when $p \gg n$ and so needs no prior variable filtering. Finally, the elastic net provides a method of obtaining approximate variable ranks in addition to selecting variables as a side effect of the classification fit. One potential weakness of the elastic net is that it will select redundant or highly related variables. As a result, it may not be able to find extremely small subsets of variables like random forests.

Another potential disadvantage of the elastic net is that although it has been shown to work well for classification, it is primarily a method for regression. For this purpose, an additional method that we call the elastic net with SVM is tested with the other methods. This is a simple adjustment. This method will be the same as the elastic net, except that the final fit will be by a 2-norm SVM rather than the fit produced by the elastic net. The generalization of this algorithm is also somewhat more difficult, but is the same as the generalization of the SVM.

2.3   RESULTS

In order to empirically assess the effectiveness of the various classifiers discussed above, the classifiers were run on some of the data sets in Ref. [2]. The tuning procedures for the elastic net and the 1-norm SVM are described below.

In the elastic net, the two parameters one needs to select are the penalty for the L2 norm and the number of steps the LARS algorithm takes on the augmented data matrix. The L2 norm penalty was chosen to be 0.01. This is a value large enough so that the influence of the ridge regression is present, but small enough to let the lasso regression dominate most of the fitting process. The number of steps was chosen based on 10-fold cross-validation error. The smallest number of steps resulting in an error within one standard error of the minimum among the first 200 steps was the value chosen. In order to ease the computational burden, 1000 variables were preselected based on the F-ratio. This has been shown not to greatly influence the results [6]. For the 1-norm SVM, the algorithm used required tuning the parameters δ and v. In contrast to Ref. [5], in which a large range of values were tested and an extremely thorough procedure was used to test each value, we tuned the parameters over a much smaller set to ease the computational burden. The values of δ and v that were tested were 1/10, 1, and 10 for each variable. Hence, nine parameter settings were tested in all. The error was evaluated using fivefold cross-validation. This gives a general idea of how the 1-norm SVM performs, but it is possible that the performance could be enhanced by a more thorough tuning procedure.

The results for varSelRF, SC.s, and NN.vs are taken from Diaz-Uriarte and Alvarez de Andres [1], as are the data sets. Figure 2.1 shows the description of the selected data sets.

Figure 2.2 shows the classification error of varSelRF, SC.s, NN.vs., the elastic net, the elastic net with SVM, and the 1-norm SVM. The first number is the classification error and the second is the ranking among the seven methods. The column entitled no info corresponds to an algorithm that simply guesses the majority class. For each method and data set, the classification error was estimated using the 0.632+ bootstrap method [10, 11] with 200 bootstrap samples. This method has been shown to produce low variance estimates of the true test error and is therefore considered a valuable method to use when the data are limited. The 0.632+ bootstrap method is in simple terms a weighted average between the training error and the test error of a classifier, where the weight depends on a parameter called the no info error. The exact procedure is given in Ref. [10].

As is shown in Fig. 2.2, all of the methods perform better than the rule that selects the most frequent class. The performances of the various methods vary significantly across the data sets. For instance, the linear separator methods (elastic net, SVM,

Data Set	No info	varSelRF s.e. = 0	varSelRF s.e. = 1	SC.s	NN.vs	1-norm SVM	Elastic Net with SVM	Elastic Net
Leukemia	0.289	0.087	0.075	0.062	0.056	0.068	**0.045**	0.053
Rank		7	6	4	3	5	1	2
Breast	0.429	0.337	0.332	**0.326**	0.337	0.342	0.343	0.332
Rank		3	2	1	3	4	5	2
Colon	0.355	0.159	0.177	**0.122**	0.158	0.155	0.139	0.147
Rank		6	7	1	5	4	2	3
Prostate	0.490	**0.061**	0.064	0.089	0.081	0.125	0.076	0.073
Rank		1	2	6	5	7	4	3
Average rank		4.25	4.25	3	4	5	3	2.5

Figure 2.2   Classification error of the methods.

and elastic net with SVM) perform extremely well on the leukemia data set. However, they do not perform exceptionally well on the other data sets. The shrunken centroids method produces the lowest error rates on two of the data sets, but does not beat the elastic net in terms of overall ranking. One clear pattern is that both the elastic net and the elastic net with SVM appear to perform better overall than the 1-norm SVM as implemented on these data sets. However, no one method dominates the others.

In terms of average ranking, we see that on these data sets, the elastic net performed the best and the 1-norm SVM performed the worst with the varSelRF methods next in line. The reason why the effectiveness of these methods vary from data set to data set is probably that the cancer groups distinguish themselves from the noncancer groups in different ways in the various data sets. If the actual relationship between the cancer patients and the noncancer patients is such that a hyperplane can actually separate their representation on a scatter plot, methods such as the elastic net and the 1-norm SVM are likely to be very effective. However, in cases where the relationship is of a different type, other methods may classify better.

Figure 2.3 shows the median number of variables each procedure selected in bootstrap samples of the data set. Again, rankings are given below the number of variables selected. Comparing Fig. 2.3 with Fig. 2.2, we see that there is a trade-off between the number of variables an algorithm uses and its classification power. The

Data Set	varSelRF s.e. = 0	varSelRF s.e. = 1	SC.s	NN.vs	1-norm SVM	Elastic Net
Leukemia	2	2	46	23	20	26
Rank	1	1	5	3	2	4
Breast	9	4	55	23	43	42
Rank	2	1	6	3	5	4
Colon	5	3	22	9	28	21
Rank	2	1	5	3	6	4
Prostate	5	2	3	6	42	38.5
Rank	3	1	2	4	6	5
Average rank	2	1	4. 5	3.25	4.75	4.25

Figure 2.3 Number of variables selected from the individual data.

varSelRF methods select an extremely small number of variables on average, far less than any of the other algorithms. Nearest neighbors + variable selection is next in line, and then the elastic net. The reason why the varSelRF methods select fewer variables is probably that they are more flexible in the way they classify. In addition, as mentioned before, varSelRF will not necessarily select redundant or highly correlated variables. This indicates that if an analyst wants an extremely small subset of variables, varSelRF could be the best algorithm to use among those discussed.

2.4 VISUALIZATION

After a method is chosen and fit, one may want to confirm that the variables selected are indeed useful and may want to understand how the chosen variables describe the data and the differences between cancer patients and noncancer patients. Such plots can also help the analyst distinguish among feature selection methods, as differences in error rates may be a result of the classifying algorithm rather than the feature selection method. If the algorithm chosen selects only two variables, this is not particularly difficult. Figures 2.4–2.6 show scatter plots of the data with the best two variables picked from varSelRF, the F-ratio, and the elastic net on the leukemia and prostate cancer data sets. The circles represent noncancer patients and the x's represent cancer patients.

These plots show that each of these three methods choose a pair of features that clearly separates the two groups. Based on the visual appearance of the plots, the elastic net seems to separate the leukemia data most clearly. A particularly interesting

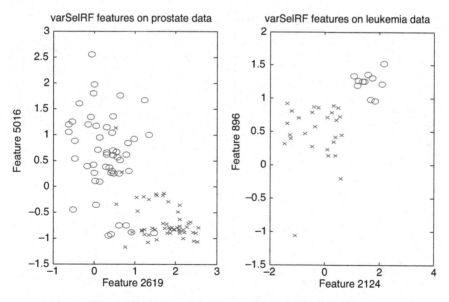

Figure 2.4 Scatter plots of features selected by varSelRF.

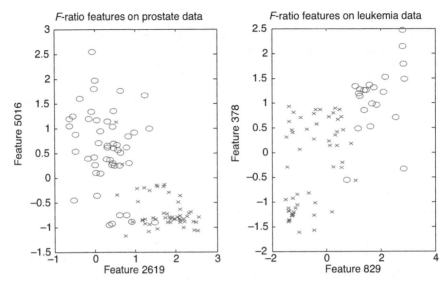

Figure 2.5   Scatter plots of features selected by F-ratio.

thing about these plots, however, is that varSelRF, the F-ratio, and the elastic net select the exact same best two features for the prostate data set. This suggests that the method of random forests seems to require fewer variables in general to make its selection. In addition, for the leukemia data, the elastic net selected one feature that varSelRF

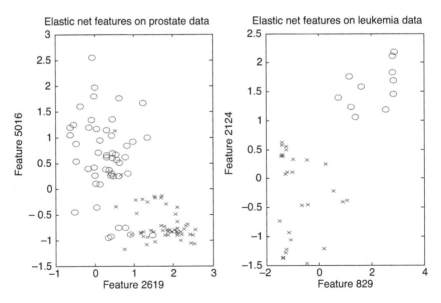

Figure 2.6   Scatter plots of features selected by elastic net.

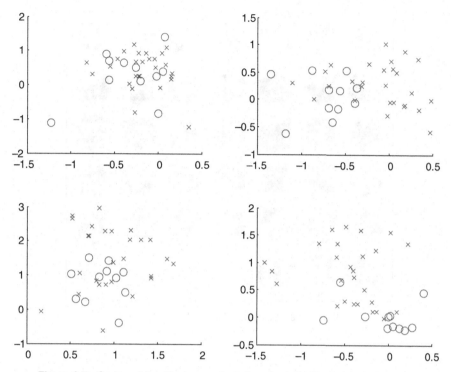

Figure 2.7 Scatter plots of features selected randomly for the leukemia data set.

selected and one feature that the F-ratio selected. For comparison in the case of the leukemia data plots, four random pairs of features were selected and are plotted in Fig. 2.7. This figure shows that the data are already reasonably well separated, but the two features selected by the algorithms above enhance this separation.

Visualization of results when there are more than three variables becomes a difficult task, but there are methods that are helpful to a certain extent. One such method is the star plot. If there are p variables, the star plot uses p axes, but these p axes are all shown in a two-dimensional plot. The idea is as follows: in a two-dimensional plotting space with a center, each axis begins at the center and ends at the end of the two-dimensional plotting space. The angles between the axes are equal. Hence, the axes for a four-dimensional star plot would be the positive x axis, the positive y axis, the negative x axis, and the negative y axis. In plotting a matrix $\mathbf{X}$, the columns represent axes and the rows represent different stars. The values of the columns of $\mathbf{X}$ are scaled so that they all lie between 0 and 1 and then these values for a particular row (star) are marked on the appropriate axis and connected to form a star.

Figures 2.8–2.14 demonstrate the use of star plots on the leukemia data sets. Figure 2.8 is a star plot of the entire data set. Observations labeled with a "1" are cancer observations and observations labeled with a "0" are noncancer observations. Note that in each of the star plots, only the last 11 stars are noncancer observations. This plot in itself is of little help. Figure 2.9 is a star plot of the entire data set, but

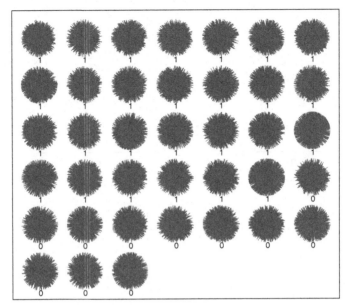

Figure 2.8 Star plots of the leukemia data set.

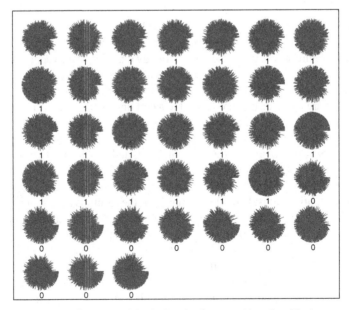

Figure 2.9 Star plots of the leukemia data set with ordered features.

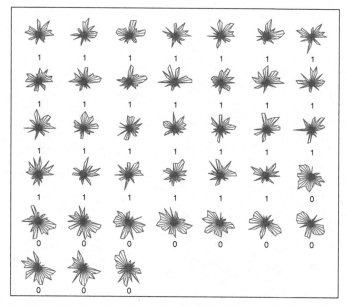

Figure 2.10 Star plots of the leukemia data set with features selected by LARS.

with variables ordered by increasing t-ratio. Hence, the t-ratio decreases as the axis moves counterclockwise from 3 o'clock. Here, there is a clear visual difference between patients with cancer and patients without cancer if the analyst looks at the portions of the stars near the positive x axis. Since variables were ranked so that there would be a visual difference, this is not completely surprising and should be interpreted with care.

Figures 2.10 and 2.11 show star plots using the best 34 variables selected by the elastic net and the F-ratio, respectively, as this is the number of variables in the tuned elastic net. For comparison, Fig. 2.12 shows star plots using 34 randomly selected variables. In all cases, variables are ordered in the same way as before. Figure 2.12 shows that the variable ordering itself demonstrates a difference between cancer and noncancer patients. One can see this difference by noting the differences between the third quadrant of the stars of patients with cancer and the stars of patients without cancer. However, the plots that use feature selection show a more drastic difference. Finally, Fig. 2.13 shows a star plot using the best 512 variables, as this is the number of variables in the optimal nearest neighbor variable selection, and Fig. 2.14 shows a star plot using a random set of 512 variables. Here there is a fairly clear difference in both plots, perhaps because 512 is a large number to begin with.

These plots do not necessarily say much about the algorithms' ability to generalize, but do indicate that the algorithms are able to select variables that distinguish the existing data well.

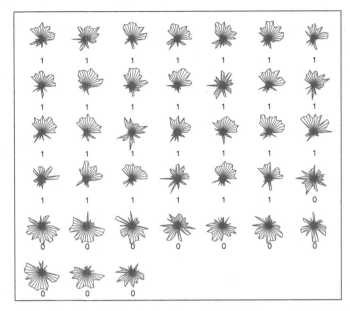

Figure 2.11 Star plots of the leukemia data set with 34 features selected by F-ratio.

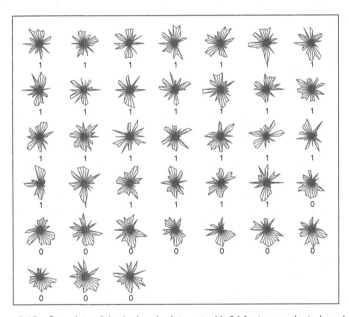

Figure 2.12 Star plots of the leukemia data set with 34 features selected randomly.

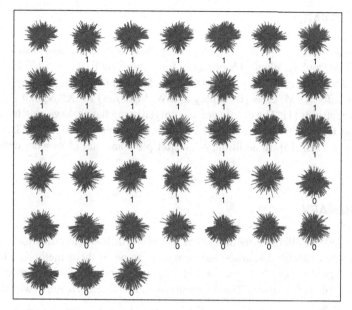

Figure 2.13 Star plots of the leukemia data set with 512 features selected by *F*-ratio.

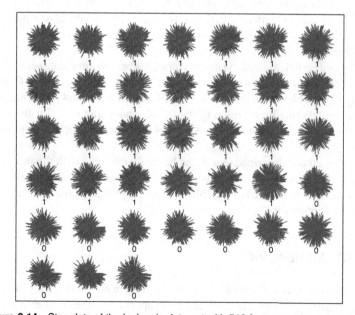

Figure 2.14 Star plots of the leukemia data set with 512 features selected randomly.

2.5 SOFTWARE

All methods from Ref. [1] were solved using R, using the packages randomForest (from A. Liaw and M. Wiener) for random forest, varSelRF (R. Diaz-Uriarte) for the varSelRF methods, e1071 (E. Dimitriadou, K. Hornik, F. Leisch, D. Meyer, and A. Weingessel) for SVM, class (B. Ripley and W. Venables) for KNN, and PAMR for shrunken centroids [1]. The elastic net was solved using the package elasticnet (from H. Zou and T. Hastie). Finally, the SVM 1-norm was implemented in MATLAB using the code from Ref. [5]. In addition, all scatter plots and star plots were done using MATLAB.

2.6 SUMMARY

There are several effective methods that can perform both classification and gene selection. This chapter discusses and compares some of these methods. However, many of these methods are qualitatively different and are difficult to compare. Different methods classify well on different data sets and select features with different characteristics.

Nearest neighbors + variable selection is the most intuitive among the methods and performs well, while selecting a moderate number of variables. This set of variables is useful in that there is a ranking among the set, so that any set size could be chosen, but may be redundant. VarSelRF methods have the advantage that they select an extremely small number of features and the interaction between these features and the response can be easily visualized with a classification tree. Meanwhile, the elastic net seems to perform extremely well, has a capability to rank features, and selects a reasonable number of features.

As discussed earlier, the method that performs the best on a new data set will depend on the characteristics of the actual data, which are not known ahead of time. One way of addressing this problem is to then try a few different methods and infer which works best. This can be done by calculating the $0.632+$ bootstrap error for the different methods and comparing them or by some sort of visualization, such as scatter plots or star plots.

All of the methods described above can be easily executed with the free software package, R, and MATLAB code exists for many of the methods as well. Several additional methods that were not described in this chapter also exist. The next section gives some references for further reading.

2.7 FURTHER READING

Many papers discuss classification and gene selection in microarray data and propose additional feature selection methods. Several papers have been written about modifications of SVMs. Zou [12] proposes the hybrid SVM, in which the 1-norm in the 1-norm SVM is replaced by a weighted 1-norm, where the weights corresponding to each variable are set to be a power of the inverse of the weights that a 2-norm SVM would produce. Meanwhile, Duan and Rajapakse [13] introduce the MSVM-RFE method, which is similar to the varSelRF methods, except features are eliminated based on their weights produced by multiple SVMs.

Approaches involving other principles exist as well. Yeung et al. [14] propose an efficient algorithm based on Bayesian model averaging. Meanwhile, Jirapech and Aitken [15] propose an evolutionary algorithm, which is a stochastic search and optimization technique motivated by the principles that drive natural selection. These methods have been shown to work well, but have not been compared with the techniques described in this chapter.

Finally, Khan et al. [16] propose a new method, Robust LARS, which attempts to make the LARS algorithm robust to outliers. LARS can be written as an algorithm that works using only location and covariance matrix estimates. Robust LARS replaces these estimates with robust counterparts.

REFERENCES

1. Diaz-Uriarte, R. and Alvarez de Andres, S. Gene selection and classification of microarray data using random forest. *BMC BioInformatics*, 7: 3, 2006.

2. http://ligarto.org/rdiaz/Papers/rfVS/randomForestVarSel.html.

3. Breiman, L. Random forests. *Machine Learning*, 45: 5–32, 2001.

4. Zhu, J., Rosset, S., Hastie, T., and Tibshirani, R.L1 norm support vector machines. Technical report, Stanford University, 2003. Available at http://citeseer.ist.psu.edu/zhu03-norm.html.

5. Fung, G. and Mangasarian, O. L. A feature selection newton method for support vector machine classification. *Computational Optimization and Applications*, 28: 185–202, 2004.

6. Tibshirani, R., Hastie, T., Narasimhan, B., and Chu, G. Diagnosis of multiple cancer types by shrunken centroids of gene expression. *Proceedings of the National Academy of Sciences of the United States of America*, 99(10): 6567–6572, 2002.

7. Zou, H., and Hastie, T. Regularization and variable selection via the elastic net. *Journal of the Royal Statistical Society B*, 62(2): 301–320, 2005.

8. Tibshirani, R. Regression shrinkage and selection via the lasso. *Journal of the Royal Statistical Society B*, 5: 267–288, 1996.

9. Efron, B., Hastie, T., Johnstone, I., and Tibshirani, R. Least angle regression. *Annals of Statistics*, 32: 407–499, 2004.

10. Efron, B. and Tibshirani, R. Improvements on cross-validation: the .632+ bootstrap method. *Journal of American Statistical Association*, 92: 548–560, 1997.

11. Ambroise, C. and McLachlan, G. J. Selection bias in gene extraction on the basis of microarray gene-expression data. *Proceedings of the National Academy of Sciences of the United States of America*, 99(10): 6562–6566, 2002.

12. Zou, H. Feature selection via a hybrid support vector machine. Technical report, University of Minnesota 2006.

13. Duan, K. and Rajapakse, J. Multiple SVM-RFE for gene selection in cancer classification with expression data. *IEEE Transactions on Nanobioscience*, 4(3): 228–234, 2005.

14. Yeung, K. Y., Bumgarner, R. E., and Raftery, A. E. Bayesian model averaging: development of an improved multi-class, gene selection and classification tool for microarray data. *Bioinformatics*, 21: 2394–2402, 2005.

15. Jirapech-Umpai, T. and Aitken, S. Feature selection and classification for microarray data analysis: Evolutionary methods for identifying predictive genes. *BMC Bioinformatics*, 6: 148, 2005.

16. Khan, J. A., Aelst, S. V., and Zamar, R. H. Robust linear model selection based on least angle regression. Working Paper, 2005.

3

ADAPTIVE KERNEL CLASSIFIERS VIA MATRIX DECOMPOSITION UPDATING FOR BIOLOGICAL DATA ANALYSIS

Hyunsoo Kim and Haesun Park

3.1 INTRODUCTION

In this chapter, efficient adaptive methods for kernelized Fisher's discriminant analysis (KDA) are presented. By formulating the KDA as a regularized minimum squared error (MSE) minimization problem, the problem is simplified to updating and downdating of a least squares problem that can be solved by either efficient complete orthogonal decomposition updating and downdating or QR decomposition updating and downdating with regularization. The downdating schemes can also be used to design efficient leave-one-out cross-validation (LOOCV) for KDA.

Fisher's linear discriminant analysis (FDA) determines a linear transformation for feature extraction that maximizes the between-class scatter and minimizes the within-class scatter [1–3]. It has been successfully applied to many pattern classification problems due to its conceptual simplicity. However, it requires one of the scatter matrices to be nonsingular, and it is applicable only to linearly separable problems. In

Machine Learning in Bioinformatics. Edited by Yan-Qing Zhang and Jagath C. Rajapakse
Copyright © 2009 by John Wiley & Sons, Inc.

order to overcome the nonsingularity restriction in undersampled high-dimensional problems, linear discriminant analysis (LDA) based on the generalized singular value decomposition (LDA/GSVD) [4, 5] or LDA based on MSE (LDA/MSE) [1, 6] can be used. Other approaches for the generalization of LDA have also been studied [7–9]. To deal with nonlinearly separable data sets, kernelized nonlinear extensions of FDA incorporating Mercer's kernel [10] have been proposed as kernelized discriminant analysis based on generalized singular value decomposition (KDA/GSVD) [11] and the minimum squared error formulation (KDA/MSE) [12].

The KDA/MSE solution can be obtained from singular value decomposition (SVD). In KDA based on regularized MSE (KDA/RMSE), a regularized version of KDA/MSE, the solution can be obtained from the QR decomposition, which in general is faster than the SVD [13]. In addition, this regularization is beneficial for designing adaptive classifiers that can efficiently perform updating and downdating of solutions when data sets are modified by incorporating new data points and deleting obsolete data points. This is especially advantageous since updating and downdating of the SVD are computationally expensive. The adaptive classifiers avoid expensive recomputation of the solution from scratch.

An efficient method for updating the decision function is required when data points are appended. For state-of-the-art support vector machines (SVM) [10, 14, 15], incremental learning has been proposed for efficient handling of very large data sets [16], where training a subset of the entire training data set and merging the support vectors found in the steps of training is repeated to train a large data set. For many practical applications that require frequent updating of decision boundary, it is desirable to develop efficient adaptive algorithms.

In the numerical linear algebra literature, updating and downdating of matrix decompositions have been intensively studied [13, 17], for example, updating and downdating of least squares solutions [18, 19], the QR decomposition [20, 21], adaptive condition estimation [22–24], and the Cholesky decomposition [25–27]. Updating and downdating of the SVD [28, 29] are computationally expensive. The ULV and URV decompositions have been shown to efficiently approximate the SVD and update/downdate the relevant fundamental subspaces [30–37].

In this chapter, efficient algorithms for adaptive nonlinear kernel discriminant analysis are reviewed. An efficient adaptive KDA algorithm, incorporating updating and downdating, is achieved by utilizing regularization and KDA based on an MSE formulation called aKDA/RMSE [38, 39]. In aKDA/RMSE, the problem is formulated as a full-rank least squares problem so that efficient QR decomposition updating and downdating can be utilized. Another method we propose is adaptive KDA/MSE (aKDA/MSE) without regularization via updating and downdating the URV and ULV decompositions. The aKDA/RMSE avoids the computationally expensive updating of an eigenvalue decomposition (EVD) by updating and downdating of the QR decomposition, but it requires an additional step for the determination of a regularization parameter. However, the aKDA/MSE does not have a regularization parameter and updates its solution by updating and downdating the URV and ULV decompositions. In addition, we show that the downdating algorithm can be used for efficient LOOCV, which plays an important role in determining model parameters. Experimental results are presented to illustrate the effective results that the proposed adaptive methods produce.

Throughout this chapter we use the following notations. A data set $A \in \mathbb{R}^{m \times n}$ with two classes is given,

$$A = \begin{bmatrix} \mathbf{a}_1^T \\ \vdots \\ \mathbf{a}_m^T \end{bmatrix} = \begin{bmatrix} A_1 \\ A_2 \end{bmatrix} \in \mathbb{R}^{m \times n},$$

where the jth row $\mathbf{a}_j^T$ of the matrix A denotes the jth data item, and the rows of submatrices A_1 and A_2 belong to classes 1 and 2, respectively. In addition, m_i denotes the number of items in class i, c_i the centroid vector, which is the average of all the data in class i, for $1 \leq i \leq 2$, and c the global centroid vector. Suppose a nonlinear feature mapping ϕ maps the input data to a feature space. The mapped jth data point in the feature space is represented as $\phi(\mathbf{a}_j)$, $1 \leq j \leq$ m, $\mathbf{c}_i^{\phi}$ denotes the centroid vector, which is the average of all the data in class i in the feature space for $1 \leq i \leq 2$, and $\mathbf{c}^{\phi}$ the global centroid vector in the feature space. Note that $\mathbf{c}_i^{\phi} \neq \phi(\mathbf{c}_i)$ and $\mathbf{c}^{\phi} \neq \phi(\mathbf{c})$, in general, since the mapping ϕ is nonlinear.

The rest of this chapter is organized as follows. A brief overview of KDA/MSE is given in Section 3.2. In Section 3.3, an adaptive KDA/MSE using updating the URV decomposition is described. An adaptive KDA/RMSE using updating the QR decomposition is reviewed in Section 3.4. In Section 3.5, we describe an efficient LOOCV method using downdating the new QR decomposition-based KDA/RMSE. In Section 3.6, the experimental results are shown. Finally, a summary is given in Section 3.7.

3.2 KERNEL DISCRIMINANT ANALYSIS BASED ON THE MINIMUM SQUARED ERROR FORMULATION

In this section, KDA based on the MSE formulation for binary classifier design in feature space [12] is briefly reviewed. Assume that a feature mapping $\phi(\cdot)$ maps the input data to a higher dimensional feature space:

$$\mathbf{a} \in \mathbb{R}^{n \times 1} \to \phi(\mathbf{a}) \in \mathbb{R}^{N \times 1}, \quad n < N.$$

Then, for the training data set $(\mathbf{a}_i, y_i)$, $1 \leq i \leq m$, in order to obtain the discriminant function for the binary classification in the feature space,

$$f(\mathbf{x}) = \mathbf{w}^T \phi(\mathbf{x}) + \beta, \tag{3.1}$$

a linear system can be built as

$$\begin{bmatrix} 1 & \phi(\mathbf{a}_1)^T \\ \vdots & \vdots \\ 1 & \phi(\mathbf{a}_m)^T \end{bmatrix} \begin{bmatrix} \beta \\ \mathbf{w} \end{bmatrix} \cong \begin{bmatrix} y_1 \mathbf{u}_{m_1} \\ y_2 \mathbf{u}_{m_2} \end{bmatrix}, \tag{3.2}$$

where y_1 and y_2 are the values that indicate the class membership of the data point $\phi(\mathbf{a}_j)$, $1 \leq j \leq m$, and $\mathbf{u}_{m_i} \in \mathbb{R}^{m_i \times 1}$ is a column vector with 1's as its elements, for $1 \leq i \leq 2$. Various pairs of numbers can be assigned to (y_1, y_2) to discriminate two classes. When N is very large, Equation 3.2 becomes an underdetermined system. In general, the problem can be reformulated by setting up Equation 3.2 as a problem of minimizing the L_2-norm of the error as

$$\min_{\beta, \mathbf{w}} \left\| \begin{bmatrix} \mathbf{u}_{m_1} & \phi(A_1) \\ \mathbf{u}_{m_2} & \phi(A_2) \end{bmatrix} \begin{bmatrix} \beta \\ \mathbf{w} \end{bmatrix} - \begin{bmatrix} y_1 \mathbf{u}_{m_1} \\ y_2 \mathbf{u}_{m_2} \end{bmatrix} \right\|_2^2 = \min_{\beta, \mathbf{w}} \left\| F \begin{bmatrix} \beta \\ \mathbf{w} \end{bmatrix} - \mathbf{y} \right\|_2^2, \qquad (3.3)$$

where

$$\phi(A_1) = \begin{bmatrix} \phi(\mathbf{a}_1)^T \\ \vdots \\ \phi(\mathbf{a}_{m_1})^T \end{bmatrix}, \quad \phi(A_2) = \begin{bmatrix} \phi(\mathbf{a}_{m_1+1})^T \\ \vdots \\ \phi(\mathbf{a}_m)^T \end{bmatrix},$$

$$F = \begin{bmatrix} \mathbf{u}_{m_1} & \phi(A_1) \\ \mathbf{u}_{m_2} & \phi(A_2) \end{bmatrix} \quad \text{and} \quad \mathbf{y} = \begin{bmatrix} y_1 \mathbf{u}_{m_1} \\ y_2 \mathbf{u}_{m_2} \end{bmatrix}.$$

The minimum norm solution for Equation 3.3 is obtained as

$$\begin{bmatrix} \beta \\ \mathbf{w} \end{bmatrix} = F^\dagger \mathbf{y}, \qquad (3.4)$$

where $F^\dagger$ is the pseudoinverse of F.

The main idea of the kernel method is that without knowing the nonlinear feature mapping, ϕ, we can work in feature space through kernel functions, as long as the problem formulation depends only on inner products between data points in feature space. This is based on the fact that for any kernel function, $\mathbf{k}$, satisfying Mercer's condition [15], there exists a mapping ϕ such that $\langle \phi(\mathbf{x}), \phi(\mathbf{y}) \rangle = \mathbf{k}(\mathbf{x}, \mathbf{y})$ where $\langle, \rangle$ is an inner product. An example of such a kernel function is the Gaussian radial basis function kernel $\mathbf{k}(\mathbf{x}, \mathbf{y}) = \exp(-\gamma \|\mathbf{x} - \mathbf{y}\|^2)$, $\gamma \in R$. For a finite data set $\{\mathbf{a}_1, \ldots, \mathbf{a}_m\}$, a kernel function $\mathbf{k}$ satisfying Mercer's condition can be rephrased as the kernel matrix $K = [\mathbf{k}(\mathbf{a}_i, \mathbf{a}_j)]_{1 \leq i,j \leq m}$ being positive semidefinite [15]. Since LDA is a special case of kernelized nonlinear discriminant analysis, the discussion in the rest of this chapter will focus on nonlinear discriminant analysis.

Although Equations 3.3, 3.4 show how the MSE method can be applied in feature space, it needs to be reformulated in terms of kernel functions when the feature mapping ϕ is unknown. Note that $\mathbf{w}$ in Equation 3.1 can be expressed as a linear combination of $\phi(\mathbf{a}_j)$'s, that is,

$$\mathbf{w} = \sum_{j=1}^m z_j \phi(\mathbf{a}_j) = [\phi(\mathbf{a}_1) \quad \cdots \quad \phi(\mathbf{a}_m)]\mathbf{z}, \quad \text{where} \quad \mathbf{z} = [z_1 \quad \cdots \quad z_m]^T, \qquad (3.5)$$

for some scalars z_j, $j = 1, \ldots, m$ [11]. Applying Equation 3.5 to Equation 3.3, we obtain

$$\min_{\beta, \mathbf{z}} \left\| [\mathbf{u} \quad \phi(A)\phi(A)^T] \begin{bmatrix} \beta \\ \mathbf{z} \end{bmatrix} - \mathbf{y} \right\|_2^2 = \min_{\beta, \mathbf{z}} \left\| P \begin{bmatrix} \beta \\ \mathbf{z} \end{bmatrix} - \mathbf{y} \right\|_2^2, \qquad (3.6)$$

where $\mathbf{u} = (1, \ldots, 1)^T \in \mathbb{R}^m$, $K_{i,j} = \mathbf{k}(\mathbf{a}_i, \mathbf{a}_j)$, and

$$P = [\mathbf{u} \quad \phi(A)\phi(A)^T] = [\mathbf{u} \quad K] \in \mathbb{R}^{m \times (m+1)}. \qquad (3.7)$$

Then, the decision rule for binary classification is given by sign $(f(\mathbf{x}))$ from KDA/MSE, where

$$f(\mathbf{x}) = \mathbf{k}(\mathbf{x}, A^T)\mathbf{z} + \beta, \quad \mathbf{k}(\mathbf{x}, A^T) = [\mathbf{k}(\mathbf{x}, \mathbf{a}_1), \ldots, \mathbf{k}(\mathbf{x}, \mathbf{a}_m)]. \qquad (3.8)$$

The minimum norm solution among all possible solutions that satisfy Equation 3.6 is

$$\begin{bmatrix} \beta \\ \mathbf{z} \end{bmatrix} = P^\dagger \mathbf{y}, \qquad (3.9)$$

where $P^\dagger$ is the pseudoinverse of P. The pseudoinverse $P^\dagger$ can be obtained as

$$P^\dagger = [V_1 \ V_2] \begin{bmatrix} \Sigma_1^{-1} & 0 \\ 0 & 0 \end{bmatrix} [U_1 \ U_2]^T = V_1 \Sigma_1^{-1} U_1^T,$$

based on the SVD of P,

$$P = [U_1 \ U_2] \begin{bmatrix} \Sigma_1 & 0 \\ 0 & 0 \end{bmatrix} [V_1 \ V_2]^T = U_1 \Sigma_1 V_1^T,$$

where $U_1 \in \mathbb{R}^{m \times r}$ satisfies $U_1^T U_1 = I_r$, $\Sigma_1 \in \mathbb{R}^{r \times r}$, is nonsingular and diagonal with positive diagonal elements in nonincreasing order, $V_1 \in \mathbb{R}^{(m+1) \times r}$ satisfies $V_1^T V_1 = I_r$, and $r = \text{rank}(P)$.

When rank $(P) = m$, its pseudoinverse can be computed much more efficiently using the QR decomposition of P^T. Let the QR decomposition of P^T be given as

$$P^T = Q \begin{bmatrix} R \\ 0_{1 \times m} \end{bmatrix},$$

where $Q \in \mathbb{R}^{(m+1) \times (m+1)}$ with $Q^T Q = I_{m+1}$ and the upper triangular matrix $R \in \mathbb{R}^{m \times m}$ is nonsingular. Then,

$$P^{\dagger} = P^T (PP^T)^{-1} = Q \begin{bmatrix} R^{-T} \\ 0 \end{bmatrix}. \tag{3.10}$$

In many applications, positive definite kernels such as the Gaussian radial basis function have been successfully utilized. Assuming no duplicated data points, positive definite kernels produce a positive definite kernel matrix K. Accordingly, rank $(P) = \text{rank}(K) = m$ for a positive definite kernel matrix K and $P^{\dagger}$ can be computed via the QR decomposition. This presents an important advantage in designing an efficient adaptive KDA algorithm based on regularized minimum squared error, called aKDA/RMSE, for binary class problems. When a new data item is added, or an existing data item is removed, the decision boundary can be computed by updating the SVD for the old data. Although updating the SVD is expensive, it is less expensive than recomputing the SVD all over. In the following, we propose two methods that are designed for fast updating of the SVD. The first is based on approximation of the SVD by complete orthogonal decompositions such as the URV and ULV decompositions, which are effective and efficient for updating and downdating. The second is based on regularization that results in a full-rank problem and therefore allows efficient QR decomposition updating and downdating.

3.3 ADAPTIVE KDA/MSE BASED ON COMPLETE ORTHOGONAL DECOMPOSITIONS

As shown in the previous section, the problem of finding the new decision function becomes that of finding the solution of the problem in Equation 3.6 when the matrix P and the right-hand side are modified. It is well known that a general solution for Equation 3.6 can be obtained by the SVD of P regardless of rank of P. However, updating and downdating of SVD incur a high computational cost. For this, complete orthogonal decompositions [13], such as the URV and ULV decompositions, have been proposed in the context of signal subspace updating and downdating (subspace tracking) [30, 31], and we now show that these decompositions can be used in the context of machine learning for separating hyperplane tracking. The difference between the URV and ULV decompositions lies in whether the middle matrix factor is upper triangular or lower triangular. Stewart's high-rank-revealing URV and ULV algorithms [30, 31] deflate the small singular values one at a time, starting with the smallest. They are optimized for numerical ranks close to full rank. Fierro and Hansen's low-rank-revealing URV and ULV algorithms [34] peel off the large singular values one by one, starting with the largest, until a small singular value is detected, which is suited for the low-rank situation. Since in many of the biological data analysis problems, the rank deficiency is not high, we used Stewart's high-rank-revealing algorithms. We focus on the URV decomposition and note that the ULV decomposition is analogous to the URV decomposition. A theoretical comparison between the URV decomposition and the ULV decomposition can be found in Refs [40, 41].

In this section, we introduce an adaptive KDA/MSE algorithm based on updating the URV decomposition, which can efficiently compute the updated solution when data points are appended or removed. In KDA/MSE, the discriminant function (3.1) is estimated by solving Equation 3.6. Let the URV decomposition of $P^T \in \mathbb{R}^{(m+1)\times m}$ be

$$P^T = \begin{bmatrix} \mathbf{u}^T \\ K^T \end{bmatrix} = URV^T,$$

where $U \in \mathbb{R}^{(m+1)\times m}$ is a matrix with orthonormal columns, $R \in \mathbb{R}^{m\times m}$ is an upper triangular matrix, and $V \in \mathbb{R}^{m\times m}$ is an orthogonal matrix. When $r = \mathrm{rank}(P)$, the solution for Equation 3.6 can be obtained by

$$\begin{bmatrix} \beta \\ \mathbf{z} \end{bmatrix} = U(:, 1 : r)R^T(1 : r, 1 : r)^{-1}V(:, 1 : r)^T\mathbf{y}. \tag{3.11}$$

Specifically, the solution can be obtained from solving a linear system

$$R^T(1 : r, 1 : r)\mathbf{w} = V(:, 1 : r)^T\mathbf{y} \tag{3.12}$$

for $\mathbf{w}$ and then computing

$$\begin{bmatrix} \beta \\ \mathbf{z} \end{bmatrix} = U(:, 1 : r)\mathbf{w}. \tag{3.13}$$

The above shows that the KDA/MSE solution of Equation 3.6 can be obtained by the URV decomposition, a linear system solver, and a matrix vector multiplication. Moreover, by updating the URV decomposition, when data points are appended or removed, we can efficiently obtain the updated solution $\tilde{\beta}$ and $\tilde{\mathbf{z}}$. It is much more efficient to update the URV decomposition of P^T than to compute it from scratch or by updating the SVD [13, 28, 29, 42].

An efficient adaptive KDA algorithm can be designed by using KDA/MSE and URV decomposition updating and downdating [30, 33]. Suppose that we have the URV decomposition for the matrix P, and now we wish to obtain the updated solutions $\tilde{\beta}$ and $\tilde{\mathbf{z}}$ after a change in the data set. When a new data point that belongs to class i is added, the matrix P and $\mathbf{y}$ are first to be modified. Because of the special structure of the kernel matrix K, a new row as well as a new column needs to be attached to P and a value of $\mathbf{y}$, that is, y_1 or y_2 for the corresponding class, needs to be inserted. For example, when a new data point $\mathbf{a}_{new}$ is added to the first class, the MSE formulation is changed to

$$\min_{\tilde{\beta},\tilde{\mathbf{z}}} \left\| \tilde{P} \begin{bmatrix} \tilde{\beta} \\ \tilde{\mathbf{z}} \end{bmatrix} - \begin{bmatrix} y_1\mathbf{u}_{m_1+1} \\ y_2\mathbf{u}_{m_2} \end{bmatrix} \right\|_2^2 \quad \text{with} \quad \tilde{\mathbf{z}} \in \mathbb{R}^{(m+1)\times 1},$$

where

$$
\tilde{P} =
\begin{bmatrix}
1 & K_{a_{\text{new}},a_{\text{new}}} & K_{a_{\text{new}},1} & \cdots & K_{a_{\text{new}},m} \\
1 & K_{1,a_{\text{new}}} & K_{1,1} & \cdots & K_{1,m} \\
\vdots & \vdots & \vdots & & \vdots \\
1 & K_{m_1,a_{\text{new}}} & K_{m_1,1} & \cdots & K_{m_1,m} \\
1 & K_{m_1+1,a_{\text{new}}} & K_{m_1+1,1} & \cdots & K_{m_1+1,m} \\
\vdots & \vdots & \vdots & & \vdots \\
1 & K_{m,a_{\text{new}}} & K_{m,1} & \cdots & K_{m,m}
\end{bmatrix},
$$

$$
K_{i,a_{\text{new}}} = \mathbf{k}(\mathbf{a}_i, \mathbf{a}_{\text{new}}), \quad \text{and} \quad K_{a_{\text{new}}j} = \mathbf{k}(\mathbf{a}_{\text{new}}, \mathbf{a}_j).
$$

When the kth data point $\mathbf{a}_j$ that belongs to the first class is deleted, the problem is to find the new solution $\tilde{\beta}$ and $\tilde{\mathbf{z}} \in \mathbb{R}^{(m-1)\times 1}$ for the problem

$$
\min_{\tilde{\beta},\tilde{\mathbf{z}}} \left\| \tilde{P} \begin{bmatrix} \tilde{\beta} \\ \tilde{\mathbf{z}} \end{bmatrix} - \begin{bmatrix} y_1 \mathbf{u}_{m_1-1} \\ y_2 \mathbf{u}_{m_2} \end{bmatrix} \right\|_2^2, \tag{3.14}
$$

where

$$
\tilde{P} =
\begin{bmatrix}
1 & K_{1,1} & \cdots & K_{1,k-1} & K_{1,k+1} & \cdots & K_{1,m} \\
\vdots & \vdots & & \vdots & \vdots & & \vdots \\
1 & K_{k-1,1} & \cdots & K_{k-1,k-1} & K_{k-1,k+1} & \cdots & K_{k-1,m} \\
1 & K_{k+1,1} & \cdots & K_{k+1,k-1} & K_{k+1,k+1} & \cdots & K_{k+1,m} \\
\vdots & \vdots & & \vdots & \vdots & & \vdots \\
1 & K_{m_1,1} & \cdots & K_{m_1,k-1} & K_{m_1,k+1} & \cdots & K_{m_1,m} \\
1 & K_{m_1+1,1} & \cdots & K_{m_1+1,k-1} & K_{m_1+1,k+1} & \cdots & K_{m_1+1,m} \\
\vdots & \vdots & & \vdots & \vdots & & \vdots \\
1 & K_{m,1} & \cdots & K_{m,k-1} & K_{m,k+1} & \cdots & K_{m,m}
\end{bmatrix}.
$$

As shown in the examples, a row and a column are appended to P when a data point is added, and a row and a column are removed from P when a data point is deleted. In aKDA/MSE, the new solution is obtained by updating or downdating the URV decomposition of P^T. For details of URV and ULV decompositions updating and downdating, refer to References [30, 31, 33, 36].

3.4 ADAPTIVE KDA/RMSE BASED ON REGULARIZATION AND QR DECOMPOSITION

In this section, we describe an adaptive KDA based on regularized MSE, which can efficiently compute the updated solution when data points are appended or removed. In general, a kernel matrix is symmetric positive semidefinite. This is evident since even the kernel matrix for a positive definite kernel becomes positive semidefinite when there are duplicated data points in the training data set.

First, we review the KDA/RMSE method that overcomes the potential rank deficiency problem by regularization. In KDA/RMSE, the discriminant function (3.1) is estimated by solving

$$\min_{\beta, \mathbf{z}} \left\| \begin{bmatrix} \mathbf{u} & K + \lambda I \end{bmatrix} \begin{bmatrix} \beta \\ \mathbf{z} \end{bmatrix} - \mathbf{y} \right\|_2^2, \tag{3.15}$$

where $\lambda > 0$ is a regularization parameter. The advantage of this formulation compared to Equation 3.6 is that the matrix

$$P_\lambda = \begin{bmatrix} \mathbf{u} & K + \lambda I \end{bmatrix} \in \mathbb{R}^{m \times (m+1)} \tag{3.16}$$

has full rank for any $\lambda > 0$ and therefore the minimum 2-norm solution for Equation 3.15 can be found by computing the QR decomposition, whereas Equation 3.6 requires a rank-revealing decomposition such as the SVD. Let the QR decomposition of P_λ^T be

$$P_\lambda^T = Q \begin{bmatrix} R \\ 0_{1 \times m} \end{bmatrix},$$

where $Q \in \mathbb{R}^{(m+1) \times (m+1)}$ is an orthogonal matrix and $R \in \mathbb{R}^{m \times m}$ is a nonsingular upper triangular matrix. Then, the solution for Equation 3.15 can be obtained by applying Equation 3.10 to P_λ, that is, by solving

$$\mathbf{y} = R^T \mathbf{r} \tag{3.17}$$

for $\mathbf{r}$ and then computing

$$\begin{bmatrix} \beta \\ \mathbf{z} \end{bmatrix} = Q \begin{bmatrix} \mathbf{r} \\ 0 \end{bmatrix}. \tag{3.18}$$

The above shows that the KDA solution of Equation 3.6 can be achieved by the QR decomposition, a linear system solver, and a matrix vector multiplication. We can efficiently obtain the updated solution $\tilde{\beta}$ and $\tilde{\mathbf{z}}$ by updating Q and R when data points are appended or removed. It is much more efficient to update the QR decomposition of P_λ^T than to compute it from scratch or by updating the SVD [13, 28, 29, 42].

We now describe aKDA/RMSE utilizing updating of the QR decomposition. Suppose that we have the QR decomposition for the matrix P_λ, and now we wish to obtain the updated solutions $\tilde{\beta}$ and $\tilde{\mathbf{z}}$ after a change in the data set. When a new data point that belongs to class i is added, the matrix P_λ and $\mathbf{y}$ are first to be modified. Because of the special structure of the kernel matrix K, a new row as well as a new column needs to be attached to P_λ and a value of $\mathbf{y}$, that is, y_1 or y_2 for the corresponding class, needs to be inserted. For example, when a new data point $\mathbf{a}_{\text{new}}$ is added to the first class, the MSE formulation is changed to

$$\min_{\tilde{\beta},\tilde{z}} \left\| \tilde{P}_\lambda \begin{bmatrix} \tilde{\beta} \\ \tilde{z} \end{bmatrix} - \begin{bmatrix} y_1 \mathbf{u}_{m_1+1} \\ y_2 \mathbf{u}_{m_2} \end{bmatrix} \right\|_2^2 \quad \text{with} \quad \tilde{z} \in \mathbb{R}^{(m+1)\times 1},$$

where

$$\tilde{P}_\lambda = \begin{bmatrix} 1 & K_{a_{\text{new}},a_{\text{new}}} + \lambda & K_{a_{\text{new}},1} & \cdots & K_{a_{\text{new}},m} \\ 1 & K_{1,a_{\text{new}}} & K_{1,1} + \lambda & \cdots & K_{1,m} \\ \vdots & \vdots & \vdots & & \vdots \\ 1 & K_{m_1,a_{\text{new}}} & K_{m_1,1} & \cdots & K_{m_1,m} \\ 1 & K_{m_1+1,a_{\text{new}}} & K_{m_1+1,1} & \cdots & K_{m_1+1,m} \\ \vdots & \vdots & \vdots & & \vdots \\ 1 & K_{m,a_{\text{new}}} & K_{m,1} & \cdots & K_{m,m} + \lambda \end{bmatrix},$$

$$K_{i,a_{\text{new}}} = \mathbf{k}(\mathbf{a}_i, \mathbf{a}_{\text{new}}), \quad \text{and} \quad K_{a_{\text{new}}j} = \mathbf{k}(\mathbf{a}_{\text{new}}, \mathbf{a}_j).$$

When the kth data point $\mathbf{a}_j$, which belongs to the first class, is deleted, the problem is to find the new solutions $\tilde{\beta}$ and $\tilde{z} \in \mathbb{R}^{(m-1)\times 1}$ for the problem

$$\min_{\tilde{\beta},\tilde{z}} \left\| \tilde{P}_\lambda \begin{bmatrix} \tilde{\beta} \\ \tilde{z} \end{bmatrix} - \begin{bmatrix} y_1 \mathbf{u}_{m_1-1} \\ y_2 \mathbf{u}_{m_2} \end{bmatrix} \right\|_2^2, \tag{3.19}$$

where

$$\tilde{P}_\lambda = \begin{bmatrix} 1 & K_{1,1} + \lambda & \cdots & K_{1,k-1} & K_{1,k+1} & \cdots & K_{1,m} \\ \vdots & \vdots & & \vdots & \vdots & & \vdots \\ 1 & K_{k-1,1} & \cdots & K_{k-1,k-1} + \lambda & K_{k-1,k+1} & \cdots & K_{k-1,m} \\ 1 & K_{k+1,1} & \cdots & K_{k+1,k-1} & K_{k+1,k+1} + \lambda & \cdots & K_{k+1,m} \\ \vdots & \vdots & & \vdots & \vdots & & \vdots \\ 1 & K_{m_1,1} & \cdots & K_{m_1,k-1} & K_{m_1,k+1} & \cdots & K_{m_1,m} \\ 1 & K_{m_1+1,1} & \cdots & K_{m_1+1,k-1} & K_{m_1+1,k+1} & \cdots & K_{m_1+1,m} \\ \vdots & \vdots & & \vdots & \vdots & & \vdots \\ 1 & K_{m,1} & \cdots & K_{m,k-1} & K_{m,k+1} & \cdots & K_{m,m} + \lambda \end{bmatrix}.$$

As shown in the examples, a row and a column are appended to P_λ when a data point is added, and a row and a column are removed from P_λ when a data point is deleted. In aKDA/RMSE, the new solution is obtained by updating or downdating the QR decomposition of P_λ^T. For details of QR decomposition updating and downdating, refer to References [13, 20, 21].

3.5 EFFICIENT LEAVE-ONE-OUT CROSS-VALIDATION FOR KERNEL DISCRIMINANT ANALYSIS BY DOWNDATING

Kernel discriminant analysis has shown excellent classification performance in many applications. The most commonly used model selection methods are k-fold CV and LOOCV. The procedure of LOOCV is as follows: given a training set of m data points, the first data point in the training set, $\mathbf{a}_1$, is left out. Then the classifier is trained on the remaining $(m-1)$ data points and tested on $\mathbf{a}_1$ producing a score, s_1, which is either 0 (incorrect) or 1 (correct). Then the first data point $\mathbf{a}_1$ is inserted back into the data set. This process is repeated until every data point in the data set has had the opportunity to be left out. The LOOCV rate is defined as the average score of all of the individual classifiers:

$$\text{LOOCV rate} = \sum_{i=1}^{m} s_i/m.$$

The LOOCV performance is a realistic indicator of performance of a classifier on unseen data and is a widely used statistical technique. The LOOCV is rarely adopted in large-scale applications since it is computationally expensive, though it has been widely studied due to its simplicity. The LOOCV rate for KDA can be efficiently computed by downdating of the QR decomposition and KDA/RMSE. The parameters β and z are first computed with the entire training data set of m items. Then for testing the effect of leaving out each data point, downdating of the QR decomposition of P_λ for KDA/RMSE is performed to obtain the new parameters. The total LOOCV rate can efficiently be computed by applying QR decomposition downdating m times to obtain each KDA/RMSE solution. This new technique efficiently computes the LOOCV rate by downdating the KDA/RMSE solution. Optimal kernel parameters can also be determined efficiently by this LOOCV algorithm.

3.6 EXPERIMENTAL RESULTS

In this section, some experimental results are presented to illustrate the effectiveness of the proposed adaptive methods for KDA. All test results were obtained using a P3 600 MHz machine with 512 MB memory, and the algorithms were implemented in MATLAB [43].

First, using a small artificial classification problem, we show that the same decision boundaries are computed by aKDA/RMSE algorithm and KDA/RMSE that recomputes the solution for each newly acquired data point rather than updating each time a data point is appended or removed. The nonlinearly separable data set consists of 12 two-dimensional data points

$$A = \begin{bmatrix} 2 & 3 & 2 & 8 & 6 & 4 & 9 & 9 & 9 & 6 & 7 & 4 \\ 7 & 6 & 2 & 1 & 4 & 8 & 5 & 9 & 4 & 9 & 4 & 4 \end{bmatrix}^T \in \mathbb{R}^{12\times2}$$

and class index

$$\mathbf{y}_s = \begin{bmatrix} -1 & -1 & -1 & -1 & -1 & 1 & 1 & 1 & 1 & 1 & 1 & 1 \end{bmatrix}^T \in \mathbb{R}^{12\times1}.$$

Figure 3.1 shows the updated decision boundaries obtained from aKDA/RMSE when a data point is removed (left panel) or appended (right panel). For the left panel, the 12th data point is removed. For the right panel, a data point $\mathbf{a}_{new} = [5\ 6]$ that belongs to the positive class is inserted. The dash-dotted contour presents a decision boundary of aKDA/RMSE and the dashed contour presents that of the KDA/RMSE where the solution vector is computed from scratch with the entire new set of data points. The contours perfectly match in spite of the different numerical pathways to solve the problems. The radial basis function (RBF) kernel $\mathbf{k}(\mathbf{a}_i^T, \mathbf{a}_j) = \exp(-\gamma\|\mathbf{a}_i - \mathbf{a}_j\|^2)$ with $\gamma = 0.1$ was used. Similarly, Fig. 3.2 shows the updated decision boundaries obtained from aKDA/MSE using the URV decomposition (aKDA/MSE-URV) when a data point is removed (left panel) or appended (right panel). The aKDA/MSE-URV generated the same decision boundary as KDA/MSE when a data point is removed or appended.

Next, we used the following data sets that were also used in Refs [44, 45] in order to confirm the performances of KDA/RMSE, KDA/MSE using the SVD (KDA/

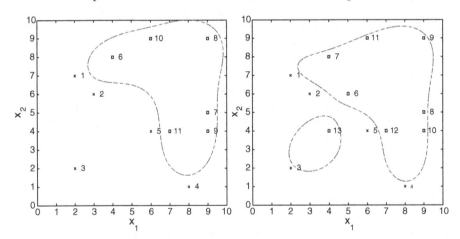

Figure 3.1   Classification results for an artificial data set of 12 items after deleting the 12th data point (left panel) and appending a data point (right panel). The dash-dotted contour presents a decision boundary of the aKDA/RMSE and the dashed contour presents that of the KDA/RMSE using the recomputed solution vector from scratch. The two lines coincide exactly since the results are identical. The radial basis function (RBF) kernel with parameter $\gamma = 0.1$ was used. The regularization parameter was set to $\lambda = 10^{-7}$.

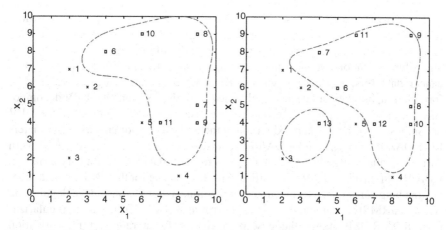

Figure 3.2 Classification results for an artificial data set of 12 items after deleting the 12th data point (left panel) and appending a data point (right panel). The dash-dotted contour presents a decision boundary of the aKDA/MSE using updating the URV decomposition and the dashed contour presents that of the KDA/MSE using the URV decomposition from scratch. The two lines coincide exactly since the results are identical. The radial basis function (RBF) kernal with parameter $\gamma = 0.1$ was used.

MSE-SVD), and KDA/MSE using the URV decomposition (KDA/MSE-URV). The central nervous system (CNS) tumors data set includes 34 samples representing four distinct morphologies: 10 classic medulloblastomas, 10 malignant gliomas, 10 rhabdoids, and 4 normal cerebella. The lung cancer data set [46] is composed of four known classes: 139 adenocarcinomas, 21 squamous cell carcinomas, 20 carcinoids, and 17 normal lung. We also used the St. Jude leukemia data set [47] for the classification of diagnostic bone marrow sample from pediatric lymphoblastic leukemia. There were 248 samples including the 6 diagnostic groups (15 BCR-ABL, 27 E2A-PBX1, 64 'Hyperdiploid > 50' chromosomes, 20 MLL rearrangements, 43 T-lineage ALL, and 79 TEL-AML1). Table 3.1 shows the five-fold CV classification

Table 3.1 Comparison of the five-fold CV errors for various kernel classifiers (KDA/RMSE using the QR decomposition, KDA/MSE using the SVD, and KDA/MSE using the URV decomposition) on three gene expression data sets

Data Set	Classifier	γ	λ	CV Error (%)
CNS tumors	KDA/RMSE	2^{-35}	2^{-5}	2.86
(34×5597)	KDA/MSE-SVD	2^{-33}		5.71
	KDA/MSE-URV	2^{-33}		5.71
Lung tumors	KDA/RMSE	2^{-34}	2^{-15}	3.01
(197×1000)	KDA/MSE-SVD	2^{-32}		3.53
	KDA/MSE-URV	2^{-32}		3.53
St. Jude leukemia	KDA/RMSE	2^{-37}	$2^{-5}, 2^{-10}, \ldots, 2^{-40}$	1.22
(248×985)	KDA/MSE-SVD	2^{-37}		1.22
	KDA/MSE-URV	2^{-37}		1.22

errors of various kernel classifiers (KDA/RMSE using the QR decomposition, KDA/MSE using the SVD, and KDA/MSE using the URV decomposition) in three gene expression data sets. We extended binary KDA algorithms to multiclass KDA algorithms with one-versus-others coding scheme to classify these multiclass cancer data sets. KDA/MSE-SVD and KDA/MSE-URV are using the same MSE formulation, but their solutions are obtained from the SVD and the URV decomposition, respectively. Their CV errors were the same in most cases since the URV decomposition is a good approximation of the SVD. The optimal kernel parameters for KDA/MSE (γ_{MSE}) and KDA/RMSE (γ_{RMSE}) were obtained from CVs with various γ values. Note that γ_{MSE} and γ_{RMSE} for each data set are not very different. KDA/RMSE produced better classification performance for the CNS tumor data set and the lung cancer data set due to regularization. As for the St. Jude leukemia data set, KDA/RMSE generated the same CV error as KDA/MSE in spite of regularization. KDA/RMSE shows the best performance for a range of regularization parameters ($\lambda \in [2^{-5}, 2^{-10}, \ldots, 2^{-35}, 2^{-40}]$). When the regularization effect is very small, KDA/MSE and its adaptive classifier aKDA/MSE-URV could be used.

For experiments regarding adaptive classifiers, we used the leukemia gene expression data set (ALLAML) [48]. The ALLAML data set contains acute lymphoblastic leukemia (ALL) that has B- and T-cell subtypes and acute myelogenous leukemia (AML) that occurs more commonly in adults than in children. This gene expression data set consists of 38 bone marrow samples (27 ALL and 11 AML) with 7129 genes. The RBF kernel was also used. In Table 3.2, the five-fold CV errors are presented. The Gaussian RBF kernel parameter γ_{RMSE} and the regularization parameter λ for KDA/RMSE were chosen by five-fold CV to obtain the optimal solution. The parameters were $\gamma_{RMSE} = 2^{-34}$ and $\lambda = 2^{-5}$, respectively. For testing adaptive classifiers, after training with 75% of the training data points by KDA, we used the remaining 25% of the training data points to obtain the final optimal classification function by using an incremental strategy where the remaining data points are inserted one by one. The experimental results illustrate that aKDA/RMSE efficiently produces the same solutions as those obtained by KDA/RMSE, which recomputes the solution from scratch each time data point is appended or removed. We also tested aKDA/MSE-URV that is an adaptive classifier for KDA/MSE-URV that is a good approximation of KDA/MSE-SVD. The RBF kernel parameter of KDA/MSE was $\gamma_{MSE} = 2^{-34}$, which is the same as that of KDA/RMSE. We obtained the

Table 3.2   Comparison of the five-fold CV errors in percent of various kernel classifiers on the ALLAML leukemia data set of size 38×7129

Classifier	Parameter	CV Error (%)
KDA/RMSE	$\gamma = 2^{-34}, \gamma = 2^{-5}$	5.0
KDA/RMSE (75%) + aKDA/RMSE (25%)	$\gamma = 2^{-34}, \gamma = 2^{-5}$	5.0
KDA/MSE-SVD	$\gamma = 2^{-34}$	10.0
KDA/MSE-URV (75%) + aKDA/MSE-URV (25%)	$\gamma = 2^{-34}$	10.0

For testing adaptive classifiers, after training 75% of the training data points, the remaining 25% of the training data points were inserted one by one.

five-fold CV error from KDA/MSE-sᴠᴅ from scratch. This result was the same as that obtained from the adaptive classifier test strategy using KDA/MSE-ᴜʀᴠ (for 75% training points) and aKDA/MSE-ᴜʀᴠ (for the remaining 25% training points). This shows that aKDA/MSE-ᴜʀᴠ is an extremely viable adaptive classifier for KDA/MSE-ᴜʀᴠ or KDA/MSE-sᴠᴅ.

In order to show the performance of LOOCV using aKDA/RMSE, we used a drug design data set used in the 2001 KDD cup data mining competition. This data set can be obtained from http://www.cs.wisc.edu/dpage/kddcup2001. It consists of 1909 compounds tested for their ability to bind to a target site on thrombin, a key receptor in blood clotting. Of these compounds, 42 are active (bind well) and the others are inactive. Each compound is described by a single feature vector comprised of a class value (A for active, I for inactive) and 139,351 binary features, which describe the three-dimensional properties of the molecule. We used a total of 142 compounds (42 active compounds and 100 inactive compounds) with 139,351 binary features. Figure 3.3 shows that aKDA/RMSE using updating of the QR decomposition can significantly reduce computational time for LOOCV rates, especially when the number of data points increases.

So far, we have dealt with aKDA and an efficient LOOCV for KDA via matrix decomposition updating and downdating. This approach can also be used to devise an adaptive least squares SVM (aLS-SVM). The original Vapnik's SVM [10] is modified so that LS-SVM [49] uses equality constraints and a squared loss function. LS-SVM is formulated as a linear Karush–Kuhn–Tucker (KKT) system with regularization that can be solved by the QR decomposition. We designed aLS-SVM by updating and downdating the QR decomposition. The detailed description of aLS-SVM and an efficient LOOCV for LS-SVM can be found in References [38, 50].

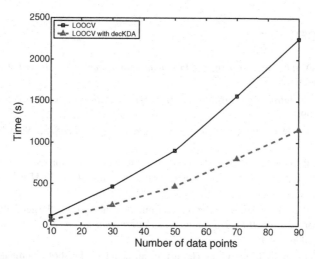

Figure 3.3 The computation time of LOOCV for different numbers of data points. The solid line represents the computation time of ordinary LOOCV and the dashed line represents that of LOOCV using decremental KDA/RMSE that is denoted as decKDA/RMSE.

3.7 SUMMARY

The implementation of an adaptive KDA requires an eigenvalue decomposition update or SVD update. Unfortunately, the SVD updating schemes require $O(m^3)$ operations, which are of the same order of magnitude computational complexity as recomputing the SVD from scratch, although there are still some gains from SVD updating [42]. The URV decomposition is a complete orthogonal decomposition [31, 32, 36] and gives a good approximation to the SVD. By formulating KDA in terms of MSE, and utilizing the updating algorithms of the URV decomposition, we have introduced two efficient adaptive KDA algorithms: aKDA/MSE-URV, which approximates the solution by efficient URV updating and downdating, and aKDA/RMSE, which incorporates regularization into the MSE formulation so that the solutions can be modified by updating/downdating of the QR decomposition with computational complexity $O(m^2)$. This method often produces better classification accuracy through regularization, although we need to also determine the regularization parameter. Efficient LOOCV algorithms using our adaptive classifiers can be devised so as to compute LOOCV rates and determine model parameters. The proposed adaptive methods are shown to produce effective results in several biological data analyses.

ACKNOWLEDGMENTS

This study is based upon work supported in part by the National Science Foundation Grants ACI-0305543 and CCF-0621889. Any opinions, findings and conclusions, or recommendations expressed herein are those of the authors and do not necessarily reflect the views of the National Science Foundation.

REFERENCES

1. Duda, R. O. and Hart, P. E. *Pattern Classification and Scene Analysis*, Wiley-Interscience, New York, 1973.

2. Fukunaga, K. *Introduction to Statistical Pattern Recognition*, 2nd edition, Academic Press, Boston, 1990.

3. Duda, R. O., Hart, P. E., and Stork, D. G. *Pattern Classification*, Wiley-Interscience, New York, 2001.

4. Howland, P., Jeon, M., and Park, H. Structure preserving dimension reduction for clustered text data based on the generalized singular value decomposition. *SIAM Journal on Matrix Analysis and Applications*, 25(1): 165–179, 2003.

5. Howland, P. and Park, H. Generalizing discriminant analysis using the generalized singular value decomposition. *IEEE Transactions on Pattern Analysis and Machine Intelligence*, 26 (8): 995–1006, 2004.

6. Koford, J. S. and Groner, G. F. The use of an adaptive threshold element to design a linear optimal pattern classifier. *IEEE Transactions on Information Theory*, IT-I2: 42–50, 1966.

7. Chen, L., Liao, H. M., Ko, M., Lin, J., and Yu, G. A new LDA-based face recognition system which can solve the small sample size problem. *Pattern Recognition*, 33: 1713–1726, 2000.

8. Yu, H. and Yang, J. A direct LDA algorithm for high-dimensional data with application to face recognition. *Pattern Recognition*, 34: 2067–2070, 2001.

9. Yang, J. and Yang, J. Y. Why can LDA be performed in PCA transformed space? *Pattern Recognition*, 36: 563–566, 2003.

10. Vapnik, V. *The Nature of Statistical Learning Theory*, Springer-Verlag, New York, 1995.

11. Park, C. H. and Park, H. Nonlinear discriminant analysis using kernel functions and the generalized singular value decomposition. *SIAM Journal on Matrix Analysis and Applications*, 27(1): 87–102, 2005.

12. Billings, S. A. and Lee, K. L. Nonlinear Fisher discriminant analysis using a minimum squared error cost function and the orthogonal least squares algorithm. *Neural Networks*, 15: 263–270, 2002.

13. Golub, G. H. and Van Loan, C. F. *Matrix Computations*, 3rd edition, Johns Hopkins University Press, Baltimore, 1996.

14. Vapnik, V. *Statistical Learning Theory*, John Wiley & Sons, New York, 1998.

15. Cristianini, N. and Shawe-Taylor, J. *Support Vector Machines and Other Kernel-Based Learning Methods*, University Press, Cambridge, 2000.

16. Syed, N. A., Liu, H., and Sung, K. K. Incremental learning with support vector machines. *Proceedings of the Workshop on Support Vector Machines at the International Joint Conference on Artificial Intelligence (IJCA1-99)*, Stockholm, Sweden, 1999.

17. Gill, P. E., Golub, G. H., Murray, W., and Saunders, M. A. Methods for modifying matrix factorizations. *Mathematics of Computation*, 28: 505–535, 1974.

18. Björck, A., Park, H., and Eldén, L. Accurate downdating of least squares solutions. *SIAM Journal of Matrix Analysis and Applications*, 15: 549–568, 1994.

19. Eldén, L. and Park, H. Block downdating of least squares solutions. *SIAM Journal of Matrix Analysis and Applications*, 15: 1018–1034, 1994.

20. Daniel, J., Gragg, W. B., Kaufman, L., and Stewart, G. W. Reorthogonalization and stable algorithms for updating the Gram–Schmidt QR factorization. *Mathematics of Computation*, 30: 772–795, 1976.

21. Yoo, K. and Park, H. Accurate downdating of a modified Gram–Schmidt QR decomposition. *Behaviour & Information Technology*, 36(1): 166–181, 1996.

22. Ferng, W. R., Gotub, G. H., and Plemmons, R. J. Adaptive Lanczos methods for recursive condition estimation. *Numerical Algorithms*, 1: 1–20, 1991.

23. Shroff, G. and Bischof, C. H. Adaptive condition estimation for rank-one updates of QR factorizations. *SIAM Journal of Matrix Analysis and Applications*, 13: 1264–1278, 1992.

24. Pierce, D. J. and Plemmons, R. J. Fast adaptive condition estimation. *SIAM Journal of Matrix Analysis and Applications*, 13: 274–291, 1992.

25. Stewart, G. W. The effects of rounding error on an algorithm for downdating a Cholesky factorization. *Journal of the Institute of Mathematics and its Applications*, 23: 203–213, 1979.

26. Bojanczyk, A.W., Brent, R. P., Van Dooren, P., and De Hoog, F. R. A note on downdating the Cholesky factorization. *SIAM Journal of Scientific and Statstical Computing*, 8: 210–221, 1987.

27. Eldén, L. and Park, H. Perturbation analysis for block downdating of a Cholesky decomposition. *Numerical Mathematics*, 68: 457–468, 1994.

28. Moonen, M. Van Dooren, P., and Vandewalle, J. A singular value decomposition updating algorithm. *SIAM Journal of Matrix Analysis and Applications*, 13: 1015–1038, 1992.

29. Van Huffel, S. and Park, H. Two-way bidiagonaltzation scheme for downdating the singular value decomposition. *Linear Algebra and Its Applications*, 222: 1–17, 1995.

30. Stewart, G. W. An updating algorithm for subspace tracking. *IEEE Transactions on Signal Processing*, 40: 1535–1541, 1992.

31. Stewart, G. W. Updating a rank-revealing ULV decomposition. *SIAM Journal of Matrix Analysis and Applications*, 14: 494–499, 1993.

32. Stewart, G. W. Updating URV decompositions in parallel. *Parallel Computing*, 20(2): 151–172, 1994.

33. Park, H. and Eldén, L. Downdating the rank-revealing URV decomposition. *SIAM Journal of Matrix Analysis and Applications*, 16(1): 138–155, 1995.

34. Fierro, R. D. and Hansen, P. C. Low-rank revealing UTV decompositions. *Numerical Algorithms*, 15: 37–55, 1997.

35. Fierro, R. D., Hansen, P. C., and Hansen, P. S. K. UTV tools: Mattab templates for rank-revealing UTV decompositions. *Numerical Algorithms*, 20: 165–194, 1999.

36. Stewart, M. and Van Dooren, P. Updating a generalized URV decomposition. *SIAM Journal of Matrix Analysis and Applications*, 22(2): 479–500, 2000.

37. Fierro, R. D. and Hansen, P. C. UTV expansion pack: special-purpose rank-revealing algorithms. *Numerical Algorithms*, 40: 47–66, 2005.

38. Kim, H. Machine Learning and Bioinformatics, Ph.D. Thesis, University of Minnesota, Twin Cities, MN, USA, 2004.

39. Kim, H., Drake, B. L., and Park, H. Adaptive nonlinear discriminant analysis by regularized minimum squared errors. *IEEE Transactions on Knowledge and Data Engineering*, 18 (5): 603–612, 2006.

40. Fierro, R. D. and Bunch, J. R. Bounding the subspaces from rank-revealing two-sided orthogonal decompositions. *SIAM Journal of Matrix Analysis Applications*, 16: 743–759, 1995.

41. Fierro, R. D. Perturbation analysis for two-sided (or complete) orthogonal decompositions. *SIAM Journal of Matrix Analysis Applications*, 17: 383–400, 1996.

42. Björck, A. *Numerical Methods for Least Square Problems*, SIAM, Philadelphia, PA, 1996.

43. MATLAB, *User's Guide*, The Math Works, Inc., Natick, MA 01760, 1992.

44. Monti, S., Tamayo, P., Mesirov, J., and Golub, T. R. Consensus clustering: a resampling-based method for class discovery and visualization of gene expression microarray data. *Machine Learning Journal*, 52(1–2): 91–118, 2003.

45. Kim, H. and Park, H. Sparse non-negative matrix factorizations via alternating non-negativity- constrained least squares. In: D.-Z. Du, Ed. *Proceedings of the IASTED International Conference on Computational and Systems Biology (CASB2006)*, pp. 95–100, November 2006.

46. Bhattacharjee, A., Richards, W. G., Staunton, J., Li, C., Monti, S., Vasa, P., Ladd, C., Beheshti, I., Bueno, R., Gillette, M., Loda, M., Weber, G., Mark, E. J., Lander, E. S., Wong, W., Johnson, B. E., Golub, T. R., Sugarbaker, D. J., and Meyerson, M. Classification of human lung carcinomas by mRNA expression profiling reveals distinct adenocarcinomas

sub-classes. *Proceedings of National Academy of Sciences of the United States of America*, 98(24): 13790–13795, 2001.

47. Yeoh, E.-J., Ross, M. E., Shurtleff, S. A., Williams, W. K., Patel, D., Mabfouz, R., Behm, F. G., Raimondi, S. C., Relling, M. V., Patel, A., Cheng, C., Campana, D., Wilkins, D., Zhou, X., Li, J., Liu, H., Pui, C.-H., Evans, W. E., Naeve, C., Wong, L., and Downing, J. R. Classification, subtype discovery, and prediction of outcome in pediatric acute lymphoblastic leukemia by gene expression profiling. *Cancer Cell*, 1(2): 133–143, 2002.

48. Golub, T. R., Slonim, D. K., Tamayo, P., Huard, C., Gaasenbeek, M., Mesirov, J. P., Cotler, H., Loh, M. L., Downing, J. R., Caligiuri, M. A., Bloomfield, C. D., and Lander, E. S. Molecular classification of cancer: class discovery and class prediction by gene expression monitoring. *Science*, 286: 531–537, 1999.

49. Suykens, J. A. K., Van Gestel, T., De Brabanter, J., De Moor, B., and Vandewalle, J. *Least Squares Support Vector Machines*, World Scientific, Singapore, 2002.

50. Kim, H. and Park, H. Incremental and decrements least squares support vector machine and its application to drug design. *Proceedings of the IEEE Computational Systems Bioinformatics Conference (CSB2004)*, pp. 656–657, August 2004.

4

BOOTSTRAPPING CONSISTENCY METHOD FOR OPTIMAL GENE SELECTION FROM MICROARRAY GENE EXPRESSION DATA FOR CLASSIFICATION PROBLEMS

Shaoning Pang, Ilkka Havukkala, Yingjie Hu, and Nikola Kasabov

4.1 INTRODUCTION

Consistency modeling for gene selection is a new topic emerging from recent cancer bioinformatics research. The result of operations such as classification, clustering, or gene selection on a training set is often found to be very different from the same operations on a testing set, presenting a serious consistency problem. In practice, the inconsistency of microarray data sets prevents many typical gene selection methods working properly for cancer diagnosis and prognosis. In an attempt to deal with this problem, we present a new concept of performance-based consistency and apply it for microarray gene selection problem by using a bootstrapping approach, with encouraging results.

Machine Learning in Bioinformatics. Edited by Yan-Qing Zhang and Jagath C. Rajapakse
Copyright © 2009 by John Wiley & Sons, Inc.

The advent of microarray technology has made it possible to monitor the expression levels for thousands of genes simultaneously, which can help clinical decision making in complex disease diagnosis and prognosis, especially for cancer classification and for predicting the clinical outcomes in response to cancer treatment. However, often only a small proportion of the genes contribute to classification, and the rest of genes are considered as noise, that is, redundant or irrelevant. Gene selection is used to eliminate the influence of such noise genes and to find out the informative genes related to disease.

4.1.1 Review of Gene Selection Methods

Selecting informative genes, as a critical step for cancer classification, has been implemented using a diversity of techniques and algorithms. Simple gene selection methods come from statistics such as t-statistics Fisher's linear discriminate, criterion, and principal component analysis (PCA) [1–4]. Statistical methods select genes by evaluating and ranking their contribution or redundancy to classification [5] and are able to filter out informative genes very quickly. Margin-based filter methods have also been introduced recently [6]. However, the performance of these methods is not satisfactory, when applied on data sets with large number of genes and small number of samples.

More sophisticated algorithms are also available, such as noise sampling method [7], Bayesian model [8, 9], significance analysis of microarrays (SAM) [10], artificial neural networks [11], support vector machines (SVM) [12], and neural fuzzy ensemble method [13]. These methods define a loss function, such as classification error, to evaluate the goodness of a candidate subset. Most are claimed to be capable of extracting out a set of highly relevant genes [14]; however, their computational cost is much higher than that in the simpler statistical methods.

A bootstrapping approach can also be used. This can select genes in a number of iterations and can use a diversity of criteria simultaneously. For example, Huerta et al. [15] proposed a GA/SVM gene selection method that achieved a very high classification accuracy (99.41%) for colon cancer data [16]. Li et al. [17] introduced a GA/KNN gene selection method that is capable of finding a set of informative genes, and the selected genes were highly repeatable. Wahde and Szallasi [18] used an evolutionary algorithm based on a gene relevance ranking and surveyed such methods [19]. The main drawbacks of the bootstrapping methods are the difficulties is developing a suitable postselection fitness function and in determining the stopping criterion.

4.1.2 Importance of Consistency for Gene Selection

For a disease microarray data set, we do not know initially which genes are truly differentially expressed for the disease. All gene selection methods seek a statistic to find out a set of genes with an expected loss of information minimized. Most previous methods work by estimating directly a "class-separability" criterion (i.e., rank of contribution to classification, or loss of classification) for a better gene selection. In a different vein, reproducibility is addressed by Mukherjee [20] as the number of common genes obtained from the statistic over a pair of subsets randomly drawn from a microarray data set under the same distribution.

Class-separability criteria approximate the "ground truth" as the class-separation status of the training set (one part of a whole data set). However, this whole data set normally is just a subset of a complete data set of disease (a data set includes all possible microarray distributions of a disease). This leads to bad reproducibility, that is, the classification system works well on the data set that it was built on, but fails on future data. Reproducibility criteria take advantage of certain properties of microarray data; thus, they do not approximate the "ground truth", but indirectly minimize the expected loss under true data generating distribution.

However, it is not clear to what extent the selected highly differentially expressed genes using common-gene reproducibility criterion are correlated to a substantially good performance of the classification of microarray data. In other words, an erroneous cancer classification may also occur using a set of genes that are selected under the criterion of common-gene reproducibility.

Consistency in terms of classification performance is addressed here to derive a gene selection model with both good class separability and good reproducibility. A bootstrapping consistency method was developed by us with the purpose of identifying a set of informative genes for achieving replicably good results in microarray data analysis.

4.1.3 Consistency Concepts for Gene Selection

Given a data set D pertaining to a bioinformatics classification task, consisting of n samples with m genes, we define D_a and D_b as two subsets of D obtained by random subsampling; these two subsets serve as training and testing data, respectively.

$$D = D_a \cup D_b \quad \text{and} \quad D_a \cap D_b = \emptyset. \tag{4.1}$$

Provided an operation function F over D and a gene selection function f_s on D_a, the fundamental consistency concept C on gene selection f_s can be defined as

$$C(F, f_s, D) = |P_a - P_b|, \tag{4.2}$$

where P_a and P_b are the outcomes of function F on D_a and D_b, respectively.

$$P_i = F(f_s(D_i), D_i) \quad \text{for} \quad i = a, b. \tag{4.3}$$

F can be any of the various data processing models, such as a common-gene computation, clustering function, feature extraction function, classification function, and so on. F determines the feature space on which the consistency is based on. Consistency is modeled on a pair of subsets (D_a and D_b) drawn from the whole microarray data set D under the same distribution.

4.1.3.1 *Common-Gene Consistency* Mukherjee et al. [21] set F as a common-gene computation; the approach is as follows. Suppose f_s is a ranking function for gene selection generating two lists of sorted genes from the two data sets. Let top-ranked genes in each case be selected and denoted by S_a and S_b. Then, the consistency in terms of common gene C_g is defined as

$$C_g(f_s, D_a, D_b) = |S_a \cap S_b|. \tag{4.4}$$

Consistency C_g in Equation 4.4 depends on the ranking function, data, and number of selected genes. Hence, a greater C_g value represents a more consistent gene selection.

4.1.3.2 *Classification Consistency* As F is assigned as a classification function, the above consistency C in Equation 4.2 is called a classification consistency C_c, where Equation 4.3 can be implemented as

$$P_a = F(f_s(D), D_a, D_b) \quad \text{and} \quad P_b = F(f_s(D), D_b, D_a). \tag{4.5}$$

Substituting Equation 4.5 into Equation 4.2, we have

$$C_c(F, f_s, D) = |F(f_s(D), D_a, D_b) - F(f_s(D), D_b, D_a)|, \tag{4.6}$$

where $f_s(D)$ specifies D as the data set for gene selection. D_a in the first term of Equation 4.5 is assigned for classifier training, and D_b is for testing. The second term of Equation 4.5 specifies a reversed training and testing position for D_a and D_b, respectively. Note that a smaller C_c value here represents a more consistent gene selection.

Figure 4.1 illustrates the procedure of computing Equation 4.5. First, the performance P_a is computed by one classification of training subset D_a; then, P_b is obtained by another classification of testing subset D_b.

Alternatively, Equation 4.7 gives another form of the performance-based consistency definition, which is obtained by switching the training and testing sets of Equation 4.5.

$$C_c(F, f_s, D) = |F(f_s(D), D_a, D_a) - F(f_s(D), D_a, D_b)|. \tag{4.7}$$

Figure 4.2 shows the procedure of computing Equation 4.7. Here, the classifier is trained on D_b, and then the performance is computed by the classifier on the other subset D_a. The important difference to the procedure in Fig. 4.1 is that the testing and training subsets are switched. Ideally, when doing the analysis, one should use both

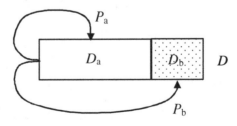

Figure 4.1 Procedure of computing consistency (Form 1).

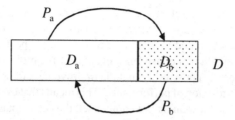

Figure 4.2 Procedure of computing consistency (Form 2).

procedures to check which data set gives better consistency when used for training. This is to safeguard the training data set against some kind of bias that may otherwise lead to suboptimal results, when the training data set is not a truly random sample of all data.

4.2 CONSISTENCY MODELING AND GENE SELECTION METHODS

4.2.1 Gene Selection in Terms of Common-Gene Consistency

Data-adaptive method (DA) is a gene selection function optimized with the consistency in terms of the number of common genes. Given a data set D with a list of genes S, the algorithm is summarized into the following steps [21]:

(1) D is resampled into a pair of D_a and D_b by using a bootstrap algorithm, each having as many samples and genes as D.

(2) Testing statistical function F_t is applied to D_a and D_b for selecting two lists of top-ranked genes denoted by S_a and S_b, respectively. F^* is defined as

$$f^* = \arg \max_{f \in \mathcal{F}} C_g(F_t, S, D), \tag{4.8}$$

where $\mathcal{F}$ is a family of test functions f optimized by the data-adaptive algorithm and S is a set of currently selected genes.

(3) The consistency C_g is calculated by Equation 4.4.

(4) Repeat steps 1–3 n times (n is a prespecified number, normally several hundreds).

4.2.2 Gene Selection in Terms of Classification Consistency

For performance-based consistency, a testing classification function F is used for performance estimation after gene selection, so that gene selection seeks f_s^* with the following consistency C_c minimized,

$$f_s^* = \arg \min_{f_s \in \mathcal{F}} C_c(F, S, D), \tag{4.9}$$

where S is a set of currently selected genes, $\mathcal{F}$ refers to a family of gene selection functions, and $C_c(\ )$ represents a classification consistency computation that has F and f_s as classification function and gene selection function, respectively.

In practice, the evaluation of consistency eventually is a multiobjective optimizing problem, because there is a possibility that the improvement of consistency might be coupled with the deterioration of performance. This means that even if the consistency of one set of genes is better than that of another set, the performance of classification P on microarray data may not be as good as we expect. In other words, it might be a case of consistently bad classification (with low classification accuracy). Therefore, a ratio of consistency to performance R is used for the purpose of optimizing these two variables simultaneously,

$$ R = \frac{C_c}{w \cdot P}, \tag{4.10} $$

where w is a predefined weight (relative importance of C_c and P) for adjusting the ratio in experiment and P is the performance evaluation on data set D.

In this sense, Equation 4.9 can be rewritten as

$$ f_s^* = \arg \min_{f_s \in \mathcal{F}} R(F, S, D), \tag{4.11} $$

where S is a set of currently selected genes. Function C_c in Equation 4.9 is replaced by a desired R to ensure a good balance between consistency and classification performance.

4.2.3 Bootstrapping Iteration for Consistent Gene Selection

The algorithm can be simply summarized into the following steps:

(1) Split all genes of data set D into N segments based on their mean value.
(2) Randomly select one gene from each of the N segments, respectively. The initial candidate gene set contains N genes and is denoted by S.
(3) Apply the operation function F (i.e., classifier) to the data containing those genes listed in S and compute the consistency C_c by Equation 4.5 or Equation 4.6.
(4) Perform gene selection function f_s on S to get a new generation of genes S' and recompute the consistency C_c'.
(5) If $C_c' > C_c$, then $C_c = C_c'$ and $S = S'$
(6) Repeat steps 3–5 until C_c becomes smaller than a predefined threshold value.
(7) Output the finally selected genes.

Algorithm 1 presents the bootstrapping consistency method for gene selection in pseudocode, where consistency and performance are optimized simultaneously.

Algorithm 1 (Bootstrapping consistency gene selection).
```
/* Initial gene selection */
Initialize the number of initial genes N;
Build gene spectrum by sorting genes in an increasing order of
   mean value;
Divide the above gene spectrum into N segments;
S ← Ø;
for each segment
   Randomly select one representative gene g;
   S = S ∪ g;
end

/*Consistency computing */
for j=1 to B   /* B is predefined resampling times *
   Partition data D into Dₐ, Db;
   Calculate Pₐ, Pb on S;
   Calculate consistency score C_cj by Equation 4.5 or Equation
      4.6;
end
Calculate consistency C = ∑_{j=1}^{B} C_cj;
Calculate classification accuracy P on D and S;
Calculate ratio R;

/* Bootstrapping gene selection */
while R>;ξ /* ξ is predefined stop criterion */
   Update S to S' by mutation or crossover operation;
   Update consistency C'_c on S';
   Update R';
   If R'>R
      S←S'; R←R';
   end
end
Output S   /* final selected informative genes */
```

The optimized gene selection is obtained by generation of optimization of consistency and classification performance. In each generation, D_a and D_b are resampled B times depending on the size of samples. For example, if the sample size of data set is larger than 30, B is set to 50, otherwise 30. Consequently, C is the mean value of the consistency scores for B rounds of computation.

4.3 CANCER DIAGNOSIS EXPERIMENTS

4.3.1 Data Sets

The proposed concept for gene selection is applied to six well-known benchmark cancer microarray data sets and one proteomics data set. Table 4.1 summarizes the seven data sets used for gene selection in the experiment.

4.3.2 Experimental Setup

As suggested in the literature for estimating generalization error [28, 29], a fivefold cross-validation scheme is used in all the experiments, except for those data sets that had originally separated training and testing sets. For each cross-validation, a totally unbiased verification scheme shown in Fig. 4.3b is used, where both gene selection and classification are working only on the training set, so that no testing information is included in any part of the cancer diagnosis modeling.

For consistency evaluation, the data set is randomly partitioned into two subsets. One subset contains two-thirds of all samples, and the other subset has the remaining one-third of samples. Using a classifier such as KNN or SVM, two classification accuracies can be computed on two subsets, respectively; the absolute difference between these two accuracies is defined as the consistency (C) in terms of classification performance (refer to Eqs. 4.5 and 4.6). After several hundred iterations, the mean value of the computed consistencies is taken as the final result.

In our example we use the above bootstrapping consistency gene selection method and Equation 4.6 for consistency evaluation. All genes of a given microarray data set (the search space) are first segmented into N segments, and N is set as 20.

Table 4.1 Cancer data sets used for testing the algorithm

Cancer	Class 1 Versus Class 2	Genes	Train Data	Test Data	Ref.[a]
Lymphoma(1)	DLBCL versus FL	7129	(58/19)77	—	[1]
Leukemia	ALL versus AML	7129	(27/11)38	34	[2]
CNS cancer	Survivor versus failure	7129	(21/39)60	—	[3]
Colon cancer	Normal versus tumor	2000	(22/40)62	—	[4]
Ovarian cancer	Cancer versus normal	15154	(91/162)253	—	[5]
Breast cancer	Relapse versus nonrelapse	24482	(34/44)78	19	[6]
Lung cancer	MPM versus ADCA	12533	(16/16)32	149	[7]

Columns for training and validation data show the total number of patients. The numbers in brackets are the ratios of the patients in the two classes.

[a]1: [22]. DLBCL, diffuse large B-cell lymphoma; FL, follicular lymphoma. http://www.genome.wi. mit.edu/MPR/lymphoma/. 2: [23]. ALL, acute lymphoblastic leukemia; AML, acute myeloid leukaemia. http://www.genome.wi.mit.edu/MPR/. 3: [24]. http://www-genome.wi.mit.edu/mpr/CNS/. 4: [16]. http:// microarray.princeton.edu/oncology/. 5: [25]. http://clinicalproteomics.steem.com/. 6: [26]. http://www.rii. com/publications/2002/vantveer.htm. 7: [27]. MPM, malignant pleural mesothelioma; ADCA, adenocarcinoma. http://www.chestsurg.org/.

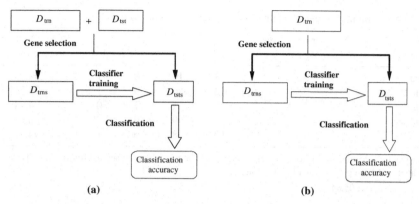

Figure 4.3 Comparison between a biased and a totally unbiased verification scheme, where D_{trn} and D_{tst} are training and testing sets, and D_{trns} and D_{tsts} are training and testing sets with selected genes, respectively. In case (a) (biased verification scheme), the testing set is used twice in gene selection and classifier procedures, which creates a bias error in the final classification results. In case (b) (the totally unbiased scheme), the testing set is independent of the gene selection and classifier training procedures.

For each fold of the given data set, the data set is initially partitioned into training and testing sets, on which the bootstrapping runs generation optimization until R becomes less than a predefined threshold ξ, and the selected informative genes are output. There are two setting choices for resampling times B. Depending on the size of data set, B is set as 50 for those data sets with more than 30 samples, and B is set as 30 for the data sets with smaller sample sizes. w in Equation 4.10 is set as 5.0, suggesting that consistency is more important than performance in the optimization. ξ is set as 0.1.

4.3.3 Results and Discussion

Experiments are presented in this section to verify the performance-based consistency gene selection method. The experiments use seven benchmark data sets, six cancer microarray data sets, and one proteomics data set and then compare them with the experimental results of these data sets reported in the original publications, in terms of the cancer diagnosis prediction accuracy (refer to the cited papers in Table 4.1).

4.3.3.1 Lymphoma Data Table 4.2 shows the bootstrapping classification results for lymphoma data and Fig. 4.4 illustrates the optimizing procedure in fivefold cross-validation, where consistency and classification accuracies are recorded at every optimizing step.

As shown in Table 4.2, the overall classification accuracy of the testing set of lymphoma data set is fairly high (greater than 95%). The number of selected informative genes is around 30, and the final calculated classification accuracy

Table 4.2   The classification validation results for lymphoma data

Lymphoma	Number of Selected Genes	TP	TN	FP	FN	Classification Accuracy (%)
Fold 1	36	8	10	0	1	94.74
Fold 2	25	12	6	0	1	94.74
Fold 3	34	11	7	1	0	94.74
Fold 4	36	10	9	0	0	100
Fold 5	32	10	8	0	1	94.74

Note that fivefold cross-validation is used for calculating classification accuracy. TP, true positive; TN, true negative; FP, false positive; and FN, false negative.
Overall accuracy: 95.84%.

is stable (94.74–100%). Moreover, the results of confusion matrix (TP, TN, FP, and FN) show that the proposed method is very effective on lymphoma data set in terms of both classification accuracy (TP and TN) and misclassification rate (FP and FN).

Figure 4.4 presents the optimizing procedure of bootstrapping consistency gene selection method. The optimized consistency is seen to decrease below 0.1, while the training classification accuracy increases above 90%. It shows that the proposed method is capable of improving consistency simultaneously with classification performance. Note that a smaller consistency value indicates a better consistency characteristic of data.

4.3.3.2 Leukemia Data   Table 4.3 and Fig. 4.5 present the classification and consistency results obtained by the described bootstrapping consistency method for leukemia data. Table 4.3 shows that the achieved classification accuracy of the testing set is about 95%, when 35 genes are used for constructing the final optimized classifier. In Fig. 4.5, after 15 rounds of optimization based on the improvement of ratio R of consistency and classification performance (refer to Eq. 4.10), the classification accuracy of the training set improves to 1 and the consistency value is reduced to 0, indicating that the maximum possible consistency is obtained.

4.3.3.3 CNS Cancer Data   Table 4.4 and Fig. 4.6 present the experimental results obtained by the described bootstrapping consistency method for CNS cancer data. Table 4.4 shows that the classification results for five folds of CNS cancer data set have high variance: the highest accuracy is 83.33%, while the lowest is only 41.67%. The overall accuracy is only 65%, which is not acceptable for solving the real clinical problem of disease diagnosis. The confusion matrix clearly shows that one misclassification rate (FN) is high, that is, the number of false negatives (FN) obtained on folds 2 and 3 is five individuals; this is larger than the TN accuracy rate.

Figure 4.6 shows that the initial consistency value of CNS cancer data set is quite high (around 0.4) and cannot be decreased in the optimizing process as much as in the previous data sets. The classification performance of training sets on folds 1–4 data rises approximately from 60% to 80%, while the consistency is decreased from 0.4 to 0.2. Although the accuracy of the fold 5 data is significantly improved, from 40% to 80%, the

Table 4.3   The classification validation result for leukemia data

Data Set	Number of Selected Genes	TP	TN	FP	FN	Classification Accuracy (%)
Leukemia	35	12	20	0	2	94.12

An independent test data set was used for validation.

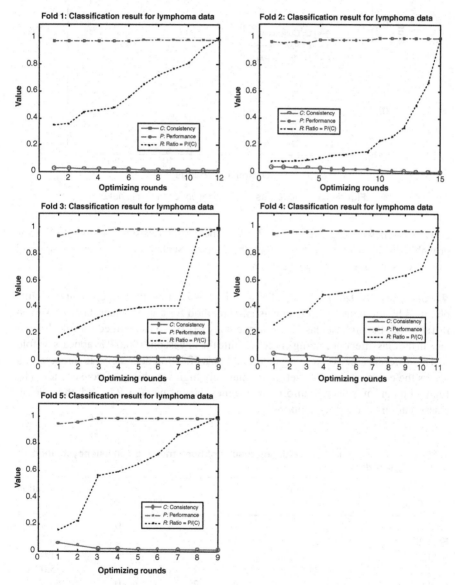

Figure 4.4   The optimization results for lymphoma data, where horizontal axis represents the optimizing rounds and vertical axis shows the results of consistency (C), classification performance (P) and the ratio (R) of consistency and performance calculated in the optimizing process. Note that accuracy P is the training classification accuracy obtained in the classifier optimizing process.

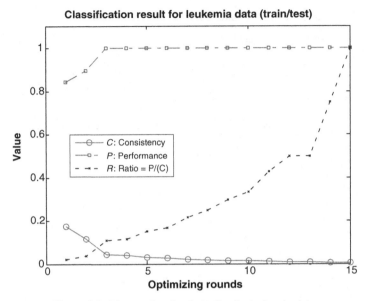

Figure 4.5   The results of optimization for leukemia data.

best consistency is still greater than 0.2, which means the consistency is not satisfactory, probably due to inherent problems with the data set. Such a situation results in the bad overall classification accuracy (65.00%).

4.3.3.4 Colon Data   Table 4.5 and Fig. 4.7 show the experimental results obtained by the bootstrapping consistency method for colon cancer data. As shown in Table 4.5, the highest classification accuracy (91.67%) is obtained for fold 1 and 4 data in the classifier optimizing process, while the lowest one (66.67%) appears on fold 3. The difference between these computed classification accuracies is large, which shows the colon cancer data set has a relatively high variability of consistency. The final number of selected informative genes is on average 23, and the overall classification accuracy is about 84%.

Table 4.4   **The classification validation results of bootstrapping consistency method for CNS cancer data**

CNS Cancer	Number of Selected Genes	TP	TN	FP	FN	Classification Accuracy (%)
Fold 1	44	9	1	2	0	83.33
Fold 2	56	4	3	0	5	58.33
Fold 3	43	3	2	2	5	41.67
Fold 4	44	7	2	3	0	75.00
Fold 5	44	6	2	4	0	66.67

Overall accuracy: 65.00%.

Table 4.5   The classification validation results of iterative bootstrapping method for colon cancer data

Colon Cancer	Number of Selected Genes	TP	TN	FP	FN	Classification Accuracy (%)
Fold 1	22	4	7	0	1	91.67
Fold 2	17	4	6	2	0	83.33
Fold 3	21	2	6	1	3	66.67
Fold 4	29	5	6	1	0	91.67
Fold 5	28	1	11	0	2	85.71

Overall accuracy: 83.81%.

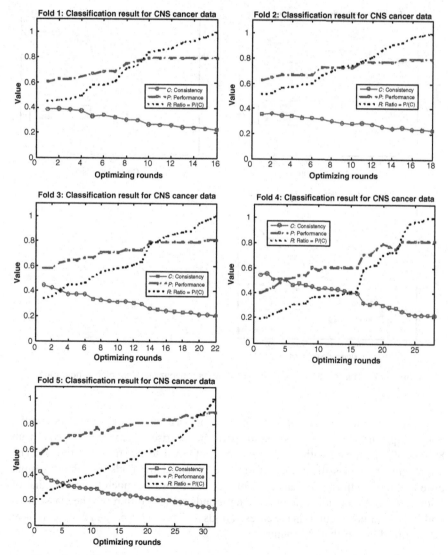

Figure 4.6   The results of iterative bootstrapping optimization for CNS cancer data.

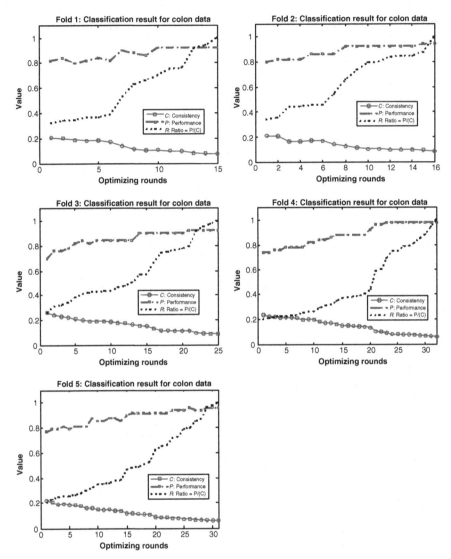

Figure 4.7   The results of iterative bootstrapping optimization for colon cancer data.

Figure 4.7 shows that both the consistency and performance improve significantly. For example, in fold 1, the classification performance rises approximately 10% (from 80% to 90%) coupled with the improvement of consistency (from 0.2 to 0.1). The improvement of classification performance obtained for folds 1–5 data is quite different, though the performance of folds 3–5 is improved much more than that of folds 1 and 2. Meanwhile, the optimizing rounds are also different. The classifier is optimized within 25 rounds in the cases of folds 3–5, but in folds 1 and 2, the classifier is optimized in less than 20 rounds.

Table 4.6   The classification validation results of iterative bootstrapping method for ovarian cancer data

Ovarian Cancer	Number of Selected Genes	TP	TN	FP	FN	Classification Accuracy (%)
Fold 1	18	25	24	0	1	98.00
Fold 2	28	31	18	1	0	98.00
Fold 3	24	33	16	1	0	98.00
Fold 4	24	34	16	0	0	100
Fold 5	34	38	15	0	0	100

Overall accuracy: 98.80%.

4.3.3.5 Ovarian Data

Table 4.6 and Fig. 4.8 give the experimental results obtained by iterative bootstrapping method for ovarian cancer data set. Table 4.5 shows the classification results based on the selected informative genes. The proposed method produces an overall accuracy of 98.80%. The difference between the highest accuracy (100%) and the lowest accuracy (98%) is only 2%. Moreover, the confusion matrix shows that both the classification accuracy rate and the misclassification rate are very good, for example, there are no misclassified samples in the cases of folds 4 and 5.

Figure 4.8 shows that both the classification performance and consistency are stable during the process of classifier optimization. It turns out that the ovarian data set has a good and little varying consistency characteristic, which leads to successful classification results for all cross-validation sets. Consequently, the improvement of consistency is less than 0.05 in all five folds.

4.3.3.6 Breast Cancer Data

Table 4.7 and Fig. 4.9 show the experimental results obtained by the bootstrapping consistency method for breast cancer data set. Table 4.6 shows that the low classification accuracy of the testing set is related to the high inconsistency characteristic of this breast cancer data set. The classification accuracy obtained by iterative bootstrapping method with 50 selected informative genes is only 63.16%, which is not very useful for identifying disease patterns in real clinical practice.

Figure 4.9 presents the relatively poor consistency and classification accuracy obtained by iterative bootstrapping method in the optimizing process. The best classification performance of the training data in gene selection procedure is 80%, when the final optimized consistency (approximately 0.2) is achieved after nine iterations.

4.3.3.7 Lung Cancer Data

Table 4.8 and Fig. 4.10 show the results obtained by the bootstrapping consistency method for lung cancer data. As shown in Table 4.7, the experimental result of lung cancer data reaches a satisfactory level in which the classification accuracy of testing set is 91.28% with 34 selected genes identified by our method.

As shown in Fig. 4.10, the classifier is optimized in nine rounds. Unlike the relatively poor consistency and classification performance in the breast cancer data

Table 4.7   The classification validation results of iterative bootstrapping method for breast cancer data

Data Set	Number of Selected Genes	TP	TN	FP	FN	Classification Accuracy(%)
Breast cancer	50	5	7	5	2	63.16

An independent test data set was used for validation.

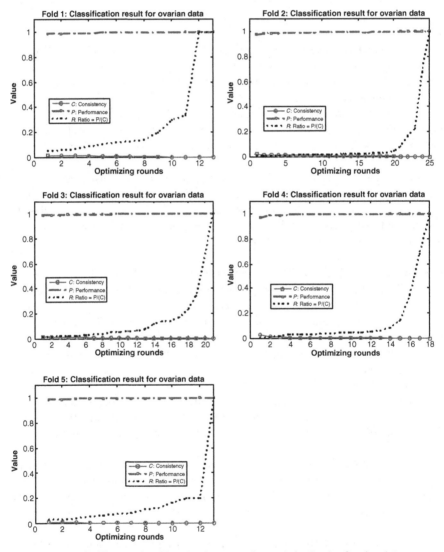

Figure 4.8   The results of iterative bootstrapping optimization for Ovarian data.

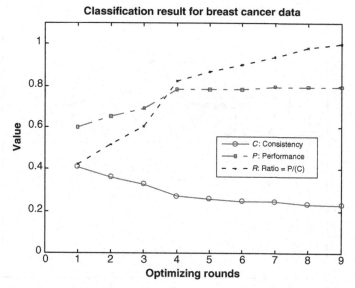

Figure 4.9 The optimizing results of iterative bootstrapping method for breast cancer data.

set, there was no difficulty in the optimizing process here, due to the already good initial consistency characteristic of lung cancer data set. It can be seen that the initial classification accuracy is greater than 90%, and the consistency calculated in the first round is about 0.1, so that it takes only nine optimizing rounds to achieve a high classification accuracy coupled with a good consistency in the training process.

4.3.4 Classification Accuracy Summary: Consistency Method Versus Publication

For clarity, the classification accuracies obtained by the presented bootstrapping consistency gene selection method are summarized and compared with the literature reported results in Table 4.9. The consistency method outperforms the published methods on four data sets, and the classification result for colon data is very close to the reported accuracy. However, the classification accuracies of two data sets (CNS,

Table 4.8 The classification results of iterative bootstrapping method for lung cancer data

Data Set	Number of Selected Genes	TP	TN	FP	FN	Classification Accuracy (%)
Lung cancer	34	121	15	0	13	91.28

An independent test data set was used for validation.

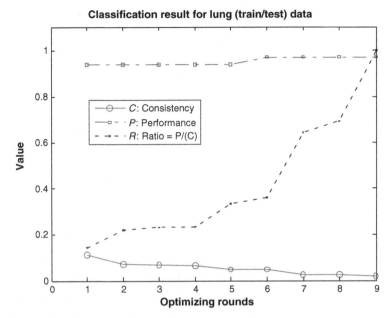

Figure 4.10   The results of iterative bootstrapping optimization for lung cancer data.

breast) are much lower than the published ones. Many published classification results are based on a biased validation scheme as given in Fig. 4.2a, which results in the experiments being unreplicable and too optimistic.

However, the experimental results obtained by the consistency method can be easily reproduced, because of the totally unbiased validation scheme of Fig. 4.2b, as applied in this study. These results suggest that a reproducible prognosis is possible for only five of the seven used benchmark data sets.

Table 4.9   Classification accuracy comparison: Consistency method results versus known results from literature

Data Set	Average Classification Accuracy	
	Consistency Method (%)	Publication (%)
Lymphoma	95.84	72.50
Leukemia	94.12	85.00
CNS cancer	65.00	83.00
Colon cancer	83.81	87.00
Ovarian cancer	98.80	97.00
Breast cancer	63.16	94.00
Lung cancer	91.28	90.00

4.4 SUMMARY

The results with the bootstrapping consistency gene selection method described in this chapter have demonstrated that the consistency concept can be used for gene selection to solve the reproducibility problem in microarray data analysis. The main contribution of the consistency method is that it ensures the reliability and generalizability of microarray data analysis experiment and improves the disease classification performance as well. In addition, because the method does not need previous knowledge about the given microarray data, it can be used as an effective tool in unknown disease diagnosis.

From the perspective of generalization error, it should be pointed out that the experimental results can be seen as totally unbiased, because the data for validation are independent and never used in the training process, that is, before the final informative genes are selected, the test data are isolated and have no correlation with these genes. Therefore, the selected informative genes are entirely fair to any given data for validation. Such a mechanism of gene selection might result in bad performance in certain microarray data sets, which is due to the special characteristics of these data sets. This makes the reported good results in some published papers for these data sets suspect, as also discussed in Ref. [30]. Recently, many papers have reported on development of guidelines and procedures for more reliable microarray profiling [31–33], reviewed existing methods [34, 35], and suggested improvements in meta-analysis [36]. However, none of these works has tackled explicitly the problem of consistency in the gene selection step, as investigated by us.

The consistency concept was investigated on six benchmark microarray data sets and one proteomic data sets. The experimental results show that the different microarray data sets have different consistency characteristics and that better consistency can lead to an unbiased and reproducible outcome with good disease prediction accuracy.

The recommended protocol for using our method is as follows:

(1) Use Equation 4.6 with your training/test sets.
(2) Run your classification algorithm of choice.
(3) Use Equation 4.7 with your training/test sets.
(4) Run your classification algorithm of choice again with same settings.
(5) Choose the results with the test/training set that gives better consistency in step 2 or step 4.
(6) Run the better model with total data or with new future data sets.

We believe our implementation of the performance-based consistency using iterative bootstrapping can provide a small set of informative genes that perform consistently with different data subsets. Compared with the traditional gene selection methods without using consistency measurement, bootstrapping consistency method can thus provide more accurate classification results. More importantly, results demonstrate that gene selection with the consistency measurement is able to enhance the reproducibility and consistency in microarray- and proteomics-based diagnosis decision systems. This is important when the classification models are used to analyze new future data sets.

ACKNOWLEDGMENT

This study was partially funded by the New Zealand Foundation for Research, Science and Technology under the grant NERF/AUTX02-01.

REFERENCES

1. Ding, C. and Peng, H. Minimum redundancy feature selection for gene expression data. *Proceedings of the IEEE Computer Society Bioinformatics Conference (CSB 2003)*, Stanford, CA, 2003.

2. Furey, T. S., Cristianini, N., Duffy, N., Bednarski, D. W., Schummer, M., and Haussler, D. Support vector machine classification and validation of cancer tissue samples using microarray expression data. *Bioinformatics*, 16(10): 906–914, 2000.

3. Jaeger, J., Sengupta, R., and Ruzzo, W. L. Improved gene selection for classification of microarrays. *Proceedings of the Pacific Symposium on Biocomputing*, 8: 53–64, 2003.

4. Tusher, V. G., Tibshirani, R., and Chu, G. Significance analysis of microarrays applied to the ionizing radiation response. *Proceedings of the National Academy of Sciences of the United States of America*, 98: 5116–5121, 2001.

5. Zhang, C., Lu, X., and Zhang, X. Significance of gene ranking for classification of microarray samples. *IEEE/ACM Transactions on Computational Biology and Bioinformatics*, 3(3): 312–320, 2006.

6. Duch, W. and Biesiada, J. Margin-based feature selection filters for microarray gene expression data. *International Journal of Information Technology and Intelligent Computing*, 1, 9–33, 2006.

7. Draghici, S., Kulaeva, O., Hoff, B., Petrov, A., Shams, S., and Tainsky, M. A. Noise sampling method: an ANOVA approach allowing robust selection of Differentially regulated genes measured by DNA microarrays. *Bioinformatics*, 19(11): 1348–1359, 2003.

8. Efron, B., Tibshirani, R., Storey, J. D., and Tusher, V. Empirical bayes analysis of a microarray experiment. *Journal of the American Statistical Association*, 96, 1151–1160, 2001.

9. Lee, K. E., Sha, N., Dougherty, E. R., Vannucci, M., and Mallick, B. K. Gene selection: a Bayesian variable selection approach. *Bioinformatics*, 19(1): 90–97, 2003.

10. Tibshirani, R. J., A simple method for assessing sample sizes in microarray experiments. *BMC Bioinformatics*, 7(106): 1–6, 2006.

11. Kasabov, N., Middlemiss, M., and Lane, T. A generic connectionist-based method for on-line feature selection and modelling with a case study of gene expression data analysis. *Conferences in Research and Practice in Information Technology Series: Proceedings of the First Asia-Pacific Bioinformatics Conference on Bioinformatics 2003*, Vol. 19, Adelaide, Australia, pp. 199–202, 2003.

12. Duan, K.-B., Rajapakse, J. C., Wang, H., and Azuaje, F. Multiple SVM-RFE for gene selection in cancer classification witn expression data. *IEEE Transactions on Nanobioscience*, 4(3): 228–234, 2005.

13. Wang, Z., Palade, V., and Xu, Y. Neuro-fuzzy ensemble approach for Microarray cancer gene expression data Analysis. *Proceedings of 2006 International Symposium on Evolving Fuzzy Systems*, pp. 241–246, 2006.

14. Wolf, L., Shashua, A., and Mukherjee, S. Selecting relevant genes with A spectral approach, CBCL Paper No. 238, Massachusetts Institute of Technology, Cambridge, MA, 2004.

15. Huerta, E. B., Duval, B., and Hao, J.-K. A hybrid GA/SVM approach for gene selection and classification of microarray data. *Lecture Notes in Computer Science*, 3907: 34–44, 2006.

16. Alon, U., Barkai, N., Notterman, D. A., Gish, K., Ybarra, S., Mack, D., and Levine, A. J. Broad patterns of gene expression revealed by clustering analysis of tumor and normal colon tissues probed by oligonucleotide Arrays. *Proceedings of the National Academy of Sciences of the United States of America*, 96(12): 6745–6750, 1999.

17. Li, L., Weinberg, C. R., Darden, T. A., and Pedersen, L. G. Gene selection for sample classification based on gene expression data: study of sensitivity to choice of parameters of the GA/KNN method. *Bioinformatics*, 17(12): 1131–1142, 2001.

18. Wahde M. and Szallasi, Z. Improving the prediction of the clinical outcome of breast cancer using evolutionary algorithms. *Soft Computing*, 10(4): 338–345, 2006.

19. Wahde, M. and Szallasi, Z. A Survey of methods for classification of gene expression data using evolutionary algorithms. *Expert Review of Molecular Diagnostics*, 6(1), 101–110, 2006.

20. Mukherjee, S., and Roberts, S. J., et al., Probabilistic consistency analysis for gene Selection. *Proceedings of the IEEE Computer Society Informatics (CSB2004)*, Stanford, CA, 2004.

21. Mukherjee, S., Roberts, S. J., and van der Laan, M. J. Data-adaptive test statistics for microarray data. *Bioinformatics*, 21 (Suppl 2): ii108–ii114, 2005.

22. Shipp, M. A., Ross, K. N., Tamayo, P., Weng, A. P., Kutok, J. L., Aguiar, R. C., Gaasenbeek, M., Angelo, M., Reich, M., Pinkus, G. S., Ray, T. S., Koval, M. A., Last, K.W., Norton, A., Lister, T. A., Mesirov, J., Neuberg, D. S., Lander, E. S., Aster, J. C., and Golub, T.R. Diffuse large B-cell lymphoma outcome prediction by gene-expression profiling and supervised machine learning. *Nature Medicine*, 8(1): 68–74, 2002. (Supplementary Information).

23. Golub, T. R. Toward a functional taxonomy of cancer. *Cancer Cell*, 6(2): 107–108, 2004.

24. Pomeroy, S. L., Tamayo, P., Gaasenbeek, M., Sturla, L. M., Angelo, M., McLaughlin, M. E., Kim, J. Y., Goumnerova, L. C., Black, P. M., Lau, C., Allen, J. C., Zagzag, D., Olson, J. M., Curran, T., Wetmore, C., Biegel, J. A., Poggio, T., Mukherjee, S., Rifkin, R., Califano, A., Stolovitzky, G., Louis, D. N., Mesirov, J. P., Lander, E. S., and Golub, T. R. Prediction of central nervous system embryonal tumour outcome based on gene expression. *Nature*, 415 (6870): 436–442, 2002.

25. Petricoin, E. F., Ardekani, A. M., Hitt, B. A., Levine, P. J., Fusaro, V. A., Steinberg, S. M., Mills, G. B., Simone, C., Fishman, D. A., Kohn, E. C., and Liotta, L. A. Use of proteomic patterns in serum to identify ovarian cancer. *Lancet*, 359: 572–577, 2002.

26. van 't Veer, L. J., Dai, H., van de Vijver, M. J., He, Y. D., Hart, A. A., Mao, M., Peterse, H. L., van der Kooy, K., Marton, M. J., Witteveen, A. T., Schreiber, G. J., Kerkhoven, R. M.,

Roberts, C., Linsley, P. S., Bernards, R., and Friend, S. H. Gene expression profiling predicts clinical outcome of breast cancer. *Nature*, 415: 530–536, 2002.

27. Gordon, G. J., Jensen, R. V., Hsiao, L. L., Gullans, S. R., Blumenstock, J. E., Ramaswamy, S., Richards, W. G., Sugarbaker, D. J., and Bueno, R. Translation of microarray data into clinically relevant cancer diagnostic tests using gene expression ratios in lung cancer and mesothelioma. *Cancer Research*, 62: 4963–4967, 2002.

28. Breiman, L. and Spector, P. Submodel selection and evaluation in regression: The X-random case. *International Statistical Review*, 60: 291–319, 1992.

29. Kohavi, R. A study of cross-validation and bootstrap for accuracy estimation and model selection, *Paper presented at the International Joint Conference on Artificial Intelligence* (IJCAI), Montreal, Quebec, Canada, 1995.

30. Ransohoff, D. F., Bias as a threat to the validity of cancer molecular-marker research. *Nature Reviews Cancer*, 5(2): 142–149, 2005.

31. Staal, F. J., Cario, G., Cazzaniga, G., Haferlach, T., Heuser, M., Hofmann, W. K., Mills, K., Schrappe, M., Stanulla, M., Wingen, L. U., van Dongen, J. J., and Schlegelberger, B. Consensus guidelines for microarray gene expression analyses in leukemia from three European leukemia networks. *Leukemia*, 20: 1385–1392.

32. Allison, D. B., Cui, X., Page, G. P., and Sabripour, M. Microarray data analysis: from disarray to consolidation and consensus. *Nature Reviews Genetics*, 7: 55–65, 2006.

33. Kawasaki, E. S. The end of the microarray tower of Babel: Will universal standards lead the way? *Journal of Biomolecular Techniques*, 17: 200–206, 2006.

34. Pham, T. D., Wells, C., and Crane, D. I. Analysis of microarray gene expression data. *Current Bioinformatics*, 1: 37–53, 2006.

35. Asyali, M., Colak, D., Demirkaya, O., and Inan, M. S. Gene expression profile classification: a review. *Current Bioinformatics*, 1: 55–73, 2006.

36. Sauerbrei, W., Hollnder, N., Riley, R. D., and Altman, D. G. Evidence-based assessment and application of prognostic markers: the Long way from single studies to meta-analysis. *Communications in Statistics — Theory and Methods*, 35: 1333–1342, 2006.

5

FUZZY GENE MINING: A FUZZY-BASED FRAMEWORK FOR CANCER MICROARRAY DATA ANALYSIS

Zhenyu Wang and Vasile Palade

5.1 INTRODUCTION

Microarray experiments have first been used for studying biological problems by a research team at Stanford University in 1995. Microarray techniques allow simultaneous measuring of the expression of thousands of genes under different experimental environments and conditions. They allow us to analyze the gene information very rapidly by managing them at one time. The gene expression profiles from particular microarray experiments have been recently used for cancer classification [1, 8, 20]. This approach promises to provide a better therapeutic assistance to cancer patients by diagnosing cancer types with improved accuracy [20]. However, the amount of data produced by this new technology is usually too large to be manually analyzed. Hence, the need to automatically analyze the microarray data offers an opportunity for machine learning methods to have a significant impact on cancer research.

Unsupervised methods such as clustering [2] and self-organizing maps (SOMs) [13] were initially used to analyze the relationships among different genes. Recently, supervised methods, such as support vector machines (SVMs) [6], multilayer

Machine Learning in Bioinformatics. Edited by Yan-Qing Zhang and Jagath C. Rajapakse
Copyright © 2009 by John Wiley & Sons, Inc.

perceptrons (MLP or NNs) [12, 19], K-nearest neighbor (KNN) method [11, 14], and decision trees (DTs) [19], have been successfully applied to classify different tissues. But, most of the current methods in microarray analysis cannot completely bring out the hidden information in the data. Meanwhile, they are generally lacking robustness with respect to noisy and missing data.

Different from black-box methods, fuzzy rule-based models not only provide good classification results, but also can easily be explained and interpreted in human understandable terms by using fuzzy rules. This provides the researchers an insight into the developed models. Meanwhile, fuzzy systems adapt numerical data (input/ output pairs) onto human linguistic terms, which offer a very good capabilities of dealing with noisy and missing data. However, defining the fuzzy rules and membership functions requires a lot of prior knowledge. This can be usually obtained from human expert, and especially in the case of large amount of gene expression data, it is not an easy task. Hybrid neuro-fuzzy (NF) models, which combine the learning ability of neural systems and the transparency of fuzzy systems, can automatically generate and adjust the membership functions and linguistic rules directly from the data.

Compared to some other benchmark problems in machine learning, microarray data sets may be problematic. The number of features (genes), usually in the range of 2000–30000, is much larger than the number of examples (usually in the range of 40–200). But, not all of these genes are needed for classification. Most genes do not influence the performance of the classification task. Taking such genes into account during classification increases the dimension of the classification problem, poses computational difficulties, and introduces unnecessary noise in the process. A major goal for diagnostic research is to develop diagnostic procedures based on inexpensive microarray that have enough probes to detect certain diseases. This requires the selection of some genes that are highly related to the particular classification problem, that is, the informative genes. This process is called gene selection [5, 14], which corresponds to feature selection from machine learning in general. Most of the current gene selection methods cannot sufficiently solve the following two questions: (1) how to estimate from numerical noisy data; and (2) how to reduce the redundancy of selected genes. In this chapter, we propose two novel fuzzy-based gene selection methods, the evolving fuzzy gene selection method (EFGS) and the fuzzy C-mean clustering-based enhanced gene selection method (FCCEGS), in order to address the above problems. EFGS combine evolutionary algorithms and fuzzy clustering methods to adjust fuzzy membership functions to represent the property of the gene data by analyzing the relationship between the trained membership functions in order to determine the rank of different genes. FCCEGS first classifies similar genes into clusters and then select the final gene subset from different clusters. By doing this, we avoid having all genes from the same pathway, with similar biological meaning. Because of the nature of the microarray gene expression data, sometimes there are several pathways involved in the perturbation, but one pathway usually has a major influence. Therefore, we also propose a new mechanism to determine the balance between the redundancy and the diversity in the selected gene subset.

Unfortunately, NF methods have suffered some well-known limitations in dealing with high dimensional data. Although some fuzzy-rule-based applications for microarray analysis have already been presented [10, 17], all these reported systems are small models and only perform well for simple data sets. Because large rule-based models imply huge computational costs, they sometimes are unacceptable in practice. In order to improve the inherent weaknesses of individual NF models, a neuro-fuzzy ensemble (NFE) model is developed in this paper. The proposed NFE model is tested on three benchmark microarray cancer data sets, namely, leukemia cancer data set, colon cancer data set, and lymphoma cancer data set.

Because a small number of available data usually cannot sufficiently represent the whole search space, traditional strategies on using different training and testing data sets do not perform well. As suggested in previous works [5, 12], we use leave-one-out-cross-validation strategy (LOOCV) to evaluate models, as described in Section 5.4.4.

The rest of the chapter is organized as follows. How to classify different types of cancer by using the microarray technology and what is the major challenge of this approach are discussed in Section 5.2. More details about two novel gene selection methods, EFGS and FCCEGS, are given in Section 5.3. What are the difficulties of using fuzzy techniques for this particular application and how to solve these problems by using a NFE model is detailed in Section 5.4. Experimental results and analytical works are presented in Section 5.4.5. Section 5.5. outlines the use of fuzzy techniques for knowledge discovery for this particular application. Some conclusions are drawn in Section 5.6.

5.2   CANCER MICROARRAY ANALYSIS

In microarray experiments, different DNA samples are fixed to a glass microscope slide, each at a predefined position in the array, known as the "gene chip". mRNAs isolated from different tissue samples, or under different conditions, labeled with two different fluorochromes (generally, the green Cy3 and the red Cy5), are then hybridized with the arrayed DNA probes on the slide. Using a fluorescent microscope and image analysis, the gene expression data (denoted as g) are measured by computing the log ratio between the intensities of the two dyes:

$$g = \log_2 \frac{\text{Int(Cy5)}}{\text{Int(Cy3)}}, \tag{5.1}$$

where Int(Cy5) is the intensity of the red color and Int(Cy3) is the intensity of the green color. Noise can be introduced during the synthesis of probes, the creation and labeling of samples, or the reading of the fluorescent signals (see Fig. 5.1).

Generally, microarray experiments can be divided into two types. One focuses on time series data that contain the gene expression data of various genes during the time span of an experiment. Another type of microarray experiment consists of gene expression data of various genes taken from different tissue samples or under different experimental conditions. Different conditions can be used to answer such questions, for example, which genes are changed under certain conditions. Meanwhile, different tissues under the same experiment conditions are helpful in the classification of

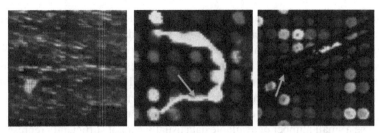

Figure 5.1 Noisy and missing experimental data are very common in microarray experiments. The left figure is an overview of a gene chip with high noise. The middle and right figures show how the noisy and missing experimental data may occur in the experiments (marked by arrows).

different types of tissues. The data from a series of n such experiments can be represented as an $n \times m$ gene expression matrix (see Table 5.1), where each row represents a sample that consists of m genes from one experiment. Each sample belongs to a certain class (cancer/noncancer).

Most analyzes on a cancer microarray data X are focused on the following two aspects:

- *Analysis Among Columns.* Let the given gene expression data be denoted as

$$D = \{G_1, \ldots, G_i, \ldots, G_m\}, \tag{5.2}$$

where each vector $G_i = (g_1, \ldots, g_n)$ denotes the different expression level of a certain position gene G_i from several repeated experiments (n is the number of experiments). Most analysis techniques used in this part are unsupervised learning methods. Cluster analysis or other statistical methods can be adopted to find out the relationship among different G_i's or the importance of genes, that is, the so-called gene clustering (GC) [18] and gene selection (GS) [5, 14]. Normally, GC and GS analyzes are used to provide the researchers with an overall picture of the microarray data. Also, good GC and GS results can be used as a nice starting point for further data analysis.

- *Analysis Among Rows.* Let us denote the given gene expression data as

$$\tilde{D} = \{(S_1, t_1), \ldots, (S_j, t_j), \ldots, (S_n, t_n)\}, \tag{5.3}$$

Table 5.1 A typical gene expression matrix X, where rows represent samples obtained under different experimental conditions and columns represent genes

	Gene 1	Gene 2	...	Gene m-1	Gene m	Class
Sample 1	165.1	276.4	...	636.6	784.9	1
Sample 2	653.6	1735.1	...	524.1	104.5	−1
...	...	...	...	...	...	...
Sample n-1	675.0	45.1	...	841.9	782.8	−1
Sample n	78.2	893.8	...	467.9	330.1	1

where an input vector $S_j = (g'_1, ..., g'_m)$ denotes a gene expression sample, m is the number of genes in this pattern, t represents the class to which the pattern belongs. The most common analysis is to first choose $\tilde{m}$ genes out of m according to certain GS algorithms. Then, select $\tilde{n}$ patterns with most informative/important $\tilde{m}$ genes to train the classifier and leave $n - \tilde{n}$ patterns (with $\tilde{m}$ genes) out to test the performance of the trained model. This method can be used to help researchers to classify whether a pattern belongs to the cancer or noncancer class. It is clear that classifying cancer microarray gene expression data is a high-dimensional low-sample problem, with highly noisy data. A typical cancer microarray data classification system is shown in Fig. 5.2.

Different from many other types of classifiers, fuzzy-based classifiers stand midway between the two options mentioned above. An NF system can classify different patterns as other classifiers do, but it can also represent the relationship between different genes correlating to final classification results by using fuzzy "IF-Then" rules.

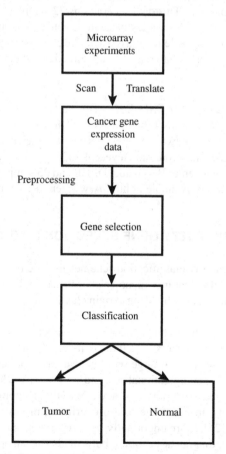

Figure 5.2 The general scheme of a cancer classification system.

5.2.1 Microarray Cancer Data Sets

In this study, the proposed models are tested on three cancer microarray gene expression data sets: leukemia cancer data set, colon cancer data set, and lymphoma cancer data set.

5.2.1.1 Colon Cancer Data Set   The data set we used here was firstly reported in Ref. [4]. The "cancer" biopsies were collected from tumors and the "normal" biopsies were collected from healthy parts of the colons of the same patients [4]. This data set contains 62 samples. There are 40 tumor samples and 22 normal samples. From about 6000 genes represented in each sample in the original data set, only 2000 genes were selected. The data set is available at http://sdmc.i2r.a-star.edu.sg/rp/ColonTumor/ColonTumor.html.

5.2.1.2 Leukemia Cancer Data Set   This data set was reported in Ref. [8]. The gene expression measurements were taken from 63 bone marrow samples and 9 peripheral blood samples [8]. This data set contains 72 samples. All samples can be divided into two subtypes: 25 samples of acute myeloid leukemia (AML) and 47 samples of acute lymphoblastic leukemia (ALL). The expression levels of 7129 genes were reported. The data set is available at http://sdmc.i2r.a-star.edu.sg/rp/Leukemia/ALLAML.html.

5.2.1.3 Lymphoma Cancer Data Set   The data set we used here was reported in Ref. [15]. This data set contains 47 samples. B-cell diffuse large cell lymphoma (B-DLCL) data set that includes two subtypes: germinal center B cell-like DLCL and active B cell-like DLCL. The expression levels of 4026 genes were reported. Twenty-four samples are germinal center B-cell-like DLCL and 23 samples are active B-cell-like DLCL. The data set is available at http://www.genome.wi.mit.edu/MPR.

5.3 THE EVOLVING FUZZY GENE SELECTION METHOD

Common cancer microarray classification schemes include two steps: first, select a subset of the most useful features (gene selection); second, classify the different samples as cancer or noncancer by using certain classification methods, according to the selected features.

Some recent researches have shown that a small number of genes is sufficient for an accurate diagnosis of most cancers even if the number of genes vary greatly between different types of diseases [24]. A large set of gene expression features will not only significantly bring higher computational cost and slow down the learning process, but will also decrease the classification accuracy due to the phenomenon known as the curse of dimensionality, in which the risk of overfitting increases as the number of selected genes grows [24]. More importantly, by using a small subset of genes, we can not only obtain a better diagnostic accuracy, but will also get an opportunity to further analyze the nature of the disease and the genetic mechanisms responsible for it.

Therefore, the microarray cancer classification problem can be formulated as a combinatorial optimization problem with two main objectives: minimizing the number of selected genes and maximizing the classification accuracy.

5.3.1   General Gene Selection Methods

The problem of feature selection received a thorough attention in pattern recognition and machine learning. The gene expression data sets are problematic in that they contain a large number of genes and, thus, methods that search over subsets of features can be prohibitively expensive. Moreover, as these data sets contain only a small number of samples, the detection of irrelevant genes can suffer from statistical instabilities. Two basic approaches, namely, the filter and wrapper methods, for feature selection are used in machine learning and information theory literature [24].

- Filter methods calculate the goodness of the proposed feature subset based on the relation of each single gene with the class label by using some simple statistical approaches. The most common way is to rank all features in terms of the values of a univariate scoring metric. The top-ranked features are selected for classification.
- In wrapper methods, a search is conducted in the space of genes. The goodness of each gene subset found is evaluated by estimating the accuracy of the specific classifier to be used, and then the classifier is trained with the found genes only.

In theory, wrapper methods should provide more accurate classification results than filter methods [5]. But, the main disadvantage of the wrapper approach is its computational cost when combined with more complex algorithms such as SVM. The wrapper approach, which is popular in many machine learning applications, is not extensively used in DNA microarray tasks. Most microarray gene selection is done by ranking genes on the basis of scores, correlation coefficients, mutual information, and sensitivity analysis.

One of the most widely used filtering gene selection methods is called signal-to-noise ratio (SNR). SNR is essentially used in Ref. [20], which also reported that the best performance was obtained with the relative class separation metric defined by

$$\text{SNR}(G_i, c) = \frac{|\mu_1 - \mu_2|}{|\delta_1 - \delta_2|}, \tag{5.4}$$

where c is the class vector, G_i is the gene expression vector (column in Table 5.1), μ_1 and μ_2 denote the mean expression level of G_i for the samples in Class 1 and Class 2, and δ_1 and δ_2 are the standard deviation of expression for the samples in Class 1 and Class 2, respectively. We then take the genes with the highest scores as our top features to be used in the classification task. The top 20 colon genes with the highest scores obtained by the SNR method are shown in Table 5.2.

SNR emphasizes that the separation between the two classes of expression data is proportional to the distance between their means [20]. Furthermore, this distance is

Table 5.2   Twenty top-ranked colon genes selected by signal-to-noise ratio method

Rank	ID	Gene Name	Description
1	1122	M22488	Human bone morphogenetic protein 1 (BMP-1) mRNA
2	738	L01664	Human eosinophil Charcot–Leyden crystal (CLC) protein mRNA, complete cds
3	1408	R72644	Choline kinase (Homo sapiens)
4	242	M58297	Zinc finger protein 42 (HUMAN)
5	1856	X80692	Homo sapiens ERK3 mRNA
6	1848	H41528	Stage V sporulation protein E
7	511	H09599	Mitochondrial import receptor MOM38 (*Neurospora crassa*)
8	1359	U26648	Human syntaxin 5 mRNA, complete cds
9	332	T72582	Glutamate receptor 5 precursor
10	1133	H47650	PTS system, sucrose-specific IIABC
11	1513	D28124	Human mRNA for unknown product, complete cds
12	197	X83535	Homo sapiens mRNA for membrane-type matrix metalloproteinase
13	1533	H63361	Translation initiation factor EIF-2B-Alpha subunit
14	1254	X54232	Human mRNA for heparan sulfate proteoglycan
15	1173	M55543	Interferon-induced guanylate-binding protein 2 (human)
16	370	Z23115	Homo sapiens bcl-xL mRNA
17	1433	X17042	Human mRNA for hematopoetic proteoglycan core protein
18	1023	T67897	Human oviductal glycoprotein mRNA, complete cds
19	124	V00522	Human mRNA encoding major histocompatibility complex gene HLA-DR beta-I
20	1183	M69296	Human estrogen receptor-related protein (variant ER from breast cancer) mRNA, complete cds

normalized by the standard deviation of the classes. A large standard deviation value implies that we find points in the group far away from the mean value and that the separation would not be strong. For example, Fig. 5.3 shows that the selected gene on the top left is unlikely to predict well because the means of the two classes are quite close; this gene cannot give us enough power to distinguish between classes. The means for the top right one and for the bottom genes are the same, but the bottom one has less variation around the mean, and hence likely to be a better gene for classification.

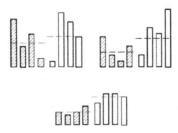

Figure 5.3   Three genes and eight samples. Samples 1–4 belong to one class and samples 5–8 belong to another class.

5.3.2 Evolving Fuzzy Gene Selection Method

Inspired from the statistical mechanism behind the SNR method, in this section, we introduce a novel fuzzy-based gene selection approach. The method first combines the evolutionary programming and fuzzy clustering methods to adjust the membership functions to represent the property of the gene vector G_i and thereafter it uses the relationship between the trained membership functions to determine the rank of different genes. The whole algorithm can be summarized as given below.

Each input gene expression data G_i, $\forall_i \ni \{1, \ldots, n\}$ is described by two linguistic values with the Gaussian membership functions set to be as follows:

$$f(x, \delta, \mu) = e^{\frac{-(x-\mu)^2}{2\delta^2}},\tag{5.5}$$

where μ and δ are the mean and the standard deviation of the function, respectively.

(1) Generate an initial population of individuals and set $k = 0$. Each individual is represented by a real-valued vector, $[\delta_{j1}, \mu_{j1}, \delta_{j2}, \mu_{j2}]$, $\forall j \in \{1, \ldots, w\}$, where μ_1 is the mean of the Gaussian membership function 1, δ_1 is the standard deviation of the Gaussian membership function 1, μ_2 is the mean of the Gaussian membership function 2, and δ_2 is the standard deviation of the Gaussian membership function 2. The initial value for δ_{j1}, μ_{j1}, δ_{j2} and μ_{j2} can be set as follows:

$$\delta_{j1} = \gamma_1 \delta_D,\tag{5.6}$$

$$\delta_{j2} = \gamma_2 \delta_D,\tag{5.7}$$

$$\mu_{j1} = \eta_1 \mu_D,\tag{5.8}$$

$$\mu_{j2} = \eta_2 \mu_D,\tag{5.9}$$

where $\gamma_1, \gamma_2, \eta_1$, and η_2 are user-defined parameters, and δ_D and μ_D are the mean and the standard deviation of the whole gene expression data set D, respectively. Some initial membership functions we used for the colon data are shown in Fig. 5.4.

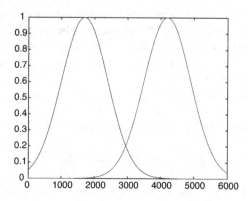

Figure 5.4 Initial membership functions of individual *i*.

(2) Each individual $[\delta_{j1}, \mu_{j1}, \delta_{j2}, \mu_{j2}]$, $\forall i \in \{1,\ldots, w\}$, creates a single offspring $[\delta'_{j1}, \mu'_{j1}, \delta'_{j2}, \mu'_{j2}]$ as given below:

$$\delta'_{j1} = \delta_{j1} + \lambda N(0,1), \tag{5.10}$$

$$\delta'_{j2} = \delta_{j2} + \lambda N(0,1), \tag{5.11}$$

$$\mu'_{j1} = \mu_g + \lambda N(0,1), \tag{5.12}$$

$$\mu'_{j2} = \mu_g + \lambda N(0,1), \tag{5.13}$$

$$\lambda'_j = \lambda_j \exp(r'N(0,1) + rN(0,1)), \tag{5.14}$$

where $N(0,1)$ denotes a normally distributed one-dimensional random variable with mean zero and variance one. δ', μ', and λ' are the parameters of the new offspring after mutation. λ is a strategy parameters in self-adaptive evolutionary algorithms [25]. The values r and r' are usually set to $\left(\sqrt{2\sqrt{n}}\right)^{-1}$ and $\left(\sqrt{2n}\right)^{-1}$.

(3) Use a simple clustering strategy to classify the pattern G_i. In our approach, this is based on a minimization of the sum of distances from each point to every center point; more details are described in Section 5.3.3. The only difference between our clustering method and the classical fuzzy C-mean clustering (FCC) method is that the membership function we adopted here is a Gaussian function.

(4) Calculate the fitness function value "*fit*" for each individual by testing how many patterns are correctly classified.

(5) Conduct pairwise comparison of the union of parents $[\delta_{j1}, \mu_{j1}, \delta_{j2}, \mu_{j2}]$ and offspring $\delta'_{j1}, \mu'_{j1}, \delta'_{j2}, \mu'_{j2}]$. For each individual, q opponents are chosen uniformly at random from all parents and offsprings. For each comparison, if the individual's fitness is not smaller than that of the opponents, it receives a $\circ$win−. Select μ individuals out of $[\delta_{j1}, \mu_{j1}, \delta_{j2}, \mu_{j2}]$ and $[\delta'_{j1}, \mu'_{j1}, \delta'_{j2}, \mu'_{j2}]$ with the largest wins to form the next generation.

(6) Stop if the halting criterion is satisfied and select the top-fitted individual to represent the final membership functions set; otherwise, $k = k+1$ and go to Step 3.

Figure 5.5 shows two trained membership functions after the evolving process had stopped. From Fig. 5.5, it can be seen that the distance between the centers of the two trained membership functions hints the distance between the means of the two classes of expression data, and the widths of the trained membership functions hint the standard deviation of the classes.

Therefore, similar to the SNR method, a good pattern implies that the two membership functions should be far from each other, and the width of each membership function should be small [21]. As shown in Equation 5.15:

$$F\_score(G_i, c) = \frac{|C_1 - C_2|}{|W_1 - W_2|}. \tag{5.15}$$

*F*_score is the score of each G_i, where c is the class vector; G_i is the gene expression vector; C_1 and C_2 are the centers of the two trained Gaussian membership functions, which can be equaled to μ_1 and μ_1; and W_1 and W_2 represent the

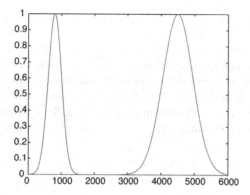

Figure 5.5 Trained membership functions of top selected individuals.

widths of the two trained Gaussian membership functions, which can be replaced by δ_1 and δ_2. The gene selection results yielded by our fuzzy gene selection method are shown in Table 5.3.

5.3.3 Enhanced Fuzzy C-Means Clustering-Based Gene Selection Method

Similar to other traditional filter methods, in the fuzzy gene selection method the selected informative genes are typically highly correlated. Highly correlated genes not

Table 5.3 Twenty top-ranked colon genes selected by fuzzy gene selection method

Rank	ID	Gene Name	Description
1	332	T72582	Glutamate receptor 5 precursor
2	30	M11799	Human MHC Class I HLA-Bw58 gene, complete cds
3	370	Z23115	Homo sapiens bcl-xL mRNA
4	104	M57710	Human IgE-binding protein (epsilon-BP) mRNA, complete cds
5	34	T48041	Human messenger RNA fragment for the beta-2 microglobulin
6	145	H48072	Cytochrome C oxidase polypeptide via liver (human)
7	18	T63508	Ferritin heavy chain (human)
8	56	H22688	Ubiquitin (human)
9	45	R36455	Nucleolar transcription factor 1
10	38	T63258	Elongation factor 1-Alpha 1 (human)
11	1122	M22488	Human bone morphogenetic protein 1 (BMP-1) mRNA
12	1173	M55543	Interferon-induced guanylate-binding protein 2 (human)
13	1408	R72644	Choline kinase (Homo sapiens)
14	242	M58297	Zinc finger protein 42 (human)
15	1856	X80692	Homo sapiens ERK3 mRNA
16	1848	H41528	Stage V sporulation protein E
17	1539	H78346	Paired box protein PAX-3
10	1133	H47650	PTS system, sucrose-specific IIABC
19	1533	H63361	Translation initiation factor EIF-2B-Alpha subunit
20	1254	X54232	Human mRNA for heparan sulfate proteoglycan

only just bring additional computational cost, but also lead to the same misclassifications. A better way is to group the genes with similar profiles, or the genes from the same pathway, and then select informative genes from these different groups to avoid redundancy [21]. In this section, we propose a FCCEGS method that improves the performance of selecting informative genes for microarray data.

Fuzzy C-means clustering allows each pattern to belong, with a certain degree of membership, to two or more clusters in the same time. Same as the other clustering methods, it is based on the minimization of the following distance function:

$$J = \sum_{i=1}^{N} \sum_{j=1}^{C} u_{ij}^{m} |x_i - c_j|^2,$$ (5.16)

where $m \geq 1$, u_{ij} is the degree of membership of pattern x_i to cluster j, c_j is the center of cluster j, C is the number of clusters, and N is the number of patterns in the data set. The algorithm is detailed below:

(1) Initialize the membership degree matrix $[u_{ij}]$ (every u_{ij} is a random number between 0 and 1) and set $k = 0$.

(2) At each step k, calculate the new cluster centers c_j according to the matrix U^k given at Step 3 [18]:

$$c_j = \frac{\sum_{i=1}^{N} u_{ij} \cdot x_i}{\sum_{i=1}^{N} u_{ij}}.$$ (5.17)

(3) Update the matrix U^k [18],

$$u_{ij}^{k+1} = \left(\frac{1}{\sum_{k=1}^{N} \left(\frac{|x_i - c_j|}{|x_i - c_j|} \right)} \right)^{\frac{2}{m-1}}$$ (5.18)

(4) If $||J(k+1) - J(k)|| \leq o$, then stop, where o is a user-defined small number.

Once we have done the FCC, we know that all the genes in one cluster show similar profiles and might be involved in the same pathway. Not only for the classification purposes, but also for the result analysis purposes, it is ideal to have genes that are still highly informative from different clusters. Generally, we believe highly correlated genes have a similar biological explanation. We prefer for our final decision model to be consisted of features with different meanings. Therefore, we would like to use more uncorrelated genes (but still informative) instead of the highly correlated top genes. As suggested in Ref. [9], we can apply a traditional gene selection method such as SNR, or the EFGS method, to select the genes in each cluster and then the final gene subset is determined by selecting a certain number ϑ of top-ranked genes from each cluster.

The value of ϑ can be set according to some statistical methods, for example,

$$\vartheta = T \times \frac{\sum_{i=1}^{C} F\_\text{score}(j)}{\sum_{i=1}^{D} F\_\text{score}_D},$$ (5.19)

where T is the maximum number of selected genes, $F\_$score (j) is the sum of the score of all gene in the clustering j, $F\_$score (D) is the sum of gene score in the whole data space D. By doing this, if there are several pathways involved in the perturbation, but one pathway has the major influence, we will probably select more genes from this pathway [9]. The gene selected by using the FCCEGS are listed in Table 5.4. The gene selection results before (left) and after (right) clustering are compared in Fig. 5.6.

Table 5.4   Seven top-ranked colon genes selected from three different clusters by using FCCEGS

Rank	ID	Gene Name	Description
Cluster A			
1	59	H22688	Ubiquitin (human)
2	29	R02593	60S Acidic Ribosomal protein P1
3	46	T61602	40S Ribosomal protein S11 (human)
4	73	H43908	transforming growth factor beta 2 precursor
5	33	T51574	40S Ribosomal protein S24
6	76	H13281	Photosystem II 44 KD reaction center protein
7	18	T63508	Ferritin heavy chain (human)
Cluster B			
1	1122	M22488	Human bone morphogenetic protein 1 (BMP-1) mRNA
2	1408	R72644	Choline kinase (Homo sapiens)
3	836	M23254	Human Ca2-activated neutral protease (CANP) large subunit mRNA, complete cds
4	1258	R67358	Map kinase phosphatase-1
5	1250	H09665	Lamin B receptor (*Gallus gallus*)
6	1856	X80692	Homo sapiens ERK3 mRNA
7	1433	X17042	Human mRNA for hematopoetic proteoglycan core protein
Cluster C			
1	334	H65823	B (human)
2	232	D31883	Human mRNA (KIAA0059) for ORF
3	490	H77348	5-Lipoxygenase-activating protein (*Macaca mulatta*)
4	115	X67325	Homo sapiens p27 mRNA
5	259	X54163	Troponin I, cardiac muscle (human); contains element MER22 repetitive element
6	54	T52015	Elongation factor 1-Gamma
7	370	Z23115	Homo sapiens bcl-xL mRNA

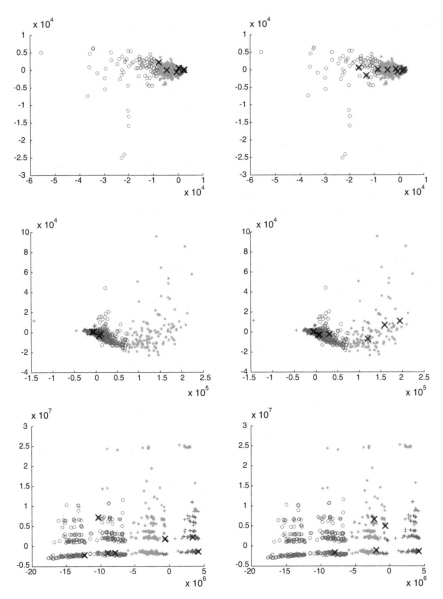

Figure 5.6 The "x" marks represent the top nine selected genes from three different data sets. The figures on the left-hand side are the gene selection results given by the traditional SNR methods, and the figures on the right-hand side are the gene selection results given by the FCCEGS. The top two figures are gene selection results for colon cancer data set, the traditional SNR method selects all genes from the right two clusters. The figures in the second row are gene selection results for leukemia cancer data set, from the left figure we can see that most of the selected genes are from one cluster. The bottom two figures are gene selection results for lymphoma cancer data set, none of the genes from the middle cluster can be selected by the traditional methods. Traditional methods are very likely to select highly correlated genes, while FCCEGS can select features from a larger space.

5.4 NEURO-FUZZY CLASSIFIER ENSEMBLE

5.4.1 Gene Subsets and Fuzzy Systems

Different from other nonlinear classification systems, fuzzy systems are rule-based systems; a small number of input will normally generate a large number of potential rules. Although not all of the potential rules are useful for final classification results and a small number of most useful rules can be selected indeed at some points, this initial large number of potential rules brings a large computational cost. For this reason, in practical applications, the feasible number of selected genes is limited. But every microarray gene database requires an optimal number of genes to represent certain properties of the data. (*Note* For different purposes, to discover different properties of the data, we may need different subsets of the data and a different minimal number of genes.) The relationship between the optimal number of informative genes and the number of fuzzy rules and fuzzy system parameters (e.g., for ANFIS) is shown in the next section.

One approach to this problem is to construct a NFE model by combining several individual NF models that are trained on the same data but using different subsets of genes.

5.4.2 Individual Neuro-Fuzzy Models

We use adaptive-network-based fuzzy inference system (ANFIS) networks to build the individual classifiers in the ensemble. ANFIS is a Sugeno-like fuzzy system in a five-layered network structure. Backpropagation algorithm is used to train the membership functions, while the least mean squares algorithm (LSE) determines the coefficients of the linear parameters in the consequent part of the rules. It has been proven that this hybrid algorithm is much more efficient than a standard gradient-based method in training the ANFIS [16].

Same as the other fuzzy-rule-based systems, ANFIS cannot easily handle high dimensional problems, as it has to deal with a large number of input nodes, rules, and, hence, consequent parameters. In an ANFIS model, if the number of inputs is N and the number of membership functions for each input is K, then the number of the fuzzy rules R is

$$R = K^N. \tag{5.20}$$

The number of the adaptive parameters P is then

$$P = K^{N+1} + K^N \times N, \tag{5.21}$$

where K^{N+1} refers to the nonlinear adaptive parameters and $K^N \times N$ to the linear adaptive parameters. For example, when the number of membership functions for each input is 3, the relationship between the inputs and rules is as shown in Table 5.5. From Equations 5.20 and 5.21 and Table 5.5, we can see that the computational cost increases

Table 5.5 **The relationship between the number of input features, the number of fuzzy rules, and the number of model parameters needed to be updated in each epoch**

NoG	Number of Rules	Number of Parameters
2	9	45
3	27	108
4	81	360
5	243	774
6	729	2484
7	2187	7008
...	...	...

NoG denotes the number of selected genes.

very quickly with the number of inputs. We simulated the ANFIS models for the presented microarray data sets on an IBM R51 laptop (CPU, PIV-1.5G; Memory, 1G). The computer reaches its limits when the number of inputs is larger than six.

The collection of well-distributed, sufficient, and accurately measured input data is however the basic requirement for obtaining an accurate model [3]. When data sets require a relatively large number of genes to represent their properties, we need to design some strategies to enable the model to accept more inputs with less computational cost. Recommended approaches include evaluate and select rules, delete antecedents, delete fuzzy sets, and others.

5.4.3 Ensemble Learning

In addition to building individual ANFIS models for the microarray data sets as described above, we designed a NFE model [22], which consists of several individual ANFIS models, to deal with these problems. Each individual model uses different subsets of genes as inputs, so that the overall model finally work with a relatively large number of genes. Meanwhile, a very good performance can be obtained by the nature of the ensemble learning itself. We set the maximum number of inputs for an individual NF network in the NFE to be four, due to the consideration of the required computational cost as discussed above. Each input is randomly selected. So, in total we use $4 \times n$ genes, where n is the number of individual NF models in the ensemble. By combining with our FCCEGS method, we will try to make sure that the inputs of one individual NF model are from different clusters. The output combination strategy of our NFE model is majority voting, where the output of the majority of individual NF networks will be considered to be the output of the ensemble. The main structure of our NFE is shown in Fig. 5.7.

The advantages offered by our NFE model are summarized below [22]:

- It allows the model to use more features when the optimal subset of genes is relatively large.

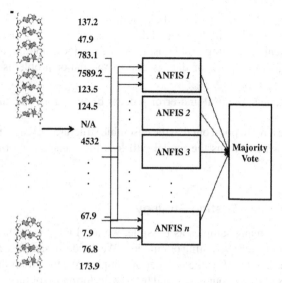

Figure 5.7 The main structure of the NFE: n individual ANFIS classifiers in the ensemble, each having R inputs, so that the overall ensemble model can use $R*n$ genes. The output of the ensemble is taken by simple majority voting (MV).

- Normally, using more than one classifier implies higher computational costs than building a single classifier with the same number of inputs. However, our NFE model will require much less computational cost than building a single large NF model, when the necessary number of inputs is relatively large. A comparison of computational costs between individual NF models and the NFE is shown in Table 5.6.

Table 5.6   Computational cost comparison between individual and ensemble NF models

| NoG | NF | | | NFE | |
	NoR	NoP		NoR	NoP
2	9	45		N/A	N/A
3	27	108		N/A	N/A
4	81	360		N/A	N/A
6	729	2484		54	216
8	6.6×10^4	7.9×10^5		162	720
12	1.7×10^7	2.7×10^8		243	1080
16	4.3×10^9	8.6×10^{10}		324	1440
20	1.1×10^{12}	2.6×10^{13}		405	1800
...	...	...		...	...

We compare the number of rules and parameters of individual NF and NFE models. Each input of the NF models has three membership functions. In this comparison, the NFE contains two individual NF models. NoG denotes the number of selected genes, NoR denotes the number of rules, and NoP denotes the number of parameters needed to be updated in each epoch.

- NFE can significantly improve the generalization ability (classification performance) compared to single NF models [23], and they can also help address three classic machine learning problems: lack of data, local optima, and representational limitations. Lack of data is one of the major problems in microarray analysis.
- NFE can relieve the trial-and-error process by tuning architectures.

However, we have to point out that the ensemble structure increases the complexity of the system, which means the model will lose the transparency of the decision-making process [7].

5.4.4 Training and Testing Strategy

In order to use as many samples as possible, we use LOOCV method or jackknife strategy for training and evaluating our models. We divide all samples, at random, into K distinct subsets, where K is the number of samples. We then train the model using $K - 1$ subsets (samples in our case), and test the performance of the Kth sample. The LOOCV accuracy is obtained by

$$\text{LOOCV}_{\text{acuracy}} = \frac{\text{Acs}}{K}, \tag{5.22}$$

where Acs is the number of correctly classified samples in K experiments. The LOOCV accuracy is strongly suggested by other researchers to be used as an evaluation measure for the microarray data classification performance. In order to compare with their work, this strategy is also adopted in our study.

5.4.5 Experimental Setup and Results

We use three important criteria for the empirical evaluation of the performances of our models:

- Number of selected genes;
- Predictive accuracy of using selected genes; and
- Extracted knowledge from the trained models.

Before running the experiments, we linearly scale all the data in the [0, 1] interval. If y is a gene expression value of a gene g, the scaled value will be

$$g(a') = \frac{y - \min(g)}{\max(g) - \min(g)} \tag{5.23}$$

where $\min(g)$ and $\max(g)$ is the minimum and maximum expression values of that gene in the database.

Each variable is described by three membership functions for both NF and NFE models, and the initial shape of the membership function is a bell-shaped function (see Fig. 5.8 top). There are five individual NF networks in our NFE model, and each NF model has four inputs. The output of the ensemble is obtained by using MV. The 20 top-ranked genes obtained by the SNR, EFGS, and FCCEGS methods were selected for classification when using the NFE model. The four top-ranked genes were selected by using the information gain (IG) [26] method and the SNR method when employing the single NF model for classification.

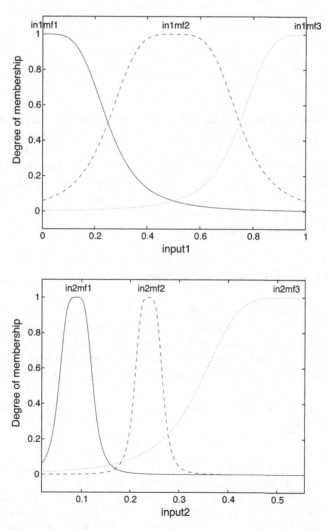

Figure 5.8 Initial membership functions (top) and adjusted membership functions (bottom) of the NFE model for colon cancer data set.

Table 5.7   Comparison of the classification performance of different classifiers and gene selection methods for leukemia cancer data set, colon cancer data set, and lymphoma cancer data set

	GS method	NoSG	Colon	Leukemia	Lymphoma
Single NF	SNR	4	90.32	83.33	89.12
Single NF	IG	4	93.55	87.5	87.23
NFE	IG	20	100	95.85	93.61
NFE	SNR	20	97.4	95.85	92.33
NFE	EFGS	20	91.94	93.06	91.49
NFE	FCCEGS	20	93.55	90.28	95.74
SVM [6]	IG	50	90.30	94.10	N/A
SVM [19]	SNR	50	65.0	59.0	76.0
	LLE	50	85.0	95.0	91.0
KNN	EA	50	75.81	72.64 [11]	74.47
C4.5	ReliefF	4–60	85.48 [26]	81.94 [26]	82.98

The performance of our NFE model was compared to that of single NF models and some other reported approaches, by using the same training and testing strategies (see Table 5.7). Our models obtained better results for colon cancer data set and Lymphoma data set, similar results for leukemia, but both NF and NFE models used less number of genes compared to other approaches. The performance of the NFE model used much better than that of single NF models for the three cancer data sets. Because of the nature of the data, the FCCEGS method performs better than the SNR and IG methods for lymphoma data set, but worse for the other data sets (see Table 5.7).

5.5 FUZZY KNOWLEDGE DISCOVERY

Different from black-box approaches, NF-based models can extract some useful knowledge from large gene expression data, for example, adjusted membership functions (Fig. 5.8), trained fuzzy rules (Table 5.8), and fuzzy decision surfaces (see Fig. 5.9). All this knowledge can be presented in a human understandable form. This seems very attractive for researchers in the area, as they can better understand the data

Table 5.8   Five rules selected from a single NF model in the ensemble of the colon cancer data set

	Descriptions of Rules
1	If (M63391 is small) and (R87126 is small), then (output is cancer)
2	If (M63391 is small) and (R87126 is medium), then (output is cancer)
3	If (M63391 is small) and (R87126 is large), then (output is normal)
7	If (M63391 is large) and (R87126 is small), then (output is cancer)
9	If (M63391 is large) and (R87126 is large), then (output is normal)

There are two membership functions for each variable.

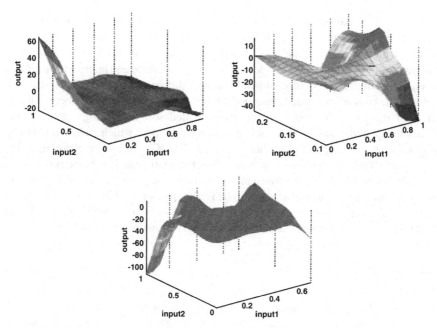

Figure 5.9 The fuzzy decision surface of the trained models when the number of selected genes $= 2$, and gene selection method $=$ IG.

or explain how the results are obtained. Meanwhile, NF-based models can also easily incorporate prior knowledge, which helps obtaining more refined models and shorten the training process.

5.6 CONCLUSION AND FUTURE WORK

In this chapter, we have proposed some fuzzy-based approaches for cancer microarray data analysis. We have first applied the fuzzy C-means for gene clustering, so that genes with similar expression profiles are grouped together. We have then used an evolutionary programming-like algorithm to estimate the membership function for each gene cluster as well as to rank genes. Finally, we have designed an ensemble of neuro-fuzzy classifiers for classifying samples into cancer or noncancer by using chosen high-ranked genes. The evolving fuzzy gene selection and fuzzy C-mean clustering based enhanced gene selection methods directly point against the two major problems of traditional gene selection methods. The neuro-fuzzy ensemble method makes the fuzzy-rule-based approach more feasible to microarray gene expression data analysis. The performances of our models are competitive. But, there still are many issues that need to be considered in future research. Fuzzy association rules offer us an insight into the interaction between genes in a human readable form. Fuzzy rules can reveal biological relevant associations between different genes, and between genes and their environment. But this depends heavily on the size of the data set.

Microarray gene data sets are usually very large, and how to understand larger number of rules becomes another prey-box problem. After combination, the NFE model becomes more complex than a single NF model and, therefore, more difficult to analyze. Although a single NF model can be easily explained and interpreted by users, an ensemble of several NF models would be more difficult to understand. The balance between classification accuracy and model interpretability should be further explored. Most of the methods used here are the so-called soft computing methods; one disadvantage of these methods is lack of theoretical support, which make our obtained performance difficult to be further analyzed. LOOCV is however not the only appropriate strategy, the bootstrap resampling technique can also be considered for such kind of problems. Fuzzy-based methods offer good potential to deal with highly noisy/missing data, which also needs to be further investigated in future research. In addition, all tested problems here are binary classification problems, we should address how to extend our model to multicategory classification problems.

REFERENCES

1. Alon, U., Barkai, N., Notterman, D. A., Gish, K., Ybarra, S., Mack, D., and Levine, A. J. Broad patterns of gene expression revealed by clustering analysis of tumor and normal colon tissues probed by oligonucleotide arrays. *Proceedings of the National Academy of Sciences of the United States of America*, 96(12): 6745–6750, 1999.

2. Baumgartner, R., Windischberger, C., and Moser, E. Quantification in functional magnetic resonance imaging: fuzzy clustering vs. correlation analysis. *Magnetic Resonance Imaging*, 16(2): 115–125, 1998.

3. Belal, S., Taktak, A., Nevill, A., Spencer, S., Roden, D., and Bevan, S. Automatic detection of distorted plethysmogram pulses in neonates and paediatric patients using an adaptive-network-based fuzzy inference system. *Artificial Intelligence in Medicine*, (24): 149–165, 2002.

4. Cho, S. and Won, H. Machine learning in DNA microarray analysis for cancer classification. *The Asia Pacific Bioinformatics Conferences*, 34: 189–198, 2003.

5. Ding, C. and Peng, H. Minimum redundancy feature selection from microarray gene expression data. *Proceedings of the IEEE Computer Society Bioinformatics Conference (CSB 2003)*, 2003, 523–528.

6. Furey, T. S., Cristianini, N., Duffy, N., Bednarski, D. W., Schummer, M., and Haussler, D. Support vector machine classification and validation of cancer tissue samples using microarray expression data. *Bioinformatics*, 16: 906–914, 2000.

7. Gabrys, B. Combining neuro-fuzzy classifiers for improved generalisation and reliability. *Proceedings of the International Joint Conference on Neural Networks (IJCNN'2002)*, Honolulu, USA, pp. 2410–2415, 2002.

8. Golub, T. R., Slonim, D. K., Tamayo, P., Huard, C., Gaasenbeek, M., Mesirov, J. P., Coller, H., Loh, M. L., Downing, J. R., Caligiuri, M. A., Bloomfield, C. D., and Lander, E. S. Molecular classification of cancer: class discovery and class prediction by gene expression monitoring. *Science*, 286: 531–537, 2002.

9. Jaeger, J., Sengupta, R., and Ruzzo, W. L. Improved gene selection for classification of microarrays. *Pacific Symposium on Biocomputing*, pp. 53–64, 2003.

10. Jiang, X. and Gruenwald, L. Microarray gene expression data association rules mining based on BSC-tree and FIS-tree. *Data & Knowledge Engineering*, 53(1): 3–29, 2005.

11. Jirapech-Umpai, T. and Aitken, S. Feature selection and classification for microarray data analysis: evolutionary methods for identifying predictive genes. *Bioinformatics*, 6: 168–174, 2005.

12. Khan, J., Wei, J. S., Ringner, M., Saal, L. H., Ladanyi, M., Westermann, F., Berthold, F., Schwab, M., Antonescu, C. R., Peterson, C., and Meltzer, P. S. Classification and diagnostic prediction of cancers using gene expression profiling and artificial neural networks. *Nature Medicine*, 7: 673–679, 2001.

13. Kohonen, T. Ed. *Self-Organizing Maps*, Springer-Verlag, New York, 1997.

14. Li, L., Weinberg, C. R., Darden, T. A., and Pedersen, L. G. Gene selection for sample classification based on gene expression data: study of sensitivity to choice of parameters of the GA/KNN method. *Bioinformatics*, 17: 1131–1142, 2001.

15. Lossos, I., Alizadeh, A., Eisen, M., Chan, W., Brown, P., Bostein, D., Staudt, L., and Levy, R. Ongoing immunoglobulin somatic mutation in germinal center B cell-like but not in activated b cell-like diffuse large cell lymphomas. *Proceedings of the National Academy of Sciences of the United States of America*, 97: 10209–10213, 2000.

16. Nauck, D. Neuro-fuzzy systems: review and prospects. *Proceedings of the Fifth European Congress on Intelligent Techniques and Soft Computing (EUFIT'97)*, 8: 1044–1053, 1997.

17. Ressom, H., Reynolds, R., and Varghese, R. Increasing the efficiency of fuzzy logic-based gene expression data analysis. *Physiological Genomics*, 13: 107–117, 2003.

18. Shannon, W., Culverhouse, R., and Duncan, J. Analyzing microarray data using cluster analysis. *Pharmacogenomics*, 4: 41–51, 2003.

19. Shi, C. and Chen, L. Feature dimension reduction for microarray data analysis using locally linear embedding. *The Asia Pacific Bioinformatics Conference*, pp. 211–217, 2005.

20. Slonim, D., Tamayo, P., Mesirov, J., Golub, T., and Lander, E. Class prediction and discovery using gene expression data. *Research in Computational Molecular Biology*, pp. 263–272, 2000.

21. Wang, Z. and Palade, V. A comprehensive fuzzy-based framework for cancer microarray data gene expression analysis. *IEEE 7th International Symposium on Bioinformatics and Bioengineering (BIBE'07)*, 2007.

22. Wang, Z., Palade, V., and Xu, Y. Neuro-fuzzy ensemble approach for microarray cancer gene expression data analysis. *Proceedings of the Second International Symposium on Evolving Fuzzy System (EFS'06)*, pp. 241–246, 2006.

23. Wang, Z., Yao, X., and Xu, Y. An improved constructive neural network ensemble approach to medical diagnoses. *Proceedings of the Fifth International Conference on Intelligent Data Engineering and Automated Learning (IDEAL'04), Lecture Notes in Computer Science, Vol. 3177*, Springer, pp. 572–577, 2004.

24. Xiong, M., Li, W., Zhao, J., Jin, L., and Boerwinkle, E. Feature (gene) selection in gene expression-based tumor classification. *Molecular Genetics and Metabolism*, 73(3): 239–247, 2001.

25. Yao, X., Liu, Y., and Liu, G. Evolutionary programming made faster. *IEEE Transactions on Evolutionary Computation*, 3(2): 82–102, 1999.

26. Yu, L. and Liu, H. Redundancy-based feature selection for microarray data. Technical report, Department of Computer Science and Engineering, Arizona State University, Tempe, AZ 2004.

6

FEATURE SELECTION FOR ENSEMBLE LEARNING AND ITS APPLICATION

Guo-Zheng Li and Jack Y. Yang

6.1 INTRODUCTION

Ensemble learning and feature selection are two hot topics in machine learning and bioinformatics community, which have been widely used to improve generalization performance of single learning machines. For ensemble learning, a good ensemble is one whose individuals are both accurate and make their errors on different parts of the input space [1]. The most popular methods for ensemble creation are bagging and boosting [2, 3]. The effectiveness of such methods comes primarily from the diversity caused by resampling the training set [4]. For feature selection, different methods can be categorized into the filter model, the wrapper model, and the embedded model [5–7], where the filter model is independent of the learning machine and both the embedded model and the wrapper model depend on the learning machine, but the embedded model has lower computational complexity than the wrapper model does and has been widely studied in recent years especially on support vector machines (SVMs) [5, 8].

Feature selection has been used to create new ensemble methods. Ho [9] proposed a random subspace method for constructing decision forests, which has been improved later [10]. Opitz [11] presented an ensemble feature selection approach based on genetic algorithm, and Oliveira et al. [12] used a multiobjective genetic algorithm. Brylla et al. [13] proposed attribute bagging based on random feature selection. Tsymbal and Cunningham [14] studied different search strategies for ensemble feature

---

Machine Learning in Bioinformatics. Edited by Yan-Qing Zhang and Jagath C. Rajapakse
Copyright © 2009 by John Wiley & Sons, Inc.

selection methods and proposed to employ genetic algorithm for ensemble feature selection [15]. Tabu search is also used to produce different feature subsets for ensemble [16]. In general, these works on ensemble feature selection create diverse individuals by generating different feature subsets, whereas bagging and boosting create diverse individuals by generating different sample subset. However, a few studied feature selection, for example, resampling method such as bagging or boosting. Therefore, Li et al. studied feature selection for bagging and its applications [17–19], which will be introduced in this chapter. Here, both the embedded feature selection model and the filter model will be employed to select optimal feature subset for individual learning machines of bagging, which is one of the main contributions of this chapter.

The second contribution is that of individual selection of ensemble learning. Zhou et al. [20] proposed a selective ensemble learning method, GASEN, where genetic algorithm was employed to select the individuals of bagging, and obtained satisfactory results. This showed that it may be better to ensemble some instead of all of them; this is also proved by Caruana et al. [21, 22]. GASEN uses genetic algorithm as an individual selection method, whose computational complexity is $O(2^n)$, n is the number of individuals; therefore, ensemble learning, selective ensemble learning, and GASEN need much computation capacity. In order to improve the efficiency of GASEN, at the same time to improve the generalization ability, mutual information-based selective ensemble (MISEN) is proposed [23], which uses mutual information as the individual selection method whose computational complexity of mutual information is $O(n^2)$, which is much lower than that of genetic algorithm. We will show how MISEN reduces the computational time and is applied to classification of brain glioma cases.

Bagging of neural networks has made a great success in the past several years, yet too many redundant features in the data sets will hurt the generalization ability of bagging: so, feature selection is employed to remove the redundant and irrelevant features [5, 7]. Although the removed features are redundant and weakly relevant, they contain useful information for improving the prediction accuracy. Multitask learning (MTL) is a method to use the redundant information, by selecting features from the discarded features to add to the target [24, 25]. MTL has been used to improve the generalization performance of several different learning machines, that is, neural networks [25], the partial least squares algorithm [26], and bagging of neural networks [27].

The previous studies on search methods for MTL are mainly on the heuristic methods [25, 26], where the number of the features selected as the input and/or target is somewhat arbitrary. When the search method is regarded as a combinational optimization problem, random search methods are also an alternative approach where genetic algorithm [28] is an easily used and powerful method; it has been used for feature selection and has obtained satisfactory results [29]. Motivated by this, Li and Lui [27] proposed a random method of genetic-algorithm-based multitask learning (GA-MTL) to improve the prediction accuracy of bagging, which will be discussed subsequently.

The rest of the chapter is arranged as follows: In Section 6.2, the embedded feature selection model with the prediction risk criteria and the filter model with the mutual information criteria for bagging are described. In Section 6.3, mutual-information-based selective ensemble is introduced. In Section 6.4, GA-MTL for improving neural networks ensemble (NNE) is presented. Then, classification of brain glioma by using the above three learning methods are performed in Section 6.5. Finally, summary is given in Section 6.6.

6.2 FEATURE SELECTION FOR INDIVIDUALS OF BAGGING

Ensemble learning methods such as bagging can effectively improve the generalization accuracy of single learning machines, this is approved and validated in many works [1–3]. Especially, on SVMs, Valentini and Dietterich concluded that reducing the error of SVMs would reduce the error of bagging of SVM [30]. At the same time, feature selection can also effectively improve the generalization ability of single learning machines, where the filter model and the embedded model are two popular methods [5–7]. Feature selection has been used in ensemble learning and has obtained some interesting results [9–15], but most of the previous works used feature selection to produce the diversity of the individuals. Therefore, combing feature selection with bagging to improve the generalization performance of single learning machines is an interesting issue. Motivated by these works, Li et al. proposed to use the embedded feature selection method with the prediction risk criteria and a filter-model-based method for bagging of SVM and presented two algorithms PRIFEB and MIFEB [17–19].

6.2.1 PRIFEB

Since the embedded model was employed on SVM and obtained satisfactory results [31], this model was firstly employed to ensemble learning, where the prediction risk criteria are used to rank the features. The prediction risk criteria were proposed by Moody and Utans [32], which evaluate each feature through estimating prediction error of the data sets when the values of all examples of this feature are replaced by their mean value.

$$S_i = \text{ERR}(\bar{x}^i) - \text{ERR}, \qquad (6.1)$$

where ERR is the training error, and ERR $(\bar{x}^i)$ is the test error for the training data set with the mean value of ith feature and defined as

$$\text{ERR}(\bar{x}^i) = \frac{1}{\ell} \sum_{j=1}^{\ell} (\tilde{y}(x_j^1, \cdots, x^i, \cdots, x_j^D) \neq y_j),$$

where ℓ and D are the number of examples S_i and features, respectively; $\bar{x}^i$ is the mean value of the ith feature; and $\tilde{y}(\ )$ is the prediction value of the jth example after the value of the ith feature is replaced by its mean value. Finally, the feature corresponding with the smallest will be deleted, because this feature causes the smallest error and is the least important one.

Input: Training data set $S_r(x^1, x^2, \cdots, x^D, C)$, Number of individuals T

Procedure:

for $k = 1 : T$

    1. Generate a training subset S_{rk} from S_r by using bootstrap sampling algorithm, the size of S_{rk} is three quarters of the size of S_r.

    2. Train an individual model L_k on the training subset S_{rk} by using the SVMs algorithm and calculate the training error ERR.

    3. Compute the prediction risk value R_i using Equation 6.1.

    4. If R_i is greater than 0, the ith feature is selected as one of the optimal features. Repeat until all the features in S_{rk} are computed.

    5. Generate the optimal training subset $S_{rk-optimal}$ from S_{rk} according to the above optimal features.

    6. Train the individual model N_k on the optimal training subset $S_{rk-optimal}$ by using SVMs.

end

Ensemble the obtained models N by the way of majority voting method for classification problems

Output: Ensemble model N

Figure 6.1   The PRIFEB approach.

In this chapter, the embedded feature selection model with the prediction risk criteria is employed to select relevant features for the individuals of bagging of SVM, which is named as PRIFEB (Prediction Risk based Feature sElection for Bagging). The basic steps of PRIFEB are described in Fig. 6.1.

6.2.2 MIFEB

The above-mentioned embedded feature selection model depends on the learning machine, whereas the filter model is independent and used frequently, where the mutual information criteria are studied widely [7]. The mutual information describes the statistical dependence of the two random features on the amount of information that one feature contains about the other, and it is a widely used information theoretic measure for the stochastic dependency of discrete random features. The mutual information between two features R and S can be defined in terms of the probability distribution of intensities as follows:

$$I(R:S) = \sum_{r \in R} \sum_{s \in S} p\{r,s\} \lg \frac{p\{r,s\}}{p\{r\}p\{s\}}, \qquad (6.2)$$

where $p\{r,s\}$ is the combined probability distribution of intensities of two features R and S. $p\{r\}$ and $p\{s\}$ are the individual probability distribution of intensities of R and S, respectively.

The mutual information criteria have been widely used in the filter feature selection model [7]; therefore, this method was employed to bagging and propose a method, MIFEB (Mutual Information based Feature sElection for Bagging) and compare it with PRIFEB to test which feature selection model will be more effective. The basic steps of MIFEB are described in Fig. 6.2.

Input: Training data set S_r, Number of individuals T

Procedure:

for $k = 1 : T$

1. Generate a training subset S_{rk} from S_r by using bootstrap sampling algorithm, the size of S_{rk} is three quarters of the size of S_r.

2. Employ the mutual information formula (Eq.6.2) on the training subset S_{rk} and the target set of S_{rk} to obtain the value vector MI .

3. Rank the vector MI in descending order and sum MI to obtain SUM(MI).

4. Select all of the first features as the optimal subset, whose total values should be greater than RMI $*$ SUM(MI). RMI is a predefined ratio which is greater than 0 but less than 1.

5. Generate the optimal training subset $S_{rk - optimal}$ from S_{rk} according to the above optimal features.

6. Train the individual model N_k on the optimal training subset $S_{rk - optimal}$ by using SVMs.

end

Ensemble the obtained models N by the way of majority voting method for classification problems .

Output: Ensemble model N.

Figure 6.2   The MIFEB approach.

6.2.3 Numerical Experiments

To perform a benchmark of the proposed algorithms, PRIFEB and MIFEB, 13 data sets are selected from the UCI machine learning repository [33] and listed in Table 6.1, in which #classes means the number of classes. We can see the number of instances ranges from hundreds to thousands and the number of features ranges from 9 to 70. To make them suitable for our algorithms, the nominal values are changed to be numerical in all data sets. Then, all the attributes are transformed into the interval of $[-1, 1]$ by an affine function.

The holdout method is used to validate the results. Experiments are repeated 50 times for each data set. The same pair of parameters for SVM, $C = 100$, $\sigma = 10$, is used and the number of individuals for bagging T is 20. The RMI for mutual information algorithm is 0.9 in our experiments. Experimental results for the accuracy obtained by the different bagging methods are shown in Table 6.2.

Table 6.2 shows that mean accuracies obtained by the bagging methods with feature selection are improved in various degrees for all the data sets that range from 0% to 6%, the mean value improved for all the data sets is 2.35% for the PRIFEB method and 2.16% for the MIFEB method. At the same time, the values of standard deviation are also reduced in some degrees.

It can also be seen that PRIFEB performs slightly better than MIFEB does for 8 out of 13 data sets, and worse for 5 data sets. Yet, in fact, the difference of the results obtained between PRIFEB and MIFEB is so slight (less than 1% for all the data sets) that we consider these two feature selection approaches perform equally.

The average numbers of features in the optimal feature subsets obtained by the different bagging methods in different data sets are shown in Table 6.3, where R_PRIFEB means the ratio of the number of the optimal subsets obtained by PRIFEB to that of the total, and R_MIFEB means the ratio of the number of the optimal subsets obtained by MIFEB to that of the total. Table 6.3 shows that all the data sets need to

Table 6.1   Properties of the used UCI data sets

Data set	#Classes	#Features	#Instances
all-bp	3	29	3772
all-hyper	5	29	3772
all-hypo	4	29	3772
backup	19	35	683
audio	24	70	226
proc-C	5	13	303
proc-H	2	13	294
soybean-l	19	34	307
statlog-h	2	13	270
glass	6	9	214
voting-records	2	16	435
Ionosphere	2	34	351
breast-cancer-W	2	9	699

Table 6.2   Statistical prediction accuracy of the bagging methods with feature selection and without feature selection for UCI data sets

Data set	PRIFEB	MIFEB	Bagging
all-bp	97.07 ± 0.45	96.86 ± 0.51	95.95 ± 0.13
all-hyper	97.87 ± 0.21	97.93 ± 0.16	96.76 ± 0.43
all-hypo	97.42 ± 0.43	97.08 ± 0.16	96.66 ± 0.61
backup	91.99 ± 1.41	91.47 ± 1.62	89.90 ± 2.06
audio	75.19 ± 3.46	75.66 ± 3.72	74.67 ± 4.23
proc-C	53.71 ± 3.67	54.16 ± 2.40	49.93 ± 3.65
proc-H	79.84 ± 2.55	79.70 ± 2.71	73.89 ± 3.52
soybean-l	85.54 ± 3.64	84.30 ± 3.38	83.25 ± 3.57
statlog-h	78.62 ± 3.30	78.98 ± 2.87	74.88 ± 3.87
glass	64.95 ± 3.74	65.78 ± 4.01	61.75 ± 5.12
voting-records	95.59 ± 1.08	94.30 ± 4.88	94.24 ± 1.40
Ionosphere	88.70 ± 2.16	88.21 ± 2.74	87.28 ± 3.07
breast-cancer-W	94.38 ± 1.02	94.01 ± 0.92	91.23 ± 1.71
Average	84.68 ± 2.08	84.49 ± 2.31	82.33 ± 2.57

remove some features in some degrees, and MIFEB can obtain slightly smaller feature subsets than PRIFEB does.

6.2.4 Summary

Feature selection for the bagging of SVMs has been studied. Experimental results for UCI data sets show that both the filter model with the mutual information criteria and the embedded model with the prediction risk criteria can improve the generalization

Table 6.3   Average number of the optimal features obtained by the bagging methods with feature selection

Data set	PRIFEB	R_PRIFEB (%)	MIFEB	R_MIFEB (%)
all-bp	23.86 ± 0.77	82.27	25.58 ± 0.24	88.21
all-hyper	24.81 ± 0.98	85.55	25.60 ± 0.27	88.28
all-hypo	18.70 ± 0.13	64.48	25.50 ± 0.31	87.93
backup	31.76 ± 0.26	90.74	25.81 ± 0.73	73.74
audio	65.71 ± 0.16	93.87	66.14 ± 0.38	94.49
proc-C	11.97 ± 0.03	92.07	9.86 ± 0.24	75.85
proc-H	11.26 ± 0.12	86.62	10.94 ± 0.27	84.15
soybean-l	31.05 ± 0.07	91.32	27.57 ± 0.74	81.09
statlog-h	11.61 ± 0.18	85.85	10.42 ± 0.28	80.15
glass	7.96 ± 0.04	88.44	7.06 ± 0.16	78.44
voting-records	13.60 ± 0.31	85.00	11.55 ± 0.34	72.19
Ionosphere	30.81 ± 0.17	90.62	30.07 ± 0.21	88.47
breast-cancer-W	7.67 ± 0.13	85.22	7.33 ± 0.09	81.44
Average	22.37 ± 0.26	86.31	21.80 ± 0.32	82.65

performance of bagging of SVM. We think that feature selection can reduce the irrelevant features and even the redundant features to improve the accuracy of single individuals, which has also been proven by the previous works. At the same time, feature selection reduces different features for different individuals and help to increase the diversity among the individuals of bagging. According to the theory [1], improving the accuracy of every individual and increasing the diversity among them will effectively improve the accuracy of an ensemble method. The above two aspects caused by feature selection will make it true that feature selection can effectively improve the generalization ability of bagging.

6.3 SELECTIVE ENSEMBLE LEARNING

Ensemble learning such as bagging and boosting has achieved great improvements for single learning machines, where these methods ensemble all of the individuals.

Recently, Zhou et al. showed that it may be better to ensemble some instead of all of the base learning machines and proposed a novel selective ensemble learning method GASEN. GASEN is based on genetic algorithm and exhibits better performance than bagging and Adaboost by using different learning machines, for example, neural networks, decision trees, k-nearest neighbor [20]. However, training of GASEN consumes a large amount of time because of the high computational complexity O (2^n) of genetic algorithm. To improve the efficiency of selecting individuals is becoming an interesting issue for selective ensemble learning.

6.3.1 MISEN

To reduce the computational time, mutual information is used instead of genetic algorithm as the individual selection method, and a mutual-information-based selective ensemble algorithm, named MISEN, is proposed [23]. Mutual information has been introduced in Section 6.2.2.

The procedure of MISEN is shown in Fig 6.3. Briefly, MISEN first employs bootstrap to generate many models and then these models are ranked by using the mutual information values calculated from the model outputs of the validation data set and the target vector of the validation data set. The best P models are selected to generate the ensemble model instead of using all the models in bagging, where P is a proportion preset by hand.

6.3.2 Numerical Experiments

MISEN is compared with GASEN and bagging of 11 data sets selected from UCI machine learning repository [33]. These data sets have been extensively used in testing the performance of diverse kinds of learning systems. To make them suitable for our algorithms, features and instances with missing values are removed and the nominal values are changed to be numerical in all the data sets. Then, all the features are transformed into the interval of $[-1, 1]$ by an affine function. The information of the UCI data sets used in our experiments is listed in Table 6.4.

Input: Training data set S, validation set V, learner L, population size T, and proportion P

Procedure:

1. for $i = 1 : T$

   Generate a subset S_i by bootstrap

   Train the learner L to generate a model N_i on S_i

   end

2. for $i = 1 : T$

   Test N_i on V and obtain output(N_i)

   Compute the mutual information value M_i between output(N_i) and the
      target vector in V by using Equation 6.2

   end

   Rank N according to M in descending order

3. Select the first INT($P * T$) models to generate the final ensemble model
   N^*

Output: Ensemble model N^*

Figure 6.3   The MISEN approach.

Table 6.4   Properties of the used UCI data sets

Data set	#Classes	#Features	#Instances
backup	19	35	683
audio	24	70	226
processed-Cleveland	5	13	303
processed-Hungarian	2	13	294
soybean-large	19	34	307
statlog-heart	2	13	270
glass	6	9	214
voting-records	2	16	435
Ionosphere	2	34	351
breast-cancer-Wisconsin	2	9	699
tic-tac-toe	2	9	958

Table 6.5 Statistical results of prediction accuracy by MISEN, GASEN, and bagging (%)

Data set	MISEN	GASEN	Bagging
backup	92.07 ± 1.17	91.85 ± 1.43	90.48 ± 2.27
audio	75.84 ± 3.99	75.62 ± 4.01	75.78 ± 3.63
processed-Cleveland	53.68 ± 2.99	53.39 ± 3.16	50.15 ± 3.30
processed-Hungarian	78.04 ± 2.56	77.63 ± 2.69	75.70 ± 3.07
soybean-large	85.04 ± 3.30	84.71 ± 1.08	83.83 ± 3.12
statlog-heart	78.50 ± 3.15	78.41 ± 2.91	75.94 ± 3.55
glass	65.19 ± 4.07	64.74 ± 4.46	61.42 ± 4.38
voting-records	94.94 ± 1.19	94.71 ± 1.23	94.12 ± 1.24
Ionosphere	89.21 ± 2.71	88.92 ± 2.72	87.55 ± 2.96
breast-cancer-Wisconsin	94.18 ± 1.11	93.69 ± 1.52	91.11 ± 1.95
tic-tac-toe	97.66 ± 0.72	97.20 ± 0.89	97.40 ± 0.97
Average	82.21 ± 2.45	81.89 ± 2.37	80.31 ± 2.78

The holdout method is used to validate the results. Experiments are repeated 50 times for each data set. According to the advices of Valentini and Dietterich [30], the same pair of parameters for SVMs, $C = 100$, $\sigma = 10$, is used and the T number of individuals for bagging is 20. The proportion P of MISEN is 75%.

The statistical prediction accuracies obtained for all data sets using MISEN, GASEN, and bagging are given in Table 6.5, which shows that the results of accuracy obtained by MISEN are slightly better than those by GASEN, and results by both MISEN and GASEN are better than those by bagging.

The computational times of MISEN and GASEN during the process of selecting individuals are given in Table 6.6, which shows that the computational time of MISEN is rather less than that of GASEN in all the 11 data sets, the ratios of the computational

Table 6.6 Average computational time of MISEN and GASEN during the process of selecting individuals

Data set	MISEN(s)	GASEN(s)	Ratio
backup	18.23 ± 1.94	33.71 ± 3.71	0.54
audio	1.99 ± 0.09	14.23 ± 0.44	0.14
processed-Cleveland	0.85 ± 0.07	4.98 ± 0.12	0.17
processed-Hungarian	0.31 ± 0.05	3.39 ± 0.09	0.09
soybean-large	3.04 ± 0.10	16.82 ± 0.70	0.18
statlog-heart	0.45 ± 0.16	5.08 ± 1.70	0.09
glass	0.87 ± 0.27	7.95 ± 2.68	0.11
voting-records	0.59 ± 0.05	3.88 ± 0.08	0.15
Ionosphere	0.61 ± 0.20	4.69 ± 1.31	0.13
breast-cancer-Wisconsin	1.69 ± 0.12	4.85 ± 0.08	0.35
tic-tac-toe	5.91 ± 0.66	10.30 ± 0.67	0.57
Average	3.14 ± 0.33	9.99 ± 1.05	0.31

time between MISEN and GASEN range from 0.09 to 0.57. The average value of ratios is 0.31; this means that MISEN uses less than one third of the computational time of GASEN for selecting individuals for all the 11 data sets.

6.3.3 Summary

This section presents individual selection methods for selective ensemble learning and proposes MISEN. MISEN further validates that selective ensemble using some of the individuals obtains better generalization performance than ensemble methods using all of the individuals. The computational complexity of mutual information is lower than that of genetic algorithm. Experimental results show that MISEN using mutual information as the individual selection method consumes less computation time than GASEN using genetic algorithm. At the same time, MISEN obtains slightly better accuracy than GASEN does. In a word, MISEN may be an alternative method in the family of selective ensemble learning.

The preset proportion P is an important parameter of MISEN, which has much effect on individual selection. However, this parameter is set by hand in this book; so, finding out the relationship between P and the generalization ability of the ensemble is an interesting issue.

6.4 MULTITASK LEARNING

Neural networks ensemble has been improved using several techniques. Although most researchers pay much attention to the diversity of NNE [4] in this section we focus on the redundant features in the data set. Instead of only removing the redundant features from the data set, MTL [24, 25] is employed to reuse the redundant information in the removed feature set to improve the prediction accuracy of NNE.

Multitask learning [24, 25] is a form of inductive transfer. It is applicable to any learning method (such as NNE) described here, which can share some of what are learned between multiple tasks. The basic idea is to use the selected features as the input feature set and to combine the target values with some of the discarded features as the target output.

There exist several heuristic search methods for MTL [25, 26]. Caruana and de Sa [25] proposed to use a filter feature selection model of the cross-entropy criteria and/or an embedded model of kernel regression to rank the features, then employ the first predefined number of features as the input feature set, and add the first predefined number of the remaining features to the target. Li et al. [26] employed clustering algorithms to select the features, which are first clustered using the Kohonen neural networks. Then the features near the center of the clusters are selected as the input feature subset. When the other unselected features are ranked by their Euclidean distance to the input, the first several features with the least distance are selected to add to the target as the output. In order to make MTL more convenient to use, a random method of MTL, GA-MTL, was proposed [27].

6.4.1 GA-MTL and Related Methods

In this section, a novel random method of GA-MTL is described, as well as the comparative methods such as H-MTL (heuristic multitask learning), GA-FS (genetic-algorithm-based feature selection), and the used NNE methods.

6.4.1.1 GA-MTL   In existing search methods [25, 26], the number of features selected for the input and/or the target is decided somewhat arbitrarily. In order to eliminate the arbitrary choice of this number, the genetic algorithm was employed to search which features to select for the input and a random method, namely, GA-MTL is proposed [27], which simultaneously selects the features for both the input and the target. The number of features for the input and the target is automatically determined by the method itself.

In GA-MTL, a binary chromosome with the same length as the feature vector is used, which equals 1 if the corresponding feature is selected as the input and 0 if the feature is selected to add to the target as the output. The fitness function is defined as

$$\text{fitness} = \frac{1}{3}E_{dr} + \frac{2}{3}E_{dv}, \tag{6.3}$$

where E_{dr} is the training error of the base learning method and E_{dv} is the prediction error for the validation data set.

The GA-MTL approach is summarized in Fig. 6.4, where the data set is divided into three parts, training set S_r, validation set S_v, and test set S_s as given in Section 6.4.2.

6.4.1.2 H-MTL   In order to compare GA-MTL with the baseline method of MTL, we use a heuristic method with an embedded feature selection method based on the

---

Input: training set S_r, validation set S_v, test set S_s, and the base learner

Procedure:

1. generate a population of weight vectors

2. evolve the population where the fitness of a weight vector w is measured
   as in Equation 6.3 on S_r and S_v

3. $w^* =$ the evolved best weight vector

4. test on S_s with features corresponding to 1's in w^* as the input and those
   to 0's to add to the target

Output: prediction accuracy of the test set S_s

Figure 6.4   The GA-MTL approach.

idea of Caruana and de Sa [25]. This is designated H-MTL. In H-MTL, the prediction risk criteria [31, 32] are used to rank the features in an ascending order, which have been introduced in Section 6.2.1. The features with zero value are removed. Then, the first quarter of the existing features are added to the target, and the last three quarters are used as the input.

6.4.1.3 *GA-FS* To show the effectiveness of MTL methods, a naive feature selection method named GA-FS is also implemented where the same genetic algorithm of GA-MTL is used for the feature selection task. The only difference between GA-MTL and GA-FS is the value of the binary chromosome; in GA-MTL, it equals 0 if the feature is selected to add to the target as the output, whereas in GA-FS, it equals 0 if the feature is removed.

6.4.1.4 *NNE* The ensemble learning method, bagging of neural networks [2], is used and the base neural networks are improved multilayer perception neural networks. These are weight-decay-based neural networks in Bayesian frame, which add a regularized term to the objective function and are to some degrees insensitive to the setting of the parameters [34].

6.4.2 Numerical Experiments

In order to validate the proposed MTL methods, three data sets, listed in Table 6.7, from the UCI machine learning repository [33] are selected.

For all the data sets, the symbols with numerical values are first replaced and then all the features are transformed into the interval of $[-1,1]$ using an affine transformation. Finally, each UCI data set is equally split into two parts according to the number of instances of each class, one part is used as training sample S_r whereas the other is used as test sample S_s. This operation is performed 50 times. For all data sets, one quarter is selected from the training sample to be used as the validation sample S_v for the genetic algorithm.

The parameters for the genetic algorithm are as follows: population size is 50, the number of generations is 20, probability of crossover is 0.6, probability of mutation is 0.001, and probability of selection of the highest-ranked individual is 0.6. The parameters for the neural networks are the same for all data sets, namely, the number of hidden nodes is 10. The number of individuals of bagging is 20.

The calculation is performed 50 times and the statistical results of the average prediction accuracy and its corresponding standard deviation using the different learning methods for each data set are listed in Table 6.8.

Table 6.7   Properties of the UCI data sets for comparison

Data set	#Features	#Instances	#Classes
ionosphere	34	351	2
proc_h	13	294	2
statlog_heart	13	270	2

Table 6.8 Statistical results of prediction accuracy by using different learning methods (%)

Data set	NNE	GA-FS	H-MTL	GA-MTL
ionosphere	87.83 ± 1.81	87.94 ± 1.99	86.23 ± 2.95	88.17 ± 2.34
proc_h	81.43 ± 2.22	80.88 ± 1.94	80.68 ± 1.67	82.11 ± 1.80
statlog_heart	80.07 ± 2.44	81.85 ± 3.15	78.53 ± 4.83	81.93 ± 3.37
average	83.11 ± 2.16	83.56 ± 2.36	81.83 ± 3.15	84.07 ± 2.50

Table 6.8 shows that (1) for two data sets, GA-FS obtained higher accuracy than the naive NNE method did, and the same result is obtained for the average values. (2) GA-MTL further improved the accuracy of GA-FS for all the three data sets, and obtained the best results among the four NNE-based learning methods. (3) H-MTL did not obtain an expected result, which was perhaps due to the fact that we did not adjust the parameters for H-MTL for the data sets.

6.4.3 Summary

Experimental results have shown that GA-MTL performed slightly better than the H-MTL, the previous heuristic multitask learning method, and the GA-FS. In this section, we discuss reasons for the same.

Why does multitask learning succeed?
In the previous study, Caruana and coworker have given their explanation [24, 25]. Here experimental results are combined with the new framework of feature selection. Yu and Liu [6] proposed to categorize the features into four classes, namely, I: irrelevant features, II: weakly relevant and redundant features, III: weakly relevant but nonredundant features, and IV: strongly relevant features, where III and IV compose the optimal feature subset and I and II should be removed using the feature selection methods. It can be found that II contains useful information. These features should not be discarded, but rather be used in the learning process. MTL is a method to use these redundant features to improve the prediction accuracy of the base learning method. This is the reason for its improved performance.

Why does genetic algorithm perform better than the heuristic method does?
The chief reason why the heuristic method did not perform better than the genetic algorithm is that the previous heuristic methods use the feature-ranking technique to select the features for the input and the target. This does not consider feature redundancy and/or feature interaction. At the same time, the predefined number of the features for the input and the target is somewhat arbitrary. This is another factor that reduces the performance of the heuristic method. However, when the genetic algorithm selects features for the input and the target, it simultaneously considers feature redundancy and/or feature interaction. So it can automatically determine the number of features for the input and the target. In fact, this result has been proved by Kudo and

Sklansky [35] in the naive feature selection field, where they compared the different complete heuristic and random algorithms and proved that the genetic algorithm has a high probability of finding better solutions than the other algorithms for feature selection problems.

6.5   THE BRAIN GLIOMA CASE

The degree of malignancy in brain glioma [36] decides the treatment, because if it is grade I or II according to Kernohan, the success rate of operation is satisfactory; otherwise, if it is grade III or IV, there will be high surgical risk and poor life quality [37] after surgery, which must be taken into account before any further decision. Now, the degree of malignancy is predicted mainly by magnetic resonance imaging (MRI) findings [38] and clinical data [39] before operations. Some features obtained manually are fuzzy values, some features are redundant, even irrelevant, which make the prediction of the degree of malignancy a hard task. Moreover, brain glioma is severe but infrequent and only a small number of neuroradiologists have the chances to accumulate enough experiences to make correct judgments. Therefore, it is worth helping the neuroradiologists to predict the degree of malignancy of tumors.

In the previous work, degree prediction of malignancy in brain glioma had been solved by Ye et al. [40], where a fuzzy max–min neural networks was used. Furthermore, Li et al. [41] employed SVMs to improve the prediction accuracy and obtained satisfactory results. Nowadays, ensemble learning is becoming a hot topic in the data mining community [1], which has been widely used to improve the generalization performance of single learning machine. Therefore, using the bagging [2] of SVM to predict the degree of malignancy is an interesting issue. Yet, there are many irrelevant and redundant features in the medical data sets; so, feature selection [5, 42] is important to improve the ensemble method.

6.5.1   The Brain Glioma Data Set

The brain glioma data set [40, 41] was gathered by neuroradiologists from HuaShan Hospital in Shanghai, China. There are more than 20 items in each case, including symptoms on different features, preoperative diagnosis made by some neuroradiologists and, without an exception, a clinical grade (the actual grade of glioma obtained from surgery). With the help of domain experts, we chose 15 features, gender, age, shape, contour, capsule of tumor, edema, mass effect (occupation), Postcontrast Enhancement, Blood Supply, Necrosis/Cyst Degeneration, Calcification, hemorrhage, signal intensity of the T1-Weighted Image, signal intensity of the T2-Weighted image, and clinical grade [39, 43]. In some cases, the value of Postcontrast enhancement is unknown. In fact, location and size also help to make the diagnosis, but their complex descriptions cannot be well modeled by our algorithms, so we did not adopt it. Except for gender, age, and clinical grade, the other items are obtained from MRI of the patient and described with uncertainty to various extents.In order to predict the degree of malignancy in brain glioma, descriptions of all features are converted into numerical values as in Ref. [41], of which the unknown value of postcontrast enhancement is defined as -1.

Originally, four grades are used to mark the degree of malignancy; we merge grade I and II into low grade and grade III and IV to high grade. According to the grade, all 280 cases of brain glioma are divided into two classes, low grade and high grade, of which 169 are of low-grade and 111 are of high-grade gliomas.

There are 126 cases containing missing values on postcontrast enhancement, and in the other subset of 154 complete cases, 85 cases are of low-grade gliomas and 69 are of high grade.

6.5.2   Classification of Brain Glioma by Using PRIFEB and MISEN

To improve prediction accuracy of degree of brain glioma, MISEN, PRIFEB, and bagging of SVM are employed to the classification of brain glioma. To compare classification ability of methods of MISEN, PRIFEB, and that of bagging of SVM and single SVM, the 10-fold cross-validation technique is used in this computation. The SVMs used in this experiment are with linear kernel and the parameter of the trade-off between the complexity and the error is $C = 100$ and the number of individuals of bagging is 20. These parameters are not the optimal, but they can make the learning methods obtain satisfying results according to the experience [4]. Computational results are shown in Table 6.9, where D280 is the total data set of 280 cases with missing values in postcontrast enhancement, whereas D154 is the data set of 154 complete cases. The results of the previous studies are also reprinted in Table 6.9, where results of MISEN are collected from Ref. [23], those of PRIFEB and bagging as well as SVM and FMMNN-FRE are from Ref. [17] and Refs [40, 41]

Table 6.9 shows that (1) the results of bagging are better than that of single SVM, which shows that bagging of SVM can really improve the prediction accuracy of single ones. (2) The results of MISEN and PRIFEB, the novel algorithm, are better than that of the general case of bagging of SVM.

Table 6.9   Results of accuracy obtained by 10-fold cross-validation method (%)

	Data set	R_{acc}	Standard deviation	Highest	Lowest
MIFEB	D280	87.54	7.73	100.00	75.00
	D154	87.03	6.57	100.00	75.00
PRIFEB	D280	87.29	7.91	97.68	75.00
	D154	86.57	8.23	100.00	73.33
Bagging of SVM	D280	86.36	7.22	96.43	75.00
	D154	86.43	8.17	100.00	73.33
SVM	D280	85.70	6.52	96.43	71.43
	D154	84.96	9.97	100.00	73.33
FMMNN-FRE	D280	83.21	5.31	89.29	75.00
	D154	86.37	8.49	100.00	73.33

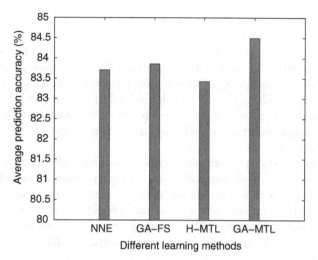

Figure 6.5 Statistical results of prediction accuracy obtained by different learning methods for the brain glioma data set.

6.5.3 Classification of Brain Glioma by Using GA-MTL

Here, bagging of neural networks with a novel algorithm GA-MTL is applied to the classification of brain glioma [44]. To compare the classification ability of the methods of GA-MTL with that of GA-FS, H-MTL, and naive NNE, the data set is equally divided into two parts according to the number of instances of each class. One part is used as the training sample, the other is used as the test sample. This operation is performed 50 times. For GA-MTL and GA-FS, one quarter of the training sample is selected to be used as the validation sample for the genetic algorithm. Statistical results for the 50 different divided copies of the glioma data set are shown in Fig. 6.5 [44].

Figure 6.5 shows that (1) The results of GA-MTL and GA-FS are better than that of NNE, which shows that feature selection can improve the prediction accuracy of NNE, and some features should be removed from or added to the target. (2) The result of GA-MTL, the novel algorithm, is better than that of H-MTL. We think that some of the reasons may be that we did not adjust the parameters for H-MTL for the glioma data set. (3) GA-MTL obtains better results than GA-FS does, which shows that the redundant features are helpful in improving the generalization performance of NNE.

It is to be noted that the improvement of prediction accuracy is although small, the improvement is still valuable. Even 1% improvement means that one person out of hundred is predicted correctly.

6.6 SUMMARY

From the above discussions, we can say that although ensemble learning have strong generalization performance, feature selection is needed to further improve the performance. First, feature selection produces different feature subsets for individuals of bagging, which helps to improve learning performance of individuals and at the same time to increase the diversity among individuals; therefore, feature selection improves generalization performance of bagging. Second, feature selection selects individuals for bagging to increase diversity among individuals and shows *many could be better than all* [20]. Third, removing redundant information could improve generalization performance of learning machines, including ensemble learning; however, redundant information could also help to improve generalization ability of ensemble learning.

The previous works on feature selection for bagging have shown ensemble learning can benefit from feature selection; of the few studies on feature selection for boosting or other popular ensemble creation methods, this is one future research direction. Another one is that to improve efficiency of feature selection; since ensemble learning needs much computation, only efficient feature selection can reduce computational complexity and simultaneously improve generalization performance of ensemble learning.

ACKNOWLEDGMENTS

The authors wish to thank Dr. Tian-Yu Liu for his contribution to numerical experiments. This work was supported in part by the Nature Science Foundation of China under Grant No. 20503015, the STCSM Project of China under Grant No. 07DZ19726, the Shanghai Rising-Star Program under Grant No. 08QA14032, and open funding by Institute of Systems Biology of Shanghai University.

REFERENCES

1. Dietterich, T. Machine-learning research: four current directions. *The AI Magazine*, 18(4): 97–136, 1998.
2. Breiman, L. Bagging predictors. *Machine Learning*, 24(2): 123–140, 1996.
3. Bauer, E. and Kohavi, R. An empirical comparison of voting classification algorithms: bagging, boosting, and variants. *Machine Learning*, 36(2): 105–139, 1999.
4. Brown, G., Wyatt, J. L., and Tino, P. Managing diversity in regression ensembles. *Journal of Machine Learning Research*, 3: 1621–1650, 2005.
5. Guyon, I. and Elisseeff, A. An introduction to variable and feature selection. *Journal of Machine Learning Research*, 3: 1157–1182, 2003.
6. Yu, L. and Liu, H. Efficient feature selection via analysis of relevance and redundancy. *Journal of Machine Learning Research*, 5(Oct): 1205–1224, 2004.
7. Liu, H. and Yu, L. Toward integrating feature selection algorithms for classification and clustering. *IEEE Transactions on Knowledge and Data Engineering*, 17(3): 1–12, 2005.
8. Lal, T. N., Chapelle, O., Weston, J., and Elisseeff, A. Embedded methods. In: I. Guyon, S. Gunn, and M. Nikravesh, Eds, *Feature Extraction, Foundations and Applications*, Springer, Physica-Verlag, 2006.

9. Ho, T. The random subspace method for construction decision forests. *IEEE Transaction Pattern Analysis and Machine Intelligence*, 20(8): 832–844, 1998.

10. Simon, G. and Horst, B. Feature selection algorithms for the generation of multiple classifier systems and their application to handwritten word recognition. *Pattern Recognition Letters*, 25(1): 1323–1336, 2004.

11. Optiz, D., Feature selection for ensembles. *Proceedings of the 16th National Conference on Artificial Intelligence (AAAI)*, pp. 379–384, 1999.

12. Oliveira, L., Sabourin, R., Bortolozzi, F., and Suen, C. Feature selection using multi-objective genetic algorithms for handwritten digit recognition. *The 16th International Conference on Pattern Recognition*, pp. 568–571, 2002.

13. Brylla, R. Gutierrez-Osunab, R., and Queka, F. Attribute bagging: improving accuracy of classifier ensembles by using random feature subsets. *Pattern Recognition*, 36(6): 1291–1302, 2003.

14. Tsymbal, A. and Cunningham, P. Search strategies for ensemble feature selection in medical diagnostics. *Proceedings of the 16th IEEE Symposium on Computer-Based Medical Systems (CBMS)*, pp. 124–129, 2003.

15. Tsymbal, A., Pechenizkiy, M., and Cunningham, P. Sequential genetic search for ensemble feature selection. *Proceedings of the International Joint Conference on Artificial Intelligence 2005 (IJCAI2005), Edinburgh, Scotland*, pp. 877–882, 2005.

16. Tahir, M. A. and Smith, J. Improving nearest neighbor classifier using tabu search and ensemble distance metrics. *Sixth International Conference on Data Mining, (ICDM 06), IEEE Press*, pp. 1086–1090, 2006.

17. Li, G.-Z., Liu, T.-Y., and Cheng, V. S. Classification of brain glioma by using SVM bagging with feature selection. *Proceedings of Workshop on Data Mining for Biomedical Application 2006 (BioDM2006). Vol. 3916*, Springer, pp. 124–130, 2006.

18. Liu, T.-Y., Li, G.-Z., Liu, Y., Wu, G., and Wang., W. Estimation of the future earthquake situation by using neural networks ensemble. *Proceedings of the International Symposium on Neural Networks, Lecture Notes in Computer Science*, Vol. 3973, Springer, pp. 1231–1236, 2006.

19. Li, G.-Z. and Liu, T.-Y. Feature selection for bagging of support vector machines. *Proceedings of 9th Biennial Pacific Rim International Conference on Artificial Intelligence (PRICAI2006) Vol. 4099*, Springer, pp. 271–277, 2006.

20. Zhou, Z.-H., Wu, J.-X., and Tang, W. Ensembling neural networks: many could be better than all. *Artificial Intelligence*, 137(1–2): 239–263, 2002.

21. Caruana, R., Niculescu-Mizil, A., Crew, G., and Ksikes, A. Ensemble selection from libraries of models. *Proceedings of the 21st International Conference on Machine Learning (ICML 2004), ACM Press*, pp. 137–144, 2004.

22. Caruana, R., Munson, A., and Niculescu-Mizil, A. Getting the most out of ensemble selection. *Sixth International Conference on Data Mining (ICDM'06), IEEE Press*, pp. 828–833, 2006.

23. Liu T.-Y. Study on feature selection issues in ensemble learning, Ph.D. Thesis, *School of Computer Engineering and Science*, Shanghai University, China, 2007.

24. Caruana, R. Multitask learning. *Machine Learning*, 28(1): 41–75, 1997.

25. Caruana, R. and de Sa, V. R. Benefiting from the variables that variable selection discards. *Journal of Machine Learning Research*, 3: 1245–1264, 2003.

26. Li, G.-Z., Yang, J. Lu, J. Lu, W.-C., and Chen, N.-Y. On multivariate calibration problems. *Proceedings of the International Symposium on Neural Networks, Lecture Notes on Computer Science* Vol. 3173, Springer, pp. 389–394, 2004.

27. Li, G.-Z. and Liu, T.-Y. Improving generalization ability of neural networks ensemble with multi-task learning. *Journal of Computational Information Systems*, 2(4): 1235–1239, 2006.

28. Goldberg, D. E. *Genetic Algorithms in Search, Optimization, and Machine Learning*, Addison Wesley, Boston, 1998.

29. Yang, J. and Honavar, V. Feature subset selection using a genetic algorithm. *IEEE Intelligent Systems*, 13: 44–49, 1998.

30. Valentini, G. and Dietterich, T. Bias-variance analysis of support vector machines for the development of SVM-based ensemble methods. *Journal of Machine Learning Research*, 5 (1): 725–775, 2004.

31. Li, G.,-Z., Yang, J., Liu, G. P., and Xue, L. Feature selection for multi-class problems using support vector machines. *Proceedings of 8th Biennial Pacific Rim International Conference on Artificial Intelligence (PRICAI2004), Vol. 3173, Springer*, pp. 292–300, 2004.

32. Moody, J. and Utans, J. Principled architecture selection for neural networks: application to corporate bond rating prediction. In: J. E. Moody, S. J. Hanson, and R. P. Lippmann, Eds, *Advances in Neural Information Processing Systems*, Morgan Kaufmann Publishers, Inc., pp. 683–690, 1992.

33. Blake, C., Keogh, E., and Merz., C. J. UCI repository of machine learning databases. Technical report, Department of Information and Computer Science, University of California, Irvine, CA, 1998. Available at http://www.ics.uci.edu/mlearn/MLRepository.htm.

34. Foresee, F. D. and Hagan, M. T. Gauss–Newton approximation to Bayesian regularization. *Proceedings of the 1997 International Joint Conference on Neural Networks*, pp. 1930–1935, 1997.

35. Kudo, M. and Sklansky, J. Comparison of algorithms that select features for pattern classifiers. *Pattern Recognition*, 33(1): 25–41, 2000.

36. Bredel, M. and Pollack, L. F. The p21-ras signal transduction pathway and growth regulation in human high-grade gliomas. *Brain Research Reviews*, 29: 232–249, 1999.

37. Wang, C., Zhang, J., Liu, A., Sun, B., and Zhao, Y. Surgical treatment of primary midbrain gliomas. *Surgical Neurology*, 53: 41–51, 2000.

38. Gonzalez, M. A. L. and Sotelo, J. Brain tumors in Mexico: characteristics and prognosis of glioblastoma. *Surgical Neurology*, 53: 157–162, 2000.

39. Chow, L. K., Gobin, Y. P., Cloughesy, T. F., Sayre, J. W., Villablanca, J. P., and Vinuela, F. Prognostic factors in recurrent glioblastoma multiforme and anaplastic astrocytoma, treated with selective intra-arterial chemotherapy. *AJNR American Journal of Neuroradiology*, 21: 471–478, 2000.

40. Ye, C.-Z., Yang, J., Geng, D.-Y., Zhou, Y., and Chen., N.-Y. Fuzzy rules to predict degree of malignancy in brain glioma. *Medical and Biological Engineering and Computing*, 40: 145–152, 2002.

41. Li, G.-Z., Yang, J., Ye, C.-Z., and Geng, D. Degree prediction of malignancy in brain glioma using support vector machines. *Computers in Biology and Medicine*, 36(3): 313–325, 2006.

42. Kohavi, R. and George, J. H. Wrappers for feature subset selection. *Artificial Intelligence*, 97: 273–324, 1997.

43. Arle, J. E., Morriss, C., Wang, Z., Zimmerman, R. A., Phillips, P. G., and Sutton, L. N. Prediction of posterior fossa tumor type in children by means of magnetic resonance image properties, spectroscopy, and neural networks. *Journal of Neurosurgery*, 86: 755–761, 1997.

44. Yang, J. Y., Li, G.-Z., Liu, L.-X., and Yang, M. Q. Classification of brain glioma by using neural networks ensemble with multi-task learning. *Proceedings of the 2007 International Conference on Bioinformatics and Computational Biology (BIOCOMP'07), CSREA Press, Nevada,* pp. 515–522, 2007.

7

SEQUENCE-BASED PREDICTION OF RESIDUE-LEVEL PROPERTIES IN PROTEINS

Shandar Ahmad, Yemlembam Hemjit Singh, Marcos J. Araúzo-Bravo, and Akinori Sarai

7.1 INTRODUCTION

Amino acid sequences carry all the information about protein structure and function. Interest in the sequence–structure relationship of proteins has existed for a very long time [1–3]. The need arose because experimentally finding the amino acid sequence of a protein or its coding DNA is much easier than determining its structure. Thus, much effort has been made to find the relationships between sequence and structure of proteins. This issue has been frequently reviewed [4–20]. Similarly, many researchers were interested in predicting biological function from sequence directly or *en route* secondary or tertiary structure [21–30]. It has now become clear that despite an intense effort to predict protein structure from the amino acid sequence, the task cannot be accomplished without error, which largely depends on the quality of an available template for which structure is known. Scientists, therefore, have focused on a much less ambitious goal of predicting the so-called one-dimensional properties of protein structure such as secondary structure, solvent accessibility, and coordination number. Similarly, biological functions for the entire protein may be difficult to find, and hence corresponding one-dimensional properties such as binding with specific ligands or

Machine Learning in Bioinformatics. Edited by Yan-Qing Zhang and Jagath C. Rajapakse
Copyright © 2009 by John Wiley & Sons, Inc.

DNA bases have been targeted for prediction. Both one-dimensional structural features of proteins and probability of binding of an amino acid with other molecules have been predicted from the information of amino acid sequence with good success [14, 31, 32] and have in many ways led the way for an eventual *ab initio* structure prediction without homology or structure models. In this chapter, we take up some of the widely studied problems of sequence-based predictions, which may be considered one-dimensional features of amino acids and, therefore, can be predicted with reasonable accuracy by looking at the sequence. We show that these predictions have computational similarities and common standards for making such sequence-based predictions need to be developed. We also discuss some of the issues of implementing prediction algorithms and measuring the performance of such prediction methods and suggest ways to solve them.

7.2   PROTEINS AS AMINO ACID SEQUENCES

Almost all the biological functions in eukaryotes are performed by proteins, which are made of only 20 types of amino acid residues connected by peptide bonds [33]. A protein is made of a number of amino acid residues, which may be less than 10 or as large as a few thousands. While writing a primary structure of a protein, amino acids are usually written in a single-letter notation [34]. Writing them in a compact way in itself requires some standards and this has led to a number of sequence formats [35]. From a computational perspective, proteins or amino acid sequences can be treated as character strings. Twenty amino acid residues are the alphabets of these strings, which may have arbitrary length and the amino acid residues may occur in any order within this sequence. Characters inside a string have certain features corresponding to their biophysical properties. Features of these amino acid residues often partially overlap, for example, both arginine and lysine residues have similar electrostatic behavior (but are different in many other ways). Properties of amino acids have been widely studied as free molecules as well as residues in the proteins, and their similarity and substitutions are of particular interest for biologists [36]. Our interest in amino acids in this chapter is limited to the preference of each amino acid to participate in a particular type of structure or its preferred role in binding sites, as discussed in the following sections. In addition, we will discuss how the preferences of each amino acids to have a structural or functional property are influenced by their "sequence neighbors" that is, the residues located in their immediate vicinity on the sequence. To this goal, we will first list up the most widely studied properties, which are known to depend on the identity of the residue and its neighbors.

7.3   RESIDUE-LEVEL PROPERTIES (RLP)

As stated above, a number of local structural and functional properties of proteins in the biological environment depend on the residues in the corresponding region of interest. It may be necessary to caution here that no major structural and functional property has so far been exclusively determined by local sequence environment (substring of amino acid sequence), primarily because in reality the amino acids that are distant on the sequence may actually be close in the three-dimensional structure and the three-dimensional organization of these residues is difficult to describe, and their exact role therefore is too complex to fully solve the problem. However, there are many properties that can be described at an individual residue level and some of them can be estimated by a fairly good confidence by just looking at the corresponding residue and its sequence neighbors. Since their estimates cannot be carried out with absolute certainty, ways of assessing their performances are needed. This will be discussed in a Section 7.8.2. Here we summarize the problems that have been solved with good estimate using these ideas. These problems range from structure-related predictions, such as secondary structure and solvent accessibility, to function prediction, such as binding site identification. We will cover the main aspects of these problems in the following sections.

7.3.1   Secondary Structure Prediction

Three-dimensional structure of a protein (i.e., a polypeptide amino acid chain), is stabilized by hydrogen bonds between atoms of different residues [37] in other types of interactions. There is a regular pattern of hydrogen bonding in proteins, which determine the pairs of residues participating in hydrogen bonds. This hydrogen bond formation, as well as the steric considerations of residues inside the proteins, imposes certain restrictions on the rotational flexibility around the peptide bond [38]. This level of organization of protein structures in terms of periodicity and pattern of hydrogen bonds leads to what we call the secondary structure of proteins (in contrast to the detailed coordinates of all atoms, which is called the tertiary structure). Eight most common secondary structures of proteins, which have been defined as a dictionary of secondary structure of proteins, are summarized in Table 7.1 [39].

Table 7.1   Definitions of eight classes of secondary structures

Secondary Structure	Notation	Hydrogen Bond Pattern
Alpha helix	H	Residue i and residue $i+4$
3–10 helix	G	Residue i and $i+3$
Pi helix	I	Residue i and $i+5$
Isolated beta bridge	B	Single pair beta-sheet hydrogen bond
Extended strand (participating in beta ladder)	E	Parallel and/or antiparallel sheet conformation (extended strand). Min length 2 residues
Bend	S	Nonhydrogen bond
Hydrogen-bonded turn	T	3-, 4-,or 5- turn types with corresponding hydrogen-bonded turns
Coil/loop or irregular structure	C	Usually irregular

Secondary structure can be assigned to each residue in a protein chain if we know the hydrogen bond patterns (which may themselves have some ambiguity because there will be no unique way to tell if a hydrogen bond exists at a given position), which comes from the three-dimensional structure. Secondary structure of proteins at each residue level is clearly the first and the most widely studied problem in the category of sequence-based predictions (See Table 7.3 and the references therein). Of the above-mentioned eight categories, most of the prediction strategies try to predict only three classes, namely, alpha (helix), beta (strand), and coil (all others), because relatively much fewer residues take other secondary structures. Some recent studies, however, have focused on the prediction of only one of these rarely occurring secondary structures [40–47]. A number of review articles have appeared on secondary structure prediction, either on secondary structure exclusively [7, 48, 49] or as part of the more general structure prediction problem [8, 13, 15, 50]. Here, we take note of only the key points of this extensively studied subject, particularly those that can be used in a general sequence-based prediction regime.

Earliest efforts to predict secondary structure of proteins from sequences were based on statistics of these residues in terms of their secondary structures relative abundance [51]. Starting with a set of 15 proteins, a simple frequency of each residue in any of the three secondary structures was counted. On the basis of these frequencies, residues were characterized for their preference or aversion to have a secondary structure. Helix and sheet breaker and former scores for some of the amino acids are listed in Table 7.2.

Typical examples from this data are that Glutamic acid (E), Alanine (A), and Leucine (L) residues prefer to form alpha helix and hence are called helix former, whereas Asparigine (N), Tyrosine (Y), and Proline (P) are helix breaker because they do not like to be part of alpha-helical structure. Similar preferences are observed for other residues in beta sheet conformation. Later evidence has shown that these preferences may substantially depend on the position of the residue within the chain as well as the overall structure of the protein [52, 53]. Further, the residue predisposition for given secondary structure are also perturbed by other residues falling in their immediate and not so immediate neighborhood, a fact used for the prediction of secondary structure by Ref. [54]. Later methods also used evolutionary information for such predictions [55]. Performances of the prominent published methods of secondary structure prediction are summarized in Table 7.3.

Starting from the simplest Chou and Fasman method [5] to the most recent method of secondary structure prediction, accuracy has gone up from 50% to as high as 80%.

Table 7.2 Helix and sheet breaker and former scores for some of the amino acids based on original [51] calculations single letter code as in [34]

Helix formers	$E = 1.37$, $A = 1.29$, $L = 1.20$, $H = 1.11$, $M = 1.07$, $Q = 1.04$, $W = 1.02$, $V = 1.02$, $F = 1.00$, $K = 0.54$, $I = 0.50$, others $= 0.00$
Helix breakers	$N = 1.00$, $Y = 1.20$, $P = 1.24$, $G = 1.38$, others $= 0.00$
Sheet formers	$M = 1.40$, $V = 1.39$, $I = 1.34$, $C = 1.09$, $Y = 1.08$, $F = 1.07$, $Q = 1.03$, $L = 1.02$, $T = 1.01$, $W = 1.00$, others $= 0.00$
Sheet breakers	$K = 1.00$, $S = 1.03$, $H = 1.04$, $N = 1.14$, $P = 1.19$, $E = 2.00$, others $= 0.00$

Table 7.3 **Methods of secondary structure prediction**

Goal	Descriptor	Model	Performance	References
H S	MSA	Bidirectional RNN	Q3 = 79.00%, C(H) = 75.6%, C(S) = 66.4%	[130]
H S C	MSA	IT, BS, and EI	Q3 = 73.50%	[145]
H S C	Evolutionary profiles	RNN	Q3 = 78.40%	131]
H S C	PSSM	Hidden NN	Q3 = 77.06%	[139]
H S C	PSSM	NN	Q3 = 79.40%	[141]
H S C	PSSM	Two-stage multiclass SVM	Q3 = 76.30%	[120]
H S C	PSSM	Critical random NN	Q3 = 81.00%	[150]
H S C		DP	Q3 = 76.70%	[155]
H S C	PSSM	MLR	Q3 = 76.40%	[158]
H S C	PSSM	Multiclass SVM	Q3 = 78.00%	[109]
H S C	—	SVM	Q3 = 78.80%	[140]
H S C	PSSM	SVM	Q3 = 78.50%	[108]
H S C	PSSM	SVM	Q3 = 77.07%	[119]
H S C	—	Pentapeptide	Q3 = 68.60%	[156]
H S C	PSI-BLAST derived profiles	Bidirectional RNN	Q3 = 78.00%	[166]
H S C	—	MLR	Q3 = 68.8% (ss), 73.7% (ms)	[138]
H S C	MSA	SVM	Q3 = 73.5 %,	[118]
-	Multiple sequence profile	FFNN	Q3 = 78.00%	[137]
H S C	MSA	NN	Q3 = 76.40%	[124, 152]
H S C	Profiles made by PSSM	NN	Q3 = 80.20%	[142]
H S C	—	NN and linear discrimination	Q3 = 76.70%	[143]
H S C	—	BS	Q3 = 68.80%	[153]
H S C	MSA	RNN	Q3 = 76.00%	[99]
H S C	PSSM	Two-stage NN	Q3 = 78.30%	[149]
H S C	—	Primary and secondary NN	Q3 = 74.60%	[157]
H S C		BS	Q3 = 72% (ms), 67% (ss)	[148]
H S C	—	NrNg	Q3 = 67.7% (ss), 72.8% (ms)	[151]
H S C	Nonintersecting local alignments	NrNg	Q3 = 71.2%.	[133]
H S C	Local pairwise alignment	DP	Q3 = 75.00%	[134]
H S C	—	BS	Q3 = 67% (ss), 72% (ms)	[148]

(continued)

Table 7.3 (*Continued*)

Goal	Descriptor	Model	Performance	References
H S C	MSA	NN	Q3 = 71.30% C(H) = 0.59 C(S) = 0.52 C(C) = 0.50	[144]
H S C	—	NN	Q3 = 67.00%	[146]
H S T	MSA	NrNg and MSA	Q3 = 72.20%	[135]
H S C	MSA	Selfoptimized	Q3 = 69.50%	[136]
H S C	MSA	Nearest neigbor	Q3 = 72.2%	[135]
H S C	MSA	NN	Q3 = 71.40%	[14]
H S C	MSA	FFNN	Q3 = 70.80%	[55]
H S C	Similarity metric[a]	NrNg	Q3 = 81.00%	[132]
H S C	MSA	NN	Q3 = 69.70%	[106]
H S C	Sequence	Probability	—	[154]
H S C	PSSM	NN	—	[147]
H S C	Sequence	Probability	Q3 = 50%	[5]

C, coil; H, helix; S, strand; T, Turn; DP, dynamic programming; FFNN, feed forward neural network; MSA, multiple sequence alignment; MLR, multiple linear regression; NN, neural network; NrNg, nearest neighbor; PSSM, position-specific scoring matrices; RNN, recurrent neural network; SVM, support vector machine; ss, single sequence; ms, multiple sequence; IT, information theory; BS, Bayesian statistics; EI, evolutionary information.

[a]Based on local structural environment scoring scheme of Bowie et al. [178] Note that PSSM, evolutionary profiles, and alignments basically present the same information but effort has been made to retain the same terminology as used by original authors.

Excluding Chou and Fasman methods [5], Sejnowski and Karplus [101, 102] methods, and Garnier [154] methods, most other secondary structure predictions utilize alignments one or another way. In that respect prediction accuracy starting from 70% of Rost and Sander method [55], there has been about 10% improvement in secondary structure prediction. Most of these improvements may be attributed to the availability of better models of relationship between the sequence space and the target property. However, it has been shown that a consensus of predictions obtained from several predictors, all designed on different models, can give the best performance. This principle has been utilized in jury-based predictions of sequence-based predictions [124, 152].

7.3.2 Solvent Accessibility Prediction

All amino acid sequences in proteins undergo process of folding, which leads to their secondary and tertiary structures. These folding patterns leave some of the residues on the surface of the so-called globule formed in this way, whereas others reside in the interior parts of this globule. Thus, in these globular proteins, amino acid residues are variously exposed to the solvent environment, under which they perform biological functions. Each of the 20 residue types has a different predisposition to remain on the surface, which is in contact with water (in the case of membrane proteins, water is substituted by lipid or other molecules). Amino acids such as Arginine, Serine, and Aspartic acids, whose free energy decreases on contact with water, are called hydrophilic, whereas the others such as Alanine, Tryptophan, and Proline, whose free energy increases upon water contact, are called hydrophobic. Many scales and methods to quantify hydrophobic behavior and solvent accessibility of amino acids in proteins have been proposed [56–62]. Among many applications of solvent accessibility, [63] used predictive methods of solvent accessibility to identify sites of deleterious mutations. Other studies have also pointed out importance of solvent accessibility in different functions of proteins [64].

Inspired by successful methods of secondary structure prediction from amino acid sequences, a successful effort to predict solvent accessibility was also reported [14]. True to the source of its inspiration, solvent-accessibility predictions started and continued for a fairly long time by predicting a category or state of burial typically characterized by their degree of burial separated by some cutoff values. Thus, for example, residues were classified into two or more states based on their solvent accessibility and treated very much like the helices and beta sheets in the secondary structure. A number of methods emerged for the prediction of solvent accessibility states or categories (see Table 7.4 and references therein).

Table 7.4   Methods of solvent accessibility predictions (burial states)

States	Descriptor	Model	Performance	References
2	PSSM	MLR	Accuracy $= 77.7\%$ (at 25% cutoff) $R = 0.548$	[158]
2 & 3	PSSM	Fuzzy k-NrNg	Accuracy $= 78.5\%$ (at 25% cutoff) Accuracy $= 64.1\%$ (at 9%; 36% cutoff)	[126]
2	PSSM and secondary structure	FFNN	MAE $= 15.90\%$, $r = 0.68$	[161]
2	PSSM	Two-stage SVM	Accuracy $= 79.4\%$ (at 16% cutoff)	[169]
2	—	Linear regression	—	[168]
2 & 3	PSSM, long-range interaction matrix, neighbor sequences	SVM	Accuracy $= 78.7\%$ (at 25% cutoff) Accuracy $= 64.5\%$ (at 9%; 36% cutoff)	[108]
2	—	NN-based regression	MAE $= 15.3$–15.8% $R = 0.64$–0.67	[159]

(continued)

Table 7.4 (*Continued*)

States	Descriptor	Model	Performance	References
2	PSSM, long-range interaction matrix, neighbor sequences	SVM	Accuracy = 78.7% (at 25% cutoff)	[121]
			Accuracy = 64.5% (at 9%; 36% cutoff)	
2	MSA	Probability profiles	Accuracy = 75.1 (at 16% cutoff for RS126 data set) R = 0.485	[165]
2	PSSM	Bidirectional (RSA) RNN	Accuracy = 77.2% (at 25% cutoff)	[69]
2	—	SVM	Accuracy = 70.1% (ss), 73.9% (ms)	[160]
2 & 3	—	NN	Accuracy = 71.1% (at 25% cutoff) r = 0.414 Accuracy = 63.0% (at 10%, 20%) R = 0.373	[102]
2 & 3	Pair-information	IT	Accuracy = 74.4% (at 25% cutoff) r = 0.47 Accuracy = 57.4% (at 9%; 36% cutoff) R = 0.41	[162]
2	MSA	MLR	Accuracy = 75.3% (at 20% cutoff) R = 0.44	[163]
2	Sequence environment	—	—	[167]
—	Single sequence	BS	—	[164]
3	MSA	NN	R = 0.54	[14]

Well-defined solvent accessibility states, as visualized in these predictions, did not exist and hence a direct prediction of solvent accessibility in terms of real values was introduced [31]. In this method, an amino acid sequence can be used to predict the actual relative solvent accessibility, which is measured in percentage, by normalizing the actual exposed surface area by the corresponding tripeptide solvent accessibility in the extended state. Several other researchers have focused on the prediction of real-valued solvent accessibility with varying degrees of success, using a number of computational methods. A nonexhaustive list of these methods is provided in Table 7.5.

Table 7.5   Real-valued prediction of solvent accessibility

Descriptor	Model	Performance Claimed	References
Single sequence	Quadratic programming	MAE = 19.43%	[127]
		r = 0.48	
PSSM	Two-stage SVR	MAE = 14.9%	[170]
		R = 0.68	
PSSM, secondary structure	FFNN	MAE = 15.9%	[161]
		R = 0.68	
PSSM	Linear regression	MAE = 15.6%	[168]
		R = 0.61–0.64	
PSSM, composition, sequence length	MLR	MAE = 16.2%–16.4%	[107]
		R = 0.64–0.66	
PSSM-derived evolutionary information	NN-based regression	MAE = 15.3–15.8%	[159]
		R = 0.64–0.67	
Single sequence	FFNN	MAE = 18.1%	[31]
		r = 0.48	

Although the complexity of models predicting solvent accessibility—both in the categories and in the real values—has grown with every new method, only small improvements could be made after first publications of either type. For example, solvent accessibility prediction proposed by Rost and Sander [14] was reported to have about 70% accuracy in two-state predictions, which has gone up to about 78% in the best available methods so far (using single-stage predictors). Multistage predictors have been shown to have slightly improved prediction, but since different data sets and cutoffs have been used in category predictions, a comparison is difficult to make. In our own view, the improvements in category-based predictions are too small to be striking and in the absense of rigorous benchmarking, such as the Critical Assessment of techniques for protein Structure Prediction (CASP) [179] they may not be treated as path breaking. It may be noted that PSSMs use basically alignment information and the number of alignments grow with a growth in the sequence data. With more and more sequence data getting available, improvement in prediction accuracy is more likely to emanate from improved alignments rather than from the complexity of models [176].

7.3.3 Other Predicted Structure Features

Secondary structure and solvent accessibility are two of the most important features of protein structures predicted from sequence information. In addition to these, other structure features such as residue–residue contacts [65–67], protein flexibility [68–69], disulfide connectivity [70–78], and disordered regions [79] have also been predicted. Most of these methods use very similar computational procedures and their success rates depend on the actual role played by the residues and their neighbors in determining the property of interest.

7.3.4 Biological Function and Binding Sites

Just as the information about the three-dimensional structure of a protein is expected to be present in the amino acid sequence itself, it is natural to expect that the biological function in general and the location of binding sites and active sites in particular may be deduced from the sequence. Computational methods employed for this prediction would be remarkably similar to those used for secondary structure and other problems. Here, we take an overview of some of these biological problems of function prediction that have been handled using the above approach with some promising degree of success.

7.3.5 DNA Binding

It is common knowledge that proteins tightly regulate the expression of biological information contained in the DNA. This control is applied by way of some protein–DNA interactions, which initiate, accelerate, suppress, or modify gene expression. Thus, there are a number of DNA-binding protein families, which perform various tasks related to the control of gene expression. The ultimate aim of the pharmaceutical and biotechnological science is to exactly learn about these processes and be able to guide this process of control in a desired way. This has far-reaching implications to the issues ranging from the disease control to genetic engineering. One of the most crucial information required to understand and control protein–DNA interactions is the knowledge of exact residues in the protein and bases of the DNA that take part in these interactions and how specifically they interact, that is, what would be the biological implication of a single or multiple mutation(s) at these locations. Role of each residue in DNA binding can be assessed by carrying out site-directed mutagenesis experiments (where residues are systematically replaced by others and their thermodynamic effects can be measured). However, it is neither practically possible nor desirable to conduct such high-throughput experiments to scan the entire genomic sequences and document the role of each one of them in DNA binding. Computational approaches are, therefore, required to predict DNA-binding sites of proteins. First alignment-free approach to predict DNA-binding sites of proteins was proposed by us [32, 80]. Some other studies have since appeared on this subject [81, 82]. Some use information from structure (in addition to the sequence and evolutionary profiles) to predict binding sites. In terms of computational description of the local sequence environment, they follow the same sequence encoding and multiple alignment profiling as in the case of secondary structure and other one-dimensional structural features. What makes the prediction of DNA binding possible? The underlying principle behind these predictions is that there are certain residue types (particularly the basic residues, Arginine and Lysine), which are overrepresented in the binding sites of DNA binding proteins. The sequence neighbors of these residues further

Table 7.6 Predictive performance of DNA binding and similar binding problems

Binding	Descriptor	Model	Performance	References
Transition		SVM, bidirectional	Accuracy = 73% (histidines), 61%	[123]
Metal		RNN	(cysteines)	
DNA	MSA with similar sequence homologues	naive Bayes classifiers	Accuracy = 8%,	[81]
			r = 0.28, spec = 44%, sens = 41%	
DNA,	Side chain pK$_a$, hydrophobicity, amino acid molecular mass	SVM	(DNA binding)	[82]
RNA			Accuracy = 70.31% sens = 69.40%, spec = 70.47% (RNA binding) Accuracy = 69.32% sens = 66.28%, spec = 69.84%	
RNA	PSSM	naive Bayes classifiers	Accuracy = 85%,	[171]
			r = 0.35, spec = 0.51, sens = 0.38 (window size = 25)	
DNA	PSSM	NN	67.10% (sens + spec)/2	[32]
Metal		NN		[173]
Metal ion		Empirical force field	Accuracy = 90–97%	[175]
Metal		MGenTHREADER	Accuracy = 94.5%	[174]
Metal	—	—	Accuracy ≥ 90%	[172]

influence this compositional bias in the binding regions, thus making it possible to predict the binding residues with a fair degree of confidence. Exact compositional biases in and around the binding residues were extensively examined before applying a predictive model of neural network in our works [80]. Predictive performance of DNA binding methods and similar problems are summarized in Table 7.6.

7.3.6 Metal Binding

Interactions between metallic ions and proteins are of interest in a number of transport, therapeutic, and pathological situations [83–89]. Sequence and structural aspects of metal-binding proteins have been reviewed by others [90–94]. One special aspect of metal binding, not shared by DNA- and RNA-binding proteins, is their competitive nature and ability for the exchange of bound metals between proteins [94]. Many metal-binding sites are rich in one or the other types of amino acids (e.g., Cystein in mercury and copper transport, Cystein, and Histidine in some transcription factors). These preferences of amino acids to interact with metals in a competitive way have been used by a number of researchers to develop methods of predicting metal-binding sites (Table 7.6). From the prediction perspective, an additional problem in metal binding is that the number of residues interacting with metals is a very small fraction of the total number in the entire protein sequence. This leaves the control and experimental data highly unbalanced and hence the development of a predictive model and evaluation of its performance becomes additionally complicated. We shall discuss these problems in a general perspective in Section 7.8.

7.3.7 Protein–Protein Interaction Sites

A sequence-based prediction of protein–protein interactions can proceed by just looking at one of the protein sequences and trying to predict what could be the potential binding sites, irrespective of the identity of the binding residues and sequences of the other protein. Defined in this way, the problem of predicting protein–protein interaction sites becomes the same as the problems being discussed here. Some recent studies in this direction have been successful in predicting protein–protein interactions with a reasonable accuracy [e.g., 177]. The methods of measuring these accuracy values have emphasized on the best and worst sets of interaction residues, and it is reported that these values can widely vary from 20% to 94% accuracy. However, protein interaction sites are highly specific to the identity of both the interacting proteins and, therefore, there remains a long way to cover in the direction of accurate prediction of protein–protein interaction sites.

7.4 RESIDUE PROPENSITIES AND CONTEXT DEPENDENCE

As is evident from the above discussions, the possibility of sequence-based predictions of any structural or functional features of a residue depends on the following two factors:

(1) Residue propensity, that is, each of the 20 amino acid types has different affinity to acquire the corresponding structural or functional trait, for example, preference to be more exposed or to be in the binding sites.

(2) Context dependence, that is, the presence of other or similar residues of particular types increasing the propensity scores for those contexts, for example, an Arginine residues may have a certain propensity to interact with DNA, but the Arginine residues surrounded by Lysine may be more likely to bind than the Arginine residues surrounded by Alanine residues.

Measure of propensity has been utilized by a number of researchers [95–97], and in a general form, propensity of a feature c, of residue i, is defined as

$$P(i) = \frac{N_c(i)/N(i)}{T_c/T}. \tag{7.1}$$

Here the numerator is the number of residues $N_c(i)$ of type i having the desired feature c relative to the total number of residues $N(i)$ of type i in the data set, whereas the denominator is the ratio of the residues of any type T_c with feature c, relative to the overall number of residues T in the data used for calculations. Thus, the numerator is a measure of relative frequency of the residue of a given type in the given state and the denominator is the measure of the same relative frequency of all residues. By this definition a propensity score greater than 1 indicates the preference of that residue to have a given property, whereas a score less than 1 represents its aversion to the same. There will be 20 propensity scores for each feature of interest and the rough average of the 20 scores should be equal to 1.

Sometimes the propensity score may be scaled by subtracting from the numerator the average value of the relative frequencies [98]. This will shift the neutral propensity scores to zero and positive and negative numbers will correspondingly denote the over- and underrepresentations of residues in the given feature.

This concept of propensity has been implemented by way of probability measures in the earlier algorithms of secondary structure prediction [5]. Chou and Fasman rely on the probability scores of amino acids to have one of the three secondary structures. These scores easily correspond to the propensity values defined above. The context dependence defined above is also taken care of in terms of second-level probabilities in these methods [54].

Thus, for the prediction of any features of a residue from a protein sequence, the first test to be conducted is the measure of propensity scores and their context values, similar to the Chou and Fasman approach. Once preliminary information on propensity scores is obtained (and if propensity scores are not too similar for all the 20 residue types), an optimization of predictive method can be carried out. Many times machine learning methods will be used for the optimization of the model [99], but before a model is selected and optimized, a good scheme of digitizing amino acids and their contexts has to be developed. We will discuss this issue of representing amino acid residues in the next section.

7.5 REPRESENTATION OF AMINO ACID RESIDUES AND THEIR SEQUENCE NEIGHBORS

We have extensively reviewed the methods of amino acid representation for machine learning algorithms in the context of solvent accessibility prediction [100]. The results obtained can be easily extended to other problems. Here we present the general aspects of sequence representation applicable to prediction methods.

In their first approach of secondary structure prediction using neural networks, Qian and Sejnowski [101] and Holley and Karplus [102] employed a 20-dimensional encoding of amino acid residues. In this method, a single residue is encoded by a 20-bit vector, in which one component is set to one and the rest to zero. This nonzero component identifies one of the 20 residues being represented. In some cases, a 21-bit vector has been used where an additional bit identifies the residue if it is on the terminal position [103]. Thus, if we expect that the context information will be supplied by n neighboring residues on either side of the target residue, we build patterns of n 20-bit vectors, each of the 20-bits allotted to a single position on the amino acid sequence.

This type of 20-bit notation is sufficient to represent any pattern made of only 20 amino acids. We have also shown that nonbinary 20-dimensional codifications in terms of BLOSUM and PAM style matrices may also be used without the loss of any information [100].

7.6 EVOLUTIONARY INFORMATION AND PROFILE-BASED PREDICTIONS

Levin et al. [104] developed a method of secondary structure prediction using sequence similarity. Although our focus in this chapter is restricted to alignment-free methods of prediction, alignments may be used for predictions by first finding sequence homologues and then compiling features of aligned sequences [105], yet we shall include the use of evolutionary information for prediction because alignment profiles serve as sequence representations in an evolutionary context.

The first use of machine learning techniques for secondary structure prediction using evolutionary profiles may be attributed to Rost and Sander [106]. The scientific explanation for a huge success of this type of representation is the fact that the positions, which remain unchanged or conserved during evolution, must carry some information about their structure and function. Unfortunately, there are a very few studies on the actual analysis of this information and on real relationship between evolutionary profiles and predicted structural or functional features (we have carried out one such study for solvent accessibility [107]). What is more frequently reported is the use of some implementations of neural network [31, 102, 106] or an SVM [108, 109] and the predicted performance thereof.

7.7 PREDICTIVE ALGORITHMS

Once an amino acid sequence or its local environment has been represented as a set of feature vectors and similarly the target property has also been encoded as a binary vector or a real value, the next important task is to find a suitable model for their relationship. In the computational sense, it is now a simple problem of finding a model that best relates a set of feature vectors to their corresponding target vectors. Various methods to achieve this have been used in the works cited above. Instead of presenting them in a chronological manner, we try to organize them in the order of complexity, which makes sense and is also likely to be useful for systematically searching for an optimum model of feature predictions.

7.7.1   Multiple Linear Regressions

A simple relationship between a pair of variables X and Y can be expressed by a linear regression

$$Y' = a + b \cdot X, \tag{7.2}$$

where a and b are the coefficients to be determined by minimizing an error function, which depends on the definition of error or maximizing an accuracy function to be selected manually X represents the predictor variables and Y is the target variable to be predicted. The task of selecting a good error function is not straightforward and some of the scores used for this purpose are presented in the subsequent section. In Equation 7.2, the coefficients a and b are called the regression coefficients and Equation 7.2 itself is called a regression relation or equation. Equation 7.2, however, assumes a linear relationship between X and Y and for this reason, it is called a linear regression equation. In reality, the relationship between X and Y may not be linear and, therefore, a linear estimate by the predictor may not work very well. Other types of regression relations can be tried in such a case.

In a sequence-based prediction, the predicted values of Y do not depend on a single variable. Instead, Y depends on variables whose number is equal to $D = (2n + 1) \cdot m$, where n is the number of residue neighbors considered important (there will be n residues on either side of the target residue, which makes the total number $2n + 1$, including the residue itself) and m is the number of bits used to represent each residue, which may be 20 (as in the case of binary representations) or more (to include additional information about the residue). If each of these D variables is denoted by X_i, where i goes from 1 to D, we can write a modified form of regression equation to replace Equation 7.2 as follows:

$$Y' = a + \sum_{i=1}^{D} b_i \cdot X_i. \tag{7.3}$$

Note that we now have D different coefficients for b_i instead of one as given in Equation 7.2. Due to the use of multiple variables and hence the multiple coefficients in this regression relation, it is called multiple linear regression (MLR). For a more formal discussion on multiple linear regressions, readers may refer to Cohen et al. [110].

An MLR method for sequence-based prediction provides a kind of baseline or minimum degree of accuracy that any other model of prediction should offer to be useful. The great advantage of MLR is that the regression coefficients may be directly interpreted in terms of role they play in determining the nature of the target property to be predicted. A method to directly obtain the role of each variable of the feature space $(X'_i s)$ in the prediction, without actually determining the regression best fit, was recently demonstrated by Wang et al. [107]. In this method, the entire n-dimensional feature space is resolved into n feature vectors for the entire data sets. Pearson's coefficient of correlation is then calculated between each of these vectors on the one hand and the target property on the other. If the target property is a multidimensional vector as in the case of secondary structure prediction or in a 3-state or 10-state prediction of ASA, correlation between pairs of vectors may be calculated such that a pair is formed by taking one vector from the feature space and the other from the target space.

7.7.2 Neural Networks

As discussed above, the actual relation between feature and target vectors may not be linear. In fact, the shape of the model, which relates the two spaces, is not known *a priori*. There are a number of ways to solve this problem. One of them is to speculate the shape of the function that approximates this relationship. This requires the estimate of the type of the function (e.g. exponential or logarithmic) or if we fix a function arbitrarily, optimization of parameters in it. For example, if we try to fit a polynomial of degree 2 for each variable in the D-dimensional feature space, the maximum number of coefficients to be optimized becomes approximately D^2. For a feature space made of 100 dimensions (using only two residue neighbors on either side), this increase in the number of variables is very rapid and may quickly outnumber the amount of data available to determine these coefficients. Further, we actually do not know what degree of polynomial will best describe the relation between features and targets. Thus, polynomial multiple regression has never been used in sequence-based predictions.

Neural network fits in a regression model without *a priori* knowledge of the nature of relation that exists between features and target vectors. This model not only provides for nonlinearity with the use of minimum number of variables, but also automatically determines the nature of function that will do the best job of prediction. What is even more powerful is the ability to optimize the model avoiding the possibility of overfitting (a situation in which model is artificially optimized for the data set used because of too many parameters optimized. Such models will underperform on a data set, which the model has not seen earlier).

There are many different types of neural networks [111]. Here we describe the most frequently used neural network called a multilayered feed-forward back-propagation neural network. This type of neural network is by far the most widely used method of prediction. Its first applications to sequence-based predictions may be attributed to almost simultaneous pair of publications [101, 102]. Computational scheme of a multilayered neural network is shown in Fig. 7.1. This model is thought to be a network connecting certain neurons (each neuron or a node is like a container having some numerical state at a given time). Neurons in one layer are connected to the other through links, which perform the task of transferring information from one layer to the other. In the foremost layer called the input layer, all the information about the features coming from the representations of residues is supplied. These values are then processed by an activation function, which determines the states of neurons in the next layer. During the learning phase the output layer receives both predictions from the previous layer and target values from an external supervisor, during the prediction

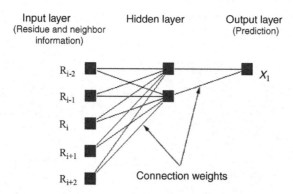

Figure 7.1   A multilayered neural network for sequence-based predictions.

phase it producers predictions for the target variable. Presence of a hidden layer between the input and the output layers and the activation function between layers provide for the nonlinearity of the function between the features sent to the input layer and the target property predicted at the output layer. The links between layers are represented by certain weight matrices, which transform the output of a layer into the input of the next layer by a process of matrix multiplication. The goal of neural network optimization then is to determine the optimal elements of these matrices. The optimized neural network is nothing but a combination of the neural network layer structure and the matrices of connection weights (one matrix connecting two successive layers). Most of the other aspects of the neural networks employed for sequence-based predictions are similar to any other multilayered neural networks, which have been extensively reviewed in the literature [112–114]. Some of the sequence-based predictors employing neural networks have been listed in Tables 7.2–7.4. A schematic representation of using evolutionary information (by position specific scoring matrices or PSSM) and neural network combination for prediction is shown in Figure 7.2.

7.7.3 Support Vector Machines

A comprehensive review of the theory and implementation of support vector machine (SVM) is provided in Ref. [115]. Here we present only the basic ideas, which are relevant for developing sequence-based predictors using SVM. In particular, we would focus on how they differ from neural networks—the most widely used model for such problems.

One of the most important differences between SVM and neural networks is that neural network looks for any hyperplane (or any arbitrary geometry) separating two classes of data with best accuracy, whereas SVM tries to find the hyperplane that has the *maximum margin* from selected data points in each class. The original idea of this type of classifier was given by Vapnik [116]. Although the initial idea was to use only a hyperplane to classify the data classes, the same was extended to a nonlinear separator to account for other kinds of data categories [117]. In a linear separator, the final

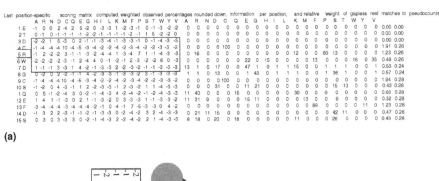

(a)

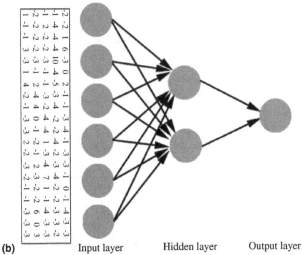

(b)   Input layer   Hidden layer   Output layer

Figure 7.2 Use of evolutionary information for predicting DNA-binding sites. (a) Rows of position specific scoring matrices selected for neural network input. (b) Network inputs of the PSSM of the target residue and its two neighboring residues on C- and N-terminals and design of neural network.

predictor is obtained by a dot product between the feature vectors (defined in the earlier sections) and the weight-like matrix in a way similar to neural networks (but without any hidden layer and with an additional requirement of maximizing the margins from the hyperplane defined by the weight space). Typical principle of SVM is shown in Fig. 7.3. The dimensionality of the weight space is the same as the feature space. The principles of training and validation will be the same as in the case of neural networks. In a nonlinear classifier, the transformation from the feature space to the target space is carried out by way of a multidimensional function called the kernel, instead of the matrix and dot product. Choice of kernel is similar to any nonlinear curve fitting between a pair of variables. Some of the most widely employed kernels are polynomial, Gaussian, and sigmoidal (see details in [115]).

Hua and Sun [118] were probably the first to apply SVM to sequence-based predictions. In their model of secondary structure prediction, the authors claimed an

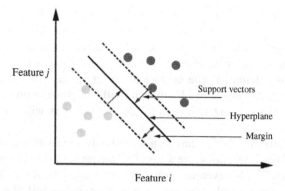

Figure 7.3   Two dimensions of SVM hyperplane for classification problem.

improvement in accuracy and generalization, using SVM over neural-network-based methods. SVM has some advantages including the effective avoidance of overfitting and the ability to handle large feature spaces. Apart from usual helix-sheet-coil-type secondary structure prediction [108, 119, 120], SVM-based predictors have also been used for solvent accessibility prediction [109, 121], beta and gamma turns [46], alpha turns [122], DNA/RNA-binding sites [82], metal binding sites [123], and so on. In most of these prediction models, SVM predictor has appeared after a neural-network-based predictor was already known, and many times comes with a claim in an improvement in accuracy. This improvement could be because of SVM's maximum margin approach that would enable the classification of those patterns in the test data sets, which did not appear in the training data, but their separating hyperplane was more accurately estimated due to the margin optimization, which is not done in the case of neural networks.

7.7.4   Multistage Methods

It has been reported that the prediction of secondary structure at one residue position may affect similar or other types of secondary structures and hence it may be possible to use the two-stage neural networks, in which the first stage is similar to the classification as explained above, and the second stage uses the first stage predictions and finds a new set of weights [124]. Similar two-stage SVMs have also been shown to improve prediction performance in the case of solvent accessibility and secondary structure [109]. We have developed multistage SVMs to integrate class and real-value predictions in solvent accessibility [125]. In general, multistage approach is likely to be effective where the structure or function of neighbors is mutually cooperative or competitive in nature, for example, when a solvent-exposed residue makes it more (or may be less) likely for its subsequent neighbor also to be solvent-exposed.

7.7.5   Other Approaches

Although the neural networks, SVMs, and linear regressions—perhaps in that order—remain the most popular approaches for sequence-based predictions, some other methods have also been applied with success (see Tables 7.3–7.6). Most noteworthy among them are "k-means clustering" [126] and "energy minimization" [127]. The former is based on the principle of Euclidean distances from hypothetical geometric centers in the feature space, from where the distance to all the data points for each class is calculated. The number of centers is the same as the number of classes and the distances may be Euclidean (with weighted or unweighted dimensions). The goal of optimization is to locate these centers with the best possible classification.

Energy minimization method has been successfully applied to solvent accessibility and is powerful because it does give a physical interpretation of the model developed for prediction [127]. However, the problem is that the solvent accessibility has an immediate parameter of solvation energy, which can be optimized to construct this model, but the application of this method to other problems such as secondary structure and binding sites is yet unclear.

Another statistical approach to sequence-based prediction is the use of Bayes method. Bayes methods use the information from the sequence and its environment to evaluate/approximate the conditional probability of the target vectors (e.g., DNA binding) using Bayes theorem [128]. These methods have recently been used to approximate DNA binding and other applications reporting similar degrees of accuracy as neural networks and SVM [81].

7.8 SPECIAL PROBLEMS

7.8.1 Data Scarcity

In the most general case, a single-residue position is represented by 20-dimensional vector. Thus, a property that is expected to depend on n-neighboring residues constitutes of at least $20 \cdot (2n + 1)$ dimensions in the feature space for a linear predictor. For a nonlinear predictor, this number will be even higher. In the case of neural networks or SVM, it leads to situations where the number of parameters to be optimized approaches the number of data points in the either or all prediction classes. Solution to this problem is either to discard information from distant neighbors (using fewer residue information) or to reduce the dimensionality of the amino acid space by optimizing the model for a single dimension and then using that representation for larger feature sets. The former was the motivation for the use of only two-neighbor information (in contrast to seven or eight neighbors in secondary structure prediction) in our DNA-binding prediction methods [32] and the later issue has been addressed in detail in our work on this subject [100].

7.8.2 Error and Accuracy Scores

The choice of a good error function is probably the most important part of optimizing and comparing the prediction methods presented above. Some of the commonly used accuracy/error scores are as follows:

(1) *Single-Residue Accuracy.* Defined as the ratio of the number of correctly classified residues in any category to the total number of predictions made.

(2) *Matthew's Correlation.* Based on the calculation of overpredictions and underpredictions in each category [129]. In a multistate prediction, each state will have its own Matthew's correlation (C_a). This is defined as

$$C_a = \frac{(p_a n_a) - (u_a o_a)}{\sqrt{(n_a + u_a)(n_a + o_a)(p_a + u_a)(p_a + o_a)}},\tag{7.4}$$

where, p, n, u, and o represent correct predictions, correct rejections, underpredictions and overpredictions, respectively, for the state represented by subscript a.

(3) *Pearson's Correlation.* Overall prediction quality (of all states) can be assessed by this index; defined as the ratio of covariance in the prediction state to the product of variance in each state.

(4) *Sensitivity.* Defined as TP/(TP + FN), where TP means true positive and FN means false negative.

(5) *Specificity.* Defined as TN/(TN + FP), where TN and FP are true negative and false positive, respectively.

(6) *Mean Absolute Error (MAE).* Defined as the absolute per residue error of prediction in a numerical property.

(7) *Mean Square Error (MSE).* Defined as the standard deviation of all the errors obtained for each residue prediction.

In addition to these scores, a special segment overlap (SOV) score is used for secondary structure prediction, which takes into account the continuous stretches of correctly predicted secondary structure elements [49].

It may be noted that the sensitivity of a two-state predictor will be 100% if all predictions were to be "positives," and specificity will be 100% if all predictions were "negatives." For this reason none of these scores alone can be utilized for comparing or optimizing predictor performance. Between the Matthew's and Pearson's correlation coefficients, the value of Matthew's correlation of each category is the same as the overall Pearson's correlation for a two-state classification, if the two states are complementary and exclusive.

7.9 CONCLUSION

A number of problems in bioinformatics require mapping of a feature space of protein sequences to a target space of structural or functional properties. Many methods to address these problems have been developed with varying degrees of success. These approaches differ in terms of their knowledge representation, modeling, and nature of target spaces. A unified approach to all these problems is needed and in this chapter we survey the current status and also try to set a stage to achieve an integrated approach to sequence-based predictions.

REFERENCES

1. Du Vigneaud, V., Resseler, C., and Trippett, S. The sequence of amino acids in oxytocin, with a proposal for the structure of oxytocin. *Journal of Biological Chemistry*, 205(2): 949–957, 1953.

2. Perutz, M. F. Relation between structure and sequence of haemoglobin. *Nature*, 194: 914–917, 1962.

3. Guzzo, A. V. The influence of amino-acid sequence on protein structure. *Biophysics Journal*, 5(6): 809–822, 1965.

4. Sternberg, M. J. and Thornton, J. M. Prediction of protein structure from amino acid sequence. *Nature*, 271(5640): 15–20, 1978.

5. Chou, P. Y. and Fasman, G. D. Prediction of the secondary structure of proteins from their amino acid sequence. *Advances Enzymology and Related Areas of Molecular Biology*, 47: 45–148, 1978.

6. Sternberg, M. J. Prediction of protein structure from amino acid sequence. *Anti-cancer Drug Design*, 1(3): 169–178, 1986.

7. Yada, R. Y., Jackman, R. L., and Nakai, S. Secondary structure prediction and determination of proteins: a review. *International Journal of Peptide and Protein Research*, 31(1): 98–108, 1988.

8. Saito, N. Principles of protein architecture. *Advances in Biophysics*, 25: 95–132, 1989.

9. Fasman, G. D. Protein conformational prediction. *Trends in Biochemical Sciences*, 14(7): 295–299, 1989.

10. Garnier, J. Protein structure prediction. *Biochimie*, 72(8): 513–524, 1990.

11. Garnier, J. and Levin, J. M. The protein structure code: What is its present status? *Computer Applications in the Biosciences*, 7(2): 133–142, 1991.

12. Benner, S. A. and Gerloff, D. L. Predicting the conformation of proteins. Man versus machine. *FEBS Letters*, 325(1–2): 29–33, 1993.

13. Benner, S. A., Jenny, T. F., Cohen, M. A., and Gonnet, G. H. Predicting the conformation of proteins from sequences. Progress and future progress. *Advances in Enzyme Regulation*, 34: 269–353, 1994.

14. Rost, B. and Sander, C. Conservation and prediction of solvent accessibility in protein families. *Proteins*, 20(3): 216–226, 1994.

15. Moult, J. The current state of the art in protein structure prediction. *Current Opinion in Biotechnology*, 7(4): 422–427, 1996.

16. Jones, D. T. Progress in protein structure prediction. *Current Opinion in Structural Biology*, 7(3): 377–387, 1997.

17. Al-Lazikani, B., Jung, J., Xiang, Z., and Honig, B. Protein structure prediction. *Current Opinion in Chemical Biology*, 5(1): 51–56, 2001.

18. Schonbrun, J., Wedemeyer, W. J., and Baker, D. Protein structure prediction in 2002. *Current Opinion in Structural Biology*, 12(3): 348–354, 2002.

19. Zhou, C., Zheng, Y., and Zhou, Y. Structure prediction of membrane proteins. *Genomics Proteomics Bioinformatics*, 2(1): 1–5, 2004.

20. Moult, J. A decade of CASP: progress, bottlenecks and prognosis in protein structure prediction. *Current Opinion in Structural Biology*, 15(3): 285–289, 2005.

21. Holt, C. and Sawyer, L. Primary and predicted secondary structures of the caseins in relation to their biological functions. *Protein Engineering*, 2(4): 251–259, 1988.

22. Hirst, J. D. and Sternberg, M. J. Prediction of structural and functional features of protein and nucleic acid sequences by artificial neural networks. *Biochemistry*, 31(32): 7211–7218, 1992.

23. Carter, J. M. Epitope prediction methods. *Methods in Molecular Biology*, 36: 193–206, 1994.

24. May, A. C., Johnson, M. S., Rufino, S. D., Wako, H., Zhu, Z. Y., Sowdhamini, R., Srinivasan, N., Rodionov, M. A., and Blundell, T. L. The recognition of protein structure and function from sequence: adding value to genome data. *Philosophical Transaction of the Royal Society of London, Series B: Biological Sciences*, 344(1310): 373–381, 1994.

25. Doerks, T., Bairoch, A., and Bork, P. Protein annotation: detective work for function prediction. *Trends in Genetics*, 14(6): 248–250, 1998.

26. Koonin, E. V., Tatusov, R. L., and Galperin, M. Y. Beyond complete genomes: from sequence to structure and function. *Current Opinion in Structural Biology*, 8(3): 355–363, 1998.

27. Lazaridis, T. and Karplus, M. Effective energy functions for protein structure prediction. *Current Opinion in Structural Biology*, 10(2): 139–145, 2000.

28. Rost, B. Prediction in 1D: secondary structure, membrane helices, and accessibility. *Methods of Biochemical Analysis*, 44: 559–587, 2003.

29. Whisstock, J. C. and Lesk, A. M. Prediction of protein function from protein sequence and structure. *Quarterly Review of Biophysics*, 36(3): 307–340, 2003.

30. Wolfson, H. J., Shatsky, M., Schneidman-Duhovny, D., Dror, O., Shulman-Peleg, A., Ma, B., and Nussinov, R. From structure to function: methods and applications. *Current Protein and Peptide Science*, 6(2): 171–183, 2005.

31. Ahmad, S., Gromiha, M. M., and Sarai, A. A real value prediction of solvent accessibility from amino acid sequence. *Proteins*, 50(4): 629–635, 2003.

32. Ahmad, S. and Sarai, A. PSSM-based prediction of DNA binding sites in proteins.. *BMC Bioinformatics*, 6: 33–35, 2005.

33. Bailey, K. and Sanger, F. The chemistry of amino acids and proteins. *Annual Review of Biochemistry*, 20: 103–130, 1951.

34. IUPAC-IUB Commission on Biochemical Nomenclature, (CBN). A one-letter notation for amino acid sequences. Archives of Biochemistry and Biophysics, 125 (3): i–v, 1968.

35. Araúzo-Bravo, M. J. and Ahmad, S. Protein sequence and structure databases: a review. *Current Analytical Chemistry*, 1(3): 355–371, 2005.

36. Betts, M. J. and Russell, R. B. Amino acid properties and consequences of substitutions: In: M. R. Barnes and I. C. Gray, Eds, *Bioinformatics for Geneticists*, Wiley, Hoboken, NJ, 2003.

37. Myers, J. K. and Pace, C. N. Hydrogen bonding stabilizes globular proteins. *Biophysical Journal*, 71(4): 2033–2039, 1996.

38. Ramachandran, G. N. and Sasisekharan, V. Conformation of polypeptides and proteins. *Advances in Protein Chemistry*, 23: 283–438, 1968.

39. Kabsch, W. and Sander, C. Dictionary of protein secondary structure: pattern recognition of hydrogen-bonded and geometrical features. *Biopolymers*, 22(12): 2577–2637, 1983.

40. Shepherd, A. J., Gorse, D., and Thornton, J. M. Prediction of the location and type of beta-turns in proteins using neural networks. *Protein Science*, 8(5): 1045–1055, 1999.

41. Pal, L. and Basu, G. Neural network prediction of 3(10)-helices in proteins. *Indian Journal of Biochemistry and Biophysics*, 38(1–2): 107–114, 2001.

42. Kaur, H. and Raghava, G. P. A neural-network based method for prediction of gamma-turns in proteins from multiple sequence alignment. *Protein Science*, 12(5): 923–929, 2003.

43. Kaur, H. and Raghava, G. P. A neural network method for prediction of beta-turn types in proteins using evolutionary information. *Bioinformatics*, 20(16): 2751–2758, 2004.

44. Kim, S. Protein beta-turn prediction using nearest-neighbor method. *Bioinformatics*, 20(1): 40–44, 2004.

45. Fuchs, P. F. and Alix, A. J. High accuracy prediction of beta-turns and their types using propensities and multiple alignments. *Proteins*, 59(4): 828–839, 2005.

46. Pham, T. H., Satou, K. and Ho, T. B. Support vector machines for prediction and analysis of beta and gamma-turns in proteins. *Journal of Bioinformatics Computational Biology*, 3 (2): 343–358, 2005.

47. Wang, Y., Xue, Z. D., Shi, X. H., and Xu, J. Prediction of pi-turns in proteins using PSI-BLAST profiles and secondary structure information. *Biochemical and Biophysical Research Communications*, 347(3): 574–580, 2006.

48. Rost, B., Sander, C. and Schneider, R. Redefining the goals of protein secondary structure prediction. *Journal of Molecular Biology*, 235(1): 13–26, 1994.

49. Rost, B. Review: protein secondary structure prediction continues to rise. *Journal of Structural Biology*, 134(2–3): 204–218, 2001.

50. Dunbrack, R. L. Jr., Sequence comparison and protein structure prediction. *Current Opinion in Structural Biology*, 16(3): 374–384, 2006.

51. Chou, P. Y. and Fasman, G. D. Conformational parameters for amino acids in helical, beta-sheet, and random coil regions calculated from proteins. *Biochemistry*, 13(2): 211–222, 1974.

52. Petukhov, M., Uegaki, K., Yumoto, N., and Serrano, L. Amino acid intrinsic α-helical propensities III: positional dependence at several positions of C terminus. *Protein Science*, 11: 766–777, 2002.

53. Costantini, S., Colonna, G., and Facchiano, A. M. Amino acid propensities for secondary structures are influenced by the protein structural class. *Biochemical and Biophysical Research Communications*, 342(2): 441–451, 2006.

54. Garnier, J., Osguthorpe, D. J., and Robson, B. Analysis of the accuracy and implications of simple methods for predicting the secondary structure of globular proteins. *Journal of Molecular Biology*, 120(1): 97–120, 1978.

55. Rost, B. and Sander, C. Prediction of protein secondary structure at better than 70% accuracy. *Journal of Molecular Biology*, 232(2): 584–599, 1993.

56. Lee, B. and Richards, F. M. The interpretation of protein structures: estimation of static accessibility. *Journal of Molecular Biology*, 55: 379–400, 1971.

57. Janin, J. Surface and inside volumes in globular proteins. *Nature*, 277: 491–492, 1979.

58. Charton, M. and Charton, B. I. The structural dependence of amino acid hydrophobicity parameters. *Journal of Theoretical Biology*, 99: 629–644, 1982.

59. Rose, G., Geselowitz, A., Lesser, G., Lee, R., and Zehfus, M. Hydrophobicity of amino acid residues in globular proteins. *Science*, 229: 834–838, 1985.

60. Wolfenden, R., Andersson, A., Cullis, P., and Southgate, C. Affinities of amino acid side chains for solvent water. *Biochemistry*, 20: 849–855, 1981.

61. Kyte, J. and Doolittle, R. A simple method for displaying the hydropathic character of a protein. *Journal of Molecular Biology*, 157: 105–132, 1982.

62. Cornette, J., Cease, K. B., Margalit, H., Spouge, J. L., Berzofsky, J. A., and DeLisi, C. Hydrophobicity scales and computational techniques for detecting amphipathic structures in proteins. *Journal of Molecular Biology*, 195: 659–685, 1987.

63. Chen, H. and Zhou, H. X. Prediction of solvent accessibility and sites of deleterious mutations from protein sequence. *Nucleic Acids Research*, 33(10): 3193–3199, 2005.

64. Rost, B., Liu, J., Nair, R., Wrzeszczynski, K. O., and Ofran, Y. Automatic prediction of protein function. *Cellular and Molecular Life Sciences*, 60(12): 2637–2650, 2003.

65. Vullo, A. and Frasconi, P. A bi-recursive neural network architecture for the prediction of protein coarse contact maps. *Proceedings of the IEEE Computer Society Bioinformatics Conference*, 1: 187–196, 2002.

66. Punta, M. and Rost, B. PROFcon: novel prediction of long-range contacts. *Bioinformatics*, 21(13): 2960–2968, 2005.

67. Sim, J., Kim, S. Y., and Lee, J. PPRODO: prediction of protein domain boundaries using neural networks. *Proteins*, 59(3): 627–632, 2005.

68. Schlessinger, A. and Rost, B. Protein flexibility and rigidity predicted from sequence. *Proteins*, 61(1): 115–126, 2005.

69. Pollastri, G., Baldi, P., Fariselli, P., and Casadio, R. Prediction of coordination number and relative solvent accessibility in proteins. *Proteins*, 47(2): 142–153, 2002.

70. Muskal, S. M., Holbrook, S. R., and Kim, S. H. Prediction of the disulfide-bonding state of cysteine in proteins. *Protein Engineering*, 3(8): 667–672, 1990.

71. Mucchielli-Giorgi, M. H., Hazout, S., and Tuffery, P. Predicting the disulfide bonding state of cysteines using protein descriptors. *Proteins*, 46(3): 243–249, 2002.

72. Martelli, P. L., Fariselli, P., and Casadio, R. Prediction of disulfide-bonded cysteines in proteomes with a hidden neural network. *Proteomics*, 4(6): 1665–1671, 2004.

73. Zhao, E., Liu, H. L., Tsai, C. H., Tsai, H. K., Chan, C. H., and Kao, C. Y. Cysteine separations profiles on protein sequences infer disulfide connectivity. *Bioinformatics*, 21(8): 1415–1420, 2005.

74. Thangudu, R. R., Vinayagam, A., Pugalenthi, G., Manonmani, A., Offmann, B., and Sowdhamini, R. Native and modeled disulfide bonds in proteins: knowledge-based approaches toward structure prediction of disulfide-rich polypeptides. *Proteins*, 58(4): 866–879, 2005.

75. Chen, B. J., Tsai, C. H., Chan, C. H., and Kao, C. Y. Disulfide connectivity prediction with 70% accuracy using two-level models. *Proteins*, 64(1): 246–252, 2006.

76. Chen, Y. C. and Hwang, J. K. Prediction of disulfide connectivity from protein sequences. *Proteins*, 61(3): 507–512, 2005.

77. Tsai, C. H., Chen, B. J., Chan, C. H., Liu, H. L., and Kao, C. Y. Improving disulfide connectivity prediction with sequential distance between oxidized cysteines. *Bioinformatics*, 21(24): 4416–4449, 2005.

78. Ferre, F. and Clote, P. DiANNA: a web server for disulfide connectivity prediction. *Nucleic Acids Research*, 33 (Web Server issue): W230–W232, 2005.

79. Yang, Z. R., Thomson, R., and McNeil, P. Esnouf RM RONN: the bio-basis function neural network technique applied to the detection of natively disordered regions in proteins. *Bioinformatics*, 21(16): 3369–3376, 2005.

80. Ahmad, S. M., Gromiha, M., and Sarai, A. Analysis and Prediction of DNA-binding proteins and their binding residues based on composition, sequence and structural information. *Bioinformatics*, 20: 477–486, 2004.

81. Yan, C., Terribilini, M., Wu, F., Jernigan, R. L., Dobbs, D., and Honavar, V. Predicting DNA-binding sites of proteins from amino acid sequence. *BMC Bioinformatics*, 7: 262–271, 2006.

82. Wang, L. and Brown, S. J. BindN: a web-based tool for efficient prediction of DNA and RNA binding sites in amino acid sequences. *Nucleic Acids Research*, 34: 243–248, 2006.

83. Bouton, C. M. and Pevsner, J. Effects of lead on gene expression. *Neurotoxicology*, 21(6): 1045–1055, 2000.

84. Hainaut, P. and Mann, K. Zinc binding and redox control of p53 structure and function. *Antioxid Redox Signal*, 3(4): 611–623, 2001.

85. Safaei, R., Holzer, A. K., Katano, K., Samimi, G., and Howell, S. B. The role of copper transporters in the development of resistance to Pt drugs. *Journal of Inorganic Biochemistry*, 98(10): 1607–1613, 2004.

86. Bush, A. I. Metal complexing agents as therapies for Alzheimer's disease. *Neurobiology of Aging*, 23(6): 1031–1038, 2002.

87. Fatemi, N. and Sarkar, B. Molecular mechanism of copper transport in Wilson disease. *Environmental Health Perspectives*, 110 (Suppl 5): 695–698, 2002.

88. Doraiswamy, P. M. and Finefrock, A. E. Metals in our minds: therapeutic implications for neurodegenerative disorders. *Lancet Neurology*, 3(7): 431–434, 2004.

89. Huffman, D. L. and O'Halloran, T. V. Function, structure, and mechanism of intracellular copper trafficking proteins. *Annual Review of Biochemistry*, 70: 677–701, 2001.

90. Massova, I., Kotra, L. P., Fridman, R., and Mobashery, S. Matrix metalloproteinases: structures, evolution, and diversification. *FASEB Journal* 12(12): 1075–1095, 1998.

91. Harrison, M. D., Jones, C. E., and Dameron, C. T. Copper chaperones: function, structure and copper-binding properties. *Journal of Biological Inorganic Chemistry*, 4(2): 145–153, 1999.

92. Cox, E. H. and McLendon, G. L. Zinc-dependent protein folding. *Current Opinion in Chemical Biology*, 4(2): 162–165, 2000.

93. Pidcock, E. and Moore, G. R. Structural characteristics of protein binding sites for calcium and lanthanide ions. *Journal of Biological Inorganic Chemistry*, 6(5–6): 479–489, 2001.

94. Opella, S. J., DeSilva, T. M., and Veglia, G. Structural biology of metal-binding sequences. *Current Opinion in Chemical Biology*, 6(2): 217–223, 2002.

95. Bradford, J. R. and Westhead, D. R. Improved prediction of protein–protein binding sites using a support vector machines approach. *Bioinformatics*, 21(8): 1487–1494, 2005.

96. Taroni, C., Jones, S. and Thornton, J. M. Analysis and prediction of carbohydrate binding sites. *Protein Engineering*, 13(2): 89–98, 2000.

97. Joughin, B. A., Tidor, B., and Yaffe, M. B. A computational method for the analysis and prediction of prote: phosphopeptide-binding sites. *Protein Science*, 14(1): 131–139, 2005.

98. Su, C. T., Chen, C. Y., and Ou, Y. Y. Protein disorder prediction by condensed PSSM considering propensity for order or disorder. *BMC Bioinformatics*, 7: 319–334, 2006.

99. Baldi, P., Brunak, S., Frasconi, P., Soda, G., and Pollastri, G. Exploiting the past and the future in protein secondary structure prediction. *Bioinformatics*, 15(11): 937–946, 1999.

100. Araúzo-Bravo, M. J., Ahmad, S., and Sarai, A. Dimensionality of amino acid space and solvent accessibility prediction with neural networks. *Computational Biology and Chemistry*, 30(2): 160–168, 2006.

101. Qian, N. and Sejnowski, T. Predicting the secondary structure of globular proteins using neural network models. *Journal of Molecular Biology*, 202: 865–884, 1988.

102. Holley, H. and Karplus, M. Protein secondary structure prediction with a neural network. *Proceedings of the National Academy of Science of the United States of America*, 86: 152–156, 1989.

103. Ahmad, S. and Gromiha, M. M. NETASA: neural network based prediction of solvent accessibility. *Bioinformatics*, 18(6): 819–824, 2002.

104. Levin, J. M., Robson, B., and Garnier, J. An algorithm for secondary structure determination in proteins based on sequence similarity. *FEBS Letters*, 205(2): 303–308, 1986.

105. Bryant, S. H. and Altschul, S. F. Statistics of sequence-structure threading. *Current Opinion in Structural Biology*, 5(2): 236–244, 1995.

106. Rost, B. and Sander, C. Improved prediction of protein secondary structure by use of sequence profiles and neural networks. *Proceedings of the National Academy of Sciences of the United States of America*, 90(16): 7558–7562, 1993.

107. Wang, J. Y., Lee, H. M., and Ahmad, S. Prediction and evolutionary information analysis of protein solvent accessibility using multiple linear regression. *Proteins*, 61(3): 481–491, 2005.

108. Kim, H. and Park, H. Protein secondary structure prediction based on an improved support vector machines approach. *Protein Engineering*, 16(8): 553–560, 2003.

109. Nguyen, M. N. and Rajapakse, J. C. Prediction of protein relative solvent accessibility with a two-stage SVM approach. *Proteins*, 59(1): 30–37, 2005.

110. Cohen, J., Cohen, P., West, S. G., and Aiken, L. S. *Applied Multiple Regression/ Correlation Analysis for the Behavioral Sciences*. 2nd edition, Lawrence Erlbaum Associates, Hillsdale, NJ, 2003.

111. Abdi, H. A neural network primer. *Journal of Biological Systems*, 2: 247–281, 1994.

112. Dayhoff, J. *Neural Network Architectures: An Introduction*, Van Nostrand Reinhold, New York, 1990.

113. Anderson, J. A. *An Introduction to Neural Networks*, The MIT Press, Cambridge, MA, 1995.

114. Braspenning, P. J. Thuisjsman, F., and Weijters, A. J. M. M. Eds. *Artificial Neural Networks: An Introduction to ANN Theory and Practice*, Springer-Verlag, Berlin, pp. 37–66, 1995.

115. Cristianini, N. and Shawe-Taylor, J. *An Introduction to Support Vector Machines (and Other Kernel-Based Learning Methods)*, Cambridge University Press, Cambridge, MA, 2000.

116. Vapnik, V. N. *The Nature of Statistical Learning Theory*, Springer, New York, 1998.

117. Boser, B. E., Guyon, I. M., and Vapnik, V. N. A training algorithm for optimal margin classifiers. In: D. Haussler, Ed., *5th Annual ACM Workshop on COLT*, ACM Press, Pittsburgh, PA. pp. 144–152, 1992.

118. Hua, S. and Sun, Z. A novel method of protein secondary structure prediction with high segment overlap measure: support vector machine approach. *Journal of Molecular Biology*, 308(2): 397–407, 2001.

119. Ward, J. J., McGuffin, L. J., Buxton, B. F., and Jones, D. T. Secondary structure prediction with support vector machines. *Bioinformatics*, 19(13): 1650–1655, 2003.

120. Nguyen, M. N. and Rajapakse, J. C. Two-stage multi-class support vector machines to protein secondary structure prediction. *Pacific Symposium on Biocomputing*, 346–357, 2005.

121. Kim, H. and Park, H. Prediction of protein relative solvent accessibility with support vector machines and long-range interaction 3D local descriptor. *Proteins*, 54(3): 557–562, 2004.

122. Wang, Y., Xue, Z., and Xu, J. Better prediction of the location of alpha-turns in proteins with support vector machine. *Proteins*, 65(1): 49–54, 2006.

123. Passerini, A., Punta, M., Ceroni, A., Rost, B., and Frasconi, P. Identifying cysteines and histidines in transition-metal-binding sites using support vector machines and neural networks. *Proteins*, 65(2): 305–316, 2006.

124. Cuff, J. A. and Barton, G. J. Application of multiple sequence alignment profiles to improve protein secondary structure prediction. *Proteins*, 40(3): 502–511, 2000.

125. Wang, J. Y., Lee, H. M., and Ahmad, S., SVM-cabins: prediction of solvent accessibility using accumulation cutoff sets. *Proteins*, 68(1): 82–91, 2007.

126. Sim, J., Kim, S. Y., and Lee, J. Prediction of protein solvent accessibility using fuzzy k-nearest neighbor method. *Bioinformatics*, 21(12): 2844–2849, 2005.

127. Xu, Z., Zhang, C., Liu, S., and Zhou, Y. QBES: predicting real values of solvent accessibility from sequences by efficient, constrained energy optimization. *Proteins*, 63: 961–966, 2006.

128. Carlin, B. P. and Louis, T. A. *Bayes and Empirical Bayes Methods for Data Analysis*, Chapman & Hall/CRC, Boca Raton, FL, 2000.

129. Matthews, B. W. Comparison of the predicted and observed secondary structure of T4 phage lysozyme. *Biochimica et Biophysica Acta*, 405(2): 442–451, 1975.

130. Pollastri, G. and McLysaght, A. Porter: a new, accurate server for protein secondary structure prediction. *Bioinformatics*, 21(8): 1719–1720, 2005.

131. Adamczak, R., Porollo, A., and Meller, J. Combining prediction of secondary structure and solvent accessibility in proteins. *Proteins*, 59(3): 467–475, 2005.

132. Yia, T. and Lander, E. S. Protein secondary structure prediction using nearest-neighbor methods. *Journal of Molecular Biology*, 232(4): 1117–1129, 1993.

133. Salamov, A. A. and Solovyev, V. V. Protein secondary structure prediction using local alignments. *Journal of Molecular Biology*, 268(1): 31–36, 1997.

134. Frishman, D. and Argos, P. Seventy-five percent accuracy in protein secondary structure prediction. *Proteins*, 27(3): 329–335, 1997.

135. Salamov, A. A. and Solovyev, V. V. Prediction of protein secondary structure by combining nearest-neighbor algorithms and multiple sequence alignments. *Journal of Molecular Biology*, 247(1): 11–15, 1995.

136. Geourjon, C. and Deléage, G. SOPMA: significant improvements in protein secondary structure prediction by consensus prediction from multiple alignments. *Computer Applications in the Biosciences*, 11(6): 681–684, 1995.

137. McGuffin, L. J., Bryson, K., and Jones, D. T. The PSIPRED protein structure prediction server. *Bioinformatics*, 16(4): 404–405, 2000.

138. Pan, X. M. Multiple linear regression for protein secondary structure prediction. *Proteins*, 43(3): 256–259, 2001.

139. Lin, K., Simossis, V. A., Taylor, W. R., and Heringa, J. A simple and fast secondary structure prediction method using hidden neural networks. *Bioinformatics*, 21(2): 152–159, 2005.

140. Hu, H. J., Pan, Y., Harrison, R., and Tai, P. C. Improved protein secondary structure prediction using support vector machine with a new encoding scheme and an advanced tertiary classifier. *IEEE Transactions on Nanobioscience*, 3(4): 265–271, 2004.

141. Wood, M. J. and Hirst, J. D. Protein secondary structure prediction with dihedral angles. *Proteins*, 59(3): 476–481, 2005.

142. Petersen, T. N., Lundegaard, C., Nielsen, M., Bohr, H., Bohr, J., Brunak, S., Gippert, G. P., and Lund, O. Prediction of protein secondary structure at 80% accuracy. *Proteins*, 41(1): 17–20, 2000.

143. Ouali, M. and King, R. D. Cascaded multiple classifiers for secondary structure prediction. *Protein Science*, 9(6): 1162–1176, 2000.

144. Riis, S. K. and Krogh, A. Improving prediction of protein secondary structure using structured neural networks and multiple sequence alignments. *Journal of Computational Biology*, 3(1): 163–183, 1996.

145. Sen, T. Z., Jernigan, L. R., Garnier, J., and Kloczkowski, A. GOR V server for protein secondary structure prediction. *Bioinformatics*, 21(11): 2787–2788, 2005.

146. Chandonia, J. M. and Karplus, M. The importance of larger data sets for protein secondary structure prediction with neural networks. *Protein Science*, 5(4): 768–774, 1996.

147. Kneller, D. G., Cohen, F. E., and Langridge, R., Improvements in protein secondary structure prediction by an enhanced neural network. *Journal of Molecular Biology*, 214(1): 171–1820, 1990.

148. Thompson, M. J. and Goldstein, R. A. Predicting protein secondary structure with probabilistic schemata of evolutionarily derived information. *Protein Science*, 6(9): 1963–1975, 1997.

149. Jones, D. T. Protein secondary structure prediction based on position-specific scoring matrices. *Journal of Molecular Biology*, 292: 195–202, 1999.

150. Kinjo, A. R. and Nishikawa, K. Predicting secondary structures, contact numbers and residue-wise contact orders of native protein structure from amino acid sequence using critical random networks. *Biophysics*, 1: 67–74, 2005.

151. Levin, J. M. CRNPRED: highly accurate prediction of one-dimensional protein structures by large-scale critical random networks.Exploring the limits of nearest neighbour secondary structure prediction. *Protein Engineering*, (7): 771–776, 1997.

152. Barton, G. J. Protein secondary structure prediction. *Current Opinion in Structural Biology*, 5(3): 372–376, 1995.

153. Schmidler, S. C., Liu, J. S., and Brutlag, D. L. Bayesian segmentation of protein secondary structure. *Journal of Computational Biology*, 7: 233–248, 2000.

154. Garnier, J., Levin, J. M., Gibrat, J. F., and Biou, V. Secondary structure prediction and protein design. *Biochemical Society Symposium*, 57: 11–24, 1990.

155. Zhao, J., Song, P. M., Fang, Q., and Luo, J. H. Protein secondary structure prediction using dynamic programming. *Acta Biochimica et Biophysica Sinica (Shanghai)*, 37(3): 167–172, 2005.

156. Figureau, A., Soto, M. A., and Toha, J. A pentapeptide-based method for protein secondary structure prediction. *Protein Engineering*, 16(2): 103–107, 2003.

157. Chandonia, J. M. and Karplus, M. New methods for accurate prediction of protein secondary structure. *Proteins*, 35(3): 293–306, 1999.

158. Qin, S., He, Y., and Pan, X. M. Predicting protein secondary structure and solvent accessibility with an improved multiple linear regression method. *Proteins*, 61(3): 473–480, 2005.

159. Adamczak, R., Porollo, A., and Meller, J. Accurate prediction of solvent accessibility using neural networks-based regression. *Proteins*, 56(4): 753–767, 2004.

160. Yuan, Z., Burrage, K., and Mattick, J. S. Prediction of protein solvent accessibility using support vector machines. *Proteins*, 48(3): 566–570, 2002.

161. Garg, A., Kaur, H., and Raghava, G. P. Real-value prediction of solvent accessibility in proteins using multiple sequence alignment and secondary structure. *Proteins*, 61(2): 318–324, 2005.

162. Naderi-Manesh, H., Sadeghi, M., Arab, S., and Moosavi Movahedi, A. A. Prediction of protein surface accessibility with information theory. *Proteins*, 42(4): 452–459, 2001.

163. Li, X. and Pan, X. M. New method for accurate prediction of solvent accessibility from protein sequence. *Proteins*, 42(1): 1–5, 2001.

164. Thompson, M. J. and Goldstein, R. A. Predicting solvent accessibility: higher accuracy using Bayesian statistics and optimized residue substitution classes. *Proteins*, 25(1): 38–47, 1996.

165. Gianese, G., Bossa, F., and Pascarella, S. Improvement in prediction of solvent accessibility by probability profiles. *Protein Engineering*, 16(12): 987–992, 2003.

166. Pollastri, G., Przybylski, D., Rost, B., and Baldi, P. Improving the prediction of protein secondary structure in three and eight classes using recurrent neural networks and profiles. *Proteins*, 47: 228–235, 2002.

167. Carugo, O. Predicting residue solvent accessibility from protein sequence by considering the sequence environment. *Protein Engineering*, 13(9): 607–609, 2000.

168. Wagner, M., Adamczak, R., Porollo, A., and Meller, J. Linear regression models for solvent accessibility prediction in proteins. *Journal of Computational Biology*, 12(3): 355–369, 2005.

169. Nguyen, M. N. and Rajapakse, J. C. Prediction of protein relative solvent accessibility with a two-stage SVM approach. Proteins, 59(1): 30–37, 2005.

170. Nguyen, M. N. and Rajapakse, J. C. Two-stage support vector regression approach for predicting accessible surface areas of amino acids. *Proteins*, 63(3): 542–550, 2006.

171. Terribilini, M., Lee, J. H., Yan, C., Jernigan, R. L., Honavar, V., and Dobbs, D. Prediction of RNA binding sites in proteins from amino acid sequence. *RNA*, 12(8): 1450–1462, 2006.

172. Gregory, D. S., Martin, A. C., Cheetham, J. C., and Rees, A. R. The prediction and characterization of metal binding sites in proteins. *Protein Engineering*, 6(1): 29–35, 1993.

173. Lin, C. T., Lin, K. L., Yang, C. H., Chung, I. F., Huang, C. D., and Yang, Y. S. Protein metal binding residue prediction based on neural networks. *International Journal of Neural System*, 15(1–2): 71–84, 2005.

174. Sodhi, J. S., Bryson, K., McGuffin, L. J., Ward, J. J., Wernisch, L., and Jones, D. T. Predicting metal-binding site residues in low-resolution structural models. *Journal of Molecular Biology*, 342(1): 307–320, 2004.

175. Schymkowitz, J. W., Rousseau, F., Martins, I. C., Ferkinghoff-Borg, J., Stricher, F., and Serrano, L. Prediction of water and metal binding sites and their affinities by using the Fold-X force field. *Proceedings of the National Academy of Sciences of the United States of America*, 102(29): 10147–10152, 2005.

176. Przybylski, D. and Rost, B. Alignments grow, secondary structure prediction improves. *Proteins*, 47(2): 228–235, 2002.

177. Ofran, Y. and Rost, B. Predicted protein–protein interaction sites from local sequence information. *FEBS Letters*, 544(1–3): 236–239, 2003.

178. Bowie J. U., Luthy R., and Eisenberg D. A method to identify protein sequences that fold into a known three-dimensional structure. *Science*, 253 (5016): 164–170, 1991.

179. Trapane, T. L. and Lattman E. E., Seventh Meeting on the Critical assessment of techniques for protein structure prediction, CASP7. *Proceedings, Proteins: Structure, Function, and Bioinformatics*, 69(S8), 1–2, 2007.

8

CONSENSUS APPROACHES TO PROTEIN STRUCTURE PREDICTION

Dongbo Bu, ShuaiCheng Li, Xin Gao, Libo Yu, Jinbo Xu, and Ming Li

8.1 INTRODUCTION

Deciphering the three-dimensional (3D) structure of a protein has been a fundamental challenge in the area of molecular biology. As genome sequencing projects proceed, thousands to millions of new genes and proteins have been sequenced at a prodigious rate. To understand the functions of these new proteins, deciphering their 3D structures is the first and key step. However, experimental techniques alone, such as X-ray crystallography and nuclear magnetic resonance spectroscopy (NMR), cannot keep up with the protein-sequencing rate, since these techniques are usually costly and time consuming. On the other hand, it is widely accepted for the structure of a protein to be almost uniquely encoded by the sequence [1]. These facts provide both motivation and possibility for computational methods to predict the structure of a protein from its sequence.

Computational methods to predict protein structures have been studied for decades with significant advancement achieved. With the enlargement of protein databases, a variety of nontrivial algorithms have been developed and have been shown to be experimentally accurate. More importantly, some communitywide competitions, such as critical assessment of structure prediction (CASP) [2–5] and LiveBench [6–8], were organized as fair and objective platforms to test and compare different prediction techniques. CASP tests are totally blind; that is, predictors are given dozens of

Machine Learning in Bioinformatics. Edited by Yan-Qing Zhang and Jagath C. Rajapakse
Copyright © 2009 by John Wiley & Sons, Inc.

sequences whose structures are still being determined or are unpublished, and prediction results are evaluated by comparing against the experimental structures. It has been shown that many prediction results in CASP are accurate enough to make rough functional inference, and others can also bring valuable insight into further studies.

Generally speaking, the traditional prediction algorithms can be categorized into three classes: homology modeling, threading, and *ab initio* [9]. The idea of homology modeling is stems from the observation that evolutionary-related proteins tend to share similar structures. Specifically, homology modeling methods attempt to detect evolutionary relationship by aligning the target protein sequence against the proteins in a database, say PDB [10], without using their structure information, and then derive coordinates of each atom based on this sequence–sequence alignment. These methods, however, work well only for proteins with a high sequence identity with some proteins in the database, which largely restricts the applicability of these methods. In contrast, a threading method evaluates how well the target sequence fits in a known structure by sequence–structure alignment. Using structure information in addition to sequence information, threading technique applies to proteins with low sequence identity. Independent of a structure database, an *ab initio* method tries to construct a protein structure to minimize an energy function, which estimates the conformational energy of a protein. The advantage of the *ab initio* method is that the possible structure space is not confined to the known structures in a structure database. Theoretically, any structural conformations can be generated and tested in an *ab initio* method. There are a large number of hybrid methods combining these three kinds, making the boundary of these three classes more and more blurred.

Despite significant progresses, protein structure prediction methods still have a number of limitations. First, each prediction method has its own strengths and limitations; therefore, no single method can reliably construct the best models for all targets. In practice, it has been noticed that a particular prediction method may favor some targets, but performs badly on others. Second, no prediction method can reliably distinguish weak hits from wrong hits, especially in the case of remote or only partial structural similarities. Consequently, although the correct model may be reported within top candidates by a prediction method, the correct model is assigned with a nonsignificant confidence score and is thus buried among false positives [11].

To overcome these shortcomings, a natural idea is integrating the strengths of different methods to obtain more accurate structures, that is, combining some weaker predictors into a stronger one. This consensus idea was first explored in early individual prediction methods, such as INBGU [12, 13] and 3D-PSSM [14]. Then, many consensus predictors (also known as meta servers) using results reported by some individual servers were developed, such as 3D-Jury [15], Pcons [3.2], Pmodeller [16], 3D-SHOTGUN [17, 18], ACE [19], and others. As suggested by recent CASP competitions, the consensus-based prediction strategies usually outperform others by generating better results.

This chapter focuses on the consensus methods for protein structure prediction. We will first introduce the fundamentals of consensus prediction problem and illustrate some popular consensus methods. These methods do not explicitly take into consideration the correlation relationship among individual prediction approaches, which sometimes make a native-like model receive less supports than some invalid ones. Here, we describe our program ACE2 as an attempt to reduce the effect of correlation among individual prediction methods. Finally, we will conclude this chapter with a prospect of the consensus methods.

8.2 FUNDAMENTALS OF CONSENSUS PREDICTION PROBLEM

8.2.1 Consensus Prediction Problem

The basic idea of consensus prediction is to combine predictions from a set of weaker methods into a more accurate one. Besides simply selecting the most reliable structure from the prediction structure set, a consensus server may also adopt a hybrid strategy since a prediction structure sometimes is only partially correct. In the hybrid strategy, a new structure is constructed by first assembling the most conserved and reliable regions from the input structures, and then refining this roughly assembled structure to a more accurate one. The consensus prediction problem can be described as follows:

Consensus Prediction Problem. *Given a query protein sequence T, each individual prediction server $S_i(1 \leq i \leq n)$ generates a set of initial model $M_i = \{m_{i,1}, m_{i,2}, \ldots, m_{i,n_i}\}$, the goal is to simply select an initial model $m_{i,j}$ or construct a new model based on predictions $\bigcup_i^n M_i$.*

Throughout this chapter, a model refers to a protein structure generated by a prediction server. In contrast to human experts, a *server* refers to an automated system that predicts a set of structures for a given protein sequence, known as the *target*.

Most consensus prediction methods adopt a three-step schema, that is, structure comparison, feature extraction, and model selection. In the structure comparison step, similarities between all pairs of the initial models are measured. Then, for each initial model, a set of variables is calculated as its features. Finally, an initial model or parts of it are selected to generate the final predictions. Usually, an additional *assessment step* is executed to evaluate the quality of these final predictions before reporting them.

8.2.2 Protein Structure Comparison

Typically, a consensus method identifies the most reliable model or regions based on the mutual structural similarity among the input models. In general, there are four types of methods to compare protein structures: global, correspondence dependent, correspondence independent, and template-based [20, 21]. A global comparison method considers all residue pairs in the two models to be compared. The most common global metric is the root-mean-square deviation (RMSD) between corresponding residues after the optimal superposition of one model onto the other. Since all residue pairs are weighted equally, a small local structural deviation may generate a large RMSD, though the global topology structures are similar. In addition, RMSD is a measurement that is bias to protein size [22].

To overcome the shortcomings of RMSD, a number of methods such as GDT [23], MaxSub [24], LG score [25], and TM score [22] have been proposed to detect the most similar substructure. Most of these methods adopt a heuristic strategy to find the substructure to optimize a similarity measure. For example, GDT tries to detect the largest set of residue pairs with a deviation less than a cutoff. MaxSub attempts to maximize the Levitt–Gerstein score [25], while LG score attempts to minimize the p-value for the distribution of S_{etr} [21]. By introducing a size-dependent scalar into the similarity measure, the average TM score of random structure pairs has no bias to protein size [22, 26].

If the residue correspondence information is unavailable *a priori*, the optimal alignment should be identified first. DALI [27] and CE citeCE identify this alignment based on the intrastructural residue–residue distance (dRMS), while structural [28] and TM align [26] are based on the interstructural residue–residue distance (cRMS). An advantage of the dRMS distance is that the need to find the optimal rigid transformation is bypassed.

The calculated structure similarities, along with the original scores assigned by the individual methods, provide each model with a set of features for the subsequent training and judging processes.

8.2.3 Protein Structure Assessment

Before returning a final prediction, it is beneficial to evaluate the overall and local quality of this prediction. Besides providing confidence measures for a whole model or its parts, the knowledge of local quality can be utilized to combine the reliable regions from different models into a hybrid model.

A variety of knowledge-based methods have been designed to assess the quality of a protein structure. Verify3D [29] derives for each residue the preference for environments, including solvent area, fraction of polar area, and local secondary structure. An unfavorable 3D-1D profile indicates the incompatibility of the sequence with the structure. Errat [30] analyzes the patterns of the nonbonded atomic interaction among nitrogen (N), carbon (C), and oxygen (O) atoms, and uses a six-dimensional normal distribution to distinguish an incorrect model from correct ones. Based on Boltzmann's principle, ProsaII [31] applies the mean field to model the energetic feature of intramolecular pairwise interaction as a function of the distance of two atoms. Then, the total pairwise interaction energy is adopted to distinguish the native-like models from incorrect ones. ProQ [16] adopts the neural network technique to predict the quality of a protein structure described by MaxSub or LG score. The structural features used in ProQ include atom–atom contacts, residue–residue contacts, surface area exposure, secondary structure agreement, globular shape, and other factors. Since it has been reported that a region with a high sequence alignment score is more likely to be correct than the regions with low scores, the sequence similarity is added to the feature set in ProQProf [32]. In FRST [33], the torsion angle potential is suggested to be a feature with the strongest correlation with model quality. In addition, WHAT_CHECK [34] and PROCHECK [35] assess the quality of a protein structure based on the stereochemistry information of this structure.

8.3 CLASSICAL ALGORITHMS FOR CONSENSUS PREDICTION

To select the best model or a part of a model from a set of candidates, many machine learning techniques have been introduced and impressive results have been achieved. Here, we review some representative methods.

8.3.1 3D-SHOTGUN: Majority-Voting Method

The 3D-SHOTGUN [17] method is reminiscent of the so-called "cooperative algorithms" in the area of computer vision. In 3D-SHOTGUN method, each initial model $m_{i,j}$ is compared with others to identify the regions of significant structural similarity. Those similar models are then superimposed upon $m_{i,j}$, thereby generating a multiple structural alignment. The "majority-voting" schema is employed to build the assembled model. That is, the coordinates of the *ith* residue in the assembled model are copied from the model having the highest frequency of occurrence in a superimposed model (see Fig. 8.1). The rationale of this strategy is that the recurring structural regions are most likely to be correct since there are more ways to be wrong than to be right.

In the feature extraction step, each assembled model $m'_{i,j}$ is assigned with a score $s'_{i,j} = \sum_{k,l} s_{k,l} \times \mathrm{sim}(m'_{i,j}, m_{k,l})$, where k and l run over all models, $s_{i,j}$ is the original score given by server S_i for model $m_{i,j}$, and $\mathrm{sim}(m'_{i,j}, m_{k,l})$ is the MaxSub similarity score. The basic idea of this step is to assign a hybrid model with a score that reflects the recurrent structural features detected among the various models.

After the assembled model set is built, a confidence score is assigned to each assembled model. Then the assembled model with the highest confidence score is selected. Unfortunately, this procedure could sometimes generate nonprotein-like models; thus, an additional refinement step is applied to eliminate the potential collisions. It is believed that the application of assembly step and refinement step makes 3D-SHOTGUN more sensitive and specific than the other consensus servers. 3D-SHOTGUN performed well in both CASP and CAFASP.

Native structure:	A	B	C	D	E	F	G	(unknown at assembly)
Initial model M_3:	A	b	C	d	E	F	-	(four native-like features)
Initial model M_2:	A	B	C	d	-	f	g	(three native-like features)
Initial model M_3:	A	b	c	D	E	-	g'	(three native-like features)
Hybrid assembly for M_1:	A_1	B_2	c_3	D_3	E_3	F_1	g'_3	(five native-like features)

Figure 8.1 Majority-voting strategy in 3D-SHOTGUN assembly process [17]. Structural features are indicated by letters (A–G). Native-like structural features are shown in uppercase. The dashes represent missing features such as gaps, and so on. The assembly for model M_1 is illustrated.

8.3.2 CAFASP-CONSENSUS, LIBELULIA and Pcons: Neural Network Methods

CAFASP-CONSENSUS [36] is the first consensus server based on the predictions reported by other individual servers. In CAFASP-CONSENSUS, the first step is to detect the similarity between predictions by checking the SCOP-ID [37] of each prediction or running a structural comparison program. Next, the related predictions are identified by counting the number of the similar models. Finally, a neural network is used to select a model based on the assigned score and the number of related models. The CAFASP-CONSENSUS group was ranked top in CAFASP2, and its success motivated the development of other consensus methods.

Pcons, the first automated consensus server, includes six individual servers (*PDBBlast, FFAS, Inbgu, GenTHREADER, Sam-T98, and 3D-PSSM*) as input. For each individual server, the top 10 hits are collected and converted into PDB models. Pcons uses LG score2 [25] to measure the similarity between a pair of models, and then calculates the following features for each model:

$$f_1 = \frac{\sum_{j=1}^{N_s} \sum_{k=1}^{N_m} \delta(i,j,k)}{N_s * N_m},$$

$$f_2 = \frac{\sum_{j=1}^{N_s} \sum_{k=1}^{N_m} \gamma(j,k) * \delta(i,j,k)}{\sum_{j=1}^{N_s} \sum_{k=1}^{N_m} \gamma(j,k)},$$

$$f_3 = \frac{\sum_{j=1}^{N_s} \delta(i,j,1)}{N_s},$$

where N_s is the number of servers, N_m is the number of models taken from a particular server, $\delta(i,j,k)$ is 1 if model i is similar to the kth model from server j and 0 otherwise, and $\gamma(i,k)$ is 1 if the kth model from server j has a score higher than a cutoff and 0 otherwise.

In brief, f_1 represents the fraction of other models having significant similarity with this model, f_2 denotes the fraction of other models similar to this model weighted by a significant method score, and f_3 is the fraction of the first-ranked models. Three additional features are calculated for the templates in the same manner. These six features, along with the scores given by each individual server, are fed into a neutral network (see Fig. 8.2). Here, the original scores are translated into a uniform score by the neutral network. It is observed that training to predict log(LGScore2) generates significantly better results than training on LGScore2 directly.

Pmodeller [16], the combination of Pcons and ProQ, assigns each model a score by a simply linear combination of the ProQ and Pcons scores as follows:

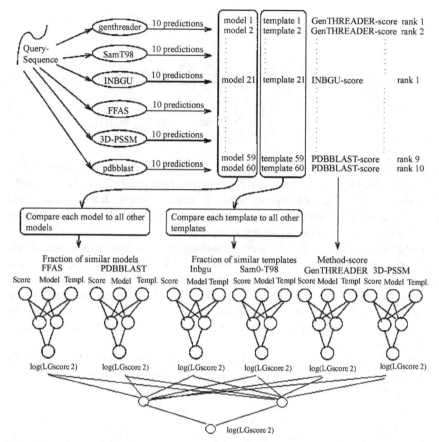

Figure 8.2   The two-layer neutral network used in Pcons. The first layer consists of only one network for each method, and its output are fed into the second layer for a normalization.

PmoddllerScore = 0.17 ProQScore + 0.85 PconsScore Pmodeller has an advantage over Pcons because a number of high-scoring false-positive models can be eliminated, resulting in a higher specificity.

LIBELLULA [38] also adopts neural network technique to evaluate fold recognition results. Besides the quality of sequence–structure alignment and compactness of the resulting models, LIBELLULA takes into account the sequence features, including the distribution of sequence-conserved positions and apolar residues.

8.3.3 ACE: An SVM Regression Method

Alignment by consensus estimator (ACE) [19] is a selection-only meta server. It can select the best model from the input set with high sensitivity and specificity by introducing three effective features and utilizing a support vector machine (SVM)-regression-based selection method.

For each model $m_{i,j}$ in the input set, the similarities with other models are calculated with MaxSub, and then two features are calculated as follows:

$$f_1(m_{i,j}) = \frac{1}{(n-1)n_i} \sum_{k=1, k\neq i}^{n} \sum_{l=1}^{n} n_i \text{sim}(m_{i,j}, m_{k,l}),$$

$$f_2(m_{i,j}) = \frac{1}{n-1} \sum_{k=1, k\neq i}^{n} \max_{l=1}^{n_i} \{\text{sim}(m_{i,j}, m_{k,l})\}.$$

Briefly, $f_1(m_{i,j})$ represents the normalized similarity with the models generated by all the servers excluding server S_i, and f_2 represents the normalized similarity with the most similar model generated by each server. In addition, another feature is calculated for each target rather than each model, which represents the prediction divergence of the individual servers for this particular target protein.

In the training step, the SVM regression technique is introduced to approximate the functional relationship between the features of a model and its structural quality. Here, the structural quality of a model is calculated by a variety of structural parameters generated by PROCHECK [35], WHAT_IF [34], and other methods. Having fixed the weights of the three features, ACE applies the combined features to evaluate the structural quality of each input model, and chooses the model with the highest score as the final prediction.

In CASP6, ACE was ranked second among 87 automatic servers, whereas Robetta ranked number first. In the case that the models predicted by the individual servers share structural similarity to each other, ACE did quite well to recognize the best prediction from the input set [5, 19].

8.3.4 Other Techniques

3D-Jury [15] is also a selection-only server. In essence, it follows an approach that performs extremely well in *ab initio* field. The experience with *ab initio* methods suggests that compared with the conformation with the lowest energy, the averages of low-energy conformations appearing most frequently are more closer to the native structure. Therefore, 3D-Jury employs the following two scoring functions to mimic this averaging step:

$$\text{3D} - \text{Jury} - \text{all}(m_{i,j}) = \frac{\displaystyle\sum_{k}^{n} \sum_{l, k\neq i \text{ or } l\neq j}^{n_k} \text{sim}(m_{i,j}, m_{k,l})}{1 + \displaystyle\sum_{k}^{n} n_k}$$

$$\text{3D} - \text{Jury} - \text{single}(m_{i,j}) = \frac{\displaystyle\sum_{k}^{n} \max_{l, k\neq i \text{ or } l\neq j}^{n_k} \text{sim}(m_{i,j}, m_{k,l})}{1 + n}$$

Here, $\text{sim}(m_{i,j}, m_{k,l})$ denotes the similarity between $m_{i,j}$ and $m_{k,l}$ calculated by MaxSub. The model with the highest score is chosen to be the final prediction. Though it seems very simple, this process can generate reasonably good predictions [15].

Other machine learning techniques were also explored in this field. For example, BPROMPT [39] adopts Bayesian belief network to construct a meta server for membrane protein prediction; JPred citeJPred employs the decision-tree technique to form a consensus method for protein secondary structure prediction.

Robetta [40] is a well-known *ab initio* prediction server that also uses consensus information and generates a new model by using fragment-assembling technique. However, due to the huge search space, Robetta cannot consistently generate good predictions, especially for a large target protein.

8.4 ACE2: AN ATTEMPT TO REDUCE SERVER CORRELATION

Although consensus servers assume that each individual server is independent, we have observed from CASP6 results that correlation exists between different servers to a large degree. The correlated servers sometimes make a native-like model receive less supports than some invalid ones. The correlation arises from the fact that many servers adopt similar techniques, including sequence alignment tools, secondary structure prediction methods, threading methods, and scoring functions.

This section presents a novel consensus method, implemented at ACE2, to reduce the impact resulting from server correlation, and performs experiments to justify our approach. Details of each step are described in the following subsections.

8.4.1 Preliminaries

Generally, two models are thought to be similar under a distance measure if their distance is below a threshold under this distance measure. The models that are similar to the native structure are called native-like and the others are invalid. Here we say two models are similar if their TM score is greater than 0.4 [22]. Alternative similarity measures such as GDT [23] gave similar results.

Given a target t_l, $1 \leq l \leq l$, a server s_i, $1 \leq i \leq u$, outputs a set of models $M_{i,l} = \{m_{i,l,q} | 1 \leq q \leq n_{i,l}\}$ as candidate structures, where $n_{i,l}$ is the number of models produced by server s_i for target t_l. Sometimes, the initial scores, which are given to model m in $M_{i,l}$ by s_i, are available. The set of models for target t_l is denoted as $M_l = \bigcup_i M_{i,l}$. A consensus server attempts to pick a model m among these candidates such that m is a structure closest to the native structure.

This section is based on the following two assumptions:

- Server s_i generates its predictions based on a confidence measure. That is, for each model $m \in M_l$, s_i has a confidence $S_{i,m,l}$.

  Usually, $S_{i,m,l}$ is approximated from the initial score given by s_i, or the similarities among different models in M_l. The underlying rationale is based on the structural clustering property, that is, among these predictions, it can be expected that the invalid models have random conformation, while the native-like models occur with higher frequency. Since the initial score sometimes is unavailable and should endure a complicated normalization first, we simply adopt the following feature as an approximation of the confidence $S_{i,m,l}$:

$$f(i,m) = \frac{1}{u-1} \sum_{j=1, j\neq i}^{u} \max_{m' \in M_{j,l}} \mathrm{sim}(m, m').$$
(8.1)

- There are some implicit latent independent servers h_j, $1 \leq j \leq v$, dominating the explicit servers s_i. Given a target t_l, h_j assigns a value $H_{j,m,l}$, $m \in M_l$, as the confidence that m is a native-like structure.

Identifying the latent independent servers is essential to reduce the negative effects of server correlations and to reduce the dimensionality of the search space, as the number of latent servers is expected to be smaller than the number of original servers. After deriving the latent servers, we can design a new and more accurate prediction server s^* by an optimal linear combination of the latent servers, which for each target t_l assigns a confidence score to each model $m \in M_l$ as follows:

$$S^*_{l,m} = \sum_{j=1}^{v} \lambda_j^* H_{j,m,l},$$
(8.2)

where λ_j^* is the weight of latent server h_j.

8.5   THE METHOD

The basic idea of our method is to reduce the negative effects caused by the correlations among prediction servers. We first employ the maximum likelihood technique to estimate the server correlations, then adopt the factor analysis technique to uncover the latent servers, and finally design a mixed ILP method to derive the optimal weights for the latent independent servers.

8.5.1   Maximum Likelihood Estimation of Server Correlations

Let the overlap set between $M_{i,l}$ and $M_{j,l}$ for a distance function dist and a threshold θ be

$$O_{i,j,l} = \{m_{i,l,q} \in M_{i,l} | \exists m_{j,l,q'} \in M_{j,l} \wedge \mathrm{dist}(m_{i,l,q}, m_{j,l,q'}) \leq \theta\}.$$

Let $o_{i,j,l} = |O_{i,j,l}|$.

For a given target, let $p_{i,j}$ be the probability of a model returned by s_i similar to that returned by server s_j. Under a reasonable assumption that targets t_l, $1 \leq l \leq \ell$ are mutually independent, the likelihood that server s_i, $1 \leq i \leq u$, generates model

$m_{i,l,q}$, $1 \le q \le n_{i,l}$, as prediction is

$$L(p_{i,j}) = \prod_{l=1}^{\ell} \binom{n_{i,l}}{o_{i,j,l}} p_{i,j}^{o_{i,j,l}} (1 - p_{i,j})^{n_{i,l} - o_{i,j,l}}.$$

Therefore, the maximum likelihood estimation of $p_{i,j}$ can be calculated as follows:

$$p_{i,j} = \frac{\sum_{l=1}^{\ell} o_{i,j,l}}{\sum_{l=1}^{\ell} n_{i,l}}. \tag{8.3}$$

In the rest of this chapter, we use P to denote the matrix $P = [p_{i,j}]_{u \times u}$.

8.5.2 Uncovering the Latent Servers

For a target t_l, let $S_{i,m,l}$ and $H_{j,m,l}$ be the confidence that model m has chosen as one of the prediction results by server s_i and h_j, respectively. Since the latent servers are mutually independent, it is reasonable to assume that $S_{i,k,l}$ is a linear combination of $H_{j,k,l}$, $1 \le j \le v$:

$$\overrightarrow{S_{i,l}} = \sum_{j=1}^{m} \lambda_{i,j} \overrightarrow{H_{j,l}}, \qquad \sum_{j=1}^{v} \lambda_{i,j} = 1, \qquad 1 \le i \le u,$$

where $\overrightarrow{S_{i,l}} = \langle S_{i,1,l}, S_{i,2,l}, \ldots, S_{i,|M_l|,l} \rangle$, $1 \le i \le u$, and $\overrightarrow{H_{j,l}} = \langle H_{j,1,l}, H_{j,2,l}, \ldots, H_{j,|M_l|,l} \rangle$, $1 \le j \le v$. Here, $\lambda_{i,j}$ is the weight, and a larger $\lambda_{i,j}$ implies a higher chance that server s_i adopts models reported by h_j.

From the correlation matrix of prediction servers s_i, factor analysis technique is employed to derive $\lambda_{i,j}$ and $\overrightarrow{H_{j,l}}$, that is, $\overrightarrow{H_{j,l}}$ can be represented to be a linear combination of $\rightarrow S_i$ as follows:

$$\overrightarrow{H_{j,l}} = \sum_{i=1}^{u} \omega_{j,i} \overrightarrow{S_{j,i,l}}, \qquad 1 \le j \le v, \qquad 1 \le l \le \ell, \tag{8.4}$$

where $\langle \omega_{j,1}, \omega_{j,2}, \cdots, \omega_{j,n} \rangle$ is an eigenvector of $P^T P$.

8.5.3 ILP Model to Weigh Latent Servers

Having derived the latent servers $h_j (1 \le j \le v)$, we can design a new and more accurate prediction server s^* as an optimal linear combination of the latent servers. For each target t_l, it assigns each predicted model $m \in M_l$ with a score as given in Equation 8.2.

To determine a reasonable setting of coefficient λ_k^*, a training process is conducted on a training data set $D = \{ < t_l, M_l^+, M_l^- >, 1 \le l \le |D| \}$, where $t_l \in T$ is a target, $M_l^+ \subseteq M_l$ denotes the set consisting of native-like models, and $M_l^- \subseteq M_l$ denotes the invalid model set. The learning process attempts to maximize the gap between scores of the native-like models and invalid ones.

More specifically, for each target t_l in the training data set, a score is assigned for each predicted model by S^*. A reasonable setting of these coefficients should assign the native-like models higher scores than those of the invalid ones. The larger the gap between the scores of native-like models and invalid models, the more the robust is this new prediction approach. In practice, "soft margin" idea is adopted to take outliers into account, that is, we try to maximize the margin while allowing errors on some samples. Formally, the learning techniques can be formulated into an ILP problem as follows:

$$\max \sum_{l=1}^{|D|} z_l. \tag{8.5}$$

$$s.t. \quad \sum_{j=1}^{v} \lambda_j^* H_{j,p,l} - \sum_{j=1}^{v} \lambda_j^* H_{j,q,l}^* \geq x_{p,q}, \qquad p \in M_l^+, q \in M_l^-, 1 \leq l \leq |D|. \tag{8.6}$$

$$\sum_{q \in M_l^-} (x_{p,q} - \epsilon)/\epsilon |M_l^-| \geq y_{p,l}, \qquad p \in M_l^+, 1 \leq l \leq |D|. \tag{8.7}$$

$$\sum_{p \in M_l^+} (1 + y_{p,l}) \geq z_l, \qquad 1 \leq l \leq |D|. \tag{8.8}$$

$$\sum_{j=1}^{v} \lambda_j^* = 1, \lambda_j \geq 0, \qquad 1 \leq j \leq v. \tag{8.9}$$

$$x_{p,q} \in \{\epsilon, -1\} \; y_{p,l} \in \{0, -1\}, \qquad z_l \in \{0, 1\}, \tag{8.10}$$

where ϵ is a parameter used as the lower bound of gap between the score of native-like structures and invalid ones.

For Equation 8.6, it is easy to see that $\sum_{j=1}^{v} \lambda_j^* H_{j,p,l} - \sum_{j=1}^{v} \lambda_j^* H_{j,q,l}^* \geq -1$. If $x_{p,q} = \epsilon$, it means that the new server s^* gives the native-like structure p a score higher than that for an invalid model q. For Equation 8.7, it is clear that

$$-1 \leq \sum_{q \in M_l^-} (x_{p,q} - \epsilon)/\epsilon |M_l^-| \leq 0$$

always holds. If $y_{p,l} = 0$, then s^* assigns a score for native-like model p higher than any invalid models for target t_l. If s^* assigns at least one of the native-like models a score higher than that for those invalid ones, then z_l will take 1 by Equation 8.8. The objective of this ILP model is to maximize the number of the targets, which has at least one native-like structure with a score higher than all the invalid models. Equation 8.9 normalizes the coefficient settings.

8.5.4 A New Prediction Server

Now, we wrap up everything to obtain a new prediction server. Given a target t^* for prediction, each server s_i produces a set of models M_i^*. The set of candidate structures is denoted as $M^* = \bigcup_i M_i^*$. For each model $m \in M^*$, a confidence can be calculated with Equation 8.1. The latent probability $H_{j,m}^* = \sum_{i=1}^{u} \omega_{j,i} S_{i,m}^*$, $1 \leq j \leq v$, is derived from Equation 8.4. Finally the consensus server produces a score for each model based on Equation 8.2, and picks the top-scored one as the final prediction.

8.6   EXPERIMENT RESULTS

8.6.1   Data Set

The biennial CASP competition provides a great opportunity to benchmark the automated structure prediction servers in the community. Unlike LiveBench tests, the CASP tests are totally blind. The native 3D structures of the CASP targets are not published before the competition. In CASP6, 87 targets, T0196–T0282, were released. Sixty-four of them are effective after some cancellations. CASP7 released 100 effective targets, T0283–T0386, for prediction. Until October 2006, 77 experimentally solved structures of these targets were published.

In this chapter, we use CASP6 data to train the ILP model and the CASP7 targets with published native structure to test the performance of our new consensus algorithm.

8.6.2   CASP6 Evaluation of ACE

ACE performed well in CASP6, ranking second in the server group after Robetta. However, some drawbacks were also observed about ACE: if the native-like predictions generated by different servers share significant structure similarity, ACE returns a good prediction; if a native-like model receives less support from individual servers than an invalid one, ACE may produce a very bad prediction [19]. To improve ACE, we first investigate this phenomena in more detail.

Table 8.1 provides some details of ACE's individual servers on a subset of CASP6 targets. In this table, if a native-like model is selected, then label 1 is filled in, otherwise

Table 8.1   Evaluation of ACE on CASP6 data set

Server	MGTH	RAPT	FUG3	PROS	ST02	SPKS	ACE
T0203	1	1	1	0	1	1	1
T0206	0	0	0	0	1	1	1
T0208	1	1	1	1	0	1	1
T0211	0	1	1	0	1	1	1
T0213	1	0	0	0	0	0	0
T0223	0	1	0	0	0	1	1
T0226	0	1	0	0	1	1	1
T0230	0	1	0	0	0	1	0
T0232	1	1	1	0	1	1	1
T0235	1	0	1	0	1	1	1
T0240	1	1	1	0	0	1	0
T0243	1	1	1	0	1	1	1
T0271	0	1	1	1	1	1	1
T0281	0	0	0	0	0	1	0
T0282	1	1	1	1	1	1	1

we set the table value to be 0. As suggested by Table 8.1, in the case that most individual servers generate native-like models, ACE usually succeeds; however, for targets T0213, T0230, and T0281, ACE fails to choose the native-like models since only one or two servers voted the optimal models. This observation implies that to improve performance, we should correctly identify the native-like model even though it receives less support than others.

8.6.3 Server Correlations and Hidden Servers

In this experiment, we investigated the correlations among six individual servers, including *mGenThreader*(MGTH), *RAPTOR*(RAPT), *FUGUE3*(FUG3), *Prospect* (PROS), *SAM-T02*(ST02), and *SparkSP3*(SPKS), and studied the relationship between the individual servers and the latent ones.

On CASP6 data set, the correlations among these individual servers are calculated according to Equation 8.3 and listed in Table 8.2. The correlation matrix is asymmetric since $o_{i,j,l}$ is not necessarily equal to $o_{i,j,l}$. As suggested by Table 8.2, most of the correlation coefficients are in the range from 0.1 to 0.4, which is prominent with respect to that there are a large number of the candidate models for each server to choose. Furthermore, some servers are correlated more tightly than others. For example, the correlation coefficient between *mGenThreader* and *RAPTOR* (0.383) is nearly twice of that between *FUGUE3* and *Prospect* (0.182). This observation suggests that these individual servers may be clustered into cliques according to their correlation factors. The servers in a small clique are underestimated according to the simple "majority voting" rule.

We then explored the relationships between the latent servers and the individual servers (see Table 8.3). As suggested by Table 8.3, some latent independent servers can be considered to be representative of some individual servers. More specifically, H_1 represents the common features shared by these individual servers; H_2 and H_4 can be reasonably considered to represent *FUGUE3* and *Prospect*; H_3 seems to be a reflection of *SAM-T02*; H_5 mainly reflects preference of *SparkSP3*; and H_6 represents *mGenThreader* and *Raptor*. Based on the eigenvalues, H_6 was eliminated since the other latent servers can explain most of the server correlation.

After deriving the optimal weight for the latent servers, we rewrite the new prediction server S^* to be a linear combination of the original individual

Table 8.2   Correlations among the individual servers

Server	MGTH	RAPT	FUG3	PROS	ST02	SPKS
MGTH	1.000	0.383	0.323	0.272	0.287	0.361
RAPT	0.378	1.000	0.367	0.300	0.300	0.381
FUG3	0.235	0.216	1.000	0.182	0.206	0.247
PROS	0.279	0.276	0.267	1.000	0.256	0.276
ST02	0.257	0.247	0.275	0.224	1.000	0.276
SPKS	0.296	0.284	0.294	0.210	0.246	1.000

Table 8.3   **Relationships between hidden servers and individual servers**

Hidden Server	H1	H2	H3	H4	H5	H6
MGTH	0.451	−0.076	0.275	0.259	−0.452	−0.666
RAPT	0.467	−0.064	0.268	0.149	−0.366	0.741
FUG3	0.355	−0.538	−0.192	−0.738	−0.010	−0.057
PROS	0.385	0.762	0.204	−0.422	0.222	−0.057
ST02	0.377	0.202	−0.864	0.259	−0.055	0.013
SPKS	0.402	−0.283	0.163	0.349	0.781	−0.022

servers as follows:

$$S^* = 0.138\text{MGTH} + 0.126\text{RAPT} + 0.084\text{FUG3} - 0.026\text{PROS} - 0.031\text{ST02}$$
$$+ 0.039\text{SPRK}. \tag{8.11}$$

Here, PROS and ST02 have a negative weight. This, however, does not mean that these servers are unimportant. In fact, as illustrated by Table 8.4, a model generated by ST02 was selected by S^* as the final prediction for target T0295. From the point of view of latent servers, a negative weight simply means that the corresponding server's contribution has been overexpressed by other individual servers that have correlation with it.

8.6.4 CASP7 Evaluation

We evaluated the sensitivity of our consensus method using the CASP7 targets with published native structure. For each of the individual prediction servers, *RAPTOR, FUGUE3, SAM-T02, Prospect, mGenThreader*, and *SparkSP3*, we downloaded its top 10 prediction models for each target. Each model is compared with the native structure by TM score [22].

Table 8.4 shows for each target the TM score of the best model at each server. Due to the length limit, we have listed only the targets for which at least one individual server predicted a model with TM score above 0.4. Here, a native-like model is listed in bold. As illustrated by Table 8.4, as long as any individual server generates a native-like prediction, ACE2 can identify it. Although ACE2 does not necessarily select the model with the highest TM score, it does always select a model with almost highest TM score.

More interestingly, in the case that a native-like model receives supports from only one or two individual servers, for example, for target T0347, only *SparkSP3* generates a good prediction, ACE2 still succeeds in selecting a correct model. For T0347, the prediction models with SCOP ID [37] "c.1.x.x" or "c.93.1.1" form a dominant cluster, while the native-like model, the second model reported by *SparkSP3*, has very few similar models. However, compared with these dominant models, the native-like model receives higher confidence from the servers with positive weight and lower confidence from the servers with negative weight; therefore, it was ranked top by

Table 8.4   TM score of the best model reported by each prediction server

Server	MGTH	RAPT	FUG3	ST02	SPSK	ACE2
T0289	**0.6010**	**0.6589**	**0.5723**	**0.5773**	**0.6435**	**0.6589**
T0292	**0.5024**	**0.4914**	**0.4902**	**0.4682**	**0.4913**	**0.4913**
T0295	0.1792	**0.6738**	**0.6752**	**0.6795**	**0.6774**	**0.6795**
T0298	**0.8354**	**0.8640**	**0.8818**	**0.8768**	**0.8914**	**0.8914**
T0301	**0.4579**	**0.5246**	**0.5041**	**0.4603**	**0.4791**	**0.5041**
T0302	**0.8760**	**0.8760**	**0.8779**	**0.8648**	**0.8761**	**0.8779**
T0303	**0.7935**	**0.8315**	**0.8350**	**0.8429**	**0.8379**	**0.8429**
T0315	**0.9576**	**0.9567**	**0.9585**	**0.9554**	**0.9604**	**0.9604**
T0316	**0.4080**	**0.4335**	**0.4131**	**0.4230**	**0.4356**	**0.4230**
T0317	**0.8413**	**0.8267**	**0.8339**	**0.8349**	**0.8507**	**0.8507**
T0318	**0.7248**	**0.8013**	**0.7868**	**0.7560**	**0.7944**	**0.7944**
T0322	**0.7025**	**0.7235**	**0.6809**	**0.7117**	**0.7092**	**0.7117**
T0323	**0.6895**	**0.6956**	**0.6848**	**0.6903**	**0.6964**	**0.6964**
T0324	**0.8276**	**0.8494**	**0.8579**	**0.8548**	**0.8712**	**0.8712**
T0325	**0.5434**	**0.6598**	**0.4230**	**0.5549**	**0.6523**	**0.6598**
T0326	**0.8498**	**0.8508**	**0.8541**	**0.8483**	**0.8498**	**0.8541**
T0327	**0.6089**	**0.6234**	**0.6247**	**0.4900**	**0.6255**	**0.6255**
T0328	**0.9274**	**0.8815**	**0.8972**	**0.9279**	**0.8884**	**0.9279**
T0329	**0.7468**	**0.8049**	**0.8055**	**0.7906**	**0.8033**	**0.8033**
T0330	**0.7332**	**0.7185**	**0.6925**	**0.7055**	**0.7033**	**0.7055**
T0331	**0.6617**	**0.7036**	**0.6321**	**0.5346**	**0.6730**	**0.6730**
T0335	**0.5849**	0.3742	**0.4185**	**0.4190**	**0.4662**	**0.4662**
T0341	**0.8137**	**0.8422**	**0.7625**	**0.8166**	**0.8480**	**0.8480**
T0342	**0.5828**	**0.5851**	**0.5947**	**0.5499**	**0.5907**	**0.5947**
T0345	**0.4354**	**0.4344**	**0.4344**	**0.4337**	**0.4332**	**0.4354**
T0347	0.1974	0.2627	0.1979	0.1523	**0.4207**	**0.4207**
T0357	0.3266	**0.4916**	**0.4186**	0.3447	**0.4659**	**0.4916**
T0362	**0.8241**	**0.8134**	**0.8307**	**0.7873**	**0.8494**	**0.8494**
T0363	0.3654	**0.4679**	0.3246	**0.5125**	**0.4680**	**0.5125**
T0364	**0.7414**	**0.7645**	**0.7264**	**0.7459**	**0.7465**	**0.7459**
T0367	**0.7082**	**0.8106**	**0.7523**	**0.5436**	**0.8383**	**0.8383**
T0369	**0.6217**	**0.6287**	**0.6155**	**0.6463**	**0.6260**	**0.6463**
T0371	**0.7469**	**0.7981**	**0.7097**	**0.7830**	**0.7733**	**0.7830**
T0375	**0.5357**	**0.5249**	**0.5273**	**0.5392**	**0.5471**	**0.5471**
T0376	**0.8222**	**0.8341**	**0.8427**	**0.6958**	**0.8517**	**0.8517**
T0380	**0.7260**	**0.8100**	**0.7802**	**0.7575**	**0.7483**	**0.7575**
T0383	0.3288	0.3957	0.2158	0.2180	**0.4258**	**0.4258**
T0384	**0.8213**	**0.8361**	**0.8346**	**0.7895**	**0.8334**	**0.8346**

Equation 8.11. Another example is T0383, which is supported by only *SparkSP3*. In this preliminary study, ACE2 has predicted successfully in the case that the "majority vote" rule fails.

Figures 8.3 and 8.4 present two good predictions made by ACE2. Figure 8.3 shows the 3D structure of T0383, for which only *SparkSP3* makes a correct prediction. In the

Figure 8.3 ACE's best prediction (left, SparkSP3's 10th model) and native structure (right) for T0383. GDT score = 0.3832, TM score = 0.4258, MaxSub score = 0.2188.

case that most of the servers output good predictions, ACE2 always chooses the best one. T0289 is such an example as shown in Fig. 8.4.

8.7 CONCLUDING REMARKS

In the field of protein folding, the state of the art is to combine the evolutionary information available from multiple sequence alignment and the structural information from the template [11]. Consensus is an effective technique to integrating the strengths of multiple prediction methods and has been proved to be powerful in CASP competitions. In fact, most of the top servers in CASP7 adopts a consensus, or a consensus–consensus [13] prediction method.

Besides the techniques introduced in this chapter, other machine learning techniques, such as SVM-Rank [41] and rank regression [42], have been proven

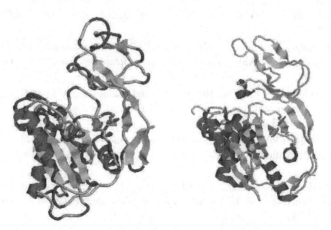

Figure 8.4 ACE's best prediction (left, RAPTOR's first model) and native structure (right) for T0289. GDT score = 0.4583, TM score = 0.6589, MaxSub-score = 0.3680.

useful in a number of areas, including information retrieval and artificial intelligence. These methods are expected to work well for protein structure prediction.

However, from our experience at CASP7 and the experiments in this chapter, it appears that the model selection-only servers that depend on pure threading servers are already at their limits. In order to further improve accuracy of protein structure computations, refinements, fragment assembly, or *ab initio* methods [40, 43] must be used.

It is also important to employ the consensus technique to predict residue contact and local structure rather than the whole model. Accurate confidences to different parts of protein models are valuable to guide the refinement process. In addition, a local quality measure can help combine the best regions from different models into a hybrid model. This strategy, in theory, can generate a model that is more accurate than all the input models.

ACKNOWLEDGMENTS

This work is supported by NSERC Grant OGP0046506 and NSF of China Grant 60496324. We would like to thank Annie Lee for her valuable discussions and suggestions.

REFERENCES

1. Braden, C. and Tooze, J. *Introduction to Protein Structure*. Garland Publishing, New York, 1999.
2. Moult, J., Hubbard, T., Fidelis, K., and Pedersen, J. Critical assessment of methods on protein structure prediction (CASP) – round III. *Proteins: Structure, Function and Genetics*, 37: 2–6, 1999.
3. Moult, J., Fidelis, K., Zemla, A., and Hubbard, T. Critical assessment of methods on protein structure prediction (CASP) - round IV. *Proteins: Structure, Function and Genetics*, 45: 2–7, 2001.
4. Moult, J., Fidelis, K., Zemla, A., and Hubbard, T. Critical assessment of methods on protein structure prediction (CASP) - round V. *Proteins: Structure, Function and Genetics*, 53: 334–339, 2003.
5. Moult, J., Fidelis, F., Rost, B., Hubbard, T., and Tramotano, A. Critical assessment of methods on protein structure prediction (CASP) - round VI. *Proteins: Structure, Function and Genetics*, 61: 3–7, 2003.
6. Bujnicki, J. M., Elofsson, A., Fischer, D., and Rychlewski, L. Livebench-2: large-scale automated evaluation of protein structure prediction servers. *Proteins*, 5: 184–191, 2001.
7. Rychlewski, L., Fischer, D., and Elofsson, A. Livebench-6: large-scale evaluation of protein structure prediction servers. *Proteins: Structure, Function and Genetics*, 53: 542–547, 2003.
8. Rychlewski, L., and Fischer, D. Livebench-8: the large-scale continuous assessment of automated protein structure prediction. *Protein Science*, 14: 240–245, 2005.

9. Baker, D. and Sali, A. Protein structure prediction and structural genomics. *Science*, 294 (5): 93–96, 2001.

10. Berman, H. M., Westbrook, J., Feng, Z., Gilliland, G., Bhat, T. N., Weissig, H., Shindyalov, I. N., and Bourne, P. E. The protein data bank. *Nucleic Acids Research*, 28: 235–242, 2000.

11. Bujnicki, J. and Fischer, D. Meta approaches to protein structure prediction. *Nucleic Acids and Molecular Biology*, 15: 23–34, 2004.

12. Fischer, D. Hybrid fold recognition: combining sequence-derived properties with evolutionary information. In: R. B. ALtman, A. K. Dunker, L. Hunter, and T. E. Klien, Eds, *Pacific Symposium on Biocomputing*, Vol. 5: World Scientific, pp. 116–127, 2000.

13. Ginalski, N. V. G., Godzik, A., and Rychlewski, L. Practical lessons from protein structure prediction. *Nucleic Acids Research*, 33(6): 1874–1891, 2005.

14. Kelley, L., MacCallum, R., and Sternberg, M. Enhanced genome annotation using structural profiles in the program 3D-PSSM. *Journal of Molecular Biology*, 299(2): 523–544, 2000.

15. Ginalski, K., Elofsson, A., Fischer, D., and Rychlewski, L. 3D-jury: a simple approach to improve protein structure predictions. *Bioinformatics*, 19(8): 1015–1018, 2003.

16. Wallner, B., Fang, H., and Elofsson, A. Automatic consensus-based fold recognition using Pcons, Proq, and Pmodeller. *Proteins: Structure, Function and Genetics*, 53: 534–541, 2003.

17. Sasson, I., and Fischer, D. Modeling three-dimensional protein structures for CASP5 using the 3D-SHOTGUN meta-predictors. *Proteins: Structure, Function and Genetics*, 53: 389–394, 2003.

18. Fischer, D. 3DS3 and 3DS5: 3D-SHOTGUN meta-predictors in CAFASP3. *Proteins: Structure, Function and Genetics*, 53: 517–523, 2003.

19. Xu, J., Yu, L., and Li, M. Consensus fold recognition by predicted model quality. *Asia-Pacific Bioinformatics Conference (APBC)*, pp. 73–83, January, 2005.

20. Cristobal, S., Zemla, A., Fischer, D., Rychlewski, L., and Elofsson, A. A study of quality measures for protein threading models. *BMC Bioinformatics*, 2(5): 1471–2105, 2001.

21. Lancia1, G. and Istrail, S. *Protein Structure Comparison: Algorithms and Applications*, Springer, Berlin/Heidelberg, 2003.

22. Zhang, Y. and Skolnick, J. Scoring function for automated assessment of protein structure template quality. *Proteins: Structure, Function, and Bioinformatics*, 57(4): 702–710, 2004.

23. Zemla, A., Venclovas, E., Moult, J., and Fidelis, K. Processing and analysis of CASP3 protein structure predictions. *Proteins: Structure, Function and Genetics*, 37: 22–29, 1999.

24. Siew, N., Elofsson, A., Rychlewski, L., and Fischer, D. Maxsub: an automated measure for the assessment of protein structure prediction quality. *Bioinformatics*, 16(9): 776–785, 2000.

25. Levitt, M. and Gerstein, M. A unified statistical framework for sequence comparison and structure. *Proceedings of the National Academy of Sciences of the United States of America*, 95: 5913–5920, 1998.

26. Zhang, Y. and Skolnick, J. TM-align: a protein structure alignment algorithm based on the TM-score. *Nucleic Acids Research*, 33(7): 2302–2309, 2005.

27. Holm, L. and Sander, C. Protein structure comparison by alignment of distance matrices. *Journal of Molecular Biology*, 233(1): 123–138, 1993.

28. Subbiah, S., Laurents, D. V., and Levitt, M. Structural similarity of DNA-binding domains of bacteriophage repressors and the globin core. *Current Biology*, 3: 141–148, 1993.

29. Luthy, R., Bowie, J. U., and Eisenberg, D. Assessment of protein models with three-dimensional profiles. *Nature*, 356(6364): 83–85, 1992.

30. Colovos, C. and Yeates, T. O. Verification of protein structures: patterns of nonbonded atomic interactions. *Protein Science*, 2: 1511–1519, 1993.

31. Sippl, M. J. Recognition of errors in three-dimensional structures of proteins. *Proteins: Structure Function and Genetics*, 17: 355–362, 1993.

32. Wallner, B. and Elofsson, A. Identification of correct regions in protein models using structural, alignment, and consensus information. *Protein Science*, 15: 900–913, 2006.

33. Silvio C. E. Tosatto. The victor/FRST function for model quality esitmation. *Journal of Computational Biology*, 12(10): 1316–1327, 2005.

34. Hooft, R. W. W., Vriend, G., Sander, C., and Abola, E. E. Errors in protein structures. *Nature*, 381: 272–272, 1996.

35. Laskowski, R. A., Macarthur, M. W., Moss, D. S., and Thornton, J. M. PROCHECK: a program to check the stereochemical quality of protein structures. *Journal of Applied Crystallography*, 26: 283–291, 1993.

36. Fischer, D., Elofsson, A., Rychlewski, L., Pazos, F., Valencia, A., Rost, B., Ortiz, A. R., and Dunbrack, R. L. CAFASP2: the second critical assessment of fully automated structure prediction methods. *Proteins: Structure, Function and Genetics*, 1: 171–183, 2001.

37. Murzin, A. G., Brenner, S. E., Hubbard, T., and Chothia, C. SCOP: a structural classification of proteins database for the investigation of sequences and structures. *Journal of Molecular Biology*, 247: 536–540, 1995.

38. Juan, D., Grana, O., Pazos, F., Fariselli, P., Casadio, R., and Valencia, A. A neural network approach to evaluate fold recognition results. *Proteins: Structure, Function and Genetics*, 50(4): 600–608, 2003.

39. Taylor, P. D., Attwood, T. K., and Flower, D. R. *Nucleic Acids Research*, 31(13): 3698–3700, 2003.

40. Simons, K. T., Strauss, C., and Baker, D. Prospects for ab initio protein structural genomics. *Journal of Molecule Biology*, 306: 1191–1199, 2001.

41. Yunbo, C. A. O., Jun, X. U., Tie-Yan, L. I. U., Hang, L. I., Yalou, H., and Hsiao-Wuen, H. Adapting ranking SVM to document retrieval. In: E. N. Efthimiadis, S. T. Dumais, D. Hawking, and Jarvelin, K.Eds, *SIGIR 2006: Proceedings of the 29th Annual International ACM SIGIR*, 2006.

42. Bura, E., and Cook, R. D. Rank estimation in reduced-rank regression. *Journal of Multivariate Analysis*, 87(1): 159–176, 2003.

43. Zhang, Y., and Skolnick, J. Automated structure prediction of weakly homologous proteins on a genomic scale. *Proceedings of the National Academy of Sciences of the United States of America*, 101(20): 7594–7599, 2004.

9

KERNEL METHODS IN PROTEIN STRUCTURE PREDICTION

Jayavardhana Gubbi, Alistair Shilton, and Marimuthu Palaniswami

9.1 INTRODUCTION

There is a significant gap between the available protein sequences and the corresponding solved protein structures. One of the reasons is the success of genome projects (e.g., Human Genome Project [1]), which has resulted in the availability of complete protein sequences. The knowledge about the three dimensional protein structure is important in understanding its function and interaction with other molecules. This will lead to the designing of better drugs. X-ray crystallography is the most popular experimental technique for solving the structure of proteins. More than 80% of the known protein structures deposited in the protein data bank (PDB) [2] is solved using protein crystallography. The diffraction data obtained from the crystal contain only the magnitude information and the phase information is absent. A computational method of obtaining the absent phases is called molecular replacement [3]. One of the steps in molecular replacement is finding similar structure in a given target sequence. The success of this depends on computationally predicting protein structure for a given sequence [4, 5]. Schwarz et al. [6] have shown that the success of molecular replacement depends on the alignment accuracy. Hence protein structure prediction has become a very attractive research problem.

In recent years, kernel methods have become increasingly popular and are being applied to various problems including bioinformatics with good results. Kernel

Machine Learning in Bioinformatics. Edited by Yan-Qing Zhang and Jagath C. Rajapakse
Copyright © 2009 by John Wiley & Sons, Inc.

methods work by implicitly (and nonlinearly) mapping data into a high dimensional feature space before applying linear techniques in this feature space. Support vector machines (SVMs) [7, 8] are one of the most popular techniques of this type and may be used for both binary classification and regression. In this chapter, we will introduce kernel methods, in particular the SVM approach. More specifically, we will discuss several protein structure prediction problems and how kernel methods may be applied to solve them. We begin by introducing different protein structures in Section 9.2. In Section 9.3, a brief introduction to support vector machines and the properties of kernels will be given. In Section 9.4, different kernels including composite kernels, homology kernels, and cluster kernels are discussed. Concluding remarks are given in Section 9.5.

9.2 PROTEIN STRUCTURES

Amino acids arrange themselves in three-dimensional space in stable thermodynamic confirmations. Such a state is called the native confirmation of the protein, and the protein is active only in this state. It is this structure that determines the final function of the protein. Most of the known structures have been determined through the use of experimental methods. About 80% of the known structures in the PDB [2] were determined using X-ray crystallography, with the remainder determined using nuclear magnetic resonance (NMR) spectroscopy techniques. Thermodynamic interactions, including hydrogen bonding, hydrophobic interactions, electrostatic interactions, and complex formation between metal ions, are the major contributors to the native confirmation.

Protein molecules are quite complex in nature and are often made up of repetitive subunits. Based on this observation, protein structures are divided into primary, secondary, tertiary, and quaternary structures. Covalently linked amino acid sequence forms the primary structure. In other words, the primary structure of the protein is defined by the sequence of amino acids covalently bounded to form the backbone of the protein. Secondary and tertiary structures are mainly due to noncovalent bonding. The backbone of the proteins is made up of periodic repetitive subunits as a result of hydrogen bonding, the sequence of which is called the secondary structure. The two most commonly occurring subunits in secondary structure are α-helix and β-pleated sheets. Other common secondary structures include 3_{10} helix, π-helix, β-bulge and β-sheet. These are categorised under coils in most secondary structure prediction methods, although eight state determination of secondary structure [9] is also addressed. The helical confirmations are very stable as the hydrogen bonds are formed parallel to helix axis within the backbone. Every turn in a helix contains about 3.6 residues. The presence of proline causes rigidity in the structure, which is one of the most important properties for disruption in stable α-helix. Groups of charged residues (due to crowding caused by bulky side chains) can also disrupt helices. Unlike α-helices, β-sheets are formed by peptide bonds perpendicular to the backbone. Hydrogen bonds can be formed between residues that are far apart in the primary structure. This is one of the chief causes of long-range interactions during protein folding. These sheets can be parallel or antiparallel depending on the direction in which alternating chains run. *Turns* help in change of direction during folding and also contribute to addition or deletion of residues in chains belonging to the same family. Glycine and proline contribute heavily to such turns.

Combinations of such secondary structures that repeat themselves are referred to as supersecondary structures. $\beta\alpha\beta$ and $\alpha\alpha$ (alpha, turn, alpha), β *turn* β (β meander), and Greek key (antiparallel β *meander*, where the chain doubles back on itself) are the common examples of supersecondary structures. Supersecondary structures repeat or combine itself to form domains (commonly called folds). Interested readers may refer to Ref. 10 for a comprehensive explanation of structures. Figure 9.1 shows schematics of primary sequence, secondary structure, supersecondary structure, and Rossmann fold.

In our discussion of secondary structures and domains, we restrict our focus to the backbone structure of the protein. Tertiary structure gives the complete three-dimensional (3D) spatial map of the protein, including locations of all atoms and metal

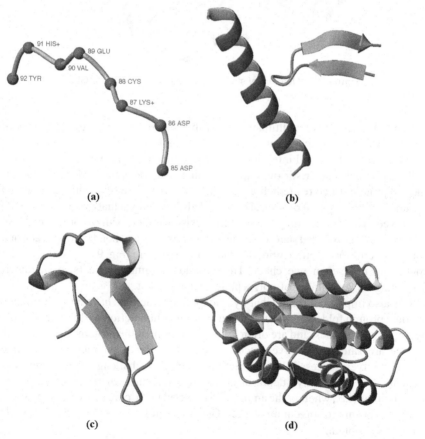

Figure 9.1 Protein structure (a) Partial primary sequence of protein 1JVW. (b) Secondary structures (α-helix and antiparallel β-strand). (c) Supersecondary structure (1JVW). (d) $\alpha + \beta$ sandwich–Rossmann fold. Figures prepared using MOLMOL [11].

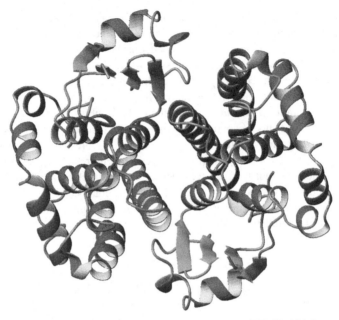

Figure 9.2 Quaternary structure of protein with PDB ID: 1GLQ.

ions in 3D space. It is the ultimate goal of the bioinformaticians to predict tertiary structure from amino acid sequence alone. Although it is known to be possible, it is very time consuming and infeasible for large numbers of proteins.

The special characteristic of cysteine forming disulfide bonds is also tackled at this stage of protein structure prediction. Disulfide bonds between cysteines are important features in the formation of several protein folds. It is known that cysteines are highly conserved in a protein family, and that they exist in either oxidized or reduced states [12–14]. In their oxidized state, cysteines often form covalent bond between each other that are referred as disulfide bridges. Quaternary structure (Fig. 9.2) refers to proteins containing more than one chain. These could be dimers, trimers, or tetrameres depending on the number of polypeptide chains present.

It is also important to note that hydrophobic interactions have a major effect on the final protein fold. Amino acids that do not favor interaction with water are called hydrophobic residues, and are often found in the core of the protein. Residues at the surface are said to be solvent accessible as they favor interaction with water. Relative solvent accessibility prediction is another challenge that involves the estimation of the relative depth of amino acid residues when compared to their surroundings (i.e., whether the amino acid is buried in the core of the protein) and the level of exposure to solvent molecules. General regression methods may be used to solve this problem.

Definitions of structure. To standardize the definitions of secondary structures, two methods are commonly used. Definition of secondary structure of proteins

(DSSP) was proposed by Kabsch and Sander [9]. It defines secondary structure, solvent accessibility, and disulfide connectivity among several other features, given the 3D atomic coordinates from the PDB. As proposed by Frishman and Argosis [15], STRIDE is another method of protein secondary structure assignment that is knowledge based.

There are three unique methods for domain and fold definitions: FSSP [16], SCOP [17], and CATH [18]. FSSP uses DALI [19] for structure–structure alignment of proteins. Structural classification of proteins (SCOP) divides the proteins into four hierarchical classes: family, superfamily, fold, and class. FSSP and SCOP use evolutionary relationships for classification. Class, architecture, topology, and homology (CATH) is another type of hierarchical classification that uses the SSAP [18] algorithm and is based on structural comparison. There is a high degree of similarity in the three databases, although there are a few disagreements.

9.3 KERNEL METHODS

Generally speaking, kernel methods are a class of machine learning algorithms based around the concept of using a nonlinear kernel function [20, 21] to achieve highly complex, nonlinear learning tasks without excessive algorithmic complexity. Specifically speaking, kernel methods use the kernel as a method of implicitly mapping data from input space to a (usually) higher dimensional feature space wherein a linear method may be used. The advantage of this approach is that the complexity of the feature map is hidden by the kernel function and need not be explicitly known.

In the case of binary classification, kernel methods work by implicitly and nonlinearly mapping input data into a (usually) higher dimensional feature space and then finding a hyperplane dividing the two classes in this feature space, which may then be used to classify other points. SVM classifiers, introduced by Vapnik [7], are binary classifiers motivated by the theory of structural risk minimization [8, 22, 23], which achieve this by maximizing the margin of separation in feature space.

9.3.1 Support Vector Classifiers

In this section we give a brief introduction to SVM classifier theory, followed by an introduction to the kernel and its properties. Support vector classifiers are a family of binary classifiers. Specifically, suppose we are given a set of objects each of which may be characterized by an observation vector $\mathbf{x}$ and is known to be classifiable in a binary fashion into either class $+1$ or class -1. Given a subset of N observations of such objects of each of known classification (called the training set), the aim of the support vector classifier is to construct a rule for classifying any object from the complete set based on its observation vector $\mathbf{x}$ alone.

Let $\mathbf{x}_i \in \mathfrak{R}^{d_L}$ be the observation vector (input) and let $y_i \in \{-1,+1\}$ be the known class (output) of the ith object in the training set. Then the training set can be denoted as follows:

$$\Theta = \{(\mathbf{x}_1, y_1), (\mathbf{x}_2, y_2), \ldots, (\mathbf{x}_N, y_N)\},$$
$$\mathbf{x}_i \in \Re^{d_L}, \tag{9.1}$$
$$y_i \in \{+1, -1\}.$$

We implicitly define a mapping function $\phi : \Re^{d_L} \rightarrow \Re^{d_H}$ from input space to feature space. Assuming that the two training classes are linearly separable in the feature space, a linear discriminant function $g : \Re^{d_L} \rightarrow \Re$ can be defined such that

$$g(x) = \mathbf{w}^T \phi(\mathbf{x}) + b,$$
$$y_i = \text{sgn}(g(\mathbf{x}_i)) \forall (\mathbf{x}_i, y_i) \in \Theta, \tag{9.2}$$

where $\mathbf{w} \in \Re^{d_H}$ and $b \in \Re$ completely specify the discriminant function g.

The equation $g(\mathbf{x}) = 0$ defines a linear decision hyperplane bisecting the two classes in feature space characterized by $\mathbf{w}$ and b. Of course, if the two classes are linearly separable in feature space, then there may exist infinitely many such decision surfaces. SVMs attempt to choose the optimal decision surface by maximizing the perpendicular distance (called the margin) between the decision surface and those training points lying closest to it (called the support vectors). For this reason, support vector classifiers are known as max-margin classifiers.

Geometrically, the margin γ in feature space is given by the projection of $\phi(\mathbf{x}_1)$ $\phi(\mathbf{x}_2)$ onto the hyperplane (refer Fig. 9.3). Hence $\gamma \propto 1/\|\mathbf{w}\|_2$. It follows that maximizing the margin is equivalent to solving

$$\min_{\mathbf{w}, b} Q(\mathbf{w}, b) = \tfrac{1}{2} \mathbf{w}^T \mathbf{w},$$
$$\text{such that} \quad y_i(\mathbf{w}^T \phi(\mathbf{x}_i) + b) \geq 1 \ \forall i. \tag{9.3}$$

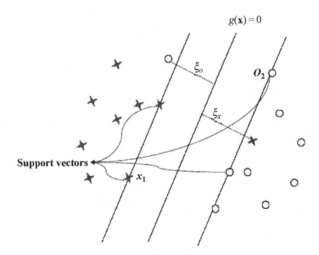

Figure 9.3 Support vector machine—an illustration.

In the nonseparable case, the strict inequalities in Equation 9.3 are relaxed by introducing a set of penalized slack variables ξ_i to give the soft margin SVM training problem:

$$\min_{\mathbf{w},b,\xi} Q(\mathbf{w}, b, \xi) = \frac{1}{2}\mathbf{w}^T\mathbf{w} + C1^T\xi,$$

$$\text{such that} \quad y_i(\mathbf{w}^T\phi(\mathbf{x}_i) + b) \geq 1 - \xi_i \; \forall i$$

$$\xi \geq \mathbf{0},$$

(9.4)

where the constant $C > 0$ controls the trade-off between the dual objectives of maximizing the margin of separation and minimizing the misclassification error. Following the usual approach, rather than solving the problem in its primal form (which is difficult because the constraint set is complex and the d_H may be high), we convert the problem into dual form, namely [7],

$$\min_{\alpha} W(\alpha) = \frac{1}{2}\alpha^T\mathbf{G}\alpha - \alpha^T 1,$$

$$\text{such that} \quad 0 \leq \alpha \leq C1$$

$$1^T\alpha = 0,$$

(9.5)

where $\mathbf{G} \in \mathfrak{R}^{N \times N}$, $G_{i,j} = y_i y_j K(\mathbf{x}_i, \mathbf{x}_j)$, and $K(\mathbf{x}_i, \mathbf{x}_j) = \phi(\mathbf{x}_i)^T \phi(\mathbf{x}_j)$. It can be shown [7] that this is a convex quadratic programming problem with no nonglobal solutions. Given α, b, it may be seen that (cf. Eq. 9.2) the final decision function is given by (denoting the unknown test vector as $\mathbf{z}$) [7, 24]:

$$g(\mathbf{z}) = \sum_{(\mathbf{x}_i, \mathbf{y}_i) \in \Theta} \alpha_i y_i K(\mathbf{x}_i, \mathbf{z}) + b.$$

The function $K(\mathbf{x},\mathbf{z}) = \phi(\mathbf{x})^T \phi(\mathbf{z})$ is called the kernel function, which plays a vital role here. This is because the kernel function K completely hides the feature map $\phi : \mathfrak{R}^{d_L} \rightarrow \mathfrak{R}^{d_H}$ in the problem. Moreover, it is well known that for any function $K : \mathfrak{R}^{d_L} \times \mathfrak{R}^{d_L} \rightarrow \mathfrak{R}$ satisfying Mercer's condition [24–26], there exists an associated feature map $\phi : \mathfrak{R}^{d_L} \rightarrow \mathfrak{R}^{d_H}$ (although calculating this map may not be a nontrivial exercise). Hence, starting with any such kernel function we may, with no knowledge of the feature map ϕ at all, optimize and use an SV-classifier based on that kernel function. This is referred to as the Kernel trick [24]. In the context of SVMs, these kernel functions are commonly called Mercer kernels, positive definite kernels, reproducing kernels, or admissible kernels in the literature.

To be specific, a positive definite (Mercer) kernel is defined as follows. Let χ be a nonempty set. A function $K : \chi \times \chi \rightarrow \mathfrak{R}$ is called a positive definite kernel if and only if K is symmetric (i.e., $K(x, x') = K(x', x) \forall x, x' \in \chi$) and

$$\sum_{j,k=1}^{n} c_j c_k K(x_j, x_k) \geq 0$$

$$\forall n \geq 1, \; c_1, c_n \in \mathfrak{R}, \; x_1, \ldots, x_n \in \chi.$$

(9.6)

If $K_1, K_2: \chi \times \chi \rightarrow \Re$ and $K_3: \Upsilon \rightarrow \Re$ are positive definite kernels over the nonempty sets χ and Υ, $\psi: \chi \rightarrow \Upsilon$, $a \in \Re^+$, $f: \chi \rightarrow \Re$, and $\mathbf{B} \in \Re^{d_L \times d_L}$ are positive semidefinite, then the following are also positive definite kernels [27]:

(1) $K(x, z) = K_1(x,z) + K_2(x,z)$

(2) $K(x, z) = aK_1\ (x,z)$

(3) $K(x, z) = K_1\ (x,z)\ K_2(x,z)$

(4) $K(x, z) = f(x)\ f(z)$

(5) $K(x, z) = K_3(\psi(x),\ \psi(z))$

(6) $K(\mathbf{x}, \mathbf{z}) = K(\mathbf{x}, \mathbf{z}) = \mathbf{x}'\mathbf{B}\mathbf{z}$ if $\chi = \Re^{d_L}$.

9.3.2 Commonly Used Kernels, Kernel Normalization, and Parameter Selection

Linear, radial basis, and polynomial kernels are commonly employed in SVMs as they give good results for most of the data and are easy to implement.

Linear kernels are the simplest of the kernels, being a simple inner product between vectors. In bioinformatics, this kernel is often modified and combined with others to formulate new kernels. It is defined by

$$K_{Lin}(\mathbf{x}, \mathbf{z}) = \mathbf{x}^T \mathbf{z}. \tag{9.7}$$

Polynomial kernels are a simple extension of the linear kernels defined by

$$K_{Pol}(\mathbf{x}, \mathbf{z}) = (1 + \mathbf{x}^T \mathbf{z})^d, \tag{9.8}$$

where $d > 1$ represents the degree of polynomial.

Gaussian radial basis function (RBF) kernels are the most popular of the kernels as they often give very good results compared to other common kernels. The Gaussian RBF kernel is defined by

$$K_{RBF}(\mathbf{x}, \mathbf{z}) = \exp\left(-\frac{\|x - z\|^2}{2\sigma^2}\right), \tag{9.9}$$

where $\sigma > 0$ acts like the variance of a standard Gaussian.

An important issue when designing more general kernels is that of kernel normalization. For optimal performance, it is important to scale the input feature values to zero mean and unit variance. This will help the classifier avoid overweighting the training samples with large numerical values, and likewise underweighting training samples with small numerical values. In bioinformatics, the following normalization function is used:

$$k'_{ij} = \frac{k_{ij}}{\sqrt{k_{ii}k_{jj}}}. \tag{9.10}$$

It should be noted that, in general, not all kernels need to be positive definite. Sigmoid kernel is an example of nonpositive definite kernel (for some values of β and θ). The sigmoid kernel is defined by

$$K_{\text{Sig}}(\mathbf{x}, \mathbf{z}) := \tanh(\beta \mathbf{x}^T \mathbf{z} + \theta), \qquad (9.11)$$

where $\beta > 0$ is called the gain and $\theta < 0$ is called the threshold. Haasdonk [28] gives a geometric interpretation of SVMs with indefinite kernel functions. Specifically, they show that SVMs constructed with indefinite kernels are optimal hyperplane classifiers utilizing minimization of distances between convex hulls in pseudo-Euclidean spaces and not max-margin classifiers as discussed earlier.

The problem of parameter selection, which encompasses both kernel parameters (e.g., σ) and the regularization parameter C, is a challenging problem as parameter selection has significant influence on the performance of the classifier. The general method for choosing these parameters is often trial and error until to minimize leave-one-out, cross-validation, or test set error. Grid-based search methods may also be used. As the number of unknown parameters increase, however, the grid becomes very complex and the computational efficiency drops significantly. In a recent work by Frohlich and Zell [29], an algorithm was proposed based on the idea of learning an online Gaussian process model of the error surface in parameter space and sampling systematically at points for which the so-called expected improvement is highest. They call the algorithm efficient parameter selection via global optimization (EPSGO). More recently, Christmann et al. [30] have compared several schemes (random search, grid search, Nelder and Mead [31], Cherkassky and Ma [32], and pattern-based search [33]) for tuning the free parameters in SVMs. They argue that there is no method that can be rated as the best, but go on to suggest that the pattern-based search is a better method in terms of error rate and computational time.

While the so-called standard kernels provide good results in many situations, there is still scope to develop new kernels for special applications such as text categorization, speech processing, and for other string-based data. Due to the fact that bioinformatics applications involve string-based prediction, a lot of interest in this area has been generated recently [20]. As most of the applications involve matching strings in bioinformatics, it is important to devise new kernels that are more appropriate for these applications. The similarity measures required for these applications are different from the similarity measures implemented by standard kernels. In areas like protein structure prediction, where defining a good similarity measure is difficult, kernel methods are very promising [34]. Hence, new kernels that separate the two classes in feature space are essential.

9.4 KERNELS IN PROTEIN STRUCTURE PREDICTION

In this section, we review some of the kernels that have been proposed for protein structure prediction and homology modeling. We discuss composite kernels, which are the most common, followed by the homology kernels and cluster kernels. Under homology kernels, Fisher kernels, mismatch kernels, dynamic programming-based kernels, and profile kernels are discussed.

9.4.1 Composite Kernels

Composite kernels are the most common type of kernels. The features extracted from the data are combined using commonly available Mercer kernels. More than one kernel is combined using properties described in Section 9.3.1 resulting in new kernels. We look at the application of these kernels to three challenging problems: secondary structure prediction, relative solvent accessibility estimation, and fold recognition.

Hua and Sun [35] have used SVMs and profiles of the multiple alignments from HSSP database as the features and reported their Q_3 score as 73.5% on the CB513 data set [36]. They used an orthogonal coding scheme for representing each amino acid. The vector contained 21 elements with 20 elements for representing each amino acid and the last feature was used to represent extension over N and C terminus. These data are mapped into feature space with the help of an RBF kernel. In 2003, Ward et al. [37] reported 77% accuracy with PSI-BLAST [38] profiles on a small set of proteins. They use polynomial kernel with evolutionary information in the form of position-specific scoring matrix (PSSM) obtained from PSI-BLAST as input. In the same year, Kim and Park [39] reported an accuracy of 76.6% of the CB513 data set using PSSM and RBF kernels. The basic ideas of these two techniques are very similar with some unique modifications. Since then, PSSM has been used as one of the standard data for other methods developed [40, 41]. The homology information may be very effectively used by employing one of the standard Mercer kernels. Hu et al. [42] reported the highest accuracy of 78.8% of a RS126 data set using a novel encoding scheme compared to others. They use hydrophobicity, BLOSUM62 matrix, and the orthogonal representation in various combinations and employ RBF kernel. More recently, we have used physicochemical properties [43], such as hydrophobicity, polarity, the effect of smaller amino acids along with statistical parameters such as Chou–Fasman parameters [44], and analyzed the effect of using such features in combination with the evolutionary features such as PSSM. We have even shown that combining the physicochemical features along with evolutionary information works better than either used separately.

Very similar developments can be observed in relative solvent accessibility prediction. The use of orthogonal coding followed by the use of PSSM has been seen over the years [45–50]. Those kernels using evolutionary information in the form of PSI-BLAST profiles (PSSM) [38] are also called *profile kernels* in literature [51]. In our recent work on relative solvent accessibility prediction [52], we have proposed a simple kernel to effectively make use of the features extracted using closure properties [53]:

$$k(\mathbf{x}, \mathbf{y}) = k_h(\mathbf{x}_h, \mathbf{y}_h)k_c(\mathbf{x}_c, \mathbf{y}_c) + k_s(\mathbf{x}_s, \mathbf{y}_s) + k_e(\mathbf{x}_e, \mathbf{y}_e), \qquad (9.12)$$

where $\mathbf{x}$ and $\mathbf{y}$ represent the input data with subscript h denoting kernel for evaluating Kyte–Doolittle hydrophobic scales [54], c for evaluating Grantham polarity values

[55], s for evaluating Chou–Fasman secondary structure conformational parameters [44], and e for evaluating evolutionary information extracted in the form of PSSM matrix obtained from PSI-BLAST. Individual kernels (k_h, k_p, k_s, k_e) are simple dot products, which were chosen ahead of other variants based on experimental results.

Ding and Dubchak [56] used amino acid composition, predicted secondary structure, hydrophobicity, van der Walls volume, polarity, and polarizability as input to the SVM for fold prediction. This was one of the earliest attempts of using composite kernels in protein structure prediction. They used it to classify protein into one of the 27 folds and showed that it performs better than neural network. In 2006, Cheng and Baldi [57] proposed a novel machine learning information retrieval method for fold recognition. They compute pairwise alignment scores with several complementary alignment tools and use them as similarity measure. Structural similarity scores of the aligned sequences, such as secondary structure, solvent accessibility, and contact map, are derived using standard kernels such as cosine, correlation, and Gaussian. The extracted data are then fed into an SVM. Interestingly, this method produces best result so far when compared with standard fold prediction schemes, including PSI-BLAST [38], HMMER [58], THREADER [59], and FUGUE [60].

Composite kernels are applied for a few other rare but important protein structure prediction problem. Zimmermann and Hansmann [61] apply a multistep SVM with PSSM as features in the first step. They report good accuracies of about 80% when compared with PSIPRED. Kuang et al. [62] propose machine learning approaches for protein backbone angle prediction. Zhang et al. [63] have used composite kernel with input as PSSM for the prediction of beta-turns with the highest accuracy of 77%.

9.4.2 Kernels for Homology Detection

It is known [64] that if the protein sequence identity of the two proteins is greater than 35%, then the proteins have similar structure. This knowledge is used in homology modeling to determine if a similar protein fold exists in the solved structure. Homology modeling involves matching of strings using certain substitution matrix and the commonly available kernels cannot be employed for this purpose. Hence, most string-based kernels for protein structure prediction were developed for homology modeling. However, this becomes challenging when the sequence identity is less than 35% (and even more so if the sequence identity is less than 25%) [64].

9.4.2.1 *Fisher Kernel*   Fisher kernels were introduced by Jaakkola et al. [65–67] to detect remote protein homologies. The basic idea of this kernel is to make use of generative models built from multiple sequence alignment. A discriminative statistical model is then employed to detect whether a given protein chain belongs to a particular family or superfamily. Hidden Markov models (HMM) are used as generative models where the sequences of a particular family are used as training vectors. Such a model yields higher probability score for that particular family compared to other families. This score is used as a similarity measure by the SVM. The procedure to compute the Fisher kernel is as follows. Given the protein family H_1 for positive model, H_0 for the negative model, and the sequence of amino acid $S = a_1, a_2, \ldots, a_n$, the likelihood ratio

is given by

$$L(S) = \log \frac{P(S|H_1)}{P(S|H_0)} + \log \frac{P(H_1)}{P(H_0)}. \tag{9.13}$$

If $L(S) > 0$, the protein sequence S is the member of the family H_1. Otherwise, it does not belong to H_1. To use this in the proposed kernel, the calculated probability model is represented as $P(S|H_1, \theta)$, where θ is the output and the transition probability of model H_1. Following the usual procedure in HMM, sufficient statistics such as observation matrix O and emission matrix E are extracted [68]. O and E are converted into the gradient vector U_s using [66, 69]

$$U_s = \frac{O_j(i)}{E_j(i)} - \sum_k O_j(k). \tag{9.14}$$

U_s is called the Fisher score and is shown to be an excellent similarity measure. The kernel function used in the protein homology detection is

$$K(\mathbf{x}, \mathbf{z}) = \exp - \frac{1}{2\sigma^2} (U_x - U_z)^T (U_x - U_z). \tag{9.15}$$

They also compare the proposed method with other popular methods—BLAST (SCOP only), BLAST (SCOP + SAMT-98 homologue), and SAMT-98—and show that the Fisher kernel outperforms the others. The main advantage of this kind of formulation is that it makes use of the prior knowledge of protein sequences in the form of probability.

9.4.2.2 *Mismatch String Kernels*

Mismatch kernels, as proposed by Leslie et al. [70], are one of the most popular string kernels in bioinformatics. We look at the formulation of these kernels and their computational complexity. Let the protein sequence be denoted by $S = a_1, a_2, \ldots, a_n$, where a_n is an amino acid picked from the set A of 20 amino acids. Let s be the substring extracted from S, where s is of length k and hence called a k-mer. s_a is the set of all k-mer substrings from S. Leslie et al. [70] define set $N_{(k,m)}(s)$, which is the set of all k-length strings in s_a that differ from s by at most m mismatches. The feature map ϕ is given as

$$\phi_{(k,m)}(s) = \begin{bmatrix} 1, & \text{if } s_a \in N_{(k,m)}(s), \\ 0, & \text{otherwise.} \end{bmatrix} \tag{9.16}$$

For the complete sequence with K k-mers, the feature map is

$$\phi_{(k,m)}(S) = \sum_{i=1}^{K} \phi_{(i,m)}(s). \tag{9.17}$$

Finally, the (k, m)-mismatch string kernel $K_{(k,m)}$ is given by the inner product:

$$K_{(k,m)}(\mathbf{x}, \mathbf{z}) = \langle \phi_{(k,m)}(\mathbf{x}), \phi_{(k,m)}(\mathbf{z}) \rangle. \tag{9.18}$$

It should be noted that if $m = 0$, Equation 9.18 reduces to k-spectrum kernel proposed by Leslie et al. [71] earlier. For efficient implementations of this kernel, the reader is referred to Refs [70, 72]. Eskin and Snir [73] design homology kernel based on the same principle as mismatch kernel and combine knowledge about local alignment, sequence homology, and predicted secondary structure. They use this kernel for protein family classification.

9.4.2.3 *Kernels Based on Dynamic Programming*   For sequence alignment, several kernels including those discussed above have shown to be useful. Due to the success of Smith–Waterman algorithm in the form of BLAST, a Smith–Waterman kernel has been developed. The kernel is based on the pairwise similarity score generated by Smith–Waterman algorithm. Just like Fisher kernels, Smith–Waterman algorithm match two proteins based on evolutionary characteristics using standard substitution matrix such as BLOSUM62. The matrix generated by Smith–Waterman scores for all the proteins is not positive semidefinite, and hence must be converted into a valid kernel by making use of an empirical kernel map to make the kernel matrix positive definite [69]. A few of the other sequence alignment kernels that have been developed can be found in Refs [53, 70, 71, 73].

We have developed a variant of the dynamic programming-based kernel for protein class prediction [74]. In Ref. [75], a dynamic time-warping kernel was developed for speech recognition. The same basic procedure for sequence alignment is used, but it is used to align the predicted structural class. Three models (α, β and $\alpha + \beta$) are constructed using separate empirically chosen substitution matrices (S_α, S_β, and $S_{\alpha+\beta}$) [74]. The input matrix In (for α class) is constructed using the following function:

$$\mathrm{In}(i,j) = S_\alpha(\mathrm{TarSS}_i, \mathrm{TemSS}_j), \tag{9.19}$$

where TarSS_i and TemSS_j, respectively, are the target and template secondary structure states of residues i and j. From the input matrix, we calculate the DP matrix D using

$$D(i,j) = \max \left\{ \begin{array}{l} D(i-1,j) + \mathrm{Gap} \\ D(i-1,j-1) + 2\mathrm{In}(i,j) \\ D(i,j-1) + \mathrm{Gap} \end{array} \right\}, \tag{9.20}$$

where Gap in the above equation is a gap penalty constant. The final kernel function is defined as follows:

$$K(\mathrm{Tar}, \mathrm{Tem}) = \frac{D(M,N)}{(M+N)}, \tag{9.21}$$

where Tar is the secondary structure of target protein sequence, Tem is the secondary structure of template protein sequence, M is length of Tar, and N is the length of Tem.

To evaluate the proposed method, we have constructed a new data set (GSPP742) using CATH version 2.6.0 (released in April 2005).[1] The data set contains 742 sequences with a maximum pairwise sequence identity of less than 25%. There are 164 protein chains belonging to class α, 223 belonging to class β, and 355 belonging to class $\alpha + \beta$. Of the total 39 topologies, there are 14 topologies in class α, 10 topologies in class β, and 15 topologies in class $\alpha + \beta$. The results are impressive, with high sensitivity of 0.77, of specificity of 0.91, and an overall accuracy of class prediction of around 87.44%.

9.4.2.4 Profile-Based Kernels   In the section on composite kernels, the use of profiles in the form of PSSM is discussed. Unlike in composite kernels where they use the profiles as features with a standard Mercer kernel, Rangwala and Karypis [76] develop profile kernels for homology detection and fold recognition. PSSM from PSI-BLAST along with position-specific frequency matrix (PSFM) are used. PSFM contains the frequencies used by PSI-BLAST to derive PSSM. The alignment score is obtained using the following function:

$$S_{X,Y}(i,j) = \sum_{k=1}^{20} \text{PSFM}_X(i,k)\text{PSSM}_Y(j,k) + \sum_{k=1}^{20} \text{PSFM}_Y(i,k)\text{PSSM}_X(j,k). \quad (9.22)$$

This similarity measure is used in the construction of window-based and local alignment kernels. In window-based kernels, the alignment score of all possible k-mers with positive alignment scores are summed. Based on the way k-mers are used, Rangwala and Karypis [76] propose three variants of window-based kernels, namely, (a) all fixed width k-mers, (b) best fixed width k-mer, and (c) best variable width k-mer. The local alignment kernel that they employ is nothing but the dynamic programming-based kernel discussed earlier in this chapter. The dynamic programming-matrix is filled with alignment scores calculated as per Equation 9.23. They test the proposed kernels on superfamily classification and fold recognition and compare their proposed ones with other popular kernels, including Fisher, mismatch and composite kernels. They report the highest ROC50[2] score of 0.904 for superfamily classification and 0.571 for fold classification using the proposed local alignment kernel.

9.4.3 Cluster Kernel for Protein Classification

Weston et al. [77] proposed two general cluster kernels: the neighborhood mismatch kernel and the bagged mismatch kernel. One of the major advantages of this kernel is the fact that the final similarity score is like BLAST family methods.

---

[1]This data set can be downloaded from http://www.ee.unimelb.edu.au/ISSNIP/downloads/.

[2]The ROC50 score is defined as the area under the ROC curve up to first 50 false positives.

In neighborhood mismatch kernel, E-value score of PSI-BLAST is used as a similarity measure. Given a target sequence x, Nbd(x) is defined as the neighborhood of sequence x whose E-value is less than a fixed threshold. Every sequence x is represented as the average of the feature vectors in its neighborhood as follows:

$$\phi_{\text{nbd}}(x) = \frac{1}{|\text{Nbd}(x)|} \sum_{x' \in \text{Nbd}(x)} \phi_{\text{orig}}(x'). \tag{9.23}$$

Kernel K_{nbd} can now be defined as

$$K_{\text{nbd}}(x, z) = \frac{\displaystyle\sum_{x' \in \text{Nbd}(x),\, z' \in \text{Nbd}(z)} K_{\text{orig}}(x', z')}{|\text{Nbd}(x)||\text{Nbd}(z)|}. \tag{9.24}$$

Bagged mismatch kernel is constructed using three steps: first, k-means is run n times, which gives the cluster assignment $c_p(x_i)$ for each i. Next, based on the fraction of times x_i and x_j fall in the same cluster, the bagged-clustering representation is constructed as follows:

$$K_{\text{bag}}(x_i, x_j) = \frac{\displaystyle\sum_p [c_p(x_i) = c_p(x_j)]}{n}. \tag{9.25}$$

Finally, given the original and the bagged kernel, the bagged mismatch kernel is represented as

$$K(\mathbf{x}, \mathbf{z}) = K_{\text{orig}}(\mathbf{x}, \mathbf{z}) \cdot K_{\text{bag}}(\mathbf{x}, \mathbf{z}), \tag{9.26}$$

where K_{orig} is the mismatch kernel. The kernel was evaluated using protein family classification problem. Kernel was compared with other kernels such as mismatch kernel, profile kernel, and graph-based kernel. Bagged mismatch kernel yielded the highest TOC50 value of 0.719, closely followed by neighborhood mismatch kernel with 0.704.

9.4.4 Kernels for Disulfide Bridge Prediction

Cheng et al. [51, 78] have applied kernel methods for disulfide bridge prediction. This is one of the most challenging problems in protein structure prediction with very limited success. PSSM was used as the input for prediction. Spectrum kernel, mismatch kernel, Fisher kernel, Smith–Waterman kernel, local alignment kernel, and profile kernel were used to find whether a disulfide bridge is present in the input protein. They reported very similar performances by all the compared kernels. The highest accuracy of 74% was obtained by mismatch kernel for SPX [51] data set whereas for SP41 [79] data set, profile kernel resulted in highest accuracy of 85%.

9.5 CONCLUSION

Kernel methods based on SVMs are becoming an increasingly popular machine learning technique in the recent years. They have been applied to various applications in computational biology with as good or even better result than other machine learning schemes. In this chapter we have given a brief overview of kernel methods and their application to protein structure prediction. Composite kernels are the most attractive kernels used with many methods developed for secondary structure, relative solvent accessibility, and fold recognition. They are shown to perform better than other machine learning methods such as the popular neural network and hidden Markov models. Support vector regression-based methods are shown to give better results for real-value solvent accessibility than any other scheme. As finding remote homology is a key challenge in protein structure prediction, several kernels are developed for string matching and shown to work well in fold recognition. Kernel methods are also applied to rare but challenging problems such as disulfide bridge prediction, backbone angle prediction, and dihedral angle prediction.

REFERENCES

1. Venter, J. C., et al. The sequence of the human genome. *Science*, 291: 1304–1351, 2001.

2. Berman, H. M., Westbrook, J., Feng, Z., Gilliland, G., Bhat, T. N., Weissig, H., Shindyalov, I. N., and Bourne, P. E. The protein data bank. *Nucleic Acids Research*, 28: 235–242, 2000.

3. Rossmann, M. G. and Blow, D. M. The detection of subunits within the crystallographic asymmetric unit. *Acta Crystallographica*, 15: 24–31, 1962.

4. Rhodes, G. *Crystallography Made Crystal Clear*, 2nd edition, Academic Press, San Diego, 2000.

5. Drenth, J. *Principles of Protein X-Ray Crystallography*, 2nd edition. Springer-Verlag, New York, 1999.

6. Schwarzenbacher, R., Godzik, A., Grzenchnik, S. K., and Jaroszewski, L. The importance of alignment accuracy for molecular replacement. *Acta Crystallographica*, D60: 1229–1236, 2004.

7. Vapnik, V. *Statistical Learning Theory*, Springer-Verlag, New York, 1995.

8. Vapnik, V., Golowich, S., and Smola, A. Support vector methods for function approximation, regression estimation, and signal processing. *Advances in Neural Information Processing Systems*, Vol. 9, MIT Press, pp. 281–287, 1997.

9. Kabsch, W. and Sander, C. Dictionary of protein secondary structure: pattern recognition of hydrogen-bonded and geometrical features. *Biopolymers*, 22: 2577–2637, 1983.

10. Campbell, M. K. and Farrell, S. O. The three-dimensional structure of proteins. *Biochemistry*, 5th edition, Thomson Brooks/Cole, pp. 80–112, 2006.

11. Koradi, R., Billeter, M., and Wüthrich, K. MOLMOL: a program for display and analysis of macromolecular structures. *Journal of Molecular Graphics*, 14: 51–55, 1996.

12. Klink, T. A., Woycechowsky, K. J., Taylor, K. M., and Raines, R. T. Contribution of disulfide bonds to the conformational stability and catalytic activity of ribonuclease A. *European Journal of Biochemistry*, 267: 566–572, 2000.

13. Fariselli, P., Riccobelli, P., and Casadio, R. Role of evolutionary information in predicting the disulfide-bonding state of cysteine in proteins. *Proteins: Structure, Function and Genetics*, 36(3): 340–346, 1999.

14. Fiser, A. and Simon, I. Predicting the oxidation state of cystines by multiple sequence alignment. *Bioinformatics*, 16(3): 251–256, 2000.

15. Frishman, D. and Argos, P. Knowledge-based secondary structure assignment. *Proteins: Structure, Function and Genetics*, 23: 566–579, 1995.

16. Holm, L. and Sander, C. The FSSP data base of structurally aligned protein fold families. *Nucleic Acids Research*, 22(17): 3600–3609, 1994.

17. Murzin, A. G., Brenner, S. E., Hubbard, T., and Chothia, C. SCOP: a structural classification of proteins database for the investigation of sequences and structures. *Journal of Molecular Biology*, 247: 536–540, 1995.

18. Orengo, C. A., Michie, A. D., Jones, S., Jones, D. T., Swindells, M. B., and Thornton, J. M. CATH- a hierarchic classification of protein domain structures. *Structure*, 5(8): 1093–1108, 1997.

19. Holm, L. and Sander, C. Touring protein fold space with DALI/FSSP. *Nucleic Acids Research*, 26: 316–319, 1998.

20. Scholkopf, B., Tsuda, K., and Vert, J.-P. *Kernel Methods in Computational Biology*, A Bradford Book, MIT Press, London, 2004.

21. Scholkopf, B. and Smola, A. J. *Learning with Kernels*, MIT Press, London, 2002.

22. Smola, A. and Schölkopf, B. A tutorial on support vector regression. Technical Report NeuroCOLT2 Technical Report Series, NC2-TR-1998-030, Royal Holloway College, University of London, UK, October 1998.

23. Burges, C. J. C. A tutorial on support vector machines for pattern recognition. *Knowledge Discovery and Data Mining*, 2(2): 121–167, 1998.

24. Scholkopf, B. and Smola, A. J. *Learning with Kernels: Support Vector Machines, Regularization, Optimization and Beyond*, MIT Press, Cambridge, MA, 2002.

25. Haussler, D.Convolution kernels on discrete structures. *Technical Report*, UCS-CRL-99(10), 1999.

26. Mercer, J., Functions of positive and negative type, and their connection with the theory of integral equations. *Transactions of the Royal Society of London*, 209(A), 1909.

27. Taylor, J. S. and Cristianini, N. *Support Vector Machines and Other Kernel-Based Learning Methods*, Cambridge University Press, 2000.

28. Haasdonk, B. Feature space interpretation of SVMs with indefinite kernels. *IEEE Transactions on Pattern Analysis and Machine Intelligence Archive*, 27(4): 482–492, 2005.

29. Frohlich, H. and Zell, A. Efficient parameter selection for support vector machines in classification and regression via model-based global optimization. *2005 Proceedings of the IEEE International Joint Conference on Neural Networks (IJCNN'05)*, Vol. 3, pp. 1431–1436, 2005.

30. Christmann, A., Lübke, K., Marin-Galiano, M., and Rüping, S. Determination of hyperparameters for kernel-based classification and regression. Technical Report, University of Dortmund, SFB-475(TR-38/2005), 2005.

31. Nelder, J. and Mead, R. A simplex method for functional minimization. *Computer Journal*, 7: 308–313, 1965.

32. Cherkassy, V. and Ma, Y. Practical selecion of SVM parameters and noise estimation for SVM regression. *Neural Networks*, 7: 113–126, 2004.

33. Momma, M. and Bennett, K. P. A pattern search method for model selection of support vector regression. *Proceedings of the Second SIAM International Conference on Data Mining*, SIAM, pp. 261–274, 2002.

34. Joachims, T., Cristianini, N., and Shawe-Taylor, J. Composite kernels for hypertext categorisation. *Proceedings of the Eighteenth International Conference on Machine Learning*, pp. 250–257, 2001.

35. Hua, S. and Sun, Z. A novel method of protein secondary structure prediction with high segment overlap measure: support vector machine approach. *Journal of Molecular Biology*, 308: 397–407, 2001.

36. Cuff, J. A. and Barton, G. J. Evaluation and improvement of multiple sequence methods for protein secondary structure prediction. *Proteins*, 34: 508–519, 1999.

37. Ward, J. J., McGuffin, L. J., Buxton, B. F., and Jonese, D. T. Secondary structure prediction with support vector machines. *Bioinformatics*, 19(13): 1650–1655, 2004.

38. Altschul, S. F., Madden, T. L., Schaffer, A. A., Zhang, J., Zhang, Z., Miller, W., and Lipman, D. J. Gapped blast and psi-blast: a new generation of protein data base search programs. *Nucleic Acid Research*, 27(17): 3389–3402, 1997.

39. Kim, H. and Park, H. Protein secondary structure prediction based on an improved support vector machines approach. *Protein Engineering*, 16(8): 553–560, 2003.

40. Nguyen, M. N. and Rajapakse, J. C. Multi-class support vector machines for protein secondary structure prediction. *Genome Informatics*, 14: 218–227, 2003.

41. Guo, J., Chen, H., Sun, Z., and Lin, Y. A novel method for protein secondary structure prediction using dual layer SVM and profiles. *Proteins: Structure, Function and Bioinformatics*, 54: 738–743, 2004.

42. Hu, H. J., Pan, Y., Harrison, R., and Tai, P. C. Improved protein secondary structure prediction using support vector machine with a new encoding scheme and an advanced tertiary classifier. *IEEE Transactions on Nanobioscience*, 3(4): 265–271, 2004.

43. Jayavardhana Rama, G. L., Palaniswami, M., Lai, D., and Parker, M. W. A study on the effect of physico-chemical properties in protein secondary structure prediction. *Applied Artificial Intelligence*, World Scientific, pp. 609–616, 2006.

44. Chou, P. Y. and Fasman, G. D. Conformational parameters for amino acids in helical, β-sheet, and random coil regions calculated for proteins. *Biochemistry*, 13(2): 211–222, 1974.

45. Yuan, Z., Burrage, K., and Mattick, J. S. Prediction of solvent accessibility using support vector machines. *Proteins: Structure, Functions and Genetics*, 48: 566–570, 2002.

46. Kim, H. and Park, H. Prediction of protein relative solvent accessibility with support vector machines and long-range interaction 3D local descriptor. *Proteins: Structure, Function, and Bioinformatics*, 54: 557–562, 2004.

47. Nguyen, M. N. and Rajapakse, J. C. Two-stage support vector machines to protein relative solvent accessibility prediction. *Proceedings of the 2004 IEEE Symposium on Computational Intelligence in Bioinformatics and Computational Biology*, pp. 67–72, 2004.

48. Nguyen, M. N. and Rajapakse, J. C. Prediction of protein relative solvent accessibility with a two stage SVM approach. *Protiens: Structure, Function and Bioinformatics*, 59: 30–37, 2006.

49. Yuan, Z. and Bailey, T. L. Prediction of protein solvent profile using SVM. *Proceedings of the 26th Annual International Conference of the IEEE EMBS*, pp. 2889–2892, 2004.

50. Yuan, Z. and Huang, B. Prediction of protein accessible surface areas by support vector regression. *Proteins: Structure, Function and Bioinformatics*, 57: 558–564, 2004.

51. Cheng, J., Saigo, H., and Baldi, P. Large-scale prediction of disulphide bridges using kernel methods, two-dimensional recursive neural networks and weighted graph matching. *Proteins: Structure, Function and Bioinformatics*, 62: 617–629, 2006.

52. Gubbi, J., Shilton, A., Parker, M., and Palaniswami, M. Real value solvent accessibility prediction using adaptive support vector regression. *Proceedings of IEEE Symposium on Computational Intelligence and Bioinformatics and Computational Biology-CIBCB*, IEEE Press, Piscataway, NJ, pp. 395–401, 2007.

53. Taylor, J. S. and Cristianini, N. *Kernel Methods for Pattern Analysis*, Cambridge University Press, 2004.

54. Kyte, J. and Doolittle, R. F. A simple method for displaying the hydropathic character of a protein. *Journal of Molecular Biology*, 157: 105–132, 1982.

55. Grantham, R. Amino acid difference formula to help explain protein evolution. *Science*, 185: 862–864, 1974.

56. Ding, C. H. Q. and Dubchak, I. Multi-class protein fold recognition using support vector machines and neural networks. *Bioinformatics*, 17: 349–358, 2001.

57. Cheng, J. and Baldi, P. A machine learning information retrieval approach to protein fold recognition. *Bioinformatics*, 22(12): 1456–1463, 2006.

58. Durbin, R., Eddy, S. R., Krogh, A., and Mitchinson, G. *Biological Sequence Analysis: Probabilistic Models of Proteins and Nucleic Acids*, Cambridge University Press, Cambridge, MA, 1998.

59. Jones, D. T. Threader: protein sequence threading by double dynamic programming. In: S. Salzberg, D. Searls, and S. Kasif, Eds, *Computational Methods in Molecular Biology*, Chapter 13, Elsevier Science, New York, 1998.

60. Shi, J., Blundell, T. L., and Mizuguchi, K. FUGUE: sequence-structure homology recognition using environment-specific substitution tables and structure-dependent gap penalties. *Journal of Molecular Biology*, 310(1): 243–257, 2001.

61. Zimmermann, O. and Hansmann, H. E. Support vector machines for prediction of dihedral angle regions. *Bioinformatics*, 22(24): 3009–3015, 2006.

62. Kuang, R., Leslie, C. S., and Yang, A.-S. Protein backbone angle prediction with machine learning approaches. *Bioinformatics*, 20(10): 1612–1621, 2004.

63. Zhang, Q., Yoon, S., and Welsh, W. J. Improved method for predicting beta-turn using support vector machines. *Bioinformatics*, 21(10): 2370–2374, 2005.

64. Rost, B. Twilight zone of protein sequence alignments. *Protein Engineering*, 12(2): 85–94, 1999.

65. Jaakkola, T., Diekhans, M., and Haussler, D. Using the Fisher kernel method to detect remote protein homologies. *Proceedings of the Seventh International Conference on Intelligent Systems for Molecular Biology*, pp. 149–158, 1999.

66. Jaakkola, T. and Haussler, D. Exploiting generative models in discriminative classifiers. *Advances in Neural Information Processing Systems*, 1999, 11, pp. 487–493.

67. Cherkassy, V. and Ma, Y. A discriminative framework for detecting remote protein homologies. *Neural Networks*, 7: 95–114, 2004.

68. Rabiner, L. R. A tutorial on hidden markov models and selected applications in speech recognition. *Proceedings of the IEEE*, 77: 257–286, 1989.

69. Noble, W. S. Support vector machine applications in computational biology. In: B. Scholkopf, K. Tsuda, and J. P. Vert Eds, *Kernel Methods in Computational Biology*, A Bradford Book, MIT Press, London, pp. 71–92, 2004.

70. Leslie, C., Eskin, E., Cohen, A., Weston, J., and Noble, W. S. Prediction of protein structural class with rough sets. *Bioinformatics*, 1(1): 1–10, 2003.

71. Leslie, C., Eskin, E., and Noble, W. S. The spectrum kernel: a string kernel for SVM protein classification. *Proceedings of the Pacific Symposium on Biocomputing*, pp. 564–575, 2002.

72. Vishwanathan, S. V. N. and Smola, A. J. Fast kernels for string and tree matching. In: B. Scholkopf, K. Tsuda, and J., P. Vert Eds, *Kernel Methods in Computational Biology*, A Bradford Book, MIT Press, London, pp. 113–130, 2004.

73. Eskin, E. and Snir, S. The homology kernel: a biologically motivated sequence embedding into Euclidean space. *Proceedings of the 2005 IEEE Symposium on Computational Intelligence in Bioinformatics and Computational Biology (CIBCB' 2005)*, 2005.

74. Gubbi, J., Shilton, A., Parker, M., and Palaniswami, M. Protein topology classification using two-stage support vector machines. *Genome Informatics*, 17(2): 259–269, 2006.

75. Shimodaira, H., Noma, K., Nakai, M., and Sagayama, S. Dynamic time-alignment kernel in support vector machine. *Advances in Neural Information Processing Systems*, Vol. 14, 2002.

76. Rangwala, H. and Karypis, G. Profile-based direct kernels for remote homology detection and fold recognition. *Bioinformatics*, 31(23): 4239–4247, 2005.

77. Weston, J., Leslie, C., Le, E., Zhou, D., Elisseeff, A., and Noble, W. S. Semi-supervised protein classification using cluster kernels. *Bioinformatics*, 21(15): 3241–3247, 2005.

78. Baldi, P., Cheng, J., and Vullo, A. Large-scale prediction of disulphide bond connectivity *Advances in Neural Information Processing Systems 17*, MIT Press, Cambridge, MA, pp. 97–104, 2005.

79. Vullo, A. and Frasconi, P. Disulfide connectivity prediction using recursive neural networks and evolutionary information. *Bioinformatics*, 10: 653–659, 2004.

10

EVOLUTIONARY GRANULAR KERNEL TREES FOR PROTEIN SUBCELLULAR LOCATION PREDICTION

Bo Jin and Yan-Qing Zhang

10.1 INTRODUCTION

The prediction of subcellular locations of proteins may help people understand proteins' functions and analyze their interactions with other molecules. Machine learning methods and computational techniques were used to predict the protein subcellular locations [1, 2]. Nakai and Kanehisa [3] presented an expert system by constructing a knowledge base and using if–then rules to predict the protein subcellular locations. The expert system works based on sequence motifs and amino acid compositions. Fourteen locations for animal cells and 17 for plant cells can be distinguished on the set of 401 eukaryotic proteins. The system can predict protein subcellular locations with the accuracy of 66% in training and 59% in testing.

Reinhardt and Hubbard [4] presented a method using neural networks to predict the subcellular locations of proteins in prokaryotic and eukaryotic cells based on their amino acid composition. The prediction accuracy can reach 81% for three possible subcellular locations for prokaryotic proteins and 66% for four locations for eukaryotic proteins.

Also based on the amino acid composition, Yuan [5] presented Markov chain models for protein subcellular location prediction. For prokaryotic proteins, the

Machine Learning in Bioinformatics. Edited by Yan-Qing Zhang and Jagath C. Rajapakse
Copyright © 2009 by John Wiley & Sons, Inc.

method can achieve a prediction accuracy of 89.1% for three subcellular locations. For eukaryotic proteins, the prediction accuracies can reach 73.0% and 78.7% within four and three location categories, respectively.

Huang and Li [6] introduced a method using a fuzzy k-nearest neighbors (k-NN) algorithm to predict protein subcellular locations from their dipeptide composition. The overall prediction accuracy is about 80% in a jackknife test. The same method is also applied to annotate six entirely sequenced proteomes, namely, *Saccharomyces cerevisiae, Caenorhabditis elegans, Drosophila melanogaster, Oryza sativa, Arabidopsis thaliana*, and a subset of all human proteins.

Besides the methods and techniques mentioned above, kernel methods such as support vector machines (SVMs) [7, 8] were also used to predict protein subcellular locations [9–11]. SVMs can effectively solve nonlinear binary classification problems with good generalization capability. SVMs have two key features: One is constructing the separating hyperplane with the maximum margin and the other is the kernel-based feature transformation. With the help of a nonlinear kernel, input data are transformed into a high-dimensional feature space where it is "easy" for the SVMs algorithm to find a hyperplane to separate data.

Hua and Sun [9] first employed SVMs for the protein subcellular location prediction. They built a group of binary SVMs models and combined outputs of SVMs with the one-versus-rest approach for multiclassification. The data features were generated based on 20 single amino acid compositions only. In the jackknife test, Hua and Sun's method achieved the overall prediction accuracy of 79.4% on the eukaryotic sequences and 91.4% on the prokaryotic sequences.

Cai et al. [10] also presented an SVM-based prediction method by incorporating the quasi-sequence-order effect. Both the amino acid composition and the sequence-order-coupling numbers were used to improve prediction quality. The experimental data set includes 2191 protein sequences of 12 groups. The prediction accuracy is 75% by using the jackknife test.

Later, Park and Kanehisa [11] presented a new method using SVMs with the one-versus-rest voting approach. Besides amino acid composition, amino acid pair and gapped amino acid pair compositions were considered in the prediction [12]. Park and Kanehisa used fivefold cross-validation to evaluate the method's performance.

In this chapter, we propose a new method to predict protein subcellular locations. The new method is designed based on SVMs with evolutionary granular kernel trees (EGKTs) [13] and the one-versus-one voting approach. Features are grouped into feature granules according to the types of amino acid compositions and the similarity of protein sequences is measured by EGKTs. The method is evaluated on the data set collected by Park and Kanehisa [11] and the simulation results show that the new method is competitive in terms of the total accuracy and the location accuracy.

The rest of the chapter is organized as follows. Granular feature transformation and EGKTs are presented in Section 10.2. Approaches to combine outputs of SVMs for multiclassification are introduced in Section 10.3. Section 10.4 shows the simulation of protein subcellular location prediction with the new method. Finally, Section 10.5 gives conclusion and outlines the future work.

10.2   GRANULAR FEATURE TRANSFORMATION AND EGKTs

The performance of SVMs is mainly affected by kernel functions. Traditional kernels, such as RBF kernels and polynomial kernels, process data as one unit in learning. With growing interests in protein sequence prediction, more powerful and flexible kernels need to be designed to handle high dimensions and incorporate prior domain knowledge. EGKTs are designed based on granular computing (GrC) [14] and evolutionary computation techniques. EGKTs can pay different attentions to feature granules and measure their similarity effectively.

10.2.1   Granular Feature Transformation

Definition 10.1 Granular feature transformation is a mapping in which a feature granule is transformed from the feature granule space into a new feature space through a function.

Here, a feature granule space is a subspace of input space and a feature granule is a vector defined in the feature granule space. Granular feature transformation may be implemented with a group of kernels on feature granules.

Definition 10.2 A granular kernel gK is a kernel that can be written in an inner product form of Equation 10.1, where φ is a mapping function from the feature granule space G to an inner product feature space and $\vec{g}, \vec{g'} \in G$.

$$gK(\vec{g}, \vec{g'}) = \langle \varphi(\vec{g}), \varphi(\vec{g'}) \rangle. \tag{10.1}$$

Just like traditional kernels, granular kernels are closed under sum, product, and multiplication with a positive constant over the granular feature spaces. Given two feature granules $\vec{g}$ and $\vec{g'}$ defined in the space G and two granular kernels gK_1 and gK_2 defined over $G \times G$, the following are granular kernel functions.

$$c \times gK_1(\vec{g}, \vec{g'}), \ c \in R^+. \tag{10.2}$$
$$gK_1(\vec{g}, \vec{g'}) + c, \ c \in R^+. \tag{10.3}$$

$$gK_1(\vec{g}, \vec{g'}) + gK_2(\vec{g}, \vec{g'}). \tag{10.4}$$

$$gK_1(\vec{g}, \vec{g'}) \times gK_2(\vec{g}, \vec{g'}). \tag{10.5}$$

10.2.2   EGKTs

An easy and effective way to design a new kernel function is combining a group of granular kernels in a tree via some simple operations such as sum and product. The following are main steps in the granular kernel tree (GKT) design.

Step 1 Generate feature granules. Features are grouped into feature granules according to some prior knowledge such as object structures, feature relationships, similarity, or functional adjacency. They may be grouped together by an automatic learning algorithm too.

Step 2 Select granular kernels. Granular kernels are selected from the candidate kernel set. Some popular traditional kernels such as RBF kernel and polynomial kernel can be chosen as granular kernels. Some kernels designed for some special problems could also be selected as granular kernels.

Step 3 Construct a tree structure. A tree structure is constructed with suitable number of layers, nodes, and connections. As in step 1, we may construct trees according to some prior knowledge or with an automatic learning algorithm.

Step 4 Select connection operations. Each connection operation can be a sum or product in GKTs. A positive connection weight may associate to each edge of the tree.

We use genetic algorithms (GAs) to optimize the parameters of GKTs and refer to such kind of evolutionary GKTs as EGKTs. In one generation, each chromosome encodes a GKT and each GKT parameter is represented by a gene. The roulette wheel method [15] is used for chromosome selection. Before selection, the best chromosome of the previous generation will replace the worst chromosome of the current generation if the best chromosome of the current generation is worse than the best chromosome of the previous generation. In selection, the sum of fitness values F_i is first calculated. A cumulative fitness $\tilde{q}_{ij}$ is then calculated for each chromosome. To select chromosomes, a random number r is generated within the range of $[0,1]$. If r is smaller than $\tilde{q}_{i1}$, then chromosome c_{i1} is selected; otherwise, chromosome c_{ij} is selected if r is in the range of $(\tilde{q}_{ij-1}, \tilde{q}_{ij}]$.

$$F_i = \sum_{j=1}^{p} f_{ij}. \tag{10.6}$$

$$\tilde{q}_{ij} = \sum_{t=1}^{j} \frac{f_{it}}{F_i}. \tag{10.7}$$

$$\tilde{q}_{ij-1} < r \le \tilde{q}_{ij}. \tag{10.8}$$

In crossover operation, two chromosomes are first randomly selected from the current generation as parents and then the crossover point is randomly chosen to separate the chromosomes. Parts of chromosomes are exchanged between two parents to generate two children. This genetic operation is equivalent to the process by which two GKTs are selected to exchange parameters. In mutation operation, some chromosomes are randomly selected and some of their genes are replaced by random values. This operation is equivalent to changing some parameters of GKTs randomly. Fivefold cross-validation is used to evaluate the system performance and the fitness for each chromosome is the mean of accuracy.

10.2.3 EGKTs Parallelization

EGKTs can be easily parallelized using the single population master–slave GAs model [16, 17], since all parameters to be optimized are independent in the EGKTs learning. As in nonparallel GAs, there is only one single population in the single population master–slave GAs model. The master node stores the population, performs genetic operations, and distributes individuals to slave nodes. Once every slave node finished the fitness calculation, they send the fitness values back to the master node. This kind of parallel model does not affect the behavior of GAs, since all individuals in the population are considered during the genetic operations.

In the parallelization of EGKTs, the master node stores the population of GKTs; executes selection, crossover, and mutation on GKTs; and then distributes the parameters of GKTs to slave nodes. Each single SVM model is trained and evaluated on one slave node. This parallel system has some characteristics. First, this is a global GAs–SVMs system, since all evaluations and operations are performed on the entire population. Second, the implementation is easy, practical, and especially suitable for SVM model selection and the training speedup of SVMs with EGKTs. The QP decomposition method can be used to speed up the GKT selection too. However, if the training data set is large, the communication costs for transferring sub-QP metaresults will be very high. In the system, the time for QP calculation in each SVM model is longer than that for the genetic operations of GAs, which generally has different magnitude. In our system, only parameters and fitness values need to be transferred between the master nodes and slave nodes. So the communication costs are small. Third, the system can be easily moved to the large distributed computing environment.

10.3 MULTICLASSIFICATION APPROACHES BASED ON SVMs OUTPUTS

SVMs, as a kind of binary classifiers, cannot directly be used for multiclassification. When using SVMs to solve a multiclassification problem, a common way [18] is first decomposing the multiclassification problem into a series of binary classification problems, building SVMs models for each of these binary classification problems, and then combining outputs of SVMs for the multiple-class prediction.

10.3.1 One-Versus-Rest

In the one-versus-rest approach, k binary SVM models are built for k classes separately in the training. The ith SVM model is built with all samples in the ith class with positive labels and all other samples with negative labels. Once an unknown sample needs to be classified, it is first predicted by these SVM models, and then classified into the class corresponding to the SVM model with the highest output value. Hua and Sun [9] first used this approach for the protein subcellular location prediction.

10.3.2 One-Versus-Rest Voting

In Ref. [11], Park and Kanehisa presented a kind of one-versus-rest voting approach for protein subcellular location prediction. Besides the amino acid composition, the amino-acid pair and gapped amino-acid pair compositions [12] were also used to generate data vectors. Based on five different types of compositions (amino acids, amino acid pairs, one-gapped amino acid pairs, two-gapped amino acid pairs, and three-gapped amino acid pairs), five groups of 12 SVMs models are built. In testing, a query protein is first classified by each group of SVMs models, which is the same as that in the one-versus-one approach. Then the final decision is made by voting among the outputs of five groups of SVMs models. In the voting, if the query protein is classified to a same location five, four, or three times, or it is classified to the same location twice and another three different locations once, it will finally be decided as belonging to this location. When the query protein is classified to two locations twice, either of these two locations could be chosen as the final decision. However, in the one-versus-rest voting approach, how to make a decision is an issue in case the query protein is classified to five different locations.

10.3.3 One-Versus-One Voting

In the one-versus-one voting approach, for a k class problem, $k(k-1)/2$ SVMs models are built where each model is built on data from two different classes. When an unknown sample needs to be classified, it is predicted by all SVMs models and $k(k-1)/2$ prediction results are generated. In the prediction with the model built on data from the ith and jth classes, if this sample is classified to the ith class, the vote for the ith class will be added by one. Otherwise, one will be added to the vote for the jth class. The sample is finally classified to the class with the maximum votes. In case two or more classes receive the same maximum votes, one of the classes with the maximum votes will be chosen randomly as the final decision. This approach is used in our method for making the final decision for protein subcellular location prediction.

10.3.4 Directed Acyclic Graph SVM

Directed acyclic graph SVM (DAGSVM) [19] is another approach to combine outputs of SVMs for the multiple-class prediction. Similar to the one-versus-one voting approach, this approach constructs models on data from any two different classes. For a k class problem, $k(k-1)/2$ SVMs models are built. Different from the one-versus-one voting approach, a rooted binary directed acyclic graph (DAG) with $k(k-1)/2$ internal nodes and k leaves is used in the prediction.

When classifying an unknown sample, the prediction starts at the root node, repeatedly moves to either the left or the right child of a node based on the node's decision, and finally reaches a leaf node that indicates the predicted class.

10.4 SIMULATIONS

10.4.1 Data Processing and Experimental Setup

The number of proteins sequences for 12 subcellular locations is listed in Table 10.1. The total number of proteins is 7579 in the data set. Each protein sequence is processed into a vector representation of 821 features, which are then grouped into three feature granules. The first feature granule contains 20 features, which are generated based on the amino acid composition together with one feature representing the sequence length. The second feature granule includes 400 features, which are generated based on the amino acid pairs without gap. The third feature granule contains another 400 features, which are generated based on the amino acid pairs with one gap. Vector features are normalized into the range of [0,1]. In the amino acid pair composition [12], two amino acids χ and β are counted together as one unit $\chi\beta$ if χ and β happen continuously in one sequence. $\chi\beta$ and $\beta\chi$ are two different pairs if χ and β are different amino acids. The gapped amino acid pair composition means some number of intervening residues can exist in the pair. The constructed GKT used to measure data similarity is shown in Fig. 10.1.

In the simulation, radial basis function (RBF) $\exp(-\gamma||\vec{x}-\vec{y}||^2)$ is chosen as the granular kernel function for each feature granule. The range of each γ is set to [0.0001–0.5]. The range of regularization parameter C of SVMs is set to [1–400]. The probability of crossover of GAs is 0.7 and the mutation ratio is 0.2. The population size is set to 100 and the number of generations is set to 50. The SVMs software used in the simulation is LibSVM [20]. The parallel system is implemented on a cluster with four head nodes and 40 computing nodes.

Table 10.1 The number of proteins for each subcellular location

Subcellular Location	No. of Entries
Nuclear	1932
Plasma membrane	1674
Cytoplasmic	1241
Extracellular	861
Mitochondrial	727
Chloroplast	671
Peroxisomal	125
Endoplasmic reticulum	114
Lysosomal	93
Vacuolar	54
Golgi apparatus	47
Cytoskeleton	40

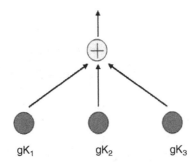

Figure 10.1 GKT used in the simulation.

The total accuracy and location accuracy are used to evaluate the method's performance, which are defined in Equations 10.9 and 10.10, respectively.

$$\text{Total accuracy} = \frac{\sum_{i=1}^{12} T}{N}. \tag{10.9}$$

$$\text{Location accuracy} = \frac{\sum_{i=1}^{12} \frac{T_i}{|S_i|}}{12}. \tag{10.10}$$

In the equations, N is the total number of protein sequences, T_i is the number of correctly predicted positive sequences in the subcellular location set S_i, and $|S_i|$ is the number of sequences in S_i.

10.4.2 Simulation Results

We compare our method with Hua and Sun's method and Park and Kanehisa's method. Three methods are tested with fivefold cross-validation and evaluated on the same data set. The prediction results are summarized in Table 10.2 and Fig. 10.2.

Table 10.2 shows that the total accuracy of our method is 77%, which is better than that of Hua and Sun's method by 7%. The location accuracy of our method can reach 56%, better than that of Hua and Sun's method by 8%. So our method is better than Hua and Sun's method in predicting every subcellular location except cytoskeleton.

Table 10.2 shows that Park and Kanehisa's method can achieve the same total accuracy as our method and a little better location accuracy than ours. However, our method has performance different from Park and Kanehisa's method in predicting different subcellular locations. In predicting mitochondrial, chloroplast, peroxisomal, and endoplasmic reticulum locations, our method has better performance than Park and Kanehisa's method. In predicting the remaining subcellular locations, Park and Kanehisa's method can achieve higher accuracies. If we consider only the first six largest location sets, the location accuracies of the three methods are 67% (Hua and Sun's method), 75% (Park and Kanehisa's method), and 76% (our method).

Table 10.2   Prediction accuracy comparison

Subcellular Location	No. of Entries (Total 7579)	Hua and Sun's Method	Park and Kanehisa's Method	Our Method
Nuclear	1932	0.84	0.90	0.89
Plasma membrane	1674	0.90	0.94	0.92
Cytoplasmic	1241	0.65	0.73	0.71
Extracellular	861	0.72	0.80	0.78
Mitochondrial	727	0.29	0.44	0.55
Chloroplast	671	0.59	0.67	0.69
Peroxisomal	125	0.13	0.20	0.26
Endoplasmic reticulum	114	0.40	0.51	0.55
Lysosomal	93	0.51	0.61	0.59
Vacuolar	54	0.13	0.28	0.30
Golgi apparatus	47	0.06	0.15	0.09
Cytoskeleton	40	0.50	0.65	0.38
Total accuracy		0.70	0.77	0.77
Location accuracy		0.48	0.57	0.56

Figure 10.2 shows that the prediction accuracies of the three methods for each subcellular location have the similar trend if we order locations according to their number of entries. All three methods achieve high accuracies ($\geq 84\%$) in predicting nuclear and plasma membrane sequences and low accuracies ($\leq 30\%$) in predicting peroxisomal, vacuolar, and Golgi apparatus sequences. Using our method, the prediction accuracies for the large location sets such as nuclear and plasma membrane

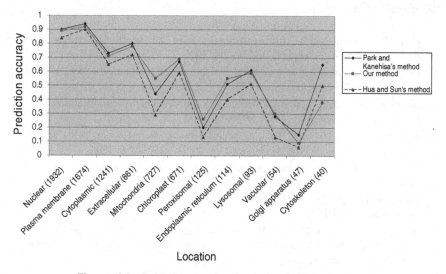

Figure 10.2 Prediction accuracy for each subcellular location.

are much better than those for the small sets such as vacuolar, Golgi apparatus, and cytoskeleton.

10.5 CONCLUSIONS AND FUTURE WORKS

In this chapter, we propose a new method using SVMs with EGKTs and the one-versus-one voting approach to predict protein subcellular locations. The new method can effectively incorporate amino acid composition information and combine binary SVMs models for protein subcellular location prediction. Simulation results show that our method is competitive in terms of total accuracy and location accuracy, especially for the large location sets. The results also prove that our method performs very well for multi-classification. As the population size is only 100 and the number of generations is only 50 in the simulation, we still have space to improve the prediction accuracy by increasing the population size and the number of generations. In future, we will also try to generate new feature granules for other types of compositions to improve the prediction accuracy. A new fitness function will be defined to evaluate the total accuracy and the location accuracy, together with the balance between the accuracy for large location sets and that for small location sets.

ACKNOWLEDGMENT

Bo Jin is supported by Molecular Basis for Disease (MBD) Doctoral Fellowship Program.

REFERENCES

1. Chou, K. C. Prediction of protein structural classes and subcellular locations. *Current Protein and Peptide Science*, 1(2): 171–208, 2000.
2. Feng, Z.-P. An overview on predicting the subcellular location of a protein. *In Silico Biology*, 2(3): 291–303, 2002.
3. Nakai, K. and Kanehisa, M. A knowledge base for predicting protein localization sites in eukaryotic cells. *Genomics*, 14(4): 897–911, 1992.
4. Reinhardt, A. and Hubbard, T. Using neural networks for prediction of the subcellular location of proteins. *Nucleic Acids Research*, 26(9): 2230–2236, 1998.
5. Yuan, Z. Prediction of protein subcellular locations using Markov chain models. *FEBS Letters*, 451(1): 23–26, 1999.
6. Huang, Y. and Li, Y. Prediction of protein subcellular locations using fuzzy k-NN method. *Bioinformatics*, 20(1): 21–28, 2004.
7. Boser, B., Guyon, I., and Vapnik, V.N. A training algorithm for optimal margin classifiers. *Proceedings of the 5th Annual Workshop on Computational Learning Theory*, ACM Press, New York, pp. 144–152, 1992.
8. Vapnik, V. N. *Statistical Learning Theory*, Wiley, New York, 1998.

9. Hua, S. and Sun, Z. Support vector machine approach for protein subcellular localization prediction. *Bioinformatics*, 17(8): 721–728, 2001.

10. Cai, Y.-D., Liu, X.-J., Xu, X.-B., and Chou, K.-C. Support vector machines for prediction of protein subcellular location by incorporating quasi-sequence-order effect. *Journal of Cellular Biochemistry*, 84(2): 343–348, 2002.

11. Park, K.-J. and Kanehisa, M. Prediction of protein subcellular locations by support vector machines using compositions of amino acids and amino acid pairs. *Bioinformatics*, 19(13): 1656–1663, 2003.

12. Chou, K. C. Using pair-coupled amino acid composition to predict protein secondary structure content. *Journal of Protein Chemistry*, 18(4): 473–480, 1999.

13. Jin, B., Zhang, Y.-Q., and Wang, B. Evolutionary granular kernel trees and applications in drug activity comparisons. *Proceedings of the IEEE Symposium on Computational Intelligence in Bioinformatics and Computational Biology*, San Diego, pp. 121–126, 2005.

14. Zadeh, L. A. Toward a theory of fuzzy information granulation and its centrality in human reasoning and fuzzy logic. *Fuzzy Sets and Systems*, 90(2): 111–127, 1997.

15. Michalewicz, Z. *Genetic Algorithms + Data Structures = Evolution Programs*, Springer-Verlag, Berlin, 1996.

16. Cantú-Paz, E. A survey of parallel genetic algorithms. *Calculateurs Paralleles. Reseaux et Systems Repartis*, 10(2): 141–171, 1998.

17. Adamidis, P. *Review of parallel genetic algorithms bibliography. Internal Technical Report,* Automation and Robotics Lab, Aristotle University of Thessaloniki, Thessaloniki, Greece, 1994.

18. Hsu, C.-W. and Lin, C.-J. A comparison of methods for multiclass support vector machines. *IEEE Transactions on Neural Networks*, 13(2): 415–425, 2002.

19. Platt, J. C., Cristianini, N., and Shawe-Taylor, J. Large margin DAGs for multiclass classification. In: S. A. Solla, Leen, T. K. and Müller, K.-R. Eds, *Advances in Neural Information Processing Systems, Vol.12*, MIT Press, Cambridge, MA, pp. 547–553, 2000.

20. Chang, C.-C. and Lin, C.-J. LIBSVM: a library for support vector machines. Available at http://www.csie.ntu.edu.tw/~cjlin/libsvm) 2001.

11

PROBABILISTIC MODELS FOR LONG-RANGE FEATURES IN BIOSEQUENCES

Li Liao

Over the last decade, bioinformatics as a rapidly emerging interdisciplinary field has witnessed unprecedented growth on many forefronts, from clustering gene expression data to inferring protein–protein interactions, from detecting homologous proteins to predicting the secondary or tertiary structures. However, because noise and errors are inevitably embedded in high-throughput data and the difficulties in data analysis are compounded by our limited understanding of underlying biological mechanisms, bioinformatics solutions to many biological problems often rely on machine learning techniques and necessarily involve statistical and probabilistic modeling. These models are built to capture and represent the characteristics of biological objects. The free parameters in the models are fixed during training of data where answers are already known and the trained models are then used to make prediction on new data. Although such paradigms of modeling seem to have a commonplace among machine learning applications, special attention is required during model building and training in order to incorporate domain-specific knowledge and details about the biological data so that the model does not become just another "black box." In this chapter, we will discuss some probabilistic models that are aimed at capturing long-range features in biosequences to achieve better predictions on the features of biosequences.

Machine Learning in Bioinformatics. Edited by Yan-Qing Zhang and Jagath C. Rajapakse
Copyright © 2009 by John Wiley & Sons, Inc.

11.1   DNA, RNA, AND PROTEINS

By biosequences we mean the sequences of DNA, RNA, and proteins—three types of macromolecules essential to the life of a cell and thereof to life of all kinds. Despite large sizes and complex structures of these macromolecules, they all are formed as chains of small units at a basic level. In DNA and RNA, these small units consist of four types of nucleotides, whereas in proteins there are 20 types of amino acids. To a large extent, it is the sequential order of how these small units are put together that determines the molecule's structures and functions. This lays down the foundation for all sequence analysis tools in bioinformatics.

Because of their different roles, questions to be asked are different for DNA, RNA, and proteins. A DNA molecule consists of two chains of nucleotides, which are folded into the famous double helix shape. People now know that genetic information of an organism is encoded in DNA molecules in terms of their nucleotide compositions. For example, the size of a human genome is about three billion nucleotides, and the number of different possible compositions is therefore equal to $4^{3,000,000,000}$. Of such a huge number of possibilities, is the human genome just a random one? The answer is no. For example, it has been found that in eukaryotic genomes, the coding regions—segments of sequences that encode proteins—tend to have more G and C types than would be expected from random sequences. Moreover, in coding regions the nucleotide C is more likely followed by a G. Naturally, such patterns have been utilized to identify coding regions. Correlations are not just observed in neighboring bases; long-range correlations in sequence composition of DNA are also observed [1–3], though their chemical and biological causes are not entirely understood as yet [4, 5]. Therefore, there is a need to detect sequence patterns, long or short, which deviate from a random (or the average) sequence composition and associate them with relevant biological functions.

The genetic information stored in the DNA is utilized in the cell via a two-step process: DNA is transcribed into messenger RNA and the messenger RNA is then translated into proteins. This process is referred to as the central dogma of molecular biology. Unlike DNA having two strands folded into a double helix, RNA molecules have a single strand of nucleotides, with the nucleotide T being substituted by uracil U, as a result of transcription. The single strand of RNA then folds onto itself by maximally pairing up C with G, and A with U. Because hydrogen bonds formed in the pairings C–G and A–U lower the free energy, the most thermodynamically stable structures correspond to those that maximally pair up C–G and A–U; and such pairings may happen across a long range of nucleotides that are separated far in the strand. It is known that the structure plays a more important role than the sequence composition in determining the functions of RNA molecules.

Proteins are the workhorse of the cell, responsible for most cellular functions, from structural support to catalytic roles as enzymes. As mentioned above, proteins are polypeptides, composed of chains of amino acid residues. Polypeptides in proteins may have 50–1000 amino acid residues, with a typical size of about 300 residues in many organisms. Proteins are synthesized according to the genes that encode them. The synthesis process is called translation, where every three nucleotides in the genes determine which amino acids are to be added to the polypeptide chain of the proteins. With a very few exceptions, the genetic code that maps the 64 possible triplets to the 20 amino acids is the same for all organisms. This universality of the genetic code makes the translation from the mRNA sequence of a gene to its protein product as trivial as following a lookup table. Consequently, if a gene is not a random sequence of nucleotides, its protein product should not be a random sequence of amino acids. As it turns out, each residue complexity of the resulting protein is much higher than that of the corresponding gene. In contrast to the double helix structure of DNA, a bigger alphabet—20 amino acids versus 4 nucleotides—and the different molecular properties possessed by these amino acid residues bestow upon proteins a very rich variety of structural conformations, from globular to transmembrane. Similar to the situations with RNA, a protein's function depends even more closely on its structure.

11.2   GENERAL TASKS IN SEQUENCE ANALYSIS

The tasks for sequence analysis can be very broad. For DNA sequences, a major task is to identify genes against noncoding regions, and this may include identifying exons/introns and splicing sites in the case of eukaryotic genomes. More recently, it has become increasingly important in sequence analysis to identify promoters in order to understand the regulation of gene expression.

Once a gene is identified, two major tasks of sequence analysis are (i) to determine its biochemical function; and (ii) to identify sequence features of biological interest, such as domains, motifs, and secondary structures, which can be responsible for the function. These two tasks, though being formulated as separate problems, are interrelated and hence are often addressed together. For example, knowing that a protein has an ATP-binding site motif in its sequence would help determine that the protein may function as an ATP-binding protein and vice versa. It is noted that these tasks are often conducted on the sequence of a protein, which is easily obtained from the gene sequence by using the genetic code.

For RNA, a major task is to predict the secondary structure, given the primary sequence. Another major task that has emerged recently, due to the discovery of the regulatory roles of some noncoding RNA, is to identify these same noncoding RNA, such as small interfering RNA (siRNA) and micro RNA (miRNA), and predict their efficacy for regulating the expression of the target genes.

11.3   COMPUTATIONAL APPROACHES AND CHALLENGES
WITH LONG-RANGE FEATURES

To address the aforementioned tasks, most sequence analysis methods take the sequences as input, and often the sequences only. Although it may sound obvious that such "sequence-only" approaches should work for DNA and RNA, it will take a major breakthrough in protein chemistry to assure that they also work for proteins. As mentioned above, what matters for DNA as a storage medium of genetic information is the sequential order of the nucleotides. For RNA, while the structures become more important, their deduction from sequences is relatively well formulated. Proteins, however, have much more diverse and complex structures, known to play key roles in determining their functions. Anfinsen's [6] thermodynamic hypothesis, supported by a series of experimental observations, established the important principle of protein chemistry that the amino acid sequence of a protein determines the protein's native conformation and the conformation determines the biological activities. This principle serves as a theoretical foundation for most sequence analyses where a common goal is to use amino acid sequences to predict protein structures and functions.

Therefore, in bioinformatics, many computational tools are developed with the objective of inferring protein functions based on the primary sequence only, without getting into molecular details. Such a strategy of relying as much as possible on the primary sequences is quite successful in some cases but not so successful in other cases. Even when molecular details are accounted for, the *ab initio* prediction of three-dimensional (3D) structure for proteins is still a very difficult and computationally demanding task, with only limited successes for small proteins. On a practical side, many details about the underlying molecular mechanisms are not directly utilized, but instead are abstracted as relationships among symbols at higher levels to be used as input to the computational models and tools. For example, amino acid substitute matrices, such as PAM [7] and BLOSUM matrices [8] used in BLAST search [9], and encapsulate a lot of information about the molecular mechanisms inside the cell, which is only manifested as how often a given amino acid may be substituted by other amino acids during evolution. There exist other matrices that are fine tuned for capturing specific sequence features [10]. Certainly, some information loss is inevitable during the abstraction and encapsulation. Yet, it presents a real challenge to balance between reducing the complexity and at the same time retaining the valuable information. Similarly important are how to extract the useful information out of biosequences, such as long-range correlations and nonlocal interactions [2, 11–15], and how to represent and integrate the information for proper use in sequence analysis.

Sequence analysis typically begins with comparisons, for example, by aligning two proteins, residue by residue, in order to reveal sequence similarity. Because significant similarity implies homology, if the function of one protein is known, then the other protein is likely to have the same function. The similarity-based homology detection approaches serve as a cornerstone for protein annotation. Similarly, sequence features or patterns that are conserved during evolution among (homologous) proteins can also be identified from alignment, and are likely to bear some biological meaning, for example, a DNA-binding domain. The accuracy of both homology detection and feature identification/extraction can be problematic when percent identity of two sequences is low (<30%), which is the case for remote homologous proteins. However, the situation can be improved by collecting aggregate statistics from a group of protein sequences, because inference drawn on weak similarities from a pairwise alignment may be strengthened by having more supporting evidences from the alignment of multiple sequences [16]. Along this line are the methods that first classify proteins into families based on functions, domains, and structures [17–19], and then transform the tasks of annotating unknown proteins into a classification problem: given a protein, which family does it belong? Here are the two typical scenarios.

Scenario 1. Given a set of proteins, classify them into families based on certain criteria such as functions, domain/motifs, or structure.

Scenario 2. Given a single query protein and a database of protein families, predict which family the query protein belongs to.

Many computational tools have been developed to meet the needs of classification and prediction. Among the methods using aggregate statistics, hidden Markov models (HMM) [20–25] are one of the most sophisticated and successful sequence analysis techniques. An HMM is a probabilistic model that is capable of capturing sequence features of meaningful biological properties. In a typical application, the model represents the sequence as a first-order Markov chain with the sequence features as building blocks, called states. Each sequence feature, that is, a state in the Markov chain, is composed of a single or few amino acids, with variations in its composition. The architecture of a model refers to how these states are arranged. If a state is represented as a vertex and the transition between two states is represented as a directed edge between the two corresponding vertices, then the model architecture can be conveniently represented as a graph, though in many cases the graph is simply a chain of vertices. Besides the architecture, an HMM is also characterized by transitional probabilities from one state to the next in the Markov chain and by the frequencies of observing different variations of the sequence feature at a given state. Since first introduced into bioinformatics around 1994, HMM has been applied with remarkable success to gene finding, protein family classification, protein secondary structure prediction, and phylogenetic analysis [20]. Also, a number of theoretic contributions to machine learning techniques have been made from HMM, and new areas for research in HMMs in biology have been opened [25, 26].

However, HMM suffers from two fundamental problems when applied to biological sequence analysis. One problem is the model's deficiency of expressiveness namely, the model cannot capture some essential features of the biological entities (protein sequences in this case) that it is intended to simulate. For example, the model assumes that the probability of having a given amino acid at any position depends only on the state at the neighboring position. Such an assumption does not hold in general and becomes more problematic when long-range correlations are significant. It is known that nonlocal interactions are necessary for protein secondary structural elements to fold into a cohesive native structure. These interactions certainly put selection pressure on sequences during evolution and manifest via correlations among amino acids. Statistical analysis of amino acid patterns in approximately 160,000 alpha-helices in experimentally determined structures revealed dipeptides, tripeptides, and tetrapeptides whose frequencies deviate most from the statistical model. Importantly, some sequences were never found in alpha-helices. For example, almost 43% of possible tetrapeptides never appear in alpha-helices [27]. While correlations and interactions across multiple residues are typically defined as nonlocal

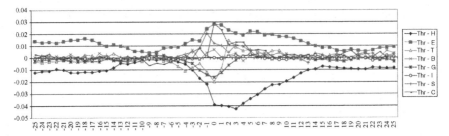

Figure 11.1 Correlation of the amino acid threonines with various secondary structures at neighboring positions.

or long-range for amino acids that have the sequence separation >6, the actual range can be longer. For example, in Fig. 11.1, amino acid threonines, as strand admirers, are shown to be correlated with various secondary structure elements at neighboring positives, and their negative correlations with alpha-helices (H) and positive correlations with strand (E) remain significant up to 20 (from −10 to 10). These statistics, as reported in Ref. [28], are collected among 282,329 amino acids in 1737 protein sequences from protein data bank (PDB) (http://www.rcsb.org/pdb/).

The other problem is about model equivalence, that is, how to determine whether two models represent the same biological entity. For example, when used for classifying proteins into families, a model represents a family of proteins. The same protein family may be represented by two apparently different HMM if constructed separately using different software systems or using different parameter settings even in the same software. It is desirable to be able to compare models for equivalenc and, even more useful, to reveal relationships among different protein families. However, it is known that model equivalence cannot be well formulated to be tractable within HMM [29].

11.4 TRANSMEMBRANE PROTEINS

Membrane proteins serve as highly active mediators between the cell and its environment or between the interior of an organelle and the cytosol. They transport specific metabolites and ions across the membrane barriers, convert energy of sunlight into chemical and electrical energies, and couple the flow of electrons to the synthesis of ATP. Furthermore, they act as signal receptors and transduce signals such as neurotransmitters, growth factors, and hormones across the membrane. On average, about 25% of the proteomes of an organism are membrane proteins. Because of their vast functional roles, membrane proteins are important targets of pharmacological agents.

Knowledge of the transmembrane helical topology—the number of membrane helices, their locations, and orientation—can help identifying binding sites and infer functions of membrane proteins. Figure 11.2 shows a type of transmembrane proteins called G-protein coupled receptor (GPCR) with seven helical transmembrane (TM)

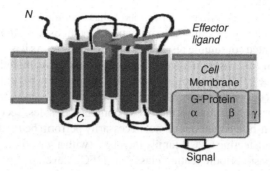

Figure 11.2   G-Protein coupled receptor. The seven black cylinders represent transmembrane helices.

domains, spanning the cell membrane, and the binding site where an effector ligand can bind and trigger the release of the G-protein coupled to the receptor from the inside. Unfortunately, membrane proteins are hard to solubilize and purify in their native conformation because they have both hydrophobic and hydrophilic regions on their surfaces and thus it is difficult to determine their structure experimentally by using X-ray crystallography or NMR spectroscopy. Although the difficulties with experimental approaches have motivated the development of various computational approaches, the computational methods face special challenges due the lack of experimentally determined structure to serve as training data.

Most of these computational approaches rely on the compositional bias of amino acids at different regions of the sequence [30, 31]. For example, there is a high propensity of hydrophobic residues in transmembrane alpha-helices due to the hydrophobic environment in lipid membranes. Because such a bias is quite noticeable and consistent, the location of transmembrane domains can often be easily identified with high accuracy even by a simple method such as applying a threshold on the hydrophobic propensity curve, calculated from a fixed size window sliding across the sequence. It is worth noting that in addition to the inherent problem stemmed from the sliding window's fixed size, prediction accuracy might have been significantly overestimated [32]. Another compositional signal in membrane proteins is the abundance of positively charged residues in the segments (loops) that are located on the cytoplasmic side of the membrane and therefore is referred to as the "positive inside rule" for predicting the orientation of a transmembrane helix [30, 31]. Unlike the hydrophobicity signal for transmembrane helices, the "positive inside rule" is a weaker signal and often confused by significant presence of positively charged residues in globular domains of the protein on the noncytoplasmic side. Consequently, it is more difficult to correctly predict the overall topology of a given protein, that is, the orientation of all transmembrane segments.

Local information was estimated to account for roughly 65% of the secondary structure formation [32]. It is therefore necessary to tap into the information that is long-range by nature, and such need seems to be a more pronounced in transmembrane proteins. The periodic domain structure in membrane proteins poses special challenges to sequence analysis methods [30, 31, 33, 34]. Since the task of analyzing whole genomes to identify a specific group of proteins has become more frequent [35], it is imperative to develop more suitable computational methods for better accuracy. Such needs have motivated continuing researches in this direction. As it is known, detection methods based on patterns/motifs are typically more sensitive, though often with low specificity as trade-off. Low specificity can be a serious drawback: for a transmembrane protein, not only it is to know which family it belongs to, say to the GPCR family but also to know the locations of these TM domains. Besides, recent studies have shown that pattern-based methods do not necessarily perform better than nonpattern-based methods in detecting GPCR family members (with a sensitivity of 78%) [36]. A method called quasi-periodic feature classifier (QFC), developed by Kim et al [37], has increased the differentiating power of its classifier by taking into account the periodic feature of transmembrane domains. Being a discriminative rather than generative model, however, the QFC method only determines whether a given protein is a GPCR, but does not predict the topological structure of transmembrane helices for the putative GPCRs.

11.4.1   Modeling Long-Range Features in Hidden Markov Models

Despite the issues with handling the long-range sequence features in HMMs in general, significant progress has been made recently in applying HMMs to classify transport membrane protein families, to identify transport proteins in newly sequenced genomes, and to predict transmembrane domains and topology [38–40]. The prediction accuracy of HMM has increased from around 75% to around 85% [39, 41]. An innovative idea proposed in Ref. [41] is to design a hidden Markov model (TMHMM) whose architecture encodes the sequence's "grammar" characteristic of transmembrane proteins, namely, the cytoplasmic and noncytoplasmic loops have to alternate with a TM in between. Further improvement in prediction accuracy was achieved in another HMM (TMMOD), which has a more refined architecture to differentiate the long loops from short loops [38, 39].

Figure 11.3 shows the architecture of TMMOD. Instead of a naive model of three states (one for each of the three features), there are four submodels designed to account

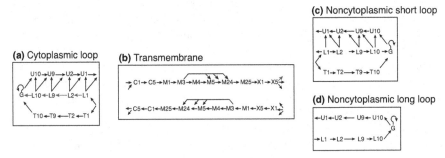

Figure 11.3   The architecture of TMMOD. Four submodels are designed to account for transmembrane domains, loops on the cytoplasmic side, and short and long loops on the noncytoplasmic side.

for transmembrane domains, loops on the cytoplasmic side, and short and long loops on the noncytoplasmic side. The design differs from TMHMM on the architecture of the submodels for loops on both sides of the membrane. The path of states (from T1 to T10), which is not present in TMHMM, is devised in TMMOD to account for loops that are longer than 20 residues so that these loops do not have to share the path of states (from L1 to L10) with loops that are 20 residues or shorter.

Therefore, the long-range effects were taken into account by introducing more hidden states in TMHMM and TMMOD. The further improvements attained in TMMOD are partly attributed to the extra states forming a bypass path in the loop submodels to accommodate the apparently different distributions for long (>20 amino acids) and short ($\leq$20 amino acids) loops. Another difference between TMHMM and TMMOD is the training procedure. TMHMM uses standard Baum–Welch algorithm [25] augmented with a specialized relabeling procedure, whereas TMMOD adopts a simple Bayesian-based approach with regularization using BLOSUM50 amino acid substitution and Dirichlet mixture prior [25].

11.4.1.1 *Coupling the Transition with Emission* The problem posed by quasi-periodic features of TM alpha-helices is that not all helices are identical, either by length or by composition. But in the model, they are collectively represented by one submodel (in Fig. 11.3b). The variations in length are accounted for by these bypass transitions, for example, there is a transition from M3 to M5 bypassing M4. However, the emission frequencies and transition probabilities in the submodel are trained by maximum likelihood optimization of all TM alpha-helices, and peculiar deviations from the average are smoothed out. One way to "memorize" is to couple the transition with the emission, that is, transition to the next state depends on the symbol emitted from the current state. Since emission and transition are orthogonal in HMMs, the coupling requires reformulation of the HMM and the associated algorithms: Viterbi decoding, forward algorithms, and Baum–Welch algorithms [25].

While coupling the transition with emission is expected to complicate the training procedures, the decoding procedure, which marks out where the TM helices are for a given sequence, is to remain unchanged in terms of the complexity. The recurrence equation used in Viterbi algorithm $v_l(i+1) = e_l(x_{i+1}) \max_k(v_k(i)\, a_{kl})$ is only slightly changed to $v_l(i+1) = e_l(x_{i+1}) \max_k(v_k(i)\, a_{kl}^i)$, where a_{kl}^i indicates the transition probability from state k to l specified by the symbol at position i, which is a simple lookup and is needed only once for all calculations in $v_l(i+1)$. The complication in general training procedure (Baum–Welch algorithm) can be largely avoided, as in TMMOD, since the training data are annotated, that is, the topology for each training sequence is known from the experiments. The Bayesian-based approach will still be invoked to incorporate Dirichlet and substitution matrix mixtures to regularize the estimated parameters. This becomes more critical to curb the potential for overfitting due to increased number of parameters. To further alleviate the problems stemming from the increased number of parameters, the coupling is limited to specific sites, for instance, the states bordering the submodels. This helps produce more accurate delimitation of the TM with the loops. Also, the coupling can be limited to groups of amino acids, for example, the hydrophobic ones as a group, instead of coupling to individual amino acids.

11.4.1.2 Feature Selection
Characterization of sequence and structural features is not only useful for improving the models but also biologically insightful.

As the hidden Markov model M for a protein family is trained to capture collective features of member proteins, it is very revealing to look at how the model would respond to individual member proteins, as a measure of how these individual member proteins deviate from the collective behaviors. One such measure is the Fisher vector score [42].

Definition. Given a hidden Markov model M and a sequence x, $P(x|M, \theta)$ is the probability that x belongs to M, where θ stands for all parameters including both emission frequencies and transition probabilities. $P(x|M, \theta)$ can be calculated by the forward algorithm. Fisher score vector is defined as the gradient $U = \nabla_\theta P(x|M, \theta)$.

Fisher score defined above was proposed for profile HMMs. A variation to the Fisher score is necessary in order to apply it to TMHMM or TMMOD. In TMMOD, unlike the task of family membership prediction using profile HMMs [42], the primary goal here is to find the most probable path of hidden states or label s for a given sequence x. Thus, the quantity of interest is not the likelihood $P(x|M, \theta)$ that the model M assigns to sequence x, but the conditional probability $P(s|x, M, \theta)$ of the label s for a given observation sequence x. Therefore, one variation to Fisher score is $U_{s|x} = \nabla_\theta P(s|x, M, \theta)$, which, after some reformulation using Bayes rule, can be calculated using the forward–backward algorithm [43]. The results demonstrate (Fig. 11.4) that the Fisher score vectors thus calculated show very distinctive features for transmembrane proteins against signal peptides. This can be useful for feature selection to reduce the dimension of the vectors and to provide sharper differentiation at the domain boundaries.

The Fisher score and the proposed variation provide a platform to apply to other classifiers. One classifier proposed is support vector machines (SVMs) [44, 45], which have been used to detect remote protein homologues [40] and other problems [46]. Figure 11.5 shows schematically how an SVM is used in tandem with TMMOD for a

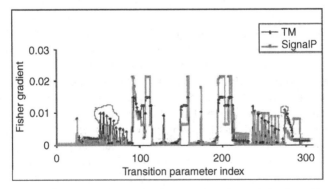

Figure 11.4 Fisher score vectors from transmembrane proteins and signal peptides. Red circles indicate distinctive features between TM and SingnalP.

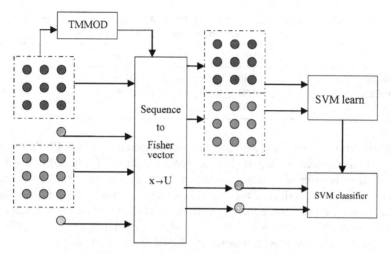

Figure 11.5 Scheme for using TMMOD and SVM in tandem.

10-fold cross-validation experiment. Sequences (blue dots for family members, green dots for nonmembers) are first converted into Fisher score vectors. The training set, with members and nonmembers mixed, are fed to SVM learn, and the testing set (light shaded blue and green dots) are fed to the trained SVM for prediction. This method has been used in Ref. [43] to distinguish transmembrane proteins from signal peptides. The performance is comparable with that in Ref. [47] where an HMM is used to model both transmembrane proteins and signal peptides.

11.4.2 Model Equivalence

As more proteins are classified into families based on functions, patterns/motifs, structures, or other criteria, it becomes critical to study the relations among these classifications, such as gene ontology (http://www.geneontology.org). Because these protein families are often classified and represented as some models, for example, HMMs, efforts have been made recently to evaluate the model equivalence, or more generally, model similarity.

Recently, a similarity metric was developed to compare two HMMs [48]. The comparison is based on the quasi-consensus sequence, which is almost the most probable sequence that an HMM can generate—finding the most probable sequence is NP-complete [29]. The metric is defined as $S = \log P(C_2|M_1)P(C_1|M_2)/P(C_2|R_1)P(C_1|R_2)$, where C_1 and C_2 are the quasi-consensus sequences for model M_1 and M_2, respectively; R_1 and R_2 are just two random reference models. If the two models M_1 and M_2 are similar, the quasi-consensus sequence C_1 from the model M_1 will have high probability of being generated by the model M_2 and vice versa. Hence, the product probability $P(C_2|M_1)P(C_1|M_2)$ is higher than that of the random reference models. The probabilities like $P(C_1|M_2)$ and $P(C_2|M_1)$ can be calculated using the standard forward algorithm [25]. The metric was tested on data collected from SUPERFAMILY and

Table 11.1   ROC scores for three approaches: COMPASS, seed sequence-based HMM comparison and quasi-consensus sequence-based HMM comparison

	COMPASS	Seed	Consensus
Symmetric	0.88	0.86	0.91
Asymmetric	0.84	0.76	0.87
E-value	0.87	–	–

improvements were achieved (see Table 11.1) as compared with COMPASS, the state-of-the-art scoring measure reported in the recent literature [10]. In addition to being used for membership identification, the similarity metric also leads to sequence alignments, as a by-product of the model comparisons, where the quasi-consensus sequences are used as a hinge to align with the seed sequences of each model using Viterbi algorithm. As shown in Table 11.1, the results reported in Ref. [47] also outperformed COMPASS as the benchmarks based on structural alignments using the PDB data. Because "direct" brute-force comparison of a pair of HMMs is NP-complete, that is, not computationally tractable [29], it is desirable to develop better heuristic solutions.

11.5 LONG-RANGE FEATURES IN RNA

The long-range features in RNA have a different nature and therefore require different models. As mentioned earlier, a major sequence analysis task with RNA is to predict the secondary structure, which minimize the free energy or equivalently[1] maximize the number of pairings of A–U and G–C. To achieve the maximum, the pairings can occur across a long range. A typical scenario is the so-called palindromic structure, as shown in Fig. 11.6. If parentheses are used to represent the base pairings, then the structure looks like a string of well-nested parentheses. A grammar different from what was used in TMHMM and TMMOD is needed to describe a "language" made up of such palindromic strings.

11.5.1 Stochastic Context-Free Grammars

In formal language theory, the Chomsky hierarchy of transformational grammars consists of four levels, from least expressive to most expressive: regular, context-free, context-sensitive, and unrestricted grammars [49, 50]. HMMs correspond to regular grammars, which cannot parse nested structures. To parse more sophisticated sequence features, grammars from a higher level in the Chomsky hierarchy are needed. For example, the palindromic structures require context-free grammars. Actually, stochastic context-free grammars (SCFG) have been developed to model the secondary structures of RNA sequences characterizing the palindromic features [51, 52].

---

[1]In addition to the canonical Watson–Crick pairs, Wobble G–U pairs also reduce the free energy.

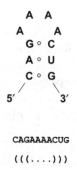

CAGAAAACUG

$(((\ldots)))$

Figure 11.6 Palindromic structure of an RNA sequence.

The SCFG-based model for predicting RNA secondary structure has a set of transformational rules that specify how a string (nonterminal) is parsed as concatenation of bases (terminals) and/or a smaller string flanked with bases on either side or both sides. In addition, for each rule the model has a probability specifying the chance of using that rule for parsing. For example, a context-free grammar allowing just the canonical Watson–Crick pairings is as follows.

$$S \rightarrow aSu|uSa|gSc|cSg|aS|uS|gS|cS|\varepsilon.$$

Then, a sequence $x =$ "guaac" can be parsed by successively applying the suitable rules given above. One parse is as follows:

$$\pi: S \rightarrow \underline{gSc} \rightarrow gu\underline{Sac} \rightarrow gua\underline{Sac} \rightarrow guaac$$
$$() \quad (()) \quad ((.)).$$

The parse corresponds to a secondary structure "((.))."
    Another possible parse is as follows:

$$\pi': S \rightarrow \underline{gS} \rightarrow gu\underline{S} \rightarrow gua\underline{S} \rightarrow guaaS \rightarrow guaacS \rightarrow guaac.$$

It corresponds to a secondary structure ". . .," which is less stable than the previous one because of fewer pairings; actually no pairing at all in this specific example. If each transformational rule is assigned a probability, for example, $p_{S \rightarrow gSc}$ for $S \rightarrow gSc$, then the joint probability $P(x, \pi)$ of parse π on sequence x is given by the product of the probabilities of the individual rules used in the parse.

$$P(x, \pi) = p_{S \rightarrow gSc} \cdot p_{S \rightarrow uSa} \cdot p_{S \rightarrow aS} \cdot p_{S \rightarrow \varepsilon}.$$

Similarly,

$$P(x, \pi') = p_{S \rightarrow gS} \cdot p_{S \rightarrow uS} \cdot p_{S \rightarrow aS} \cdot p_{S \rightarrow aS} \cdot p_{S \rightarrow cS} \cdot p_{S \rightarrow \varepsilon}.$$

If probabilities assigned to rules that have a nonterminal S flanked by bases on both sides, such as $S \rightarrow gSc$, are higher than the probabilities assigned to rules that have S flanked by a base on just the left side, such as S → gS, then obviously parse π is more probable than π' for the given sequence x. Correspondingly, the secondary structure "((.))" is more probable than "....." Since there may be more than one parse that can give the same secondary structure τ, the joint probability of structure τ for a given sequence x is the sum of all possible parses.

$$P(x, \tau) = \sum_{\pi} P(x, \pi).$$

In practice, these probabilities are assigned by training the model on a set of sequences whose secondary structures are determined experimentally. A common procedure for the parameter estimation in training such probabilistic modes is maximum likelihood (ML). Let $\theta = \{p_1, \ldots, p_n\}$ be the set of parameters (i.e., p_i is the probability for the ith rule), and $D = \{(x_1, \tau_1), \ldots, (x_m, \tau_m)\}$ be the set of training sequences and their known secondary structures. The ML procedure will pick θ that maximize the joint probabilities on the training set.

$$\theta_{ML} = \text{argmax}_\theta \prod_{i=1 \text{ to } m} P(x_i, \tau_i; \theta).$$

Despite the elegant formulation and an efficient parsing algorithm and training procedure, the SCFG-based models are not performing as well as those thermodynamic models where more parameters are introduced for capturing detailed characteristics. The popular programs of this kind include MFold [53], ViennaRNA [54], and RDFolder [55].

In a recent work, a new probabilistic model, CONTRAfold, was proposed [56]. The model generalizes SCFG by using discriminative training (i.e., use both positive and negative examples) and feature-rich scoring. CONTRAfold converts an SCFG into a conditional log-linear model (CLLM) where most of the features found in typical thermodynamic modes can be incorporated. Specifically, to follow the treatment in Ref. [56], for a particular parse σ of a sequence x, let $F(x, \sigma) \in \mathbb{R}^n$ be an n-dimensional feature vector (where n is the number of rules in the grammar), whose ith dimension, $F_i(x, \sigma)$, indicates the number of times the ith transformation rule is used in parse σ. Furthermore, let p_i denote the probability for the ith transformation rule. We rewrite the joint likelihood of the sequence x and its parse σ in log-linear form as

$$P(x, \sigma) = \prod_i p_i^{F_i(x,\sigma)} = \exp\left(\ln\left(\prod_i p_i^{F_i(x,\sigma)}\right)\right)$$
$$= \exp\left(\sum_i F_i(x, \sigma) \ln p_i\right) = \exp(\mathbf{w}^T \mathbf{F}(x, \sigma)),$$

where $w_i = \ln p_i$. For sequence x and secondary structure τ, the joint probability can now be rewritten as

$$P(x, \tau) = \sum_\sigma \exp(\mathbf{w}^T \mathbf{F}(x, \sigma)).$$

Such formulation allows for convenient inclusion of features from thermodynamic models, by using F to represent the occurrence frequencies and w the weight scores.

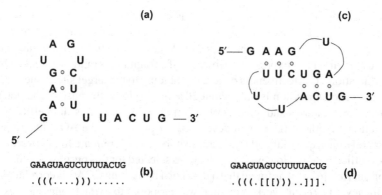

Figure 11.7   Pseudoknot. (**a**) When only nested pairings are allowed. (**b**) The unfolded sequence aligned with parenthesis notation. (**c**) A pseudoknot is formed. (**d**) The unfolded sequence aligned with augmented parenthesis notation.

The features in CONTRAfold include hairpin lengths, bulge loop lengths, internal loop lengths, internal loop asymmetry, affine multibranch loop scoring, and so on. In a set of cross-validation experiments, CONTRAfold not only outperformed these SCFG-based models, but also consistently surpassed the physics-based methods.

There is another type of RNA secondary structures, called pseudoknots, in which the pairings of bases occur not only across some distance, but more distinctively, in a nonnested manner. An example is shown in Fig. 11.7. The nonnested correlations presented in pseudoknots are more difficult to model than the nested correlations. The SCFG-based models cannot capture pseudoknots. While a general context-sensitive grammar can describe the pseudoknotted secondary structures, the parsing of such grammars is computationally intractable [25]. Many efforts have thus been made in developments of computational methods that can handle pseudoknots both effectively and efficiently [57–61]. In Ref. [57], a formal transformational grammar was proposed that encompasses the context-free grammars. The pseudoknot grammar avoids the use of general context-sensitive rules by introducing a small set of auxiliary symbols that are used to form pseudoknots by reordering strings generated by the rest of the part of the grammar that is context-free. A parsing algorithm for the grammar was implemented using dynamic programming in Ref. [62] with a complexity of $O(L^6)$ for an RNA sequence of length L. In a more recent work [63], a special subclass of tree adjoining grammars (TAG) was proposed to model a simple type of pseudoknots, which can represent most known pseudoknot structures in the RNA. In Chomsky hierarchy, the expressiveness of TAGs is in between that of context-free and context-sensitive grammars. A parsing algorithm with time complexity of $O(L^4)$ was given in Ref. [63].

11.5.2 Structural Profiling

Secondary structure may determine the function of an RNA molecule and sometimes affect how efficiently an RNA molecule can fulfill its function. Recently, the regulatory effects of some noncoding RNA have drawn a lot of attention. For example, antisense oligonucleotides (AO), typically 15–30 bases long, can bind to mRNA sequences at specific locations determined by Watson–Crick base pairing rules, and as a result can inhibit gene expression. However, if a fragment is randomly picked based on hybridization, its effectiveness of actually binding to the target site to successfully inhibit the gene expression is only about 20% *in vivo* [64]. Because of the potential therapeutic applications and the cost of experimentally designing high efficacy AO, computational methods have been developed for predicting the AO efficacy. These methods rely on sequential and structural features to differentiate high efficacy AOs from low efficacy AOs. However, most features selected for this purpose are local. For example, in Ref. [65] it was found that a motif of 4 bases long CCAC is correlated with high efficacy. In a recent work, a method was proposed to extract long features. The method works as "reversing" the process of folding the sequence into a secondary structure. Specifically, a profile vector is created by scanning the secondary structure of an AO or its target. It starts with an initial score, say zero. At each base, a score is determined according to whether the base is unpaired, paired with a base downstream, or paired with a base upstream. The cumulated score, that is, the score for the current base added to the score of the previous base, is recorded for the current position. The cumulative scores form a score landscape, which is used to characterize the secondary structure somewhat globally. A simple scoring scheme proposed in Ref. [66] is as follows:

$$\text{score}(a) = +1 \ \text{if } a \text{ is paired downstream,}$$
$$-1 \ \text{if } a \text{ is paired upstream,}$$
$$0 \ \text{if } a \text{ is unpaired,}$$

where $a \in \{A, C, G, U\}$. A refined scoring scheme can be devised to account for the varied strength of different pairing bases, for instance, higher score is assigned to G–C pair than A–U pair. The structural profiles were then concatenated with the profiles based on short local features. These profiles were then used in supervised learning with an SVM. The results from a series of cross-validation experiments indicate significant improvements in the prediction accuracy.

11.6 SUMMARY

In this chapter, we discussed some probabilistic models for analyzing biosequences, and have shown how some features of long-range significance can be incorporated to improve the performance. We particularly discussed HMMs applied to transmembrane proteins and SCFGs applied to RNA folding as two examples. The lessons learned can be summarized in the following four aspects.

Model Architecture. By architecture, it can mean the hidden states and how they are connected in the HMMs; or it can also mean the production rules transforming nonterminals to terminals in the SCFG. This is a key part and is concerned with the model expressiveness, that is, the capability of capturing and representing the targeted features present in the data.

Learning Algorithms. Another essential step in modeling is to fix the free parameters in the models based on training data. Experiences indicate that it is critical to aid the general learning procedures with regularization to allow for incorporation of prior knowledge.

Hybrid Models. Many problems cannot be properly solved by a single model or models of a single type and therefore require hybrid models. For example, a hybrid model may combine an HMM with an SUM, with a phylogenetic tree, or with a neural network. These hybrid models should be combined with other types of machine learning methods to suit specific needs.

Model Equivalence. Objects of interest in biosequences (e.g., proteins) are often classified into families based on biochemical functions and other properties; members in a family share common functions and properties, which are collectively captured and represented as models. On the empirical side, model equivalence and model similarity can tell how these families are related. On the theoretical side, this is concerned with metalearning; the ability to evaluate model equivalence can aid model design.

Because of the extensive existence of the long-range features and their roles in determining and affecting the functions and structures of biosequences, there is a noticeable trend in recent development of machine learning methods in bioinformatics to incorporate long-range information, with necessary modifications to the model architecture and learning algorithms. Indeed, what is reported here represents only a small part of what is already in the literature dealing with long-range effects, but it shows the improvements that can be achieved by treating long-range features, however ad hoc they may be. As our understanding of underlying biological mechanisms deepens, it is expected that these long-range effects will be taken into account more adequately and systematically.

REFERENCES

1. Li, W. and Kaneko, K. Europhysics. *Letters*, 17: 655–660, 1992.
2. Peng, C. K., Buldyrev, S. V., Goldberger, A. L., Havlin, S., Sciortino, F., Simons, M., and Stanley, H. E. Long-range correlations in nucleotide sequence. *Nature*, 356: 168–171, 1992.
3. Buldyrev, S. V. A., Goldberger, L., Havlin, S., Mantegna, R. N., Matsa, M. E., Peng, C.-K., Simons, M., and Stanley, H. E. Long-range correlation properties of coding and noncoding DNA sequences: GenBank analysis. *Physical Review E*, 51: 5083–5091, 1995.
4. Larhammer, D. and Chartzidimitriou-Dreismann, C. A. Biological origins of long-range correlations and compositional variations in DNA. *Nucleic Acids Research*, 22: 5167–5170, 1993.
5. Messer, P. W., Arndt, P. F., and Lassig, M. Solvable sequence evolution models and genomic correlations. *Physical Review Letters*, 94: 138103–138104, 2005.

6. Anfinsen, C. B. Principles that govern the folding of protein chains. *Science*, 181: 223–230, 1973.

7. Dayhoff, M. O., Schwartz, R. M., and Orcutt, B. C. A model of evolutionary change in proteins. *Atlas of Protein Sequence and Structure*, 5: 345–352, 1978.

8. Henikoff, S., and Henikoff, J. G. Amino acid substitution matrices from protein blocks. *Proceedings of the National Academy of Sciences*, 89: 915–919, 1992.

9. Altschul, S. F., Gish, W., Miller, W., Myers, E. W., and Lipman, D. J. Basic local alignment search tool. *Journal of Molecular Biology*, 215: 403–410, 1990.

10. Mittelman, D., Sadreyev, R., and Grishin, N. Probabilistic scoring measures for profile–profile comparison yield more accurate short seed alignments. *Bioinformatics*, 19: 1531–1539, 2003.

11. Abkevich, V. I., Gutin, A. M., and Shakhnovich, E. I. Impact of local and non-local interactions on thermodynamics and kinetics of protein folding. *Journal of Molecular Biology*, 252: 460–471, 1995.

12. Audit, B., Vaillant, C., Arneodo, A., d'Aubenton-Carafa, Y., and Thermes, C. Long-range correlations between DNA bending sites: relation to the structure and dynamics of nucleosomes. *Journal of Molecular Biology*, 316: 903–918, 2002.

13. Klein-Seetharaman, J., Oikawa, M., Grimshaw, S. B., Wirmer, J., Duchardt, E., Ueda, T., Imoto, T., Smith, L. J., Dobson, C. M., and Schwalbe, H. Long-range interactions within a nonnative protein. *Science*, 295: 1719–1723, 2002.

14. Muller, T., Rahmann, S., and Rehmsmeier, M. Non-symmetric score matrices and the detection of homologous transmembrane proteins. *Bioinformatics*, 17: S182, 2001.

15. Paulsen, I. T., Nguyen, L., Sliwinski, M. K., Rabus, R., and Saier, M. H., Jr. Microbial genome analyses: comparative transport capabilities in eighteen prokaryotes. *Journal of Molecular Biology*, 301: 75–100, 2000.

16. Orengo, C. A., Michie, A. D., Jones, S., Jones, D., Swindells, T., and Thornton, J. M. CATH—a hierarchic classification of protein domain structures. *Structure*, 5: 1093–1108, 1997.

17. Finn, R. D., Mistry, J., Schuster-Bockler, B., Griffiths-Jones, S., Hollich, V., Lassmann, S., Moxon, T., Marshall, M., Khanna, A., Durbin, R., Eddy, S. R., Sonnhammer, E. L., and Bateman, S. A. Pfam: clans, web tools and services. *Nucleic Acids Research*, 34: D247–D251, 2006.

18. Hubbard, T. J., Ailey, B., Brenner, S. E., Murzin, A. G., and Chothia, C. SCOP: a structural classification of proteins database. *Nucleic Acids Research*, 27(1999): 254–256, 1999.

19. Hulo, N., Bairoch, A., Bulliard, V., Cerutti, L., De Castro, E., Langendijk-Genevaux, P. S., Pagni, M., and Sigrist, C. J. A. The PROSITE database. *Nucleic Acids Research*, 34: D227–D230, 2006.

20. Krogh, A., Brown, M., Mian, I. S., Sjolander, K., and Haussler, D. Hidden Markov models in computational biology: applications to protein modeling. *Journal of Molecular Biology*, 235: 1501–1531, 1994.

21. Burge, C. and Karlin, S. Prediction of complete gene structures in human genomic DNA. *Journal of Molecular Biology*, 268: 78–94, 1997.

22. Gough, J., Karplus, K., Hughey, R., and Chothia, C. Assignment of homology to genome sequences using a library of hidden Markov models that represent all proteins of known structure. *Journal of Molecular Biology*, 313: 903–919, 2001.

23. Ito, H., Amari, S.-I., and Kobayashi, K. Identifiability of hidden Markov information sources and their minimum degrees of freedom. *IEEE Transactions on Information Theory*, 38: 324–333, 1992.

24. Park, J., Karplus, K., Barrett, C., Hughey, R., Haussler, D., Hubbard, T., and Chothia, C. Sequence comparisons using multiple sequences detect three times as many remote homologues as pairwise methods. *Journal of Molecular Biology*, 284: 1201–1210, 1998.

25. Durbin, R., Eddy, S., Krogh, A., and Mitchison, G. *Biological Sequence Analysis*, Cambridge University Press, Cambridge, UK, 1998.

26. Birney, E. Hidden Markov models in biological sequence analysis. *IBM Journal of Research and Development*, 45: 449–454, 2001.

27. Acevedo, O. E. and Lareo, L. R. Amino acid propensities revisited. *OMICS: A Journal of Integrative Biology*, 9: 391–399, 2005.

28. Malkov, A., Zivkovic, M. V., Beljanski, M. V., and Zaric, S. D.Correlations of amino acids with secondary structure types: connection with amino acid structure, 2005. Available at http://arxiv.org/ftp/q-bio/papers/0505/0505046.pdf.

29. Lyngso, R. B. and Pedersen, C. N. S. Complexity of comparing hidden Markov models. In: E. Eades and T. Takaoka, Eds, *International Symposium on Symbolic and Algebraic Computation, Lecture Notes in Computer Science,* pp. 416–428, 2001.

30. von Heijne, G. The distribution of positively charged residues in bacterial inner membrane proteins correlates with the transmemebrane topology. *EMBO Journal*, 5: 3021–3027, 1986.

31. von Heijne, G. Membrane protein structure prediction. Hydrophobicity analysis and the positive-inside rule. *Journal of Molecular Biology*, 225: 487–494, 1992.

32. Rost, B., Fariselli, P., and Casadio, R. Topology prediction for helical transmembrane proteins at 86% accuracy. *Protein Science*, 5: 1704–1718, 1996.

33. Casadio, R., Fariselli, P., Taroni, C., and Compiani, M. A predictor of transmembrane alpha-helix domains of proteins based on neural networks. *European Biophysics Journal*, 24: 165–178, 1996.

34. Ren, Q., Kang, K. H., and Paulsen, I. T. TransportDB: a relational database of cellular membrane transport systems. *Nucleic Acids Research*, 32: D284–D288, 2004.

35. Hill, C. A., Fox, A. N., Pitts, R. J., Kent, L. B., Tan, P. L., Chrystal, M. A., Cravchik, A., Collins, F. H., Robertson, H. M., and Zwiebel, L. J. G-Protein-coupled receptors in anopheles gambiae. *Science*, 298: 176–178, 2002.

36. Rost, B. Prediction in 1D: secondary structure, membrane helices, and accessibility. P. E. Bourne and H. Weissig, Eds, *Structural Bioinformatics*, Wiley-Liss, New York, 2003.

37. Kim, J., Moriyama, E. N., Warr, C. G., Clyne, P. J., Carlson, J. R., Identification of novel multi-transmembrane proteins from genomic databases using quasi-periodic structural properties. *Bioinformatics*, 16: 767–775, 2000.

38. Kahsay, R., Liao, L., and Gao, G. An improved hidden Markov model for transmembrane topology prediction. *Proceedings of the 16th IEEE International Conference on Tools with Artificial Intelligence*, Florida, pp. 634–639, 2004.

39. Kahsay, R., Gao, G., and Liao, L. An improved hidden Markov models for transmembrane protein detection and topology prediction and its applications to complete genomes. *Bioinformatics*, 21: 1853–1858, 2005.

40. Liao, L. and Noble, W. S. Combining pairwise sequence similarity and support vector machines for remote protein homology detection. *Proceedings of the Sixth International Conference on Research in Computational Molecular Biology (RECOMB)*, 2002.

41. Krogh, A., Larsson, B., von Heijne, G., and Sonnhammer, E. L. L. Predicting transmembrane protein topology with a hidden Markov model: application to complete genomes. *Journal of Molecular Biology*, 305: 567–580, 2001.

42. Jaakkola, T., Diekhans, M., and Haussler, D. Using the Fisher kernel method to detect remote protein homologies. *Proceedings of the Seventh International Conference on Intelligent Systems for Molecular Biology*, AAAI Press, Menlo Park, CA, pp. 149–158, 1999.

43. Kaysay, R., Gao, G., and Liao, L. Discriminating transmembrane proteins from signal peptides using SVM–Fisher approach. *Proceedings of the 4th International Conference on Machine Learning and Applications*, California, pp. 151–155, 2005.

44. Vapnik, V. *The Nature of Statistical Learning Theory*, Springer-Verlag, New York, 1995.

45. Cristianini, N. and Shawe-Taylor, J. *Introduction to Support Vector Machines and Other Kernel-Based Learning Methods*, Cambridge University Press, Cambridge, UK, 2000.

46. Scholkopf, B., Tsuda, K., and Vert, J.-P. *Kernel Methods in Computational Biology (Computational Molecular Biology)*, MIT Press, Cambridge, Massachusetts, 2004.

47. Käll, L., Krogh, A., and Sonnhammer, E. L. L. A combined transmembrane topology and signal peptide prediction method. *Journal of Molecular Biology*, 338: 1027–1036, 2004.

48. Kahsay, R. Y., Wang, G., Gao, G., Liao, L., and Dunbrack, R. Quasi-consensus-based comparison of profile hidden Markov models for protein sequences. *Bioinformatics*, 21: 2287–2293, 2005.

49. Chomsky, N. Tree models for the description of language. *IRE Transactions on Information Theory*, 2: 113–124, 1956.

50. Chomsky, N. On certain formal properties of grammars. *Information and Control*, 2(1959): 137–167, 1959.

51. Eddy, S. and Durbin, R. RNA sequence analysis using covariance models. *Nucleic Acids Research*, 22: 2079–2088, 1994.

52. Sakakibara, Y., Brown, M., Hughey, R., Mian, I. S., Sjolander, K., Underwood, R. C., and Haussler, D. Stochastic context-free grammars for tRNA modeling. *Nucleic Acids Research*, 22: 5112–5120, 1994.

53. Zuker, M. Mfold web server for nucleic acid folding and hybridization prediction. *Nucleic Acids Research*, 31: 3406–3415, 2003.

54. Hofacker, I. L., Fontana, W., Stadler, P. F., Bonhoeffer, L. S., and Schuster, P. Fast folding and comparison of RNA secondary structures (The Vienna RNA Package). *Monatshefte fur Chemie*, 125: 167–188, 1994.

55. Ying, X., Luo, H., Luo, J., and Li, W. RDfolder: a web server for prediction of RNA secondary structure. *Nucleic Acids Research*, 32: W150–W153, 2004.

56. Do, C. B., Woods, D. A., and Batzoglou, S. CONTRAfold: RNA secondary structure prediction without physics-based models. *Bioinformatics*, 14: e90–e98, 2006.

57. Rivas, E. and Eddy, S. R. The language of RNA: a formal grammar that includes pseudoknots. *Bioinformatics*, 16: 334–340, 2000.

58. Lyngso, R. B. and Pedersen, C. N. S. RNA pseudoknot prediction in energy-based models. *Journal of Computational Biology*, 7: 409–427, 2000.

59. Cai, L., Malmberg, R. L., and Wu, Y. Stochastic modeling of RNA pseudoknotted structures: a grammatical approach. *Bioinformatics*, 19: i66–i73, 2003.

60. Ieong, S., Kao, M.-Y., Lam, T.-W., Sung, W.-K., and Yiu, S.-M. Predicting RNA secondary structures with arbitrary pseudoknots by maximizing the number of stacking pairs. *Journal of Computational Biology*, 10(6): 981–995, 2003.

61. Ruan, J., Stormo, G. D., and Zhang, W. An iterated loop matching approach to the prediction of RNA secondary structures with pseudoknots. *Bioinformatics*, 20: 58–66, 2004.

62. Rivas, E. and Eddy, S. R. A dynamic programming algorithm for RNA structure prediction including pseudoknots. *Journal of Molecular Biology*, 285: 2053–2068, 1999.

63. Matsui, H., Sato, K., and Sabakibara, Y. Pair stochastic tree adjoining grammars for aligning and predicting pseudoknot RNA structures. *Bioinformatics*, 321: 2611–2617, 2005.

64. Myers, K. and Dean, N. Sensible use of antisense: how to use oligonucleotides as research tools. *Trends in Pharmacological Sciences*, 21: 19–23, 2003.

65. Matveeva, O. V., Tsodikov, A. D., Giddings, M., Freier, S. M., Wyatt, J. R., Spiridonov, A. N., Shabalina, S. A., Gesteland, R. F., and Atkins, J. F. Identification of sequence motifs in oligonucleotides whose presence is correlated with antisense activity. *Nucleic Acids Research*, 28: 2862–2865, 2000.

66. Craig, R. and Liao, L. Prediction of antisense oligonucleotide efficacy using local and global structure information with support vector machines. *Proceedings of the Fifth International Conference on Machine Learning and Applications*, Orlando, FL, 2006.

12

NEIGHBORHOOD PROFILE SEARCH FOR MOTIF REFINEMENT

Chandan K. Reddy, Yao-Chung Weng, and Hsiao-Dong Chiang

12.1 INTRODUCTION

Recent developments in DNA sequencing have allowed biologists to obtain complete genomes of several species. However, knowledge of the sequence does not imply the understanding of how genes interact and regulate one another in the genome. Many transcription factor binding sites are highly conserved throughout the sequences and the discovery of the location of such binding sites plays an important role in understanding gene interactions and gene regulation. Biological experiments such as DNA footprinting, gel shift analysis, and others [1] are tedious and time consuming. To complement these traditional experimental techniques, scientists have begun to develop computational methods to identify these regulatory sites.

In this chapter, we develop a new computational approach to find the regulatory sites in an efficient manner [2]. Although there are several variations of the motif finding algorithms, the problem studied in this chapter is defined as follows: Without any previous knowledge of the consensus pattern, discover all the occurrences of the motifs and then recover a pattern for which all of these instances are within a given number of mutations (or substitutions) [3]. Usually, every instance of a motif will have the same length but slightly different sequence compositions. In this chapter, we consider a precise version of the motif discovery problem in computational biology as discussed in Refs [4, 5]. The planted (l,d) motif problem [5] considered here is

Machine Learning in Bioinformatics. Edited by Yan-Qing Zhang and Jagath C. Rajapakse
Copyright © 2009 by John Wiley & Sons, Inc.

GAATTCATACCAGATCAC CGGATTCCCGA CTCCAAATGTGTCCCCCTCACAC

TCCC CCGATTACCGT CTTCTGCTCTTAGACCACTCTACCCTATTCCCCACACT

CACCGGAGCCAAAGCCGCGGCCCTTCCGTT CCGATTACCGA AAAGACCCCA

CCCGTAGGTGGCAAGCTAGCTTAAGTAACGCCACT TCGATTAACGA GGAAA

AATACATAACTGA CCTATTATCGA GTTCAGATCAAGGTCAGGAACAAAGAA

ACA CCGATTACCGT AACCGTAAGATAATGGTATCGATACGTAGACAGTTTA

Figure 12.1 Synthetic DNA sequences containing some instances of the pattern "CCGATT-ACCGA" with a maximum number of two mutations. The motifs in each sequence are highlighted in the box. We have a (11,2) motif where 11 is the length of the motif and 2 is the number of mutations allowed.

described as follows: Suppose there is a fixed but unknown nucleotide sequence M (the *motif*) of length l. The problem is to determine M, given t sequences with t_i being the length of the ith sequence and each one containing a planted variant of M. More precisely, each such planted variant is a substring that is M with exactly d point substitutions (see Fig. 12.1). More details about the complexity of the motif finding problem are given in Ref. [3]. A detailed assessment of different motif finding algorithms was published recently in Ref. [6].

Despite the significant amount of literature available on the motif finding problem, many do not exploit the probabilistic models used for motif refinement [7, 8]. More details on the estimates of the hardness of this problem without any complex information like overlapping motifs and background distribution are given in Ref. [9]. We provide a novel optimization framework for refining motifs based on TRansformation Under STability-reTaining Equilibria CHaracterization (TRUST-TECH) methodology that can systematically explore subspace and perform neighborhood search effectively [10]. The rest of this chapter is organized as follows: Section 12.2 gives some relevant background of the existing approaches used for finding motifs. Section 12.3 describes the problem formulation and the expectation-maximization (EM) algorithm in detail. Section 12.4 discusses our new framework and Section 12.5 details our implementation. Section 12.6 gives the experimental results obtained from running our algorithm on synthetic and real data sets. Finally, Section 12.7 concludes our discussion with future research directions.

12.2 RELEVANT BACKGROUND

Existing approaches used to solve the motif finding problem can be classified into two main categories [11]. The first group of algorithms utilizes a generative probabilistic representation of nucleotide positions to discover a consensus DNA pattern that maximizes the likelihood score. In this approach, the original problem of finding the best consensus pattern is formulated as finding the global maximum of a continuous nonconvex function. The main advantage of this approach is that the generated profiles are highly representative of the signals being determined [12]. The disadvantage, however, is that the determination of the "best" motif cannot be guaranteed and is often a very difficult problem since finding global maximum of any continuous nonconvex function is a challenging task. Current algorithms converge to the nearest local optimum instead of the global solution. Gibbs sampling [13], MEME [8], greedy CONSENSUS algorithm [14], and HMM-based methods [15] belong to this category.

The second group uses patterns with "mismatch representation", which defines a signal to be a consensus pattern and allows up to a certain number of mismatches to occur in each instance of the pattern. The goal of these algorithms is to recover the consensus pattern with the highest number of instances. These methods view the representation of the signals as discrete and the main advantage of these algorithms is that they can guarantee that the highest scoring pattern will be the global optimum for any scoring function. The disadvantage, however, is that consensus patterns are not as expressive of the DNA signal as the profile representations. Recent approaches within this framework include projection methods [4, 16], string-based methods [5], pattern-branching [17], MULTIPROFILER [18], suffix trees [19], and other branch and bound approaches [20, 11]. Theoretically, the best approach for finding the consensus pattern is an exhaustive pattern-driven search [21]. Since, the pattern search space grows exponentially, this approach is not feasible.

In summary, the consensus model is not a very good description of the functional sites. Some positions in a biologically functional site are much conserved than others and consensus model cannot represent these characteristics. The profile model, on the other hand, can reflect these characteristics in a better manner by treating each position differently with a different letter distribution. A hybrid approach could potentially combine the expressiveness of the profile representation with convergence guarantees of the consensus pattern. An example of a hybrid approach is the random projection [4] algorithm followed by the EM algorithm [8]. It uses a global solver to obtain promising alignments in the discrete pattern space followed by further local solver refinements in continuous space [22, 23]. Currently, only a few algorithms take advantage of a combined discrete and continuous space search [4, 11, 16]. In this chapter, we consider the profile representation of the motif and a new hybrid algorithm is developed to escape out of the local maximum of the likelihood surface obtained in this profile space. Some motivations to develop this new hybrid algorithm are as follows:

- A motif refinement stage is vital and popularly used by many pattern-based algorithms (like PROJECTION, MITRA, and others), which try to find optimal motifs.
- The traditional EM algorithm used in the context of motif finding converges very quickly to the nearest local optimal solution (within 5–8 iterations) [24].
- There are many other promising local optimal solutions in the close vicinity of the profiles obtained from the global methods.

In spite of the importance given to obtaining a global optimal solution in the context of motif finding, little work has been done in the direction of finding such solutions [25, 26]. There are several proposed methods to escape out of the local optimal solution to find better solutions in machine learning [27] and the optimisation [28] related problems. Most of them are stochastic in nature and usually rely on perturbing either the data or the hypothesis. These stochastic perturbation algorithms are inefficient because they sometimes miss a neighborhood solution or obtain an already existing solution.

In this chapter, we develop a TRUST-TECH-based expectation-maximization (TT-EM) algorithm and apply to the motif finding problem [10]. It has the capability to search for alignments corresponding to Tier-1 local maxima in the profile space in a tier-by-tier manner systematically. This effective search is achieved by transforming the original optimization problem into a gradient system with certain properties and obtaining dynamical and topological properties of the gradient system corresponding to the nonlinear likelihood surface. Our method has been successfully used for obtaining the parameter estimates of finite mixture models [29]. The underlying theoretical details of our method are described in Refs. [30, 31].

12.3   PROFILE MODEL AND THE EM ALGORITHM

In this section, we will describe our problem formulation and the details of the EM algorithm in the context of the motif finding problem. In the next section, we will describe some details of the dynamical system of the log-likelihood function that enables us to search for the nearby local optimal solutions.

We will now transform the problem of finding the best possible motif into a problem of finding the global maximum of a highly nonlinear log-likelihood scoring function obtained from its profile representation. The log-likelihood surface is made of $3l$ variables, which are treated as the unknown parameters that are to be estimated. Here, we will describe these parameters ($Q_{k,j}$) and construct the scoring function in terms of these parameters.

Some promising initial alignments are obtained by applying projection methods or random starts on the entire data set. Typically, random starts are used because they are cost-efficient. The most promising sets of alignments are considered for further processing. These initial alignments are then converted into a profile representation. Let t be the total number of sequences and $S = \{S_1, S_2, \ldots, S_t\}$ be the set of t sequences. Let P be a single alignment containing the set of segments $\{P_1, P_2, \ldots, P_t\}$. l is the length of the consensus pattern. For further discussion, we use the following variables:

$$i = 1, \ldots, t \qquad \text{for } t \text{ sequences}$$

$$k = 1, \ldots, l \qquad \text{for positions within an } l\text{-mer}$$

$$j \in \{A, T, G, C\} \quad \text{for each nucleotide}$$

The count matrix can be constructed from the given alignments as shown in Table 12.1. We define $C_{0,j}$ to be the overall background count of each nucleotide in all of the sequences. Similarly, $C_{k,j}$ is the count of each nucleotide in the kth

Table 12.1 A count of nucleotides _A_, _T_, _G_, _C_ at each position _K_ = 1...<i>l</i> in all the sequences of the data set

j	$k=0$	$k=1$	$k=2$	$k=3$	$k=4$	...	$k=l$
A	$C_{0,1}$	$C_{1,1}$	$C_{2,1}$	$C_{3,1}$	$C_{4,1}$	...	$C_{l,1}$
T	$C_{0,2}$	$C_{1,2}$	$C_{2,2}$	$C_{3,2}$	$C_{4,2}$	...	$C_{l,2}$
G	$C_{0,3}$	$C_{1,3}$	$C_{2,3}$	$C_{3,3}$	$C_{4,3}$	...	$C_{l,3}$
C	$C_{0,4}$	$C_{1,4}$	$C_{2,4}$	$C_{3,4}$	$C_{4,4}$	...	$C_{l,4}$

$K=0$ denotes the background count.

position (of the l-mer) in all the segments in P.

$$Q_{0,j} = \frac{C_{0,j}}{\sum\limits_{J \in \{A,T,G,C\}} C_{0,j}}. \qquad (12.1)$$

$$Q_{k,j} = \frac{C_{k,j} + b_j}{d + \sum\limits_{J \in \{A,T,G,C\}} b_J}. \qquad (12.2)$$

Equation 12.1 shows the background frequency of each nucleotide, where b_j (and b_J) is known as the Laplacian or Bayesian correction and is equal to $d*Q_{0,j}$ and d is a constant usually set to unity. Equation 12.2 gives the weight assigned to the type of nucleotide at the kth position of the motif.

A position-specific scoring matrix (PSSM) can be constructed from one set of instances in a given set of t sequences. In this model, every position of the motif is described as a probability distribution over the allowed alphabets. This model assumes that the alphabets are independently and identically distributed (i.i.d.) at each position and the starting positions at each site. It also assumes that these alphabets contribute additively to the total activity and hence the probability of an alignment matrix of n instances is determined by a multinomial distribution.

From Equations 12.1 and 12.2, it is obvious that the following relationship holds

$$\sum_{j \in \{A,T,G,C\}} Q_{k,j} = 1 \quad \forall k = 0, 1, 2, \ldots, l. \qquad (12.3)$$

For a given k value in Equation 12.3, each Q can be represented in terms of the other three variables. Since the length of the motif is l, the final objective function (i.e., the likelihood score) would contain $3l$ independent variables. It should be noted that even if there are $4l$ variables in total, the parameter space will contain only $3l$ independent variables because of the constraints obtained from Equation 12.3. Thus, the constraints help in reducing the dimensionality of the search problem.

To obtain the likelihood score, every possible l-mer in each of the t sequences must be examined. This is done so by multiplying the respective $Q_{i,j}/Q_{0,j}$ dictated by the nucleotides and their respective positions within the l-mer. Only the highest scoring l-mer in each sequence is noted and kept as part of the alignment. The total score is the sum of all the best (logarithmic) scores in each sequence.

$$A(Q) = \sum_{i=1}^{t} \log(A)_i = \sum_{i=1}^{t} \log\left(\prod_{k=1}^{l} \frac{Q_{k,j}}{Q_b}\right)_i = \sum_{i=1}^{t} \sum_{k=1}^{l} \log(Q'_{k,j})_i, \qquad (12.4)$$

where $Q_{k,j}/Q_b$ represents the ratio of the nucleotide probability to the corresponding background probability. $\text{Log}(A)_i$ is the score at each individual ith sequence. In Equation 12.4, we see that A is composed of the product of the weights for each individual position k. We consider this to be the likelihood score, which we would like to maximize. $A(Q)$ is the nonconvex $3l$ dimensional continuous function for which the global maximum corresponds to the best possible motif in the data set. More accurate motif models proposed in the literature [8] were not used here because they require significant additional calculations for fitting the parameters from the sequence data.

One of the primary tools for performing refinement of the candidate motifs by improving this likelihood score is the EM algorithm. We used the EM algorithm that has been proposed in Ref. [7]. Since the position of the motif occurrence in each sequence is not fixed *a priori*, summing of all possible locations of motif instances becomes computationally tedious. EM algorithm overcomes this problem by iteratively seeking a better likelihood model and converges linearly to a locally maximum likelihood model from a given initial guess. Hence, this EM-based refinement algorithm will seek a matrix model that locally maximizes the likelihood ratio. In other words, it tries to converge to a motif model that can explain the instances much better than the background model alone.

EM refinement performed at the end of a combinatorial approach has the disadvantage of converging to a local optimal solution. Many promising solutions might be in the close neighborhood in the model space. EM algorithm cannot obtain these models and thus outputs a suboptimal solution. To avoid this problem, our method improves this procedure for refining the motifs by understanding the details of the stability boundaries of the likelihood function and deterministically tries to escape the convergence region of the EM algorithm.

12.4 NOVEL FRAMEWORK

In this section, we will describe the TT-EM algorithm that combines the advantages of both stability regions and the EM algorithm. Our framework consists of the following three stages:

- *Global stage* in which the promising solutions in the entire search space are obtained.
- *Refinement stage* (or *local stage*) where a local method is applied to the solutions obtained in the previous stage in order to refine the profiles.
- *Neighborhood-search stage* where the exit points are computed and the Tier-1 and Tier-2 solutions are explored systematically.

In the global stage, a branch and bound search is performed on the entire data set. All of the profiles that do not meet a certain threshold (in terms of a given scoring function) are eliminated in this stage. Some promising initial alignments are obtained by applying these methods (like projection methods) on the entire data set. Most promising set of alignments is considered for further processing. The promising patterns obtained are transformed into profiles and local improvements are made to these profiles at the refinement stage. The consensus pattern is obtained from each nucleotide that corresponds to the largest value in each column of the PSSM. The $3l$ variables chosen are the nucleotides that correspond to those that are not present in the consensus pattern. Because of the probability constraints discussed in the previous section, the largest weight can be represented in terms of the other three variables.

12.4.1 Hessian Matrix and Dynamical System for the Scoring Function

In order to present our algorithm, we have to define a dynamical system corresponding to the log-likelihood function and the PSSM. The key contribution of our work is the development of this nonlinear dynamical system that will enable us to realize the geometric and dynamic nature of the likelihood surface by allowing us to understand the topology and convergence behavior of any given subspace on the surface. We construct the following *gradient system* in order to locate critical points of the objective function (12.4):

$$\dot{Q}(t) = -\nabla A(Q). \tag{12.5}$$

One can realize that this transformation preserves all of the critical points [30]. Now, we will present the details of the construction of the gradient system and the Hessian. In order to reduce the dominance of one variable over the other, the values of each of the nucleotides that belong to the consensus pattern at the position k will be represented in terms of the other three nucleotides in that particular column. Let P_{ik} denote the kth position in the segment P_i. This will also minimize the dominance of the eigenvector directions when the Hessian is obtained. The variables in the scoring function are transformed into new variables described in Table 12.2. Thus,

Table 12.2 A count of nucleotides $j \in \{A, T, G, C\}$ at each position $k = 1 \ldots l$ in all the sequences of the data set

j	$k=b$	$k=1$	$k=2$	$K=3$	$k=4$	$\ldots$	$k=1$
A	b_A	w_1	C_2	w_7	w_{10}	$\ldots$	w_{3l-2}
T	b_T	w_2	w_4	w_8	C_4	$\ldots$	w_{3l-1}
G	b_G	C_1	w_5	w_9	w_{11}	$\ldots$	C_l
C	b_C	w_3	w_6	C_3	w_{12}	$\ldots$	w_{3l}

C_k is the kth nucleotide of the consensus pattern, which represents the nucleotide with the highest value in that column. Let the consensus pattern be $GACT\ldots G$ and b_j be the background.

Equation 12.4 can be rewritten in terms of the $3l$ variables as follows:

$$A(Q) = \sum_{i=1}^{t} \sum_{k=1}^{l} \log f_{ik}(w_{3k-2}, w_{3k-1}, w_{3k})_i, \tag{12.6}$$

where f_{ik} can take the values $\{w_{3k-2}, w_{3k-1}, w_{3k}, 1 - (w_{3k-2} + w_{3k-1} + w_{3k})\}$ depending on the P_{ik} value.

The first derivative of the scoring function is a one-dimensional vector with $3l$ elements.

$$\nabla A = \begin{bmatrix} \dfrac{\partial A}{\partial w_1} & \dfrac{\partial A}{\partial w_2} & \dfrac{\partial A}{\partial w_3} & \cdots & \dfrac{\partial A}{\partial w_{3l}} \end{bmatrix}^T, \tag{12.7}$$

and each partial derivative is given by

$$\frac{\partial A}{\partial w_p} = \sum_{i=1}^{t} \frac{\dfrac{\partial f_{ip}}{\partial w_p}}{f_{ik}(w_{3k-2}, w_{3k-1}, w_{3k})} \tag{12.8}$$

$$\forall p = 1, 2, \ldots, 3l \quad \text{and} \quad k = \text{round}(p/3) + 1.$$

The Hessian $\nabla^2 A$ is a block diagonal matrix of block size 3×3. For a given sequence, the entries of the 3×3 block will be the same if that nucleotide belongs to the consensus pattern (C_k). This nonlinear transformation will preserve all the critical points on the likelihood surface. The theoretical details of the proposed method and their advantages are published in Ref. [30]. If we can identify all the decomposition points on the stability boundary of a given local maximum, then we will be able to find all the tier-1 local maxima. A Tier-1 local maximum is defined as the new local maximum that is connected to the original local maximum through one decomposition point. However, finding all of the decomposition points is computationally intractable and hence we have adopted a heuristic by generating the eigenvector directions of the PSSM at the local maximum. Also, for such a complicated likelihood function, it is not efficient to compute all decomposition points on the stability boundary. Hence, one can obtain new local maxima by obtaining the *exit points* instead of the decomposition points. The point along a particular direction where the function has the lowest value starting from the given local maximum is called the *exit point*.

12.4.2 TRUST-TECH-Based Framework

To solve Equation 12.4, current algorithms begin at random initial alignment positions and attempt to converge to an alignment of l-mers in all of the sequences that maximize the objective function. In other words, the l-mer whose $\log(A)_i$ is the highest (with a given PSSM) is noted in every sequence as part of the current alignment. During the maximization of $A(Q)$ function, the probability of weight matrix and hence the corresponding alignments of l-mers are updated. This occurs iteratively until the PSSM converges to the local optimal solution. The consensus pattern is obtained from the nucleotide with the largest weight in each position (column) of the PSSM. This converged PSSM and the set of alignments correspond to a local optimal solution. The neighborhood-search stage where the neighborhood of the original solution is explored in a systematic manner is as follows:

Input: Local maximum A.
Output: Best local maximum in the neighborhood region.
Algorithm:

Step 1: Construct the PSSM for the alignments corresponding to the local maximum (A) using Equations 12.1 and 12.2.

Step 2: Calculate the eigenvectors of the Hessian matrix for this PSSM.

Step 3: Find exit points (e_{1i}) on the practical stability boundary along each eigen-vector direction.

Step 4: For each exit point, the corresponding tier-1 local maxima (a_{1i}) are obtained by applying the EM algorithm after the ascent step.

Step 5: Repeat the above procedure for promising tier-1 solutions to obtain tier-2 neighborhood local maxima (a_{2j}).

Step 6: Return the solution that gives the maximum likelihood score among $\{A, a_{1i}, a_{2j}\}$.

Figure 12.2 illustrates the TT-EM method. To escape this local optimal solution, our approach requires the computation of a Hessian matrix (i.e., the matrix of second derivatives) and the $3l$ eigenvectors of the Hessian. These directions were chosen

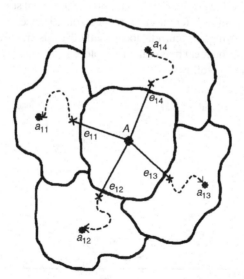

Figure 12.2   Diagram illustrates the TRUST-TECH method of escaping from the original solution (A) to the neighborhood local optimal solutions (a_{1i}) through the corresponding exit points (e_{1i}). The dotted lines indicate the local convergence of the EM algorithm.

as a general heuristic and are not problem dependent. Depending on the data set that is being worked on, one can obtain even more promising directions. The main reasons for choosing the eigenvectors of the Hessian as search directions are as follows:

- Computing the eigenvectors of the Hessian is related to finding the directions with extreme values of the second derivatives, that is, directions of extreme normal-to-isosurface change.
- The eigenvectors of the Hessian will form the basis vectors for the search directions. Any other search direction can be obtained by a linear combination of these directions.
- This will make our algorithm deterministic since the eigenvector directions are always unique.

The value of the objective function is evaluated along these eigenvector directions with some small step-size increments. Since the starting position is a local optimal solution, one will see a steady decline in the function value during the initial steps; we call this the *descent stage*. Since the Hessian is obtained only once during the entire procedure, it is more efficient compared to Newton's method where an approximate Hessian is obtained for every iteration. After a certain number of evaluations, there may be an increase in the value indicating that the stability boundary is reached. The point along this direction intersecting the stability boundary is called the *exit point*. Once the exit point has been reached, few more evaluations are made in the direction of the same eigenvector to improve the chances of reaching a new convergence region. This procedure is clearly shown in Fig. 12.3. Applying the local method directly from the exit point may give the original local maximum. The ascent stage is used to ensure that the new guess is in a different convergence zone. Hence, given the best local maximum obtained using any current local methods, this framework allows us to systematically escape the local maximum to explore surrounding local maxima. The complete algorithm is as follows:

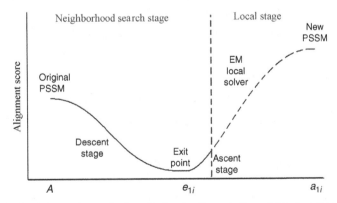

Figure 12.3 A summary of escaping from the local optimum to the neighborhood local optimum. Observe the corresponding trend of $A(Q)$ at each step.

Input: The DNA sequences, length of the motif (l), number of Mutations (d)
Output: Motif(s)
Algorithm:
Step 1: Given the sequences, apply random projection algorithm to obtain different set of alignments.
Step 2: Choose the promising buckets and apply EM algorithm to refine these alignments.
Step 3: Apply the TT-EM method to obtain nearby promising local optimal solutions.
Step 4: Report the consensus pattern that corresponds to the best alignments and their corresponding PSSM.

The new framework can be treated as a hybrid approach between global and local methods. It differs from traditional local methods in the ability to explore multiple local solutions in the neighborhood region in a systematic and effective manner. It differs from global methods in working completely in the profile space and searching a subspace efficiently in a deterministic manner. However, the main difference of this work compared to the algorithm presented in the previous chapter is that the global method is performed in discrete space and the local method is performed in the continuous space. In other words, both the global and local methods do not optimize the same function. In such cases, it is even more important to search for neighborhood local optimal solutions in the continuous space.

12.5 IMPLEMENTATION DETAILS

Our program was implemented on Red Hat Linux version 9 and runs on a Pentium IV 2.8 GHz machine. The core algorithm that we have implemented is TT_EM described in Algorithm 1. TT_EM obtains the initial alignments and the original data sequences along with the length of the motif. This procedure constructs the PSSM, performs EM refinement, and then computes the tier-1 and tier-2 solutions by calling the procedure *Next_Tier*. The eigenvectors of the Hessian were computed using the source code obtained from Ref. [32]. *Next_Tier* takes a PSSM as an input and computes an array of PSSMs corresponding to the next-tier local maxima using the TT-EM methodology.

Algorithm 1 Motif $TT\_EM(init\_aligns,seqs,l)$

```
PSSM = Construct_PSSM(init_aligns)
New_PSSM = Apply_EM(PSSM, seqs)
TIER1 = Next_Tier(seqs, New_PSSM, 1)
for i = 1 to 31 do
  if TIER1[i] <> zeros(41) then
     TIER2[i][] = Next_Tier(seqs, TIER1[i], 1)
  end if
end for
Return best(PSSM, TIER1, TIER2)
```

Given a set of initial alignments, Algorithm 1 will find the best possible motif in the profile space in a tier-by-tier manner. For implementation considerations, we have shown only for two tiers. Initially, a PSSM is computed using *construct_PSSM* from the given alignments. The procedure *Apply_EM* will return a new PSSM that corresponds to the alignments obtained after the EM algorithm has been applied to the initial PSSM. The details of the procedure *Next_Tier* are given in Algorithm 2. From a given local solution (or PSSM), *Next_Tier* will compute all the $3l$ new *PSSMs* corresponding to the tier-1 local optimal solutions. The second-tier patterns are obtained by calling the *Next_Tier* from the first-tier solutions. Sometimes, new PSSMs might not be obtained for certain search directions. In those cases, a zero vector of length $4l$ is returned. Only those new PSSMs which do not have this value will be used for any further processing. Finally, the pattern with the highest score among all thé PSSMs is returned.

Algorithm 2 PSSMs[] *Next_Tier*(*seqs, PSSM,*l)

```
Score = eval(PSSM)
Hess = Construct_Hessian(PSSM)
Eig[] = Compute_EigVec(Hess)
MAX_Iter = 100
for k = 1 to 3l do
  PSSMs[k] = PSSM      Count = 0
  Old_Score = Score      ep_reached = FALSE
  while(! ep_reached) && (Count < MAX_Iter) do
    PSSMs[k] = update(PSSMs[k],Eig[k],step)
    Count = Count + 1
    New_Score = eval(PSSMs[k])
    if (New_Score > Old_Score) then
      ep_reached = TRUE
    end if
    Old_Score = New_Score
  end while
  if count < MAX_Iter then
    PSSMs[k] = update(PSSMs[k],Eig[k],ASC)
    PSSMs[k] = Apply_EM(PSSMs[k],Seqs)
  else
    PSSMs[k] = zeros(4l)
  end if
end for
Return PSSMs[]
```

The procedure *Next_Tier* takes a PSSM, applies the TT-EM method, and computes an array of PSSMs that corresponds to the next tier local optimal solutions. The procedure *eval* evaluates the scoring function for the PSSM using Equation 12.4. The procedures *Construct_Hessian* and *Compute_EigVec* compute the Hessian matrix and the eigenvectors, respectively. *MAX_iter* indicates the maximum number of uphill evaluations that are required along each of the eigenvector directions. The neighborhood PSSMs will be stored in an array variable *PSSMs* (this is a vector). The original PSSM is updated with a small step until an exit point is reached or the number of iterations exceeds the *MAX_Iter* value. Choosing an optimal step size is a heuristic. We choose the step size to be the average of the step values taken by the EM algorithm during its convergence. If the exit point is reached along a particular direction, few more function evaluations are made to ensure that the PSSM has exited the original convergence region and has entered a new one. The EM algorithm is then used during this ascent stage to obtain a new PSSM. For completeness, the entire algorithm has been shown in this section. However, during the implementation, several heuristics have been applied to reduce the running time of the algorithm. For example, if the first-tier solution is not very promising, it will not be considered for obtaining the corresponding second-tier solutions.

The initial alignments are converted into the profile space and a PSSM is constructed. The PSSM is updated (using the EM algorithm) until the alignments converge to a local optimal solution. The TT-EM methodology is then employed to escape from this local optimal solution to compute nearby first-tier local optimal solutions. This process is then repeated on promising first-tier solutions to obtain second-tier solutions. As shown in Fig. 12.2, from the original local optimal solution, various exit points and their corresponding new local optimal solutions are computed along each eigenvector direction. Sometimes, two directions may yield the same local optimal solution. This can be avoided by computing the decomposition point corresponding to the exit point on the stability boundary [33]. There can be many exit points, but there will only be a unique decomposition point corresponding to the new local minimum. For computational efficiency, the TT-EM approach is only applied to promising initial alignments (i.e., random starts with higher likelihood score). Therefore, a threshold $A(Q)$ score is determined by the average of the three best first-tier scores after 10–15 random starts; any current and first-tier solution with scores greater than the threshold is considered for further analysis. Additional random starts are carried out in order to aggregate at least 10 first-tier solutions. The TT-EM method is repeated on all first-tier solutions above a certain threshold to obtain second-tier local optimal solutions.

12.6 EXPERIMENTAL RESULTS

Experiments were performed on both synthetic data and real data. Two different methods were used in the global stage: random start and random projection. The main purpose of our work is not to demonstrate that our algorithm can outperform the existing motif finding algorithms. Rather, the main work here focuses on improving the results that are obtained from other efficient algorithms. We have chosen to demonstrate the performance of our algorithm on the results obtained from the random projection method, which is a powerful global method that has outperformed other traditional motif finding approaches such as MEME, Gibbs sampling, WINNOWER, SP-STAR, and others [4]. Since the comparison was already published, we mainly

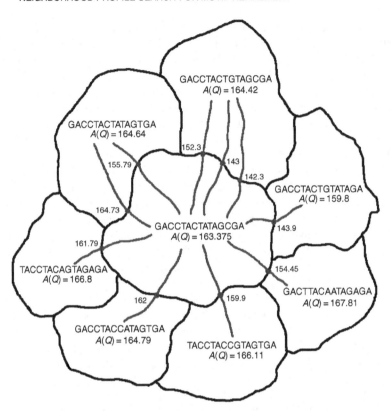

Figure 12.4 Two-dimensional illustration of first-tier improvements in a 3*l*-dimensional objective function. The original local maximum has a score of 163.375. The various tier-1 solutions are plotted and the one with highest score (167.81) is chosen.

focus on the performance improvements of our algorithm as compared to the random projection algorithm. For the random start experiment, a total of N random numbers between 1 and $(t - l + 1)$ corresponding to initial set of alignments are generated. We then proceeded to evaluate our TT-EM methodology from these alignments.

Figure 12.4 shows the tier-1 solutions obtained from a given consensus pattern. Since the exit points are being used instead of decomposition points, our method might sometimes find the same local optimal solution obtained before. As seen from the figure, the tier-1 solutions can differ from the original pattern by more than just one nucleotide position. Also, the function value at the exit point is much higher than the original value.

12.6.1 Synthetic Data Sets

The synthetic data sets were generated by implanting some motif instances into $t = 20$ sequences each of length $t_i = 600$. Let m correspond to one full random projection + EM cycle. We have set $m = 1$ to demonstrate the efficiency of our approach. We compared the performance coefficient (PC), which gives a measure of the average performance of our implementation compared to that of the random projection

Table 12.3 The results of performance coefficient with $m = 1$ for synthetically generated sequences

Motif (l,d)	PC obtained Using Random Projection	PC obtained Using TT-EM method
(11,2)	20	20
(15,4)	14.875	17
(20,6)	12.667	18

The likelihood scores are not normalized and the perfect score is 20 since there are 20 sequences.

algorithm. The PC is given by

$$PC = \frac{|K \cap P|}{|K \cup P|},$$ (12.9)

where K is the set of the residue positions of the planted motif instances, and P is the corresponding set of positions predicted by the algorithm. Table 12.3 gives an overview of the performance of our method compared to the random projection algorithm on the (l,d) motif problem for different l and d values.

Our results show that by branching out and discovering multiple local optimal solutions, higher m values are not needed. A higher m value corresponds to more computational time because projecting the l-mers into k-sized buckets is a time-consuming task. Using our approach, we can replace the need for randomly projecting l-mers repeatedly in an effort to converge to a global optimum by deterministically and systematically searching the solution space modeled by our dynamical system and improving the quality of the existing solutions. We see that for higher length motifs, the improvements are more significant.

As opposed to stochastic processes such as mutations in genetic algorithms, our approach eliminates the stochastic nature and obtains the nearby local optimal solutions systematically. Figure 12.5 shows the performance of the TRUST-TECH approach on synthetic data for different (l,d) motifs. The average scores of the 10 best solutions obtained from random starts and their corresponding improvements in tier-1 and tier-2 are reported. One can see that the improvements become more prominent as the length of the motif is increased. Table 12.4 shows the best and worst of these top 10

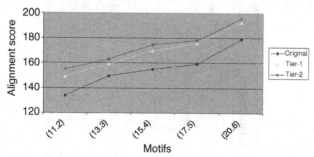

Figure 12.5 The average scores with the corresponding first-tier and second-tier improvements on synthetic data using the random starts with exit-point approach with different (l,d) motifs.

Table 12.4   The consensus patterns and their corresponding scores of the original local optimal solution obtained from multiple random starts on the synthetic data

(l,d)	Initial Pattern	Score	First-Tier Pattern	Score	Second-Tier Pattern	Score
(11,2)	AACGGTCGCAG	125.1	CCCGGTCGCTG	147.1	CCCGGGAGCTG	153.3
(11,2)	ATACCAGTTAC	145.7	ATACCAGTTTC	151.3	ATACCAGGGTC	153.6
(13,3)	CTACGGTCGTCTT	142.6	CCACGGTTGTCTC	157.8	CCTCGGGTTTGTC	158.7
(13,3)	GACGCTAGGGGGT	158.3	GAGGCTGGGCAGT	161.7	GACCTTGGGTATT	165.8
(15,4)	CCGAAAAGAGTCCGA	147.5	CCGCAATGACTGGGT	169.1	CCGAAAGGACTGCGT	176.2
(15,4)	TGGGTGATGCCTATG	164.6	TGGGTGATGCCTATG	166.7	TGAGAGATGCCTATG	170.4
(17,5)	TTGTAGCAAAGGCTAAA	143.3	CAGTAGCAAAGACTACC	173.3	CAGTAGCAAAGACTTCC	175.8
(17,5)	ATCGCGAAAGGTTGTGG	174.1	ATCGCGAAAGGATGTGG	176.7	ATTGCGAAAGAATGTGG	178.3
(20,6)	CTGGTGATTGAGATCATCAT	165.9	CAGATGGTTGAGATCACCTT	186.9	CATTTAGCTGAGTTCACCTT	194.9
(20,6)	GGTCACTTAGTGGCGCCATG	216.3	GGTCACTTAGTGGCGCCATG	218.8	CGTCACTTAGTCGCGCCATG	219.7

The best first-tier and second-tier optimal patterns and their corresponding scores are also reported.

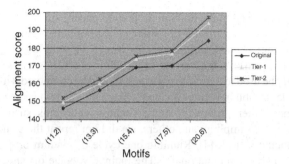

Figure 12.6 The average scores with the corresponding first-tier and second-tier improvements on synthetic data using the random projection with exit-point approach with different (*l*, *d*) motifs.

random starts along with the consensus pattern and the alignment scores. We see that for higher length motifs, the improvements are more significant.

With a few modifications, more experiments were conducted using the random projection method. The random projection method eliminates nonpromising regions in the search space and gives a number of promising sets of initial patterns. EM refinement is applied only to the promising initial patterns. Due to the robustness of the results, the TT-EM method is employed only to the top five local maxima. The TT-EM method is again repeated on the top-scoring first-tier solutions to arrive at the second-tier solutions. Figure 12.6 shows the average alignment scores of the best random projection alignments and their corresponding improvements in tier-1 and tier-2 are reported. In general, the improvements in the first-tier solutions are more significant than the improvements in the second-tier solutions.

12.6.2 Real Data Sets

Table 12.5 shows the results of the TT-EM methodology for real biological sequences. We have chosen $l = 20$ and $d = 2$. 't' indicates the number of sequences in the real data. For the biological samples taken from Refs. [4,17], the value m once again is the average number of random projection + EM cycles required to discover the motif. All other parameter values (such as projection size $k = 7$ and threshold $s = 4$) are chosen to

Table 12.5 Results of TT-EM method for biological samples

Sequence	Sample Size	t	Best (20,2) Motif	Reference Motif
E. coli CRP	1890	18	TGTGAAATAGATCACATTTT	TGTGANNNNGNTCACA
Preproinsulin	7689	4	GGAAATTGCAGCCTCAGCCC	CCTCAGCCC
DHFR	800	4	CTGCAATTTCGCGCCAAACT	ATTTCNNGCCA
Metallothionein	6823	4	CCCTCTGCGCCCGGACCGGT	TGCRCYCGG
c-fos	3695	5	CCATATTAGGACATCTGCGT	CCATATTAGAGACTCT
Yeast ECB	5000	5	GTATTTCCCGTTTAGGAAAA	TTTCCCNNTNAGGAAA

The real motifs were obtained in all the six cases using the TT-EM framework.

be the same as those used in the random projection paper [4]. All of the motifs were recovered with $m = 1$ using the TT-EM strategy. Without our algorithm, the random projection algorithm needed *multiple cycles* ($m = 8$ in some cases and $m = 15$ in others) in order to retrieve the correct motif. This elucidates the fact that global methods can only be used to a certain extent and should be combined with refined local heuristics in order to obtain better efficiency. Since the random projection algorithm has outperformed other prominent motif finding algorithms such as SP-STAR, WINNOWER, Gibbs sampling, and others, we did not repeat the same experiments that were conducted in Ref. [4]. Running one cycle of random projection + EM is much more expensive computationally. The main advantage of our strategy comes from the deterministic nature of our algorithm in refining motifs.

Let the cost of applying EM algorithm for a given bucket be f and let the average number of buckets for a given projection be b. Then, the running time of the TT-EM method will be $O(cbf)$ where c is a constant that is linear in lth length of the motif. If there were m projections, then the cost of the random projection algorithm using restarts will be $O(mbf)$. The two main advantages of using TT-EM strategy compared to random projection algorithm are as follows:

- It avoids multiple random projections, which often provide similar optimal motifs.
- It provides multiple optimal solutions in a promising region of a given bucket as opposed to a single solution provided by random projection algorithm.

12.7 CONCLUDING DISCUSSION

The TT-EM framework proposed in this chapter broadens the search region in order to obtain an improved solution that may potentially correspond to a better motif. In most of the profile-based algorithms, EM is used to obtain the nearest local optimum from a given starting point. In our approach, we obtain promising results by computing multiple local optimal solutions in a tier-by-tier manner. We have shown on both real and synthetic data sets that beginning from the EM-converged solution, the TT-EM approach is capable of searching in the neighborhood regions for another solution with an improved likelihood score. This will often translate into finding a pattern with less hamming distance from the resulting alignments in each sequence. Our approach has demonstrated an improvement in the score on all data sets that it was tested on. One of the primary advantages of the TT-EM methodology is that it can be used with different global and local methods. The main contribution of our work is to demonstrate the capability of this hybrid EM algorithm in the context of the motif finding problem. Our approach can potentially use any global method and improve its results efficiently. From our results, we observe that motif refinement stage plays a vital role and can yield accurate results deterministically. In the future, we would like to continue our work by combining other global methods available in the literature with existing local solvers such as EM or GibbsDNA that work in continuous space. By following the example of Ref. [6], we may improve the chances of finding more promising patterns by combining our algorithm with different global and local methods.

REFERENCES

1. Dofcherty, K. *Gene Transcription DNA Binding Proteins: Essential Techniques*, Wiley, New York, 1997.

2. Reddy, C. K., Weng, Y. C., and Chiang, H. D. Refining motifs by improving information content scores using neighborhood profile search. *Algorithms for Molecular Biology*, 1 (23): 1–14, 2006.

3. Pevzner, P. Finding Signals in DNA. *Computational Molecular Biology: An Algorithmic Approach*, MIT Press, pp. 133–152, 2000.

4. Buhler, J. and Tompa, M. Finding motifs using random projections. *Proceedings of the Fifth Annual International Conference on Research in Computational Molecular Biology*, pp. 69–76, 2001.

5. Pevzner, P. and Sze, S.-H. Combinatorial approaches to finding subtle signals in DNA sequences. *The Eighth International Conference on Intelligent Systems for Molecular Biology*, pp. 269–278, 2000.

6. Tompa, M., Li, N., Bailey, T. L., Church, G. M., De Moor, B., Eskin, E., Favorov, A. V., Frith, M. C., Fu, Y., Kent, W. J., Makeev, V. J., Mironov, A. A., Noble, W. S., Pavesi, G., Pesole, G., Regnier, M., Simonis, N., Sinha, S., Thijs, G., Van Helden, J. Vandenbogaert, M., Weng, Z., Workman, C. Ye, C., and Zhu, Z. Assessing computational tools for the discovery of transcription factor binding sites. *Nature Biotechnology*, 23 (1): 137–144, 2005.

7. Lawrence, C. E. and Reilly, A. A. An expectation maximization (EM) algorithm for the identification and characterization of common sites in unaligned biopolymer sequences. *Proteins: Structure, Function and Genetics*, 7: 41–51, 1990.

8. Bailey, T. and Elkan, C. Fitting a mixture model by expectation maximization to discover motifs in biopolymers. *The First International Conference on Intelligent Systems for Molecular Biology*, pp. 28–36, 1994.

9. Waterman, M., Arratia, R., and Galas, E. Pattern recognition in several sequences: consensus and alignment. *Mathematical Biology*, 46: 515–527, 1984.

10. Reddy, C. K. TRUST-TECH based methods for optimization and learning, Ph. D. thesis, Cornell University, Ithaca, NY, 2007.

11. Eskin, E. From profiles to patterns and back again: a branch and bound algorithm for finding near optimal motif profiles, *Proceedings of the Eighth Annual International Conference on Research in Computational Molecular Biology*, pp. 115–124, 2004.

12. Durbin, R., Eddy, S. R., Krogh, A., and Mitchison, G. *Biological Sequence Analysis: Probabilistic Models of Proteins and Nucleic Acids*, Cambridge University Press, Cambridge, UK, 1999.

13. Lawrence, C., Altschul, S., Boguski, M., Liu, J., Neuwald, A., and Wootton, J. Detecting subtle sequence signals: a gibbs sampling strategy for multiple alignment. *Science*, 262: 208–214, 1993.

14. Hertz, G. and Stormo, G. Identifying DNA and protein patterns with statistically significant alignments of multiple sequences. *Bioinformatics*, 15 (7–8): 563–577, 1999.

15. Eddy, S. R. Profile hidden Markov models. *Bioinformatics*, 14 (9): 755–763, 1998.

16. Raphael, B., Liu, L. T., and Varghese, G. A uniform projection method for motif discovery in DNA sequences. *IEEE Transactions on Computational Biology and Bioinformatics*, 1 (2): 91–94, 2004.

17. Price, A., Ramabhadran, S., and Pevzner, P. A. Finding subtle motifs by branching from sample strings. *Bioinformatics*, 1 (1): 1–7, 2003.

18. Keich, U. and Pevzner, P. Finding motifs in the twilight zone. *Bioinformatics*, 18: 1374–1381, 2002.

19. Sagot, M. Spelling approximate or repeated motifs using a suffix tree. *Lecture Notes in Computer Science*, 1380: 111–127, 1998.

20. Eskin, E. and Pevzner, P. A. Finding composite regulatory patterns in DNA sequences. *Bioinformatics*, 18(1): 354–363, 2002.

21. Brazma, A., Jonassen, I., Eidhammer, I., and Gilbert, D. Approaches to the automatic discovery of patterns in biosequences. *Journal of Computational Biology*, 5 (2): 279–305, 1998.

22. Barash, Y., Bejerano, G., and Friedman, N. A simple hyper-geometric approach for discovering putative transcription factor binding sites. *Proceedings of the First International Workshop on Algorithms in Bioinformatics*, 2001.

23. Segal, E., Barash, Y., Simon, I., Friedman, N., and Koller, D. From promoter sequence to expression: a probabilistic framework. *Proceedings of the Sixth Annual International Conference on Computational Biology*, pp. 263–272, 2002.

24. Blekas, K., Fotiadis, D., and Likas, A. Greedy mixture learning for multiple motif discovery in biological sequences. *Bioinformatics*, 19 (5): 607–617, 2003.

25. Xing, E., Wu, W., Jordan, M. I., and Karp, R. LOGOS: a modular Bayesian model for de novo motif detection. *Journal of Bioinformatics and Computational Biology*, 2 (1): 127–154, 2004.

26. Fogel, G. B., Weekes, D. G., Varga, G., Dow, E. R., Harlow, H. B., Onyia, J. E., and Su1, C. Discovery of sequence motifs related to coexpression of genes using evolutionary computation. *Nucleic Acids Research*, 32 (13): 3826–3835, 2004.

27. Elidan, G., Ninio, M., Friedman, N., and Schuurmans, D. Data perturbation for escaping local maxima in learning. *Proceedings of the Eighteenth National Conference on Artificial Intelligence*, pp. 132–139, 2002.

28. Cetin, B. C., Barhen, J., and Burdick, J. W. Terminal repeller unconstrained subenergy tunneling (TRUST) for fast global optimisation. *Journal of Optimization Theory and Applications*, 77 (1): 97–126, 1993.

29. Reddy, C. K., Chiang, H. D., and Rajaratnam, B. TRUST-TECH based expectation maximization for learning finite mixture models. *IEEE Transactions on Pattern Analysis and Machine Intelligence*, 30 (7): 1146–1157, 2008.

30. Chiang, H. D. and Chu, C. C. A systematic search method for obtaining multiple local optimal solutions of nonlinear programming problems. *IEEE Transactions on Circuits and Systems I: Fundamental Theory and Applications*, 43 (2): 99–109, 1996.

31. Lee, J. and Chiang, H. D. A dynamical trajectory-based methodology for systematically computing multiple optimal solutions of general nonlinear programming problems. *IEEE Transactions on Automatic Control*, 49 (6): 888–899, 2004.

32. Press, W. H., Teukolsky, S. A., Vetterling, W. T., and Flannery, B. P. *Numerical Recipes in C: The Art of Scientific Computing*, Cambridge University Press, Cambridge, UK, 1992.

33. Reddy, C. K. and Chiang, H. D. A stability boundary based method for finding decomposition points on potential energy surfaces. *Journal of Computational Biology*, 13 (3): 745–766, 2006.

13

MARKOV/NEURAL MODEL FOR EUKARYOTIC PROMOTER RECOGNITION

Jagath C. Rajapakse and Sy Loi Ho

The past decade has seen an explosion of life sciences data, posing incredible challenges to computational scientists to mine these databases and discover relevant biological knowledge leading to new discoveries in biology and medicine. Data mining and knowledge discovery from biological databases are key aspects in bioinformatics computation.

This chapter deals with discovering promoter regions from eukaryotic DNA sequences. The present approach encodes the inputs to the neural networks for the recognition of promoters, using Markov models. The Markov models represent the biological knowledge of the motifs, namely, TATA box and and Inr regions, of the promoter regions. These inputs are then processed by neural networks to extract more complex and distant interactions of bases. The predictions made by neural networks are later combined to vote for the promoter region. The present approach showed promising results in detecting promoter regions and may be extended to extract other signals or motifs from biological sequences.

13.1 INTRODUCTION

The field of computational biology, or more broadly bioinformatics, has been swift to attract much interest and concern from life sciences since its inception a few years ago.

Machine Learning in Bioinformatics. Edited by Yan-Qing Zhang and Jagath C. Rajapakse
Copyright © 2009 by John Wiley & Sons, Inc.

In early days, an application of computational biology could be simply a creation or maintenance of a database to store biological information, such as nucleotide or amino acid sequences. Many efforts in computational biology have since been focused on discovering the characteristics of biological data, such as DNA sequences, genes, and so on. Because most of the hereditary information of a living cell is encoded in its genomic DNA, the investigation of DNA; genes, and their relationships is of paramount importance to life sciences. The Human Genome Project (HGP) has sequenced a massive outpouring of genomic data that requires developing new algorithms, mathematical formulae, and statistical techniques to assess connections among the elements of large biological data sets.

The prediction of gene structure poses a major challenge to genome analysis since it lays the foundation for many subsequent characterization of genes. There is no exact efficient mechanism found yet on which a live cell relies to detect the coding regions in a DNA sequence. DNA sequences themselves do not provide knowledge of functional features embedded in the gene structure. Therefore, gene structure prediction based on biological knowledge alone is impossible. We cannot expect the full identification or knowledge about the gene structure in the near future by using pure experimental approaches because of their capital-intensive and time-consuming nature, given the enormous volume of sequence data. Therefore, automatic annotation of genomic sequences requires fast and reliable analyses methods, especially computer-based methods. This chapter deals with a computational technique that automatically identifies promoter regions in genomic sequences, an issue of vital importance for gene structure prediction efforts.

13.1.1 Promoters

Living organisms carry their genetic information in the form of DNA molecules. Each DNA molecule contains genes that decide the structural components of cells, tissues, and enzymes for biochemical reactions essential for its survival and functioning. The fundamental dogma of molecular biology is that DNA produces RNA, which, in turn, produces protein. The process of transcription, whereby genes in the DNA are transcribed into a corresponding RNA product, is an essential process in gene expression. The failure of transcription to occur obviously renders all the other steps that follow the production of the initial RNA transcript in eukaryotes, such as RNA splicing, transport to the cytoplasm, or translation into protein, redundant or faulty.

A promoter is typically referred to as a DNA region around a transcription initiation site (TSS), which is biologically the most important signal controlling and regulating the transcriptional initiation of the gene immediately downstream [20]. A gene has at least one promoter [5]. Promoters have very complex sequence structures reflecting the complexity of protein–DNA interactions during which various transcription factors (TFs) bind to promoter regions [1]. Promoter regions can be dispersed or overlapped, largely populating in about 1 Kb region upstream and surrounding the TSS. Their functions can be either positive or negative and are often context dependent. Promoters are therefore recognized by specific proteins and differ from other sections of DNA that are transcribed or translated.

Eukaryotes have three different RNA polymerase promoters that are responsible for transcribing different subsets of genes: RNA polymerase I transcribes genes encoding ribosomal RNA; RNA polymerase II transcribes genes encoding mRNA and certain small nuclear RNAs; RNA polymerase III transcribes genes encoding tRNAs and other small RNAs [13]. This chapter is devoted to the detection of type II RNA polymerase. An RNA polymerase II promoter is of particular interest because it not only transcribes into heterogeneous RNA, but also its regulation is the most complex of the three types [9].

The typical structure of an RNA polymerase II promoter located in the region [−50, +50], with respect to a TSS contains multiple binding sites for TFs that occur in a specific context. TF binding sites are typically 5–15 bp long [5, 13]. The nucleotide specificity at different positions within the sites varies. The TSS of the RNA polymerase II promoter is usually identified by an Inr and/or by a TATA box. The TATA box is a binding site usually conserved at −25 bp upstream of TSS in metazoans. Although the transcription process is not prevented when the TATA box is mutated, the start site of transcription varies from its unusual precise location, confirming the role of the TATA box as crucial positioning component of the core promoter [19]. Around the TSS, there is a loosely conserved initiator region, abbreviated by the Inr, which overlaps with TSS. Though the Inr is a much weaker signal than the TATA box, it is one determinant of promoter strength. Mutational analyses have shown that the Inr is important for directing the synthesis of properly initiated transcripts of all RNA polymerase II promoters [6]. In the absence of TATA box, the Inr can also determine the location of TSS [5]. Also, there exist many regions giving exact matches to the TATA box and the Inr, which do not represent true promoter regions. The RNA polymerase II promoters seem to obey a principle of "mix and match," that is, any combination of the promoter elements may contribute to promoter function, but none of them alone is essential for promoter function.

13.1.2 Previous Work

The problem of promoter recognition has concentrated on the TSS detection, which mainly locates the beginning of a gene instead of seeking the regulatory elements or promoter regions. It requires recognizing individual binding sites by using some overall descriptions of how these sites are spatially arranged [5].

From a number of computational techniques and algorithms recently developed for the promoter recognition, so far only the applications of probabilistic models [11, 12] and neural networks [16] have reported fair degrees of successes [5]. It is fundamental that these methods scan the TSS by shifting a window of a fixed length along the sequence. Since the promoter activity depends closely on how different TFs bind to the promoter region, in principle, the binding sites can be determined based on the contextual information.

Probabilistic models focus on the homology search in the vicinity of the promoter, for example, to recognize distinct motifs, by estimating position-specific probabilities of the elements in a promoter [1]. A recently developed probabilistic model, based on interpolated Markov chain models, was proposed to extend the range of contextual dependencies by accounting for nonfixed length subsequences whose frequencies were reliably estimated from the data set [11, 12].

Neural networks attempt to receive inputs from a neighborhood window of nucleotides and learn complicated interactions of nucleotides in regulatory regions by finding arbitrarily complex nonlinear mappings [16]. An evolutionary algorithm finding straightforward motifs in the promoter region has been applied to locate potentially important patterns by inspection [4]. Though substantial improvements in computational modeling for recognition of promoters have been reported in the recent literature, the accuracies are far from satisfactory.

13.1.3 Motivation

Accurate promoter recognition requires handling underlying correlation between nucleotides at nonfixed positions or detecting relevant features surrounding the promoter region. Based on the hypothesis that different positions within the promoter region make independent contributions to binding, the low-order probabilistic models give poor results in known cases as, for instance, the existence of multiisoforms of proteins, leading to different classes of sites, or alternative protein conformation induced by the DNA structure, leading to correlated preferences at different positions [5]. Although many researchers have previously identified the potential of higher-order probabilistic models in sequence modeling, their implementation has practically been prohibitive because of the need of large numbers of training samples and due to the compute-intensive nature of the algorithms. Probably, nonlinear methods are more capable of handling the problems of high-order correlations.

Neural network approaches receive inputs from a neighborhood window of nucleotides and are capable of learning complex interactions of elements in the input window by finding arbitrarily complex nonlinear mapping. However, because of receiving mainly low-level inputs—a string of nucleotides, they incorporate no explicit biological features. Relevant features of the promoter regions were localized by using statistic analyses to determine genetic contexts [18], but the exact location of TSSs could not be predicted by these methods.

This chapter describes a novel Markov/neural hybrid approach to the promoter recognition by introducing a novel encoding scheme for inputs to the neural networks, using lower-order Markov chains. The lower-order Markov models is used to incorporate *a priori* biological knowledge of the context of promoters; the neural networks combine the outputs from Markov chains to derive long-range and complex interactions among nucleotides, which are related to the functions of promoters. The significant improvements of accuracies of the promoter recognition with the present Markov encoding suggest that the lower-order Markov chains correctly represent the features relevant to biological phenomena and the present Markov/neural hybrid provides a potential model for representing promoters in genomic sequences.

13.1.4 Organization

The remaining of the chapter is organized as follows: Section 13.2 describes the novel hybrid approach using Markov models and neural networks for recognition of eukaryotic promoters; Section 13.3 presents the experiments with the present approach on a human promoter data set of 565 promoter sequences. A comparison among neural network models, such as multilayer perceptron (MLP), time-delay neural network (TDNN), and time-delay neural network with shortcuts (HBNN), and a comparison of the present approach with the promoter recognition program, NNPP 2.1 [16] are also given in this section. Finally, Section 13.4 presents some conclusions.

13.2 METHODS

The present model consists of three modules working in cascade, as illustrated in Fig. 13.1.

First, the model scans the TSS by shifting a window of fixed length along the given DNA sequence. A scanning window of length 100 bp [−50, 50] of DNA sequences is considered for the promoter, presuming that the scanning point representing the candidate TSS position is relatively at +1. From the scanning window, two segments correspondingly representing the TATA box and the Inr of the promoter candidate are extracted: as seen in Fig. 13.2, the TATA box segment is chosen as a subsequence of 30 bp from −40 to −10 and the Inr segment is chosen as a subsequence of 25 bp from −14 to 11. The segment sizes are selected so that the conserved motifs for both binding sites are included. Further information relating to our selection of the segment locations can be found in Refs. [1, 9, 12, 16].

13.2.1 Markov Chain Models

We model two DNA segments of the promoter, the TATA box and the Inr, and their concatenation separately by three Markov chain models where the observed state variables are elements drawn from the alphabet Ω_{DNA} of four bases: A, T, G, and C. The Markov chain is defined by a number of states that is equal to the number of nucleotides in the sequence, where each state variable of the model corresponds to a nucleotide. Consider a sequence $(s_1, s_2, \ldots, s_l)$ of length l, modeled by a Markov chain, where the nucleotide $s_i \in \Omega_{DNA}$ is a realization of the ith state variable of the Markov chain. Except from state i to state $i+1$, there is no transition from state i to the other states. The model serially travels from one state to the next while emitting letters from the alphabet Ω_{DNA} in which each state is characterized by a position-specific probability parameter. In previous works [2, 17], the Markov chain model was referred to as weight array matrix model.

If the Markov chain, say M, has an order k, the likelihood of the sequence is given by

$$P(s_1, s_2, \ldots, s_l | M) = \prod_{i=1}^{l} P_i(s_i), \qquad (13.1)$$

where the Markovian probability $P_i(s_i) = P(s_i | s_{i-1}, s_{i-2}, \ldots, s_{i-k})$ denotes how conditionally the appearance of the nucleotide at location i depends on its k

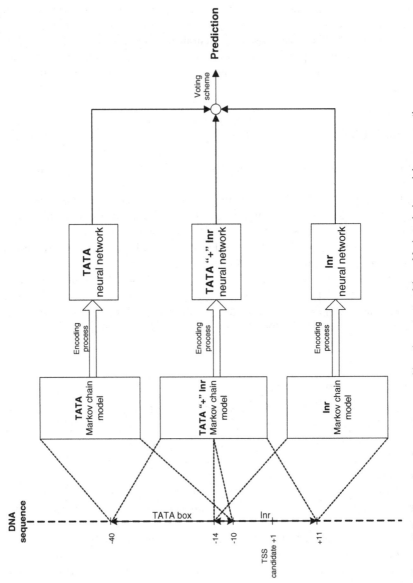

Figure 13.1 Illustration of promoter recognition: the outputs of three Markov chain models, representing TATA box, concatenation of TATA box and Inr models, and Inr models, are fed to three neural networks whose outputs are combined using a voting scheme for final prediction.

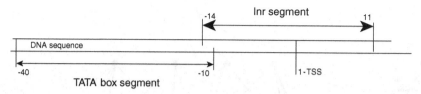

Figure 13.2   The representation of TSS by a promoter consisting of TATA box and Inr segments; TSS is considered to be at site +1.

predecessors. In case $i \leq k$, lower-order dependencies are appropriately used. Such a model is characterized by a set of parameters, $\{P_i(s) | s \in \Omega_{\text{DNA}}, i = 1, 2, \ldots, l\}$. That is, we need to maintain 4^{k+1} parameters at each state to represent the segment.

There are three independent Markov chain models in our approach: two of these models represent the TATA box and Inr segments, whereas the last model represents the concatenation of these two segments, that is, the nucleotides of the interval $[-40, -10]$ plus the nucleotides of the interval $[-14, +11]$, to capture long-range contextual correlations around the promoter region. Higher-order Markov chains is preferred to capture all nucleotide interactions in the promoter regions. However, estimation of 4^{k+1} Markovian parameters at each state limits the order of the Markov chain with the amount of the available training data. In addition, with increasing order k, some subsequences of length k and $k + 1$ might not appear in the sequences of the training set, for which the likelihood of any sequences containing these subsequences is set to zero. Therefore, the second order ($k = 2$) is chosen for the TATA box model and the first order ($k = 1$) is chosen for the Inr model because of the weak consensus of the Inr [5]; thus, the order of the concatenation model is chosen accordingly, the second order for the nucleotides belonging to the TATA box segment and the first order for those belonging to the Inr segment.

13.2.2   Inputs to Neural Networks

In majority of neural network applications, DNA sequences need to be converted into digits for inputs. The orthogonal coding method using 4-bit code per base has been widely chosen to avoid algebraic dependencies among nucleotides [16], which encodes each nucleotide s by a block of four binary digits, say $\mathbf{b}_s$, where only one digit is set to 1 and the remaining digits are set to 0; for instance, $\mathbf{b}_A = (1,0,0,0)$, $\mathbf{b}_C = (0,1,0,0)$, $\mathbf{b}_G = (0,0,1,0)$, and $\mathbf{b}_T = (0,0,0,1)$. Generally, given a sequence $(s_1, s_2, \ldots, s_l)$ of length l, where $s_i \in \Omega_{\text{DNA}}$, the encoding process results in a vector $(\mathbf{b}_{s_1}, \mathbf{b}_{s_2}, \ldots, \mathbf{b}_{s_l})$ of $4\,l$ elements.

One disadvantage of the orthogonal encoding is that it incorporates no explicit biological features. We suggest that contextual relations should be used at the input of the neural networks to reflect the biological characteristics of the input sequences. Therefore, a Markov encoding method is proposed to combine the parameters of Markov chain models and the orthogonal coding, as illustrated in Fig. 13.3. Given the sequence $(s_1, s_2, \ldots, s_l)$, the outputs from a Markov chain model's parameters of kth

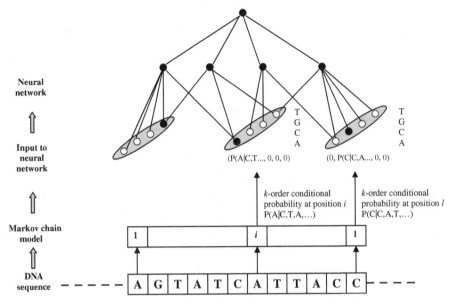

Figure 13.3   Illustration of the proposed Markov input encoding of inputs to a neural network.

order $(P_1(s_1), P_2(s_2), \ldots, P_l(s_l))$ and the orthogonal vector $(\mathbf{b}_{s_1}, \mathbf{b}_{s_2}, \ldots, \mathbf{b}_{s_l})$, the present encoding results in a vector of $4\,l$ elements as inputs: $(\mathbf{c}_{s_1}, \mathbf{c}_{s_2}, \ldots, \mathbf{c}_{s_l})$ where $\mathbf{c}_{s_i} = P_i(s_i)\mathbf{b}_{s_i}$ for $1 \le i \le l$. Due to the use of Markov chains, the present encoding method allows for the incorporation of homology present in potential promoter sites into neural network-based methods.

13.2.3 Neural Networks

Assume that an MLP network has one hidden layer of m units and one output unit, the output of the network is given by

$$y = f\left(\sum_{k=1}^{m} w_k f_k\left(\sum_{j=1}^{l} w_{kj} x_j\right)\right),\tag{13.2}$$

where f_k, $k = 1, 2, \ldots, m$, and f denote the activation functions of the hidden-layer neurons and the output neuron, respectively; w_k, $k = 1, 2, \ldots, m$, and w_{kj}, $k = 1, 2, \ldots, m$, $j = 1, 2, \ldots, l$ denote the weights connected to the output neuron and to the hidden-layer neurons, respectively.

In addition to the classical MLP, other neural network architectures were attempted for the proposed component modules, including a TDNN [7] and a hybrid of HBNN (direct connections from the input to the output units). The TDNN is a multilayer feedforward neural network whose hidden neurons and/or output neurons are replicated across time. The hidden neurons of the TDNN are connected to a limited number of input neurons which represent a consecutive pattern in the input sequence. These neurons have a receptive field, comprising a set of weights connecting one hidden neuron to a subset of input neurons, which allows them to be sensitive to a specific area of the input sequence. The hidden layer contains several copies of receptive fields such that the various replicas of a hidden neuron apply the same set of weights to each position exactly one. During training, in order to force weights in all the copies of a receptive field to have the same values, the weight update is averaged all over the copies of a weight [7, 16]. The hidden neurons are fully connected to the output layer. For promoter modeling, the TDNN is devised to capture TF binding sites that may occur at nonfixed positions and recognize underlying features even when they appear at different relative positions [16].

In order to further explore relevant feature signatures on the TF binding sites, the HBNN that extends the TDNN with direct connections from the input to the output neurons was tested. The direct links add a linear component to the network; the output of the MLP network with direct connections to the output is given by

$$ y = f\left(\sum_{k=1}^{m} w_k f_k\left(\sum_{j=1}^{l} w_{kj} x_j \right) + \sum_{j=1}^{l} v_j x_j \right), \tag{13.3}$$

where $v_j, j = 1, 2, \ldots, l$, denote the shortcuts from the inputs connected to the output neuron. The shortcuts v_j are updated by using a modification of the TDNN training algorithm, which considers partial derivative of the error function with respect to v_j.

For all the three network models, the output activation functions were selected to be a unipolar sigmoidal: $f(u) = \alpha/(1 + e^{-\beta u})$. The hidden-layer activation functions were taken in the form of hyperbolic tangent sigmoidals: $f_k(u) = \alpha_k(e^{\beta_k u} - e^{-\beta_k u})/(e^{\beta_k u} + e^{-\beta_k u})$, for all k. The weights of the neural networks were learned with the standard backpropagation algorithm [7]. The performances of these neural network architectures were compared and the comparison analysis will be presented in Section 13.3.

13.2.4   Higher-Order Markov Models of Promoters

As described in the previous section, low-order Markov chains allow for the incorporation of the prior knowledge, such as the homology of TATA box and the Inr-conserved motifs. However, these models are sufficient only in representing nucleotide dependencies in the adjacent positions in the promoter regions because of low-order Markovian assumption. In the case of data sets of small sequences, the direct implementation of high-order Markov models is not feasible. The basic idea behind our approach is that neural networks are so designed to take the Markovian parameters as input that with nonlinear mapping capabilities, higher-order dependencies in the sequences are effectively captured.

Ohler et al. [11] applied an interpolation technique to the Markov chain model in order to show that the higher-order dependencies are approximated by rational interpolation; given a subsequence $s_1^i = (s_1, s_2, \ldots, s_i)$,

$$P(s_i|s_1^{i-1}) \approx \frac{\sum_{k=0}^{i-1} a_k g_k(s_{i-k}^{i-1}) \hat{P}(s_i|s_{i-k}^{i-1})}{\sum_{k=0}^{i-1} a_k g_k(s_{i-k}^{i-1})}, \tag{13.4}$$

where $\hat{P}(\cdot)$s denote the empirical probabilities obtained from the maximum likelihood (ML) estimates from the training data sets (see Eq. 13.5) in which coefficient a_ks denote real values such that $\sum_{k=0}^{i-1} a_k = 1$, and g_ks denote functions that represent the relationships of different-order contextual interactions among variables s_k, $k = 1$, $2, \ldots, i$ [11]. The result is that by replacing conditional probabilities with the probabilities conditioned with a less number of elements, the high-order conditional probabilities are approximated favorably with polynomials of sufficient orders, by taking inputs as the low-order conditional probabilities [8].

Following Pinkus [15], the output of the neural network is capable of approximating the input–output relationship with a sufficiently large higher-order polynomial. And by application of Equation 13.4, the neural network output, y, represents a highe-rorder Markov model that takes care of all the conditional interactions among all the elements in the input sequence. In other words, by nonlinearly combining the nucleotide interactions at TF binding sites, which are given by lower-order Markovian probabilities, the neural networks are capable of capturing distant nonlinear contextual relationships.

13.3 EXPERIMENTS AND RESULTS

In this section, we provide experimental results with some benchmark data to evaluate the performance of our method and to compare the results with the previous method.

13.3.1 Data Sets

We carried out experiments with the human promoter data set provided by Reese et al. [23]. The data set was used for the training and evaluation of promoter prediction programs in human DNA, which consist essentially of three parts: promoter sequences, CDS (coding) sequences, and noncoding (intron) sequences. Each part contains five sequence files to be used with fivefold cross-validation. The promoters were extracted from the eukaryotic promoter database (EPD) release 50 (575 sequences) [21]. Entries with less than 40 bp upstream and/or 5 bp downstream were discarded, leaving 565 entries. Of these, 250 bp upstream and 50 bp downstream were extracted, resulting in 300 bp long sequences that may have flanking N regions because of the lack of data in the beginning and/or end of the promoter region. The false set contains coding and noncoding sequences from 1998 Genie data set [22]. The exons were concatenated to form single CDS sequences. The sequences were cut consecutively into 300 bp long nonoverlapping sequences. Shorter sequences and remaining sequences at the end were discarded. Finally, 565 true promoters, 890 CDS sequences, and 4345 intron sequences were included for the evaluation.

13.3.2 Training

First, the training sequences were aligned to obtain the ML estimations of Markov model parameters. The estimates of the kth-order Markov model, in our case $\hat{P}_i$s, are given by the ratios of the frequencies, determined from all partial sequences of $k+1$ elements at i and k elements at $i-1$ positions:

$$\hat{P}_i(s_i) = \frac{\#\left(s_{i-k}^i\right)}{\#\left(s_{i-k}^{i-1}\right)},\tag{13.5}$$

where $\#(\cdot)$ represents the frequency of its argument in the training data set. Smaller k's were appropriately used at the boundaries of the sequences, or when $k \le i$. A small number was assigned to those sequences that were not encountered in the training set to avoid zero values at the denominator. Only the promoter sequences were considered in training the Markov models.

Once the Markov chain parameters were learned, the training sequences were applied again to the model and the Markovian probabilities were used as inputs to the neural network for training the neural networks. The desired outputs of the neural network were set to either 0.9 or 0.1 to represent the true or false site at the output, correspondingly. In order to avoid overfitting, the number of the hidden neurons was determined such that the total number of free parameters, that is, the weights and biases, should be about ϵ times the number of the training sequences, where ϵ denotes the fraction of detection errors permitted in testing [7]. These two criteria were used to set low bound and high bound of the number of hidden neurons. The standard online error backpropagation algorithm was used to train the neural network [7] with momentum term initially set to 0.01 and learning rate initially set to 0.1 for every hidden and output neurons and updated iteratively [14]. The parameters of the activation functions were empirically set: $\alpha = 1.0$, $\beta = 1.716$, $\alpha_k = 1.0$, and $\beta_k = 1.0$. The weights of neural networks were initialized in $[-1,1]$ using Nguyen–Widrow method [10].

The present method of promoter recognition comprises four steps: (1) The TATA box and the Inr segments are extracted from the scanning window and fed to the corresponding Markov chain models; (2) each model, with its Markovian parameters, assigns a weight to each nucleotide at each position of the input sequence according to the nucleotide distribution in the positional context; (3) the parameters from these Markov models are fed into one of the three neural networks whose outputs are later compared with thresholds selected in the interval $[0.1, 0.9]$; and (4) a voting procedure makes a decision on the majority vote.

13.3.3 Results

Experiments were carried out on different neural networks separately with orthogonal encoding and with Markov encoding; the comparisons were based on two measures: true positive and false positive rates. By changing thresholds of the prediction, we obtained a range of true positives and false positives, which led to the receiver operating characteristics (ROC) giving true positives versus false positives [3].

All neural network configurations were tested by using the Stuttgart Neural Network Simulator Software toolkit [24], publicly available from the University of Stuttgart, Germany. Each component neural network was tested empirically with different numbers of neurons in the hidden layer for best recognition rate. Because of the small size of the data sets and small lengths of the TATA box and the Inr segments, we examined the following numbers of hidden neurons: 2, 3, 5, 7, 9, 12, 15, 17, and 20. We found that 7 hidden neurons for the TATA's MLP network, 7 hidden neurons for the Inr's MLP network, 15 hidden neurons for the concatenation of TATA box and Inr's MLP networks performed best. Also, we empirically determined that a receptive field size of 17 bp was the best choice for the TDNNs and the HBNNs.

Performance comparisons of different neural networks using the present encoding for detecting different motifs are given in Figs 13.4 – 13.6 with the reference to MLPs using orthogonal encoding. The HBNN showed the best performance for the recognition of the TATA box segment, which also supports the hypothesis of the highly conserved nature of the TATA box [5]. The MLP demonstrated a better accuracy than

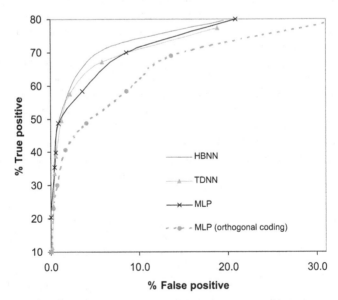

Figure 13.4 ROC curves measuring performances of several neural network models for the recognition of TATA box with the present input encoding method and comparing with MLP using orthogonal encoding.

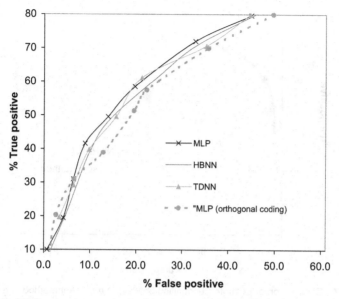

Figure 13.5 ROC curves measuring performances of several neural network models for the recognition of Inr with the present input encoding method and comparing with the MLP using orthogonal encoding.

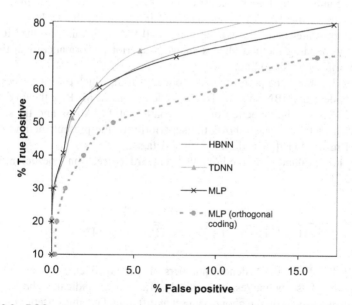

Figure 13.6 ROC curves measuring performances of several neural network models for the recognition of the promoter region [−40, +11], considering TSS at site +1, with the present input encoding method and comparing with MLP using orthogonal encoding.

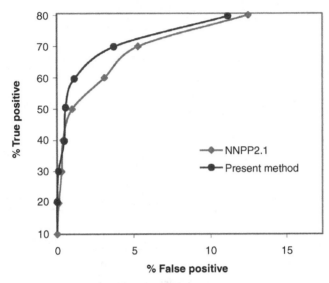

Figure 13.7   Comparison of promoter recognition between the present method and NNPP 2.1 method [16], which uses a TDNN with orthogonal encoding for inputs.

the TDNN in detecting the Inr feature. The result agrees with the previous observation that the Inr is a much weaker signal than the TATA box and TDNN and HBNN are good at detecting underlying features at different relative positions. In recognition of the concatenation of the TATA box segment and Inr segment, the TDNN and HBNN clearly demonstrated better accuracies than the MLP. Also, in this case, the MLP using the present encoding method always showed superior performance than the MLP using the orthogonal encoding.

Figure 13.7 shows the performance of our approach, which uses three best neural network models: the HBNN for the TATA box detection, the MLP for the Inr detection, and the TDNN for the detection of concatenation of TATA box and Inr segments. Comparison of ROC curves shows the superiority of the present method over the NNPP 2.1 method [16], tested on the same data set.

To further evaluate our method, the standard correlation coefficient, r, was computed [3]:

$$r = \frac{(TP \times TN) - (FN \times FP)}{\sqrt{(TP + FN) \times (TN + FP) \times (TP + FP) \times (TN + FN)}}, \quad (13.6)$$

where TP, TN, FP, and FN denote numbers of true positives, true negatives, false positives, and false negatives, respectively. Table 13.1 indicates the correlation coefficients measured on the above data set at different TP values. Better correlation coefficient of 0.69 on average was given by our method compared to the rate of 0.65 by NNPP 2.1.

Table 13.1 Comparison of correlation coefficients averaged over fivefold cross-validated sets obtained by the present method and NNPP 2.1 method

%True Positives	NNPP 2.1	Present Method
30	0.50	0.52
40	0.60	0.57
50	0.65	0.65
60	0.61	0.69
70	0.58	0.65
80	0.52	0.53
90	–	0.39

"–" indicates that the NNPP 2.1 was unable to achieve the corresponding value of true positives.

In order to directly compare the present method with a recent promoter recognition system based on evolutionary computations [9], a data set containing only promoter sequences and coding sequences of Reese's data set was used for training and testing. The data set of 565 promoter sequences and 890 coding sequences was first arranged to facilitate fivefold cross-validation such that each of the five test sets contains 452 promoters and 712 false promoters for training and the remaining 113 promoters and 178 false promoters for testing. Table 13.2 shows the results of fivefold cross-validation for the five test sets. As seen, the method performed comparably to the evolutionary computation method on average of the data sets. It should be noted that Howard and Benson [9] did not use the part of introns in the experiments that may lower the performance of their method, as the introns are more random and complex than both the promoters and exons, and they seem to not to carry any genetic information.

Table 13.2 Comparison of the present method and the evolutionary computation method [9] based on the results of fivefold cross-validation tests

		TP	TN	FP	FN
Test 1	Evolutionary method	92	159	19	21
	Present method	92	156	22	21
Test 2	Evolutionary method	87	165	13	26
	Present method	87	166	12	26
Test 3	Evolutionary method	86	168	10	27
	Present method	86	169	9	27
Test 4	Evolutionary method	91	167	11	22
	Present method	91	166	12	22
Test 5	Evolutionary method	90	160	18	23
	Present method	90	160	18	23
Σ	Evolutionary method	446	819	71	119
	Present method	446	817	73	119

The thresholds of three neural networks were set to give the corresponding TP numbers as that reported by the evolutionary computation method for each test.

13.4 CONCLUSION

This chapter introduced a hybrid approach for identifying TSSs in RNA polymerase II promoters, using Markov chains and neural networks. The approach made use of an encoding method based on Markov chains to represent DNA sequences, which allows for the incorporation of biological knowledge of regulatory elements. Lower-order Markov models used at the input stage of our approach capture nucleotide distributions in the coding and noncoding regions surrounding the TSS.

The present method used first- and second-order Markov models at the first stage and then fed the conditional probabilities into neural networks, which realizes higher-order Markov models. This approach renders an efficient and practically feasible method of implementing higher-order Markov models of promoters. The neural networks account for distant and nonlinear interactions of nucleotides in the promoter regulation regions by arbitrary complex mappings of low-order Markovian probabilities. Earlier, though higher-order Markov models have often been touted as more accurate models of promoters, their direct implementation has been practically prohibitive because of the difficulties in estimating the large number of parameters with the limited amount of the available training data.

Our experiments demonstrated that the structure of present model allows for incorporating the homology of potential promoter sites into the neural network-based methods, resulting in reduced rate of false recognitions of promoters. The present higher-order Markov model implemented using neural networks showed better performance than both low-order Markov model-based approaches and neural network approaches in the identification of the binding sites in the promoter region of the tested data sets.

Though the present model showed promise in recognition of promoter sites, our comparison was limited. The parameters of the neural networks used in the experiments were set empirically, though their optimization may lead to further improvement of accuracies. Our method is not only promising for recognition of promoters, but also suggests a framework for realizing higher-order Markov models with neural networks, which is worthwhile for further investigation as most signals or motifs in biological sequences can be characterized by features with elements having complex interactions.

REFERENCES

1. Bajic, V. B., Seah, S. H., Chong, A., Krishnan, S. P. T., Koh, J. L. Y., and Brusic, V. Computer model for recognition of functional transcription start sites in RNA polymerase II promoters of vertebrates. *Journal of Molecular Graphics and Modelling*, 21: 323–332, 2003.

2. Burge, C. and Karlin, S. Prediction of complete gene structures in human genomic DNA. *Journal of Molecular Biology*, 268: 78–94, 1997.

3. Burset, M. and Guigo, R. Evaluation of gene structure prediction programs. *Genomics*, 34: 353–367, 1996.

4. Corne, D., Meade, A., and Sibly, R. Evolving core promoter signal motifs. *Proceedings of the 2001 Congress on Evolutionary Computation, IEEE Press*, pp. 1162–1169, 2001.

5. Fickett, J. W. and Hatzigeorgious, A. G. Eukaryotic promoter recognition. *Genome Research*, 861–878, 1997.

6. Goodrich, J. A., Cutler, G., and Tjian, R. Contacts in context: promoter specificity and macromolecular interactions in transcription. *Cell*, 84: 825–830, 1996.

7. Haykin, S. *Neural Networks: A Comprehensive Foundation*, Prentice-Hall, Upper Saddle River, NJ, 1999.

8. Ho, S. L. and Rajapakse, J. C. Splice site detection with a higher-order Markov model implemented on a neural network. *Genome Informatics*, 14: 64–72, 2003.

9. Howard, D. and Benson, K. Evolutionary computation method for promoter site prediction in DNA. *Genetic and Evolutionary Computation Conference*, Chicago, 1690–1701, 2003.

10. Nguyen, D. and Widrow, B. Improving the learning speed of 2-layer neural networks by choosing initial values of the adaptive weights. *Proceedings of the International Joint Conference on Neural Networks*, San Diego, 3: 21–26, 1990.

11. Ohler, U., Harback, S., Niemann, H., Noth, E., and Rubin, G. M. Joint modeling of DNA sequence and physical properties to improve eukaryotic promoter recognition. *Bioinformatics*, 17: 199–206, 2001.

12. Ohler, U., Niemann, H., Liao, G., and Reese, M. G. Interpolated Markov chains for eukaryotic promoter recognition. *Bioinformatics*, 15: 362–369, 1999.

13. Pedersen, A. G., Baldi, P., Chauvin, Y., and Brunak, S. The biology of eukaryotic promoter prediction: a review. *Computer& Chemistry*, 23: 191–207, 1999.

14. Plagianakos, V. P., Magoulas, G. D., and Vrahatis, M. N. Learning rate adaptation in stochastic gradient descent. In: N. Hadjisavvas and P. Pardalos, Eds, *Advances in Convex Analysis and Global Optimization, Noncovex Optimization and its Applications*, Kluwer Academic Publishers, Dordrecht, pp. 433–444, 2001.

15. Pinkus, A. Approximation theory of the MLP model in neural networks. *Acta Numerica*, 143–195, 1999.

16. Reese, M. G. Application of a time-delay neural network to promoter annotation in the *Drosophila melanogaster* genome. *Computer Chem*, 26: 51–56, 2001.

17. Salzberg, S. L., Delcher, A. L., Fasman, K., and Henderson, J. A decision tree system for finding genes in DNA. *Journal of Computational Biology*, 5: 667–680, 1998.

18. Scherf, M., Klingenhoff, A., and Werner, T. Highly specific localization of promoter regions in large genomic sequences by PromoterInspector: a novel analysis approach. *Journal of Molecular Biology*, 297: 599–606, 2000.

19. Wang, Y. and Stumph, W. RNA polymerase II/III transcription specificity determined by TATA box orientation. *Proceedings of the National Academy of Sciences*, 92, 8606–8610, 1995.

20. Zhang, M. Q. Computational methods for promoter prediction. In: T. Jiang, Y. Xu, and M. Q. Zhang, Eds, *Current Topics in Computational Molecular Biology*, MIT Press, Cambridge, MA, pp. 249–267, 2002.

21. EPD. http://www.epd.isb-sib.ch/

22. GenieData. http://www.fruitfly.org/seq_tools/datasets/Human/

23. PromoterData. http://www.fruitfly.org/seq_tools/datasets/Human/promoter/

24. SNNS. http://www.ra.informatik.uni-tuebingen. de/SNNS/

14

EUKARYOTIC PROMOTER DETECTION BASED ON WORD AND SEQUENCE FEATURE SELECTION AND COMBINATION

Xudong Xie, Shuanhu Wu, and Hong Yan

14.1 INTRODUCTION

With the publication and preliminary analysis of the human genome sequence [1, 2], which marks a significant milestone in the field of biology, it has become important to characterize and functionally annotate genes based on the collected large-scale sequences. In the past decade, a number of reliable methods have been developed for protein-coding regions prediction [3]. However, for regulatory regions of genes, although a number of algorithms have been proposed [4–11], accurate promoter identification remains a challenge. A promoter is the region of a genomic sequence that is close to a gene's transcription start site (TSS), and it largely controls the biological activation of the gene [12]. Based on the characteristics of promoters, the intrinsic relations between them and the corresponding genes they control can be discovered, and this is useful for our understanding of gene transcription. Therefore, promoter identification is a fundamental and important step in gene annotation.

To discriminate a promoter region from nonpromoter regions, with the quantity of the latter being overwhelmingly larger than the former, many different features are

Machine Learning in Bioinformatics. Edited by Yan-Qing Zhang and Jagath C. Rajapakse
Copyright © 2009 by John Wiley & Sons, Inc.

considered, such as CpG islands [9, 10], TATA boxes [11, 13], CAAT boxes [11, 13], some specific transcription factor binding sites (TFBS) [11, 13, 14], pentamer matrices [15], and oligonucleotides [16]. Also, various pattern recognition technologies have been adopted for classification, for example, neural networks [9, 11, 13, 15], linear and quadratic discriminant analyses [10, 14], interpolated Markov model [11], independent component analysis (ICA) [17, 18], and nonnegative matrix factorization (NMF) [18]. Although current algorithms perform much better than the earlier attempts, according to experimental results and analyses in Ref. 19, it remains an open issue to select appropriate biological signals to be implemented in promoter prediction programs. In fact, none of these signals can cover all promoter representations, and each feature abstracted from promoter sequences has its own limitations.

The underlying principles of existing algorithms for promoter region recognition are that the properties of the promoter regions are different from those of other functional regions and can be subdivided into three main categories: (1) search by signal, (2) search by content, and (3) search by CpG island. "Search by signals" techniques are based on the identification of putative transcriptional patterns such as the TATA box and the CAAT box in DNA sequences, but these patterns cannot be the only determinants of the promoter function. For instance, it was found in one study that applying Bucher's TATA box weight matrix to a set of mammalian nonpromoter DNA sequences resulted in an average of one predicted TATA box every 120 bps [20]. This means that the application of some known transcriptional motifs to promoter prediction introduces many false positives. "Search by content" techniques are often based on the difference in the local base and local word composition between regulatory and nonregulatory DNA regions. Those classes of algorithms assume that the difference is caused by the presence of transcriptional signals, such as the binding motifs for transcriptional regulators in the promoter regions. This concept was explored by analyzing the most frequent hexamers (differential hexamer frequency) [21], other variant-length motifs, and short words [16, 22]. Searching by CpG island techniques is based on the fact that most human promoters are correlated with CpG islands, and many genes are recognized and validated successfully by using CpG island as gene markers [1, 2]. CpG islands are found near gene starts in approximately half mammalian promoters and are estimated to be associated with ~60% of human promoters [22]. Therefore, it is a good indicator for the presence of promoters. Algorithms such as the CpG-promoter [23], CpGProd [24], and first exon finder (FirstEF) [10] make use of the information of CpG islands. Nevertheless, we must bear in mind that not all the human promoters are related to CpG islands, and from this point of view, there are at least ~40% false predictions and correct predictions are limited and cannot exceed 60%, if we make prediction based on CpG islands alone.

In this chapter, we first propose a eukaryotic promoter prediction algorithm, namely PromoterExplorer I, which is based on relative entropy and information content. Then, another algorithm, PromoterExplorer II, is introduced, which combines different kinds of features as the input and adopts a cascade adaptive boosting (AdaBoost)-based learning procedure to select features and perform classification. The outputs of these two methods are finally combined to build a more reliable system, namely

PromoterExplorer III. We evaluate the performance of the proposed algorithms for promoter detection using different DNA sequences. Consistent and promising results have been obtained, which show that our method can greatly improve the promoter identification performance and that it outperforms other methods such as Promoter-Inspector [16], Dragon Promoter Finder (DPF) [15], and FirstEF [10].

14.2 PROMOTER PREDICTION BASED ON RELATIVE ENTROPY AND INFORMATION CONTENT

The underlying principle for promoter recognition is based on the fact that the properties of the promoter regions are different from other functional regions in DNA sequences. Many features may be associated with promoter sequences and functions. They include core promoter elements such as TATA boxes, CAAT boxes and transcription initiation sites (INR), CpG islands, secondary structure elements such as the HIV-1 TAR regions [25], and three-dimensional structures such as the curved DNA sequence [26]. Although most of these elements can be detected by means of computer-assisted sequence analysis, none of them are really promoter specific, and they can be found frequently outside promoter regions. Therefore, it is important to combine them to distinguish promoters from other DNA sequences, such as exon, intron, and 3'UTR.

The number and distribution of words of length k or k-word ($k > 3$) in a DNA sequence can have biological significance. Some particularly important k-words are useful for analyzing particular genomic subsequences. For example, four-word frequencies can be used to quantify the differences between *Escherichia coli* promoter sequences and "average" genomic DNA, and coding and noncoding DNA can be distinguishable in terms of their pentamer (five-word) and hexamer (six-word) distributions [27]. Here, we wish to find k-words that distinguish promoter sequence regions from other regions. Therefore, we attempt to select the most effective k-words that are overrepresented within the promoter regions compared with other DNA sequence regions and can help identify DNA "signals" required for promoter functions. Thus, the focus of our work in this section is on the following issues: (1) the relationship between word length and discriminability of promoter regions and other regions in the DNA sequences, (2) how to select the words of fixed length with the highest discriminability, and (3) how to use the selected features to build a classifier to predict promoter regions. We tackle the former two issues using the Kullback–Leibler divergence [28] and the last issue by the word information content of position distribution [29].

14.2.1 Word Selection Based on Relative Entropy

The relative entropy can be interpreted as a distance, using the Kullback–Leibler divergence. Let $p^k_{promoter}$ and $p^k_{nonpromoter}$ be the probability density functions of words in promoter sequences and in nonpromoter sequences for fixed word length k that has 4^k words in total, where $p^k_{nonpromoter}$ comes from one of the three nonpromoter regions:

exon, intron and $3'$-UTR. The Kullback–Leibler divergence is defined as follows [29]:

$$\delta(p^k_{\text{promoter}}, p^k_{\text{nonpromoter}}) = \sum_{i=1}^{4^k} p^k_{\text{promoter}}(i) \log \frac{p^k_{\text{promoter}}(i)}{p^k_{\text{nonpromoter}}(i)}. \tag{14.1}$$

The Kullback–Leibler divergence can be considered as a distance between the two probability densities, because it is always nonnegative and zero if, and only if, the two distributions are identical. Our aim is to select one group of the most effective words that can distinguish promoter sequences and nonpromoter sequences and at the same time make the distance between promoter sequences and nonpromoter sequences maximal. This group of words can be obtained by maximizing the following criterion function:

$$\begin{aligned}
S &= \underset{\{i|i\in\{1,2,\dots,4^k\}\}}{\arg} \left\{ \max \delta(p^k_{\text{promoter}}, p^k_{\text{nonpromoter}}) \right\} \\
&= \underset{\{i|i\in\{1,2,\dots,4^k\}\}}{\arg} \left\{ \max \sum_i p^k_{\text{promoter}}(i) \log \frac{p^k_{\text{promoter}}(i)}{p^k_{\text{nonpromoter}}(i)} \right\} \\
&= \{i | p^k_{\text{promoter}}(i) > p^k_{\text{nonpromoter}}(i)\},
\end{aligned} \tag{14.2}$$

where S represents the set of indexes of all the words that satisfy the condition in the equation. Often in practice, the number of words in S is still large and needs to be further reduced, especially for a large word length k. We can achieve it by simply sorting $\left\{ p^k_{\text{promoter}}(i) \log \frac{p^k_{\text{promoter}}(i)}{p^k_{\text{nonpromoter}}(i)}, \ i \in S \right\}$ in descending order and then selecting the desirable number of words. This way, we can guarantee that the selected words are of most discriminability relative to other selections. The purpose here is to select the over-represented words within scattered promoter regions over a noisy background of nonpromoters, so we use the nonsymmetric Kullback–Leibler divergence in Equation 14.1 instead of the average Kullback–Leibler divergence that is symmetric.

Our promoter prediction system, namely, PromoterExplorer I, consists of three classifiers, so it needs three different groups of words to distinguish promoter regions from exons, introns, and 3-UTR regions. These three groups of k-words can be obtained from available training sets, and we will discuss them in detail in the following sections.

14.2.2 Promoter Prediction Based on Information Content

By using the proposed method described above, three different groups of k-words, which are of most discriminablility between promoter and exon regions, promoter and intron regions, promoter and 3UTR regions, respectively, can be obtained according to the Kullback–Leibler divergence. We can consider each word group as unknown binding site motif sets that appear more frequently in promoter regions than in nonpromoter regions. Also, we must bear in mind that each of these three assumed binding site motif sets is position specific and of different properties from that of corresponding nonpromoter regions. We represent these properties by their positional weight matrices (PWM), and two PWMs are computed from a training set for each of the three classifiers, each corresponding to a different training set. For instance, for a promoter–exon classifier, two PWMs are generated in terms of a promoter training set and an exon training set, respectively. A total of six PWMs are calculated for our prediction system. Each element in PWM represents a probability distribution of some selected word at a corresponding position in the training sequence. For example, for a training set with sequence length 250, 1024 words, each with length 6, a PWM is generated with the size 1024×255.

For each classifier in our prediction system, we can obtain one group of selected k-words W_k and two corresponding PWMs, $\text{PWM}_{\text{promoter}}$, computed by a promoter training set, and $\text{PWM}_{\text{nonpromoter}}$, computed by some nonpromoter training set (one of exon, intron, and 3-UTR training sets). When an unknown sequence is input into a prediction system, two scores can be calculated by the corresponding two PWMs and a classifying result is determined by these two scores whether or not this sequence is a promoter region. The prediction system assigns a sequence to the class promoter only if all three classifiers decide that the sequence belongs to this class. Let $p_{i,j}$ be an element of PWM at the ith row and the jth column that represents the probability of the ith selected k-word at position j estimated from the training data set and $S = c_1 c_2, \ldots, c_{L-k+2} c_{L-k+1}$ be the unknown input words sequence (L is the training/unknown input DNA sequence, and k is the word length). Then a score can be calculated by the following information content [29]:

$$\text{Score} = \sum_{\substack{c_j \in W_k \\ j=1,2,\ldots,L-k+1}} -p_{i_{c_j}, j}\log(1 - p_{i_{c_j}, j}), \qquad (14.3)$$

where i_{c_j} represents the row number of word c_j in PWM. Note that we adopt the form of $-p\log(1 - p)$ instead of $-p\log p$ to compute the score. This is because $-\log p$ will be smaller when p becomes larger. Since the words in each selected word group are dominant in the promoter region, we can expect that larger scores would be obtained when an unknown input sequence belongs to a promoter region, and a smaller score would be obtained when an unknown input sequence belongs to a nonpromoter region. Based on this motivation, an unknown input sequence is predicated as the promoter in one of the three classifiers if the following two conditions are satisfied simultaneously:

$$\frac{\text{Score}_{\text{promoter}}}{\text{Score}_{\text{nonpromoter}}} > T_1 \quad \text{and} \quad \text{Score}_{\text{promoter}} > T_2, \qquad (14.4)$$

where $\text{Score}_{\text{promoter}}$ and $\text{Score}_{\text{nonpromoter}}$ are calculated by the selected words and the corresponding two PWMs, $\text{PWM}_{\text{promoter}}$ and $\text{PWM}_{\text{nonpromoter}}$ according to Eq. 14.3, and T_1 and T_2 are two thresholds that can be chosen optimally using training sequence sets. Different from other promoter prediction systems that adopt more complex classifiers, for example, artificial neural networks [9], relevance vector machines [30], and quadratic discriminant analyses [10], our classifier is intuitive and efficient.

Traditional classifiers built for promoter recognition often tend to find compromised solutions that may results in too many false positives.

14.3 PROMOTER PREDICTION BASED ON THE ADABOOST ALGORITHM

In Section 14.2, we propose a promoter prediction algorithm, PromoterExplorer I. This method uses the selected words to distinguish promoter regions, and classifiers based on information content are designed. In this section, we describe another method, PromoterExplorer II, which adopts different kinds of features, and a cascade AdaBoost-based learning procedure is used for feature selection and promoter classification [31, 32].

14.3.1 Feature Extraction from DNA Sequences

In PromoterExplorer II, we consider three different kinds of features: the local distribution of pentamers, positional CpG island features, and digitized DNA sequence. These are described in the following sections.

14.3.1.1 *Local Distribution of Pentamers* PromoterInspector is designed based on the context technique. For two sets of oligonucleotides, which are promoter related and nonpromoter related, respectively, several wildcards at multiple positions are introduced to reduce the effect of mismatching. The number of wildcards and the length of the elements in the oligonucleotide sets are optimized for classification and then used for testing. In DPF, pentamers (five-word) are used as input features. The pentamers, which most significantly contribute to the separation between the promoter and nonpromoter regions, are selected, and a PWM is used to represent the positional distribution of pentamers. From the two methods discussed above, we can argue that (1) the discrimination information between promoters and nonpromoters may be concealed in a certain set of combinations of nucleotides; in other words, the local structure plays an important role, and (2) the frequencies of occurrences of these combinations are related to their positions considered in a promoter sequence.

In PromoterExplorer II, we also select pentamers as input features. For an input DNA sequence, a set of pentamers a_i, $i = 1, 2, \ldots, W$, can be obtained, where the maximal value of W is $4^5 = 1024$. To select the most informative pentamers for discriminating promoters and nonpromoters, we consider the posterior probability of I given a_i, $P(I|a_i)$, where I is an indicator that equals 1 when the input sequence is a promoter, otherwise $I = 0$. If $P(I = 1|a_i) > P(I = 0|a_i)$, the input sequence should be a promoter with a higher probability, and vice versa. Define

$$\eta = \frac{P(I = 1|a_i)}{P(I = 0|a_i)}, \quad i = 1, 2, \ldots, W, \tag{14.5}$$

and compute the value of η for each pentamer. According to the Baye's theorem, we have

$$P(I = 1|a_i) = \frac{P(a_i|I = 1)P(I = 1)}{P(a_i)}, \quad i = 1, 2, \dots W, \qquad (14.6)$$

and

$$P(I = 0|a_i) = \frac{P(a_i|I = 0)P(I = 0)}{P(a_i)}, \quad i = 1, 2, \dots, W. \qquad (14.7)$$

From Equations 14.5 to 14.7, we can obtain that

$$\eta = \frac{P(a_i|I = 1)P(I = 1)}{P(a_i|I = 0)P(I = 0)}, \quad i = 1, 2, \dots, W. \qquad (14.8)$$

Assuming that $P(I = 1)$ and $P(I = 0)$ are constant, we define η as follows:

$$\eta = \frac{P(a_i|I = 1)}{P(a_i|I = 0)}, \quad i = 1, 2, \dots, W. \qquad (14.9)$$

The pentamers are then ranked according to their η values. The 250 pentamers with the highest values are selected to form a pentamer set Pset. Here, 250 pentamers are considered because their η values are larger than 1.5. In Table 14.1, the first 10 ranked pentamers and their corresponding η values are shown. The pentamers that include a TATA box or a CAAT box are also tabulated.

Table 14.1 Some selected pentamers in the Pset

Rank	Pentamer	η
1	CGGCG	9.11
2	GCGCG	7.71
3	GCGGC	7.43
4	CGCGG	7.35
5	CGCCG	7.10
6	CGCGC	7.00
7	CCGCG	6.94
8	TCGCG	6.43
9	GGCGG	6.37
10	CGCGA	6.31
171	CCAAT	2.02
223	CAATC	1.59
236	TATAA	1.52

For each pentamer included in Pset, DPF uses PWMs to represent their positional distributions. This method can describe the distribution of each pentamer exactly at each position in a DNA sequence. However, because there are a limited number of promoters for training, that is, from a hundred to several thousands (here we have 1024 pentamer patterns), the PWMs based on statistics may not be reliable. To solve this small sample-size problem, two modifications are made in our algorithm. First, all pentamers in Pset are considered as one class, and the others as another class. In other words, 1024 pentamer patterns are converted into two kinds of patterns: pentamers in Pset and pentamers not in Pset. Second, the pentamer at each position in a DNA sequence and those within its neighborhood are considered. A window of 51 bp moves across the sequence at an interval of 1 bp, and the number of pentamers in Pset within this window is taken as a feature at the center of the window. Therefore, for a DNA sequence with a length l, the number of features, which represent local distributions of pentamers in Pset, is $l - 4$.

14.3.1.2 *Positional CpG Island Features*   CpG islands are regions of a DNA sequence near and in the promoter of a mammalian gene where a large concentration of phosphodiester-linked cytosine (C) and guanine (G) pairs exists. The usual formal definition of a CpG island is a region with at least 200 bp, with a GC percentage greater than 50%, and with an observed/expected CpG ratio greater than 0.6 [33]. Because CpG islands can be found around gene starts in approximately half mammalian promoters and are estimated to be associated with ~60% of human promoters [22], CpG islands can be used to locate promoters across genomes [9, 10, 12]. In fact, from Table 14.1, we can find that most pentamers in Pset with high η values are also G + C rich. The most widely used CpG island features are the GC percentage (GCp) and the observed/expected CpG ratio (o/e), which are defined as follows:

$$GCp = P(C) + P(G), \tag{14.10}$$

and

$$o/e = \frac{P(CG)}{P(C) \times P(G)}, \tag{14.11}$$

where $P(CG)$, $P(C)$, and $P(G)$ are percentages of CG, C, and G in a DNA sequence, respectively.

GCp and *o/e* are two global features for G + C- rich or G + C-related promoters. However, for the promoters that are G + C poor, CpG island features cannot be used to predict the position of a promoter. It is a reasonable assumption that there are some short regions that are G + C rich, even in a G + C poor promoter sequence. These regions can then be used for promoter identification. In other words, if we consider GCp and *o/e* as a sequence of local features instead of global features, more promoters can be found based on CpG islands. Similar to pentamer feature extraction, a sliding window of 51 bp in length is used, and *GCp* and *o/e* are calculated for each window.

Then, for an l-length DNA sequence, the number of extracted positional CpG island features is $2l$.

14.3.1.3 Digitized DNA Sequence

Besides the local distribution of penta-mers and the positional CpG island features, we also adopt the digitized DNA sequence as an input feature. This is because there may be some intrinsic features in the original sequence that are useful for discriminating promoters and nonpromoters and are still unknown. In fact, some of the early research on promoter identification is based on digitized DNA sequence [34], and various digitization algorithms have been proposed [35, 36]. In our method, each nucleotide is represented using a single integer, as given by $A = 0$, $T = 1$, $G = 2$, and $C = 3$.

From the discussion above, we can see that for an l-length input DNA sequence, the number of extracted features, including the local distribution of pentamers, positional CpG island features, and digitized DNA sequence, is $l - 4 + 2l + l = 4l - 4$. These features are concatenated to form a high-dimensional vector, and then a cascade AdaBoost learning algorithm is used for feature selection and classifier training for promoter identification.

14.3.2 Feature Selection and Classifier Training with Adaboost

Boosting is a supervised learning algorithm used to improve the accuracy of any given learning algorithm. In this algorithm, a classifier with accuracy based on the training set greater than an average performance is created, and then new component classifiers are added to form an ensemble whose joint decision rule has an arbitrarily high level of accuracy in the training set [37]. In such a case, we say that the classification performance has been "boosted." In general, the algorithm trains successive compo-nent classifiers with a subset of the entire training data that is "most informative" given the current set of component classifiers [37].

AdaBoost is a boosting algorithm that runs a given weak learner several times on slightly altered training data and combines the hypotheses to one final hypothesis to achieve greater accuracy than the weak learner's hypothesis would have [38]. The main idea of AdaBoost is that each example of the training set should act in a different role for discrimination at different training stages. Those examples that can be easily recognized should be considered less in the training that follows, while those examples that are incorrectly classified in the previous rounds should be paid more attention to. In this way, the weak learner is forced to focus on the more informative or the "difficult" examples of the training set. The importance of each example is represented by a weight. At the beginning, all the weights are equal, and they are then adaptively adjusted according to the classification results based on a hypothesis in every round. The final hypothesis is a combination of the hypotheses of all rounds, namely a weighted majority vote, where hypotheses with lower classification errors have higher weights [38].

As discussed in Section 14.3.1, for an input DNA sequence with length l, the number of features extracted is $N = 4l - 4$, for example, if $l = 250$, then $N = 996$. We assume that only a small number of these features are necessary to form an effective, strong classifier. We, therefore, define our weak classifier as follows:

$$h_j(\mathbf{X}) = \begin{cases} 1 & \text{if } x_j > \theta_j \\ -1 & \text{otherwise} \end{cases} \quad j = 1, 2, \ldots, N, \quad (14.12)$$

where $\mathbf{X}$ is an input feature vector, x_j is the jth feature of $\mathbf{X}$, and θ_j is a threshold. Suppose we have a set of training samples $(\mathbf{X}_1, y_1), \ldots, (\mathbf{X}_m, y_m)$, where $\mathbf{X}_i \in \mathbf{X}$ and $y_i \in \mathbf{Y} = \{1, -1\}$ ("1" denotes positive examples, and "−1" is used for negative examples). In order to create a strong classifier, the following procedure is used:

1. Initialize the weights for each training example:

$$w_{1,i} = \begin{cases} \dfrac{1}{2N^+} & y_i = 1 \\[2mm] \dfrac{1}{2N^-} & y_i = -1 \end{cases} \quad i = 1, 2, \ldots, m, \quad (14.13)$$

where N^+ and N^- are the number of positives and negatives, respectively.

2. For $t = 1, \ldots, T$

   (a) For each feature x_j, train a classifier $h_j(\mathbf{X})$, which implies selecting the optimal θ_j to produce the lowest error. The error for the classifier h_j considers all the input samples with the condition of θ_j, which is defined as $\varepsilon_j = \sum_{i=1}^{m} w_t [y_i \neq h_j(X_i)]$.

   (b) Find the classifier $h_t: \mathbf{X} \to \{1, -1\}$ that minimizes the error with respect to the distribution w_t: $h_t = \arg\min_{h_j \in H} \varepsilon_j = \sum_{i=1}^{m} w_t [y_i \neq h_j(X_i)]$. Here, $\varepsilon_t = \min_{h_j \in H} \varepsilon_j$ should be larger than 0.5.

   (c) Update the weights of the samples for the next round $w_{t+1,i} = w_{t,i} \beta_t^{1-e_i}$, where $e_i = 0$ if sample $\mathbf{X}_i$ is correctly classified, otherwise $e_i = 1$, and $\beta_t = \varepsilon_t / (1 - \varepsilon_t)$.

   (d) Normalize the weights to make w_{t+1} a probability distribution:

$$w_{t+1,i} = \frac{w_{t+1}}{\displaystyle\sum_{j=1}^{m} w_{t+1, j}}.$$

After T iterations, the resulting strong classifier is

$$h(\mathbf{X}) = \begin{cases} 1 & \displaystyle\sum_{t=1}^{T} \alpha_t h_t(X) \geq \frac{1}{2} \sum_{t=1}^{T} \alpha_t \\ -1 & \text{otherwise} \end{cases} \quad (14.14)$$

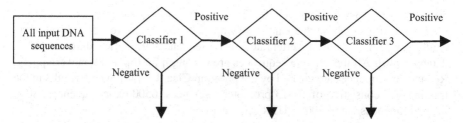

Figure 14.1   Promoter identification based on a cascade AdaBoost classifier.

where $\alpha_t = \log(1/\beta_t)$. The procedure described above not only selects features that will produce the lowest error ε_t when used as weak classifiers, but also trains the weak classifiers and the combined strong classifiers; that is, the optimal values of θ_j, w_t, and α_t are determined based on the training set.

For an input DNA sequence, the number of nonpromoter segments is much larger than the number of promoters. Therefore, it is desirable to remove early on as many nonpromoter segments from consideration as possible. Then, we can cascade our classifiers to filter out most of the nonpromoters, as shown in Fig. 14.1.

From Fig. 14.1, we can see that a number of strong classifiers are used. In the early stage, few features, or weak classifiers, are considered, which can rapidly filter out most of the nonpromoters while retaining most of the promoters. In the later stages, increasingly more complex features are adopted. For each stage, only the positive samples (promoters) and the negative samples (nonpromoters) that are incorrectly classified in the previous stage are used for training. This structure can speed up the detection and can also reduce the effect of a heavy imbalance between the number of promoters and nonpromoters training samples. In our method, a five-layer cascade is used and for each strong classifier, the numbers of weak classifiers used are 10, 20, 50, 100, and 200, respectively.

14.4   EXPERIMENT RESULTS

In this section, we will evaluate the performance of our proposed algorithms, PromoterExplorer I and PromoterExplorer II, for promoter identification based on different databases. These two algorithms adopt different mechanisms for training, that is, PromoterExplorer I designs three classifiers to separate promoters from exons, introns, and 3-UTR regions, while PromoterExplorer II identifies promoters from nonpromoters. Therefore, we describe the training procedures for these two methods, respectively.

14.4.1   Training Sequence Sets

14.4.1.1   *Training for PromoterExplorer I*   For PromoterExplorer I, the vertebrate promoter training sequence set comes from the database of transcription start sites (DBTSS) [39]. For each sequence, a section is taken from 200 bp upstream to

50 bp downstream of the TSS. Vertebrate exon and intron sequences are extracted from the Exon/Intron database that can be downloaded from Web site http://hsc.utoledo. edu/bioinfo/eid/index.html. Vertebrate 3′-UTR sequences are extracted from the UTR database [40]. All the training sequences are of length 250 bp and nonoverlapping. Redundant sequences are cleared by the program CleanUp [41], which results in the training sets consisting of 10513 promoter sequences, 6500 exon sequences, 8000 intron sequences, and 7500 3′-UTR sequences.

14.4.1.2 Training for PromoterExplorer II The training set used for PromoterExplorer II is from the Eukaryotic Promoter Database (EPD), Release 86 [42]. The positive samples are 2426 promoter sequences in EPD, which are from 200 bp upstream to 50 bp downstream of the TSS. The negative samples are randomly extracted sequences of 250 bp, which are out of the range $[-1000, 1000]$ relative to the TSS locations. Here, we do not divide the negative samples into different classes according to their categories, that is, intron, exon, or 3′-UTR, but combine them for training. Furthermore, we extract these nonpromoter sequences from the EPD instead of using annotated intron, exon, or 3′-UTR sets. This can ensure that the training sequences and the testing set are close. In fact, for a DNA sequence to be analyzed, the positions of intron, exon, promoter, and 3′-UTR are unknown, and also the lengths of these patterns vary. Therefore, for a segment to be classified, the characters may be from different kinds of patterns. Certainly, this mechanism cannot ensure that there are no partial promoter segments in the negative samples. However, considering the very small percentage of promoters in a DNA sequence and the large number of negative samples for training, that is, 11,515, the useful and effective discriminating features can still be extracted based on the training set for classification. In our experiments, all training sequences are constructed only by A, T, G, and C; in other words, the sequence that includes the letter N is excluded.

14.4.2 Testing Based on Large Genomic Sequence

14.4.2.1 Word Length and Discriminability Analysis In this section, we first illustrate the relationship between word length and discriminability by experiments, and then a selection strategy is proposed for PromoterExplorer I. In our experiments, word length k ranges from 4 to 7. We do not choose longer words since the number of words is very large and is not practical in real application. First, we calculate the word probability distributions for different word lengths ($k = 4$–7) and the corresponding training sequence set, and then the maximum distance between the promoter words and the nonpromoter words are estimated according to the Kullback–Leibler divergence described in Section 14.2.1. The computation results are provided in Table 14.2. We can conclude from Table 14.2 that the longer the word, the larger the discriminability, but the number of words would increase exponentially. For example, the total number of words for $k = 7$ is $4^7 = 16,384$ and for $k = 6$ is $4^6 = 4096$. This means that the number of selected words will also increase exponentially and that will increase the computational burden and decrease the system performance. In addition, with the increase in word length, most words will not appear simultaneously in a

Table 14.2 The maximum Kullback–Leibler distance between the promoter words and the nonpromoter words for different word lengths

	Kullback–Leibler distance		
Word length	Promoter versus exon	Promoter versus intron	Promoter versus 3'UTR
$k = 4$	0.290781	0.776157	0.583311
$k = 5$	0.349264	0.929367	0.686333
$k = 6$	0.406076	1.098631	0.786895
$k = 7$	0.470600	1.437328	0.928344

training/unknown input sequence since these sequences always take limited length; for example, there are at most 244 words in a 250 bp sequence length for word length $k = 7$. Therefore, we choose $k = 6$ in our system in consideration of the system speed and the limitation of the training sequence sets. To further decrease the computational burden, we select 1024 words for each promoter sensor in our prediction system for which the Kullback-Leibler distances are 0.397643, 1.051998, and 0.755872, respectively, that is, 98%, 95.8%, and 96.1% of the corresponding maximum Kullback–Leibler distance. For the selected words, two PWMs are calculated for each promoter sensor through corresponding training sequence sets, and two threshold parameters of each promoter sensor are optimized by randomly selecting testing sequences from training sequence sets and predefined accuracy parameters.

14.4.2.2 *Large Genomic Sequence Analysis and Comparisons*   The testing databases include six Genbank genomic sequences and human chromosome 22 (http://www.sanger.ac.uk/HGP/Chr22). When testing is performed, an input DNA sequence is divided into a set of segments of 250 bp that overlap each other with a 10-bp shift. As described in Section 14.3.1, features including the local distribution of pentamers, the positional CpG island features, and the digitized DNA sequence are obtained, followed by a cascade AdaBoost for classification. From Equation 14.14, if the final output is larger than zero, a TSS candidate is marked. Those TSS candidates that have no more than 1000 nucleotides apart from their closest neighboring prediction should be merged into a cluster. Then a new TSS prediction is used to represent this cluster, which is obtained by averaging all the TSS candidates within the cluster. To fairly compare the performance of PromoterExplorer I and PromoterExplorer II with other methods, for DPF, the minimum length of gaps in the output predictions is set as 1000, and a similar merging mechanism is also adopted for other methods such as PromoterInspector, FirstEF, and DragonGSF [9]. Combining the outputs of PromoterExplorer I and PromoterExplorer II, followed by a merging mechanism, we can obtain a sequence of new results. This method is called PromoterExplorer III, which is also evaluated in the following sections. Similar to the criteria proposed in Ref. 19, when one or more predictions fall in the region [−2000, +2000] relative to the reference TSS location, a true positive is counted, otherwise the predictions are denoted as false positives. When the known gene is missed by this count, it represents a false negative. The effect of different distance criteria for promoter prediction will be

discussed in the following section. Sensitivity (S_n) and specificity (S_p) are two criteria widely used to evaluate the performance of promoter prediction program, which are defined as follows:

$$S_n = \frac{TP}{TP + FN},$$ (14.15)

$$S_p = \frac{TP}{TP + FP},$$ (14.16)

where *TP*, *FP* and *FN* denote the numbers of true positives, false positives and false negatives, respectively. Generally, the larger the value of S_n, the more false positives are reported, and the smaller the value of S_p. It is a trade-off to balance S_n and S_p. For DPF, the values of S_n can be preset, which is used to control the predictions. In our algorithm, the sensitivity can be modulated by the number of TSS candidates within a cluster. For each cluster to be merged, if the number of TSS candidates within this cluster is larger than a threshold, the merged TSS prediction is considered a true prediction; otherwise, the cluster is removed from the output. Various thresholds will result in different outputs. The larger the threshold value, the fewer the false positives, and also the fewer the true positives predicted.

Experiment Results Based on Genbank   We evaluate the performance of the proposed methods using six Genbank genomic sequences with a total length of 1.38 Mb and 35 known TSSs in these sequences. In Refs 15 and 16, PromoterInspector and DPF outperform other methods based on this set. In another study [19], DragonGSF and FirstEF achieved the best performance in the test. Thus, in this chapter, we compare our methods with these four methods. Figure 14.2 shows the sensitivity-specificity curves of these seven algorithms. In Fig. 14.2a, the distance criterion is set as [−2000, 2000], and in Fig. 14.2b, the criterion [−500, 500] is adopted.

We can see that when a large distance criterion is used, DragonGSF achieves the best performance, and PromoterExplorer I performs better than PromoterExplorer II. When a smaller distance criterion is adopted, PromoterExplorer II outperforms PromoterExplorer I. In the case of the distance criterion [−2000, 2000] being adopted, the average distances between the predicted TSS and the real TSS for Promoter-Explorer I, PromoterExplorer II, PromoterExplorer III, PromoterInspector, DPF, DragonGSF, and FirstEF are 551 (Sn = 45.7%), 440 (Sn = 45.7%), 451 (Sn = 48.6%), 586 (Sn = 40.0%), 509 (Sn = 57.1%), 571 (Sn = 45.7%), and 865 (Sn = 74.3%), respectively. This shows that PromoterExplorer I has higher detection precision, while PromoterExplorer II has the highest locational precision. Combining them, we can see that in these two cases PromoterExplorer III can improve the performance to some extent.

Experiment Results for Human Chromosome 22   Next, we evaluate PromoterExplorer on Release 3 of the human chromosome 22, which includes 34,748,585 base pairs and 393 known genes. The annotation data were produced

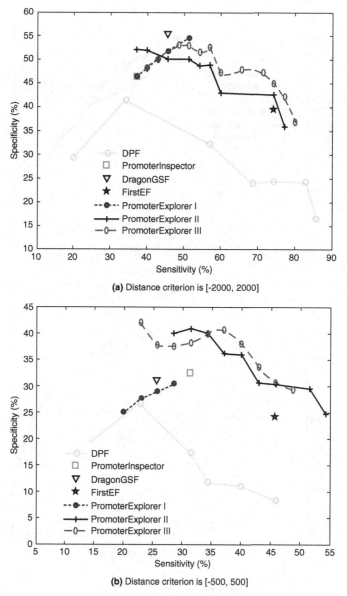

(a) Distance criterion is [-2000, 2000]

(b) Distance criterion is [-500, 500]

Figure 14.2   The sensitivity-specificity curves based on Genbank.

by the Chromosome 22 Gene Annotation Group at the Sanger Center. The comparative experiment results are shown in Fig. 14.3. Similar to the observation in previous section, PromoterExplorer I performs better than PromoterExplorer II in case of a large distance criterion, but the latter outperforms the former with a small distance criterion. When the distance criterion is [−2000, 2000], the average distances between

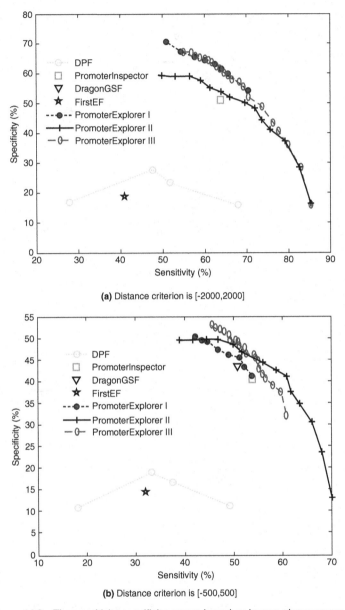

(a) Distance criterion is [-2000,2000]

(b) Distance criterion is [-500,500]

Figure 14.3 The sensitivity-specificity curves based on human chromosome 22.

the predicted TSS and the real TSS for PromoterExplorer I, PromoterExplorer II, PromoterExplorer III, PromoterInspector, DPF, DragonGSF, and FirstEF are 375 ($S_n = 64.1$), 306 ($S_n = 63.9$), 340 ($S_n = 64.1$), 351 ($S_n = 63.6$), 401 ($S_n = 67.9$), 376 ($S_n = 66.2$), and 371 ($S_n = 41.2$), respectively.

14.5 CONCLUSIONS

Computational prediction of eukaryotic promoters from the nucleotide is one of the most important problems in sequence analysis, but it is also a very difficult one. Although a number of algorithms have been proposed, most of them suffer from low sensitivity or too many false positives. In this chapter, we have proposed two effective promoter identification algorithms, namely, PromoterExplorer I and PromoterExplorer II, respectively. PromoterExplorer I focuses on the selection of the most effective words for different functional regions in DNA sequences. A new feature selection strategy based on the Kullback–Leibler divergence is proposed, and a new promoter prediction system that makes use of the position-specific information content is developed. In PromoterExplorer II, different kinds of features, that is, the local distribution of pentamers, positional CpG island features, and digitized DNA sequence, are extracted and combined. Then a cascade AdaBoost algorithm is adopted to perform feature selection and classifier training. The outputs of PromoterExplorer I and PromoterExplorer II are combined to obtain a more reliable detection result. This new classifier-integrated system is called PromoterExplorer III.

Our proposed algorithms are tested on large-scale DNA sequences from different databases. The test results show that PromoterExplorer I has higher detection precision, while PromoterExplorer II has the highest locational precision. Furthermore, PromoterExplorer III can further improve the performance. Our algorithms have favorable performances compared to several best known existing techniques. In other words, our methods not only achieve a balance between the sensitivity and the specificity of the predictions, but also have the capability of more exact TSS localization. Therefore these methods can be used to detect unknown prompter locations in a new DNA sequence.

ACKNOWLEDGMENTS

This work is supported by a grant from Hong Kong Research Grant Council (Project CityU 122607) and a grant from City University of Hong Kong (Project 9610034).

REFERENCES

1. Lander, E. S., Linton, L. M., Birren, B., et al. Initial sequencing and analysis of the human genome. *Nature*, 409: 860–921, 2001.
2. Venter, J. C., Adams, M. D., and Myers, E. W., et al. The sequence of the human genome. *Science*, 291: 1304–1351, 2001.
3. Claverie, J. M. Computational methods for the identification of genes in vertebrate genomic sequences. *Human Molecular Genetics*, 6: 1735–1744, 1997.
4. Kuo, M. D., Zhang, F., and Brunkhors, A. *E. coli* promoter prediction using feed-forward neural networks. *Proceedings of the 28th Annual International Conference of Engineering in Medicine and Biology Society*, pp. 2025–2027, 2006.

5. Zhu, H. M. and Wang, J. X. Predicting Eukaryotic promoter using both interpolated Markov chains and time-delay neural networks. *Proceedings of the International Conference on Machine Learning and Cybernetics,* pp. 4262–4267, 2006.

6. Huang, Y. F. and Wang, C. M. Integration of knowledge-discovery and artificial-intelligence approaches for promoter recognition in DNA sequences. *Proceedings of the Third International Conference on Information Technology and Applications,* pp. 459–464, 2005.

7. Rajapakse, J. C. and Ho, L. S. Markov encoding for detecting signals in genomic sequences. *IEEE/ACM Transactions on Computational Biology and Bioinformatics,* 2(2): 131–142, 2005.

8. Mehta, T. K., Hoque, M. O., Ugarte, R., Rahman, M. H., Kraus, E., Montgomery R., Melancon, K., Sidransky, D., and Rabb H. Quantitative detection of promoter hypermethylation as a biomarker of acute kidney injury during transplantation. *Transplantation Proceedings,* 38(10): 3420–3426, 2006.

9. Bajic, V. B. and Seah, S. H. Dragon gene start finder: an advanced system for finding approximate locations of the start of gene transcriptional units. *Genome Research,* 13: 1923–1929, 2003.

10. Davuluri, R. V., Grosse, I., and Zhang, M. Q. Computational identification of promoters and first exons in the human genome. *Nature Genetics,* 29: 412–417, 2001.

11. Ohler, U., Liao, G. C., Niemann, H., and Rubin, G. M. Computational analysis of core promoters in the Drosophila genome. *Genome Biology,* 3(12): 2002, RESEARCH0087.

12. Pedersen, A. G., Baldi, P., Chauvin, Y., and Brunak, S. The biology of eukaryotic promoter prediction: a review. *Computers & Chemistry,* 23: 191–207, 1999.

13. Knudsen, S. Promoter 2.0: for the recognition of PoIII promoter sequences. *Bioinformatics,* 15: 356–361, 1999.

14. Solovyev, V. V. and Shahmuradov, I. A. PromH: promoters identification using orthologous genomic sequences. *Nucleic Acids Research,* 31: 3540–3545, 2003.

15. Bajic, V. B., Chong, A., Seah, S. H., and Brusic, V. An intelligent system for vertebrate promoter recognition. *IEEE Intelligent Systems Magazine,* 17(4): 64–70, 2002.

16. Scherf, M., Klingenhoff, A., and Werner, T. Highly specific localization of promoter regions in large genomic sequences by PromoterInspector: a novel context analysis approach *Journal of Molecular Biology,* 297: 599–606, 2000.

17. Matsuyama, Y. and Kawamura, R. Promoter recognition for *E. coli* DNA segments by independent component analysis. *Proceedings of the Computational Systems Bioinformatics Conference,* pp. 686–691, 2004.

18. Hiisila, H. and Bingham, E. Dependencies between transcription factor binding sites: comparison between ICA, NMF, PLSA and frequent sets. *Proceedings of the IEEE International Conference on Data Mining,* 4: 114–121, 2004.

19. Bajic, V. B., Tan, S. L., Suzuki, Y., and Sugano, S. Promoter prediction analysis on the whole human genome. *Nature Biotechnology,* 22: 1467–1473, 2004.

20. Prestridge, D. S. and Burks, C. The density of transcriptional elements in promoter and non-promoter sequences. *Human Molecular Genetics,* 2: 1449–1453, 1993.

21. Hutchinson, G. B. The prediction of vertebrate promoter regions using differential hexamer frequency analysis. *Computer Applications in the Biosciences,* 12: 391–398, 1996.

22. Cross, S. H., Clark, V. H., and Bird, A. P. Isolation of CpG islands from large genomic clones. *Nucleic Acids Research,* 27: 2099–2107, 1999.

23. Ioshikhes, I. and Zhang, M. Q. Large-scale human promoter mapping using CpG islands. *Nature Genetics,* 26(1): 61–63, 2000.

24. Ponger, L. and Mouchiroud, D. CpGProD: identifying CpG islands associated with transcription start sites in large genomic mammalian sequence. *Bioinformatics*, 18(4): 631–633, 2002.

25. Bohjanen, P. R., Liu, Y., and GarciaBlanco, M. A. TAR RNA decoys inhibit Tat-activated HIV-1 transcription after preinitiation complex formation. *Nucleic Acids Research*, 25: 4481–4486, 1997.

26. Kim, J. Klooster, S. and Shapiro, D. J. Intrinsically bent DNA in a Eukaryotic transcription factor recognition sequence potentiates transcription activation. *Journal of Biological Chemistry*, 270: 1282–1288, 1995.

27. Claverie, J.-M. K-tuple frequency analysis: from intron/exon discrimination to T-cell epitope mapping. *Methods in Enzymology*, 183: 237–252, 1990.

28. Cover, T. M. and Thomas, J. A. In: D. L. Schilling, Ed., *Elements of Information Theory*. Wiley, New York, 1991.

29. Wu, S., Xie, X., Liew, A. W. C., and Yan, H. Eukaryotic promoter prediction based on relative entropy and positional information. *Physical Review* E, 75:041908, 2007.

30. Down, T. A. and Hubbard, T. J. Computational detection and location of transcription start sites in mammalian genomic DNA. *Genome Research*, 12: 458–461, 2002.

31. Xie, X., Wu, S., Lam, K. M., and Yan, H. An effective promoter detection method using the AdaBoost algorithm. In: D., Sankoff, L., Wang, and F., Chin, Eds, *Proceedings of the 5th Asia-Pacific Bioinformatics Conference*, 2007, Imperial College Press, Beijing, pp. 37–46.

32. Xie, X., Wu, S., Lam, K. M., and Yan, H. PromoterExplorer: an effective promoter identification method based on the AdaBoost algorithm. *Bioinformatics*, 22: 2722–2728, 2006.

33. Gardiner-Garden, M. and Frommer, M. CpG islands in vertebrate genomes. *Journal of molecular biology*, 196(2): 261–282, 1987.

34. Mahadevan, I. and Ghosh, I. Analysis of *E. coli* promoter structures using neural networks. *Nucleic Acids Research*, 22: 2158–2165, 1994.

35. Demeler, B. and Zhou, G. W. Neural network optimization for *E. coli* promoter prediction. *Nucleic Acids Research*, 19: 1593–1599, 1991.

36. Parbhane, R. V., Tambe, S. S., and Kulkarni, B. D. ANN modeling of DNA sequences: new strategies using DNA shape code. *Computers & Chemistry*, 24(6): 699–711, 2000.

37. Duda, R. O., Hart, P. E. and Stork, D. G. *Pattern Classification*, 2nd edition, Wiley, New York, 2001.

38. Freund, Y. and Schapire, R. E. A decision-theoretic generalization of on-line learning and an application to boosting. *Journal of Computer and System Sciences*, 55(1): 119–139, 1997.

39. Suzuki, Y., Yamashita, R., Nakai, K., and Sugano, S. DBTSS: DataBase of human transcriptional start sites and full-length cDNAs. *Nucleic Acids Research*, 30, 328–331, 2002.

40. Pesole, G., *et al.* UTRdb and UTRsite: specialized databases of sequences and functional elements of 5′and 3′untranslated regions of eukaryotic mRNAs. Update 2002. *Nucleic Acids Research*, 30, 335–340, 2002.

41. Grillo, G., Attimonelli, M., Liuni, S., and Pesole, G. CLEANUP: a fast computer program for removing redundancies from nucleotide sequence databases. *Computer Applications in the Biosciences*, 12(1): 1–8, 1996.

42. Schmid, C. D., Périer, R., Praz, V., and Bucher, P. EPD in its twentieth year: towards complete promoter coverage of selected model organisms. *Nucleic Acids Research*, 34: 82–85, 2006.

15

FEATURE CHARACTERIZATION AND TESTING OF BIDIRECTIONAL PROMOTERS IN THE HUMAN GENOME— SIGNIFICANCE AND APPLICATIONS IN HUMAN GENOME RESEARCH

Mary Q. Yang, David C. King, and Laura L. Elnitski

15.1 INTRODUCTION

The mechanisms of regulation of gene expression in the human genome are not well characterized. Our understanding relies greatly on our ability to identify prospective regulatory regions, and to identify them with precision. It turns out that candidate regulatory regions can be identified by searching for genes arranged in a "head-to-head" configuration. The designation of the 5′ and 3′ ends of a gene, from start-to-stop or head-to-tail, indicates that the head-to-head arrangement places the transcription

Machine Learning in Bioinformatics. Edited by Yan-Qing Zhang and Jagath C. Rajapakse
Copyright © 2009 by John Wiley & Sons, Inc.

start sites (TSSs) of two genes in proximity. Because genes are transcribed in the $5'–3'$ direction (downstream) by RNA polymerase, these adjacent genes can produce products in the $3'$ direction, without any interference of each other. The site where the RNA polymerase initially binds is a region of the DNA called a promoter; since transcription proceeds in the $5' \rightarrow 3'$ direction, the promoter must be located upstream of the $5'$ end of the gene. Two genes that have $5'$ ends located fairly close together, say, within 1000 bps, and are transcribed in opposite directions are said to be in a head-to-head configuration. The significance of this configuration is that one or more regulatory elements will be located in the stretch between the $5'$ end of one gene and the $5'$ end of the other. This stretch is known as a *bidirectional promoter*, because it influences expression of the two genes simultaneously.

Bidirectional promoters appear frequently in the human genome [9] and help to regulate DNA repair, mitochondrial, and other processes. Most early instances of bidirectional promoters were discovered in the course of investigating individual genes [9], but recent work by Trinklein et al. [9] resulted in a substantial increase in the number of known bidirectional promoters.

Spliced expressed sequence tags (ESTs) [2], which are short DNA sequences (usually 200–500 bps) obtained by sequencing one or both ends of a transcript of an expressed gene, constitute a large and intricate data set that can be used for detecting bidirectional promoters. However, working with EST data presents a challenge, as the EST database may be highly redundant and may also contain overlapping ESTs. To deal with these problems, we have developed an algorithm to identify bidirectional promoters that are present in the current gene annotations. For instance, these resources include the UCSC list of Known Genes [4], spliced EST data [2], and GenBank mRNA data [2]. The algorithm is capable of handling redundant ESTs, and also those ESTs that overlap or disagree in orientation. By combining data from the three sources, the algorithm evaluates the evidence that a candidate region is in fact a bidirectional promoter. If sufficient evidence exists, the assignment can be based on EST data alone. If not, it looks for supporting evidence by examining Known Gene and mRNA data.

This analysis identified thousands of new candidate bidirectional promoters. In doing so, the $5'$ ends of many known human genes were corroborated, which also provided validation of the method. Other data revealed new $5'$ exons that could be assigned to previously characterized genes, and in some cases, novel genes were found. The fact that our algorithm extracts significantly more bidirectional promoters than were previously known raises the question as to whether these are in fact valid promoter regions, and what biological details can be found from a more comprehensive data set. Subsequent sections are devoted to the discussion of the data and algorithm, new biological results, and ongoing classification strategies and how they provide insight into contemporary approaches to studying promoter regions by computational means.

The prediction of promoter regions remains a difficult challenge in unannotated DNA. The DNA within promoters contains a combination of mostly weak signals, including TATA box motifs, sporadic sequence conservation, and some repetitive elements. Stronger signals emerge from CpG islands, but these are not diagnostic for promoters alone. We describe a scoring approach that works very well for bidirectional promoters and shows that they can be differentiated from other genomic regions by computational means.

15.2 DATA AND ALGORITHM

The data that we use derive from three sources, which are the UCSC Human Genome Browser list of Known Genes [4], GenBank mRNA data [2], and spliced EST data from the GenBank dbEST database [2].

The algorithm for extracting bidirectional promoters is as follows:

I. **Known Gene Analysis**. Known Genes that overlap and have the same orientation are clustered; these clusters are defined by the further 5′ and 3′ ends of any gene in the cluster. The region between the 5′ ends of the two gene clusters is classified as a bidirectional promoter if the following conditions are satisfied:

   – The 5′ ends of the two gene clusters are adjacent to one another, and the two arrows that define the 5′ → 3′ direction for each gene cluster point away from each other.

   – The 5′ ends of the two gene clusters are separated by no more than 1000 base pairs.

   – There are no other gene clusters between the 5′ ends of the two gene clusters.

II. **EST Analysis**. ESTs were assessed for confidence in their orientation using the "ESTOrientInfo" table from the UCSC Genome Browser, which gives a measure of reliability of the orientation of the EST based on all overlapping transcripts from the region. Those with no score were excluded due to low confidence in their orientation. Once the orientation was confirmed, all ESTs were compared to the "intronEST" table to verify agreement; this table lists the intronic orientation for each intron of a spliced EST based on the presence of consensus splice sites.

   ESTs that overlap and have the same orientation are then clustered; these clusters are defined by the furthest 5′ and 3′ ends of any gene in the cluster. Candidate bidirectional promoter regions are formed by pairing an EST cluster with either another EST cluster or a Known Gene cluster, such that the two clusters are in a head-to-head configuration. The candidate bidirectional promoter is rejected if the two clusters overlap, or if the 5′ ends of the two clusters are separated by more than 1000 base pairs. The candidate bidirectional promoters are then classified using a decision tree [12], as shown in Fig. 15.1. The tree either rejects the candidate bidirectional promoter, or assigns it a class label "EST-Li," where i is an integer between 1 and 10. To streamline the notation, in the sequel we truncate the leading "EST-L" from the class label, so that the class label is just an integer between 1 and 10. The class label carries two pieces of information:

   – It gives a confidence level that the candidate is in fact a bidirectional promoter. The confidence level is an integer between 1 and 5, where 1 represents the lowest confidence level and 5 the highest. The confidence level can be obtained from the class label via

$$\text{confidence level} = 5 - \left\lfloor \frac{\text{class label} - 1}{2} \right\rfloor. \tag{15.1}$$

– It indicates whether the candidate bidirectional promoter is contained within a Known Gene, or not. Odd-numbered class labels indicate that the candidate bidirectional promoter is contained within a Known Gene, whereas even-numbered class labels indicate that it is not.

The classification proceeds as follows:

1. Candidate bidirectional promoters enter at the top of the tree in Fig. 15.1. If the candidate bidirectional promoter is contained within a Known Gene, and there exists base pairs of the candidate bidirectional promoter that are more than 1000 base pairs away from the 5' ends of the Known Gene in which the candidate bidirectional promoter is contained, then the candidate bidirectional promoter is rejected; otherwise, we proceed to step 2 (the next level of the tree).

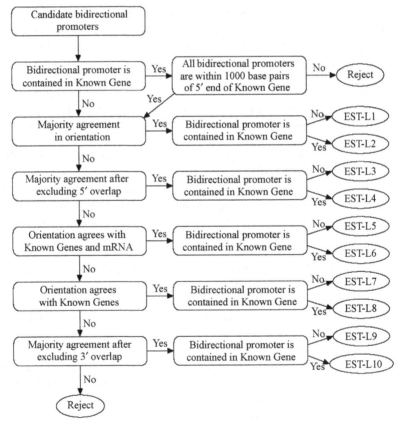

Figure 15.1 Decision tree for classifying candidate bidirectional promoter regions flanked by at least one EST cluster.

2. If the EST cluster(s) flanking the candidate bidirectional promoter satisfies the condition that the number of ESTs that overlap the cluster and disagree in orientation with the cluster is smaller than the number of ESTs comprising the cluster, then we say there is "majority agreement in orientation," and the candidate bidirectional promoter is classified to class 1 or class 2, depending on whether it is contained within a Known Gene; otherwise, we proceed to step 3 (the next level of the tree).

3. If the EST cluster(s) flanking the candidate bidirectional promoter satisfies the condition that after disregarding ESTs that disagree in orientation with the cluster and overlap the 5' end of the cluster by no more than 1000 base pairs, the number of ESTs that overlap the cluster and disagree in orientation with the cluster is smaller than the number of ESTs comprising the cluster, then we say there is "majority agreement after excluding 5' overlap," and the candidate bidirectional promoter is classified to class 3 or class 4, depending on whether it is contained within a Known Gene; otherwise, we proceed to step 4 (the next level of the tree).

4. If the EST cluster(s) flanking the candidate bidirectional promoter agrees in orientation with Known Genes and mRNA transcripts, then the candidate bidirectional promoter is classified to class 5 or class 6, depending on whether it is contained within a Known Gene; otherwise, we proceed to step 5 (the next level of the tree).

5. If the EST cluster(s) flanking the candidate bidirectional promoter agrees in orientation with Known Genes, then the candidate bidirectional promoter is classified to class 7 or class 8, depending on whether it is contained within a Known Gene; otherwise, we proceed to step 6 (the next level of the tree).

6. If the EST cluster(s) flanking the candidate bidirectional promoter exhibits majority agreement (as in step 2) after excluding ESTs that overlap the 3' end of the cluster, then the candidate bidirectional promoter is classified to class 9 or class 10, depending on whether it is contained within a Known Gene; otherwise, the candidate bidirectional promoter is rejected.

The set of bidirectional promoters extracted by the algorithm consists of those extracted in step I, which are precisely those flanked by Known Gene clusters, along with those extracted in step II, which are precisely those flanked either by the two EST clusters, or by one EST cluster and one Known Gene cluster, and furthermore are not rejected by the decision tree.

Evidence that the extracted regions indeed serve as promoters can be obtained by looking for two features in the extracted regions that are associated with promoters.

- The presence of experimentally validated TAF250 (or TAF1) binding sites: a high percentage of TAF250 binding sites coincide with other markers of promoter regions [6].
- The presence of CpG islands.

Table 15.1   Verification of regulatory regions by TAF250 and CpG overlap

Leaf	Gene Pairs	Valid TAF250 (%)	CpG Island (%)	Dual TAF250 (%)
KG	1006	74.55	90.15	71.37
EST-L1	2083	53.77	72.11	50.17
EST-L2	240	50.83	80.41	49.58
EST-L3	225	50.22	73.78	47.11
EST-L4	173	61.85	79.19	58.96
EST-L5	184	37.50	47.80	35.32
EST-L6	103	61.17	67.96	57.28
EST-L7	21	42.86	66.67	33.33
EST-L8	24	29.16	58.33	25.00
EST-L9	363	66.92	83.27	60.84
EST-L10	54	53.70	74.07	48.15
Overall (EST)	3470	52.30	70.40	48.85

For extracted regions that are flanked by two Known Gene clusters (those extracted in step I), 74% overlapped a valid TAF250 binding site and 90% overlapped a CpG island, whereas for extracted regions that are flanked either by the two EST clusters or by one EST cluster and one Known Gene cluster (those extracted in step II), 52% overlapped a valid TAF250 binding site and 70% overlapped a CpG island [12]. A summary of the percentages of extracted regions with valid TAF250 binding sites and/or CpG islands for each class is given in Table 15.1.

Evidence that the extracted regions are in fact bidirectional promoters was obtained by dividing the extracted region into two halves and looking for experimentally validated TAF250 binding sites in each half. The last column of Table 15.1 gives the percentage of extracted regions with TAF250 binding sites in each half.

15.3   BIOLOGICAL RESULTS

The algorithm identified 1006 bidirectional promoters flanked by two Known Genes and 159 bidirectional promoters flanked by one Known Gene and one EST cluster (this situation is illustrated in Fig. 15.2c). Of 5575 candidate bidirectional promoters flanked by two EST clusters, 2105 were rejected by the algorithm. Of the remaining 3470 identified bidirectional promoters [12],

- 2876 were supported by downstream sequences overlapping additional ESTs, mRNA data, or Known Gene data.
- 594 were located in Known Genes; these alternative promoters direct transcription of both a shorter form of the gene G in which they are embedded and a gene that has the opposite orientation to that of G (this situation is illustrated in Fig. 15.2b).

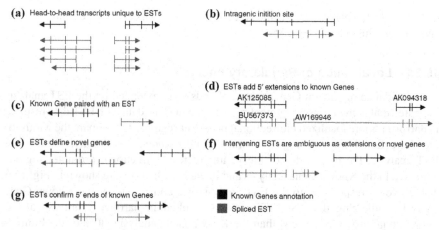

Figure 15.2 Possible configurations of Known Genes and spliced ESTs.

15.3.1 Identification of Novel Genes and Exons

For each Known Gene G that is not in a head-to-head configuration with another Known Gene, let E be the closest EST to G that is in a head-to-head configuration with G. If the 5′ end of E is no more than 1000 base pairs away from the 5′ end of G, then

- If E overlaps a downstream Known Gene G_2 having the same orientation as E, then E is considered to be an extension of the 5′ end of G_2
- If E overlaps a downstream gene G_2 having the opposite orientation to E, then E is considered to be a novel gene.
- If E does not overlap any Known Gene, but one or more downstream Known Genes have the same orientation as E, then E could either be a 5′ extension or a novel gene (this situation is illustrated in Fig. 15.2f). These ESTs require further investigation as well as experimental verification to determine if they represent 5′ extensions or novel genes.

New functional elements identified in this analysis included novel 5′ exons for characterized human genes (this situation is illustrated in Fig. 15.2d). For instance, the EST AW169946 extended the 5′ end of gene AK094318 by 144,000 base pairs to create a new transcription initiation site adjacent to the neighboring gene AK125085.

In addition to the extension of characterized genes, this analysis identified novel transcripts. These transcripts were absent from the list of Known Gene annotations and therefore were only detected by the EST analysis (this situation is illustrated in Fig. 15.2e)). These transcripts were spliced although, their protein-coding potential was not always obvious.

Of the 3470 pairings of EST clusters in a head-to-head configuration, 40% represented extensions of the 5′ ends of Known Genes and 43% represented novel transcripts [12]. ESTs that confirmed the 5' ends of Known Genes were abundant (this situation is illustrated in Fig. 15.2g).

15.3.2   Localization of Regulatory Intervals

The abundance of Known Genes whose 5' ends were extended by the EST analysis indicated that in many cases augmenting the Known Gene data with EST data resulted in narrower, more localized bidirectional promoter regions. To compare the widths of the bidirectional promoter regions extracted in the Known Gene analysis and in the EST analysis, the percentiles of the widths of the bidirectional promoter regions extracted in the Known Gene analysis and in the EST analysis are shown in Fig. 15.3 The curve corresponding to the EST analysis lies above that for the Known Gene analysis, indicating that the EST analysis resulted in narrower, more localized bidirectional promoter regions than the Known Gene analysis; 80% of the bidirectional promoters identified by the EST analysis were 300 base pairs or less, whereas 80% of the bidirectional promoters identified by the Known Gene analysis were 550 base pairs or less [12].

15.3.3   Coordinately Regulated Expression Groups

We looked for evidence of common regulatory patterns revealed by microarray expression profiles among 16,078 Known Genes [14]. For each Known Gene, a cluster was formed consisting of that Known Gene, along with the 500 Known Genes with the most similar coexpression profiles according to the GNF expression data [7]. The association rate, defined as the proportion of genes in the same cluster that are

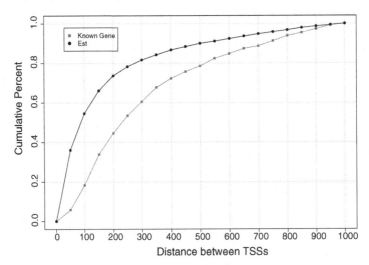

Figure 15.3   Percentiles of the widths of the bidirectional promoter regions extracted in the Known Gene analysis and in the EST analysis.

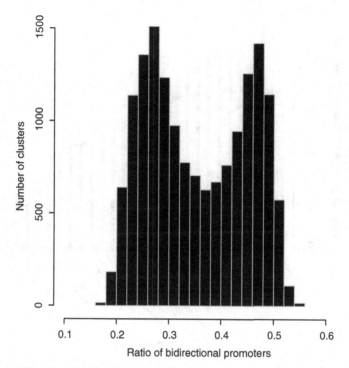

Figure 15.4 Histogram of proportions of genes with bidirectional promoters after clustering genes with similar expression profiles.

regulated by bidirectional promoters, was then calculated for each cluster; it ranged from a low of 0.16 to a high of 0.56. A histogram of the association rates, shown in Fig. 15.4, reveals a bimodal distribution. Genes with the highest rates clustered with other genes regulated by bidirectional promoters at a ratio of 2 : 1. The difference between the clusters obtained and those that would be expected by chance was statistically significant. Thus, there was strong evidence of coordinated expression among subsets of genes in a head-to-head configuration.

15.3.4 Prevalence of Bidirectional Promoters in Biological Pathways

Bidirectional promoters are known to regulate a few categories of genes [1, 15]. Using the 26 biological pathway genes from the Reactome project [5], we examined additional biological categories for enrichment of bidirectional promoters. Compared to the human genome average in which 31% of genes contained bidirectional promoters, 13 Reactome pathways had a ratio of bidirectional promoters significantly larger than 31%, as shown in Fig. 15.5. For example, the percentage of bidirectional promoters in the Influenza, HIV infection, and DNA repair pathways were respectively 48%, 42%, and 40%; these values yielded respective p-values of

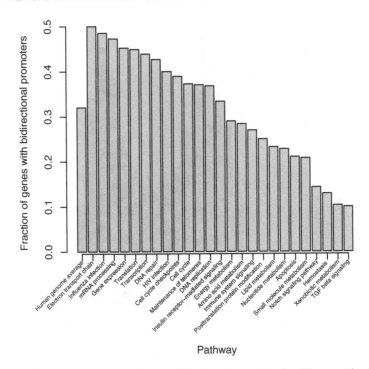

Figure 15.5 Fraction of genes regulated by bidirectional promoters for different pathways. The first bar gives the average for the human genome, which is approximately 0.31.

0.04, 0.04, and 0.09 in a Chi-square test, indicating a statistically significant enrichment of bidirectional promoters as compared to the genome average. These results suggest that bidirectional promoters could provide potential therapeutic targets for disease intervention [12].

15.4   FEATURE-BASED CLASSIFICATION OF HUMAN BIDIRECTIONAL PROMOTERS

15.4.1   Classification Strategy

The classification of regulatory regions, *de novo*, has received much attention due to the completion of several mammalian genomes. Semi-successful approaches utilize individual characteristics such as sequence conservation, phylogenetic models of substitution patterns in DNA, sequence composition, or the presence of clustered transcription factor binding sites. Improved performance is achievable when multiple characteristics are included. Combining features of conservation, composition, and substitution patterns, the regulatory potential score (RP) approach [3] enables classification of functional regions from nonfunctional ones. From its inception, the regulatory potential score has proven effective in predicting functional elements in the human genome [11]. The basic premise in this procedure transforms columns of nucleotides from a multiple-sequence alignment into a reduced alphabet of symbols. The symbols encode combinations of nucleotides, including types of substitutions at each position, and strings of letters. Additionally, alignment gaps are considered as a fifth letter in the DNA. These features are necessary for constructing an optimized alignment over sequences of any given length. The absence of gaps provides vital evidence of evolutionary pressure to maintain the information content within a column of aligned sequences.

As the available multiple alignments gain additional sequences, more efficient handling procedures are necessary to evaluate the data set. For instance, the number of possible alignment columns increases exponentially with the number of sequences in the alignment. Taylor et al. [8] have developed a method for encoding multiple alignment columns into a reduced alphabet that enables detection of regulatory features. This approach, termed ESPERR (evolutionary and sequence pattern extraction through reduced representations), has proven effective for alignments of seven mammalian sequences (human, chimpanzee, macaque, mouse, rat, cow, and dog). These sequences were extracted as aligned bases corresponding to the positions of a training set of known regulatory regions in the human genome [3]. The 97 regions in the training set comprise a mixture of regulatory elements with positively acting influences on gene expression levels. They are present in promoter, enhancer, and intronic locations. For comparison, negative training data contain aligned sequences of ancestral repeats. These ancient relics of mobile genetic elements are no longer actively moving about the genome, and thus are excellent models of neutrally evolving DNA sequences.

A variable order markov Model (VOMM), developed by Taylor et al. [8], having a maximal order of 2 was fitted to this data, resulting in the identification of a 17-symbol alphabet. A log-odds classifier was then constructed based on this model, where the log-odds scores were calculated using

$$S(a) = \sum_{i=1}^{n} \log \frac{P_{\text{SEG}}(\alpha_i | \alpha_{i-1}\alpha_{i-2}\ldots\alpha_1)}{P_{\text{AR}}(\alpha_i | \alpha_{i-1}\alpha_{i-2}\ldots\alpha_1)}. \tag{15.2}$$

In Equation 15.2, SEG represents the DNA segment a, AR represents the ancestral repeat segment, n is the number of base pairs in the multiple alignment, and i is the position.

15.4.2 Regulatory Potential as a Means of Discriminating Functional Classes

Given that RP scores can be used to predict regulatory regions [8], we decided to study the relationship between RP scores and functional regions. Specifically, we examined the distribution of RP scores within each of the following functional classes [13]:

- *Bidirectional Promoters.* This class was divided into three subclasses according to whether the bidirectional promoter was identified on the basis of Known

Gene, mRNA, or EST data; these classes are identified as BP-KG, BP-mRNA, and BP-EST, respectively.

- *Nonbidirectional Promoters.* These are regions in which the promoter length was limited by the tail end of a neighboring gene. The genes that define a nonbidirectional promoter are in a head-to-tail configuration, with the promoter region for the second gene falling between them.

- *Unbounded Promoters.* These are promoters with no proximal neighbors to place limits on their upstream boundaries. This class was divided into three subclasses, UBP1000, UBP5000, and UBP10000, which consist of promoters that have no upstream gene within 1000, 5000, and 10,000 base pairs, respectively.

- *Coding Regions.* These are among the most conserved regions in the human genome.

- *Nonpromoter Regions.* These are regions between the 3′ ends of two neighboring genes that are in a tail-to-tail configuration. It is true that regulatory elements do exist at the 3′ ends of genes, which regulate transcription in an antisense direction. However, these regulatory elements are relatively rare, and thus will not significantly affect our analysis.

- *Ancestral repeat sequences.*

The cumulative distribution function (cdf) of the RP score for each of these functional classes is shown in Fig. 15.6a. The cdfs are observed to cluster into three groups: the leftmost group, corresponding to the lowest RP scores, consists of ancestral repeats; the middle group consists of nonpromoters and unbounded promoters; and the rightmost group, corresponding to the highest RP scores, consists of bidirectional promoters (all three classes), nonbidirectional promoters, and coding regions.

The RP score cdfs for bidirectional promoters, nonpromoters, and unbounded promoters are plotted in Fig. 15.6a; the distinctness of these cdfs suggests that it may be possible to design an RP-score-based classifier to discriminate between bidirectional promoters and nonpromoters/unbounded promoters.

15.4.3 Prediction of Bidirectional Promoters from RP Scores

On the basis of Fig. 15.6a,b, it is apparent that bidirectional promoter regions tend to have higher RP scores than either nonpromoter or unbounded promoter regions. Another way to see this is to plot the class-conditional density functions $p(x|C)$, where x is the RP score and C is a functional class; this is simply the probability density function of RP scores, restricted to the functional class C. Given the class-conditional density functions $p(x|C_1)$ and $p(x|C_2)$ for classes C_1 and C_2, respectively, we can construct a likelihood ratio classifier that maps an RP score x to a functional class using the rule [13]:

$$\text{If } \frac{p(x|C_1)}{p(x|C_2)} \begin{cases} > \mu & \text{Decide class } C_1 \\ < \mu & \text{Decide class } C_2. \end{cases} \tag{15.3}$$

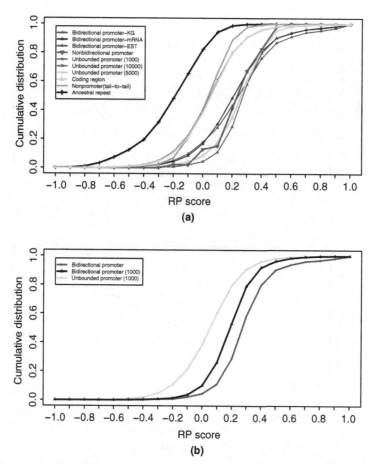

Figure 15.6   (a) Cumulative distribution function of RP scores for various functional regions. (b) Cumulative distribution function of RP score for bidirectional promoters and unbounded promoters.

The performance of this classifier for different values of the threshold μ is summarized by a receiver operating characteristic (ROC), which is a plot of sensitivity against $(1 - \text{specificity})$. We constructed two such classifiers: one to discriminate bidirectional promoters from nonpromoters, and the other to discriminate bidirectional promoters from unbounded promoters.

15.4.3.1   *Distinguishing Bidirectional Promoters from Nonpromoters*

The class-conditional probability distributions $p(x|\text{BP})$ and $p(x|\text{NP})$ are shown in Fig. 15.7a (here BP denotes the class of bidirectional promoters, and NP denotes the class of nonpromoters, that is, the intervening region between two genes in a tail-to-tail configuration). The corresponding ROC curve is shown in Fig. 15.7b. A maximum likelihood classification rule (obtained by setting $\mu = 1$ in the likelihood ratio classifier

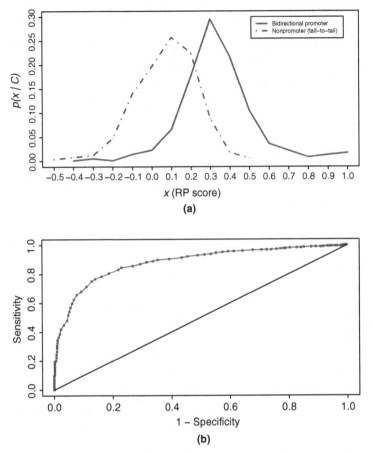

Figure 15.7   (a) Class-conditional probability density functions $p(x|BP)$ (bidirectional promoters) and $p(x|NP)$ (nonpromoters). (b) ROC for classifier that discriminates bidirectional promoters from nonpromoters.

(Eq. 15.3)) yielded a test set accuracy of 74%, a specificity of 92% (relatively high), and a sensitivity of 65% (relatively low), as shown in Fig. 15.2.

The ROC curve reveals that the sensitivity can be boosted above 80% by trading off for a specificity below 80% [13].

15.4.3.2 Distinguishing Bidirectional Promoters from Unbounded Promoters   The class-conditional probability distributions $p(x|BP)$ and $p(x|UBP1000)$ are shown in Fig. 15.8a (here BP denotes the class of bidirectional promoters, and UBP1000 denotes the class of promoters with no upstream gene within 1000 base pairs). The corresponding ROC curve is shown in Fig. 15.8b. A maximum likelihood classification rule (obtained by setting $\mu = 1$ in the likelihood

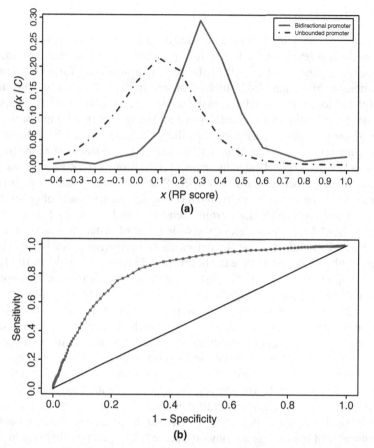

Figure 15.8   (a) Class-conditional probability density functions $p(x|BP)$ (bidirectional promoters) and $p(x|UBP1000)$ (unbounded promoters). (b) ROC for classifier that discriminates bidirectional promoters from unbounded promoters.

ratio classifier (Eq. 15.3)) yielded a test set accuracy of 80%, a specificity of 81% (relatively high), and a sensitivity of 67% (relatively low), as shown in Table 15.2. The ROC curve reveals that the sensitivity can be boosted above 80% by trading off for a specificity below 75% [13].

Table 15.2   Performance of classifiers for test data

Classifier	Accuracy (%)	Sensitivity (%)	Specificity (%)
Bidirectional promoter versus nonpromoter	74.54	65.53	92.16
Bidirectional promoter versus unbounded promoter	80.37	67.94	81.1

15.4.4 Fixed Distances

The analysis of unbounded promoters raised the question, "what distance accurately captures the full length of the functional promoter?" Currently, the best answer is "whichever distance contains all of the sequence necessary for activity under a broad number of conditions." In lieu of an optimal distance, a conventional approach is adopted that standardizes the region of a promoter into a fixed distance. For instance, by defining all promoters as a 550 bp region, they all uniformly begin 500 bp upstream of the transcription start site of the gene and end 50 bp downstream of that same point. A similar approach has been taken to standardize the length of bidirectional promoters. Under the current definition, a 1000 bp distance is the maximal distance between the two TSSs that flank the promoter region. This definition has support from empirical data showing an enrichment of genes that are located within 1000 bp distance from each other and transcribed from opposite strands of DNA [10]. Nevertheless, the wide range of distances observed between TSSs (Fig. 15.3) indicates that the definition is qualitative instead of mechanism derived. Rather than standardize the bidirectional promoter lengths to 1000 bp, we propose to represent them by the distance between the two TSSs. This approach relies on defining promoter regions by feature-based boundaries instead of using fixed distances. In working toward the optimal set of feature classifiers for bidirectional promoters, we discovered that the approach of standardizing the length of promoter regions reduces the quality of functional information in our training set. When fixed distances are used to define the promoter regions, the robustness of our separation is lost (Fig. 15.6b). The loss proves that the act of standardizing bidirectional promoters to 1000 bp regions has detrimental consequences on their classification. As a result, the overall density of high scores in the bidirectional promoter set decreased, as did the appearance of the highest scoring elements. The consequences of the change are obvious, since the largest possible size in this set was 1000 bp by definition. Therefore, we know that standardizing the set leads to an increase of nearly 600 bp in the average size of the functional regions. The increased presence of low scoring elements in the data set reduces the ability of the classifier to accurately discriminate functional regions. The result is reproducible in comparisons against unbounded and head-to tail promoter regions as well.

15.5 DISCUSSION

This analysis of bidirectional promoters resulted in the identification of thousands of new candidate head-to-head gene pairs. It also corroborated the 5' ends of many known human genes, which provided the added benefit of validating the effectiveness of using spliced ESTs to find bidirectional promoters. The new data add to our knowledge of exons at the furthest 5' end of genes. By extending the position of the "head" of the gene further upstream, these new exons show that even previously identified genes may have newly identified bidirectional promoters. The class of novel genes regulated by bidirectional promoters provides a rich data set for experimental evaluation of their function. Biological findings such as the coordinated expression of genes by bidirectional promoters open new avenues for understanding regulated gene expression. This is especially useful because these promoter regions are bounded on both ends and therefore are definable in length. Finally, the fact that bidirectional promoters participate in many pathways indicates that we have only begun to understand their diverse roles in regulating the human genome. Better characteristics of their features will enable better understanding of their biological roles. Bidirectional promoters provide a nice model for testing machine learning approaches to promoter detection. In evaluating our data set, we are able to show that RP scores provide good separation between tail-to-tail regions and bidirectional promoters. Thus, the presence of an exon on both sides of the region is not sufficient to give a strong RP score. Furthermore, RP scores provided good separation between bidirectional promoters and unbounded promoters. This difference suggests that there could be fundamental differences in the composition of these two types of promoters. In the course of this analysis, we discovered that standardizing the lengths of promoters for machine learning purposes is not the best approach. The biological distance between TSSs of genes represents a better classifier feature than the standardized promoter distance, perhaps because it retains the most concentrated signal over the given amount of sequence.

ACKNOWLEDGMENTS

We gratefully acknowledge discussions with faculty of National Human Genome Research Institute for improvement of this study. This study was supported by the Intramural Research Program of the National Human Genome Research Institute, National Institutes of Health.

REFERENCES

1. Adachi, N. and Lieber, M. R. Bidirectional gene organization: a common architectural feature of the human genome. *Cell*, 109(7): 807–809, 2002.
2. Benson, D. A., Karsch-Mizrachi, I., Lipman, D. J., Ostell, J., and Wheeler, D. L. GenBank: update. *Nucleic Acids Research*, 32: D23–D26, 2004.
3. Elnitski, L., Hardison, R. C., Li, J., Yang, S., Kolbe, D., Eswara, P., O'Connor, M. J., Schwartz, S., Miller, W., and Chiaromonte, F. Distinguishing regulatory DNA from neutral sites. *Genome Research*, 13: 64–72, 2003.
4. Hsu, F., Kent, W. J., Clawson, H., Kuhn, R. M., Diekhans, M., and Haussler, D. The *UCSC* Known Genes. *Bioinformatics*, 22(9): 1036–1046, 2006.
5. Joshi-Tope, G., Gillespie, M., Vastrik, I., D'Eustachio, P., Schmidt, E., de Bono, B., Jassal, B., Gopinath, G. R., Wu, L., Matthews, S., Birney, L. E., and Stein, L. Reactome: a knowledgebase of biological pathways. *Nucleic Acids Research*, 33: D428–D32, 2005.

6. Kim, T. H., Barrera, L. O., Zheng, M., Qu, C., Singer, M. A., Richmond, T. A., Wu, Y., Green, R. D., and Ren, B. A high-resolution map of active promoters in the human genome. *Nature*, 436: 876–880, 2005.

7. Su, A. I., Wiltshire, T., Batalov, S., Lapp, H., Ching, K. A., Block, D., Zhang, J., Soden, R., Hayakawa, M., Kreiman, G., Cooke, M. P., Walker, J. R., and Hogenesch, J. B. A gene atlas of the mouse and human protein-encoding transcriptomes. *Proceedings of the National Academy of Sciences*, 101(16): 6062–6067, 2004.

8. Taylor, J., Tyekucheva, S., King, D. C., Hardison, R. C., Miller, W., and Chiaromonte, F. ESPERR: learning strong and weak signals in genomic sequence alignments to identify functional elements. *Genome Research*, 16: 1596–1604, 2006,

9. Trinklein, N. D., Aldred, S. I. F., Hartman, S. J., Schroeder, D. I., Otillar, R. P., and Myers, R. M. An abundance of bidirectional promoters in the human genome. *Genome Research*, 14(1): 62–66, 2004.

10. Trinklein, N. D., Aldred, S. J. F., Saldanha, A. J., and Myers, R. M. Identification and functional analysis of human transcriptional promoters. *Genome Research*, 13(2): 308–312, 2003.

11. Wang, H., Zhang, Y., Cheng, Y., Zhou, Y., King, D. C., Taylor, J., Chiaromonte, F., Kasturi, J., Petrykowska, H., Gibb, B., Dorman, C., Miller, W., Dore, L. C., Welch, J., Weiss, M. J., and Hardison, R. C. Experimental validation of predicted mammalian erythroid *cis*-regulatory modules. *Genome Research*, 16: 1480–1492, 2006.

12. Yang, M. Q. and Elnitski, L. A computational study of bidirectional promoters in the human genome. *Bioinformatics Research and Applications, Spring Lecture Series: Notes in Bioinformatics*, 361–371, 2007.

13. Yang, M. Q. and Elnitski, L. Orthology and multiple class prediction of functional elements in the human genome. *Proceedings of the International Conference on Bioinformatics and Computational Biology*, 2007.

14. Yang, M. Q., Koehly, L. and Elnitski, L. Comprehensive annotation of human bidirectional promoters identifies co-regulatory relationships among somatic breast and ovarian cancer genes. *PLoS Computational Biology*, 3(4): 2007.

15. Zhao, Q., Wang, J., Levichkin, I. V., Stasinopoulos, S., Ryan, M. T., and Hoogenraad, N. J. A mitochondrial specific stress response in mammalian cells. *The EMBO Journal*, 21: 4411–4419, 2002.

16

SUPERVISED LEARNING METHODS FOR MicroRNA STUDIES

Byoung-Tak Zhang and Jin-Wu Nam

MicroRNAs (miRNAs) are a large functional family of small noncoding (nc) RNAs that play a role as posttranscriptional regulators by repressing the translation of mRNA. Recently, it has also been reported that miRNAs are involved in cancer-related processes functioning themselves as oncogenes or tumor suppressors. Therefore, miRNAs are important for investigating the regulation of gene expression as well as for understanding the pathogenesis of specific diseases. In this chapter, we identify the computational issues involved with miRNA studies and review machine learning methods proposed for solving them. In particular, we classify supervised learning methods into three major categories: kernel methods, probabilistic graphical models, and evolutionary algorithms. We then describe their application to the prediction of miRNAs and their target genes. We also discuss the emerging issues in miRNA research and suggest some new directions of future research from the machine learning point of view.

16.1 ISSUES IN CURRENT miRNA STUDIES

16.1.1 Definition of miRNAs and Their Functions

Large fractions of what was until recently considered *junk* DNA in eukaryotes are indeed turned out to be transcribed, and many researchers now believe that they may

Machine Learning in Bioinformatics. Edited by Yan-Qing Zhang and Jagath C. Rajapakse
Copyright © 2009 by John Wiley & Sons, Inc.

play a role as fundamental for understanding genomes as that of dark matter for understanding cosmological phenomena [1]. Recently, the truth that small noncoding RNAs—most of which sit in *junk* DNA—control much of posttranscriptional gene regulation has changed the concept for eukaryotic regulatory systems and finally has forced reconsideration of the original "central dogma" of gene information flow. Among various small RNAs, those of ~22 nucleotides, known as microRNAs, are positioning as key molecules.

MicroRNAs, constituting a large family of noncoding small RNAs, directly take part in posttranscriptional regulation (Fig. 16.1). MiRNAs are defined as single-stranded RNAs of ~22 nucleotides (nt) in length (range 19–25 nt) generated from endogenous transcripts that can form local hairpin structures [3]. They act by binding to the complementary sites on the 3′ untranslated region (UTR) of the target gene

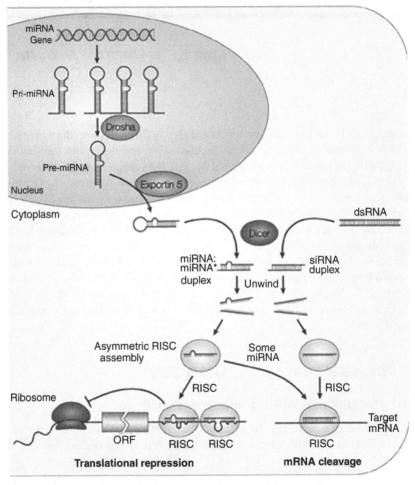

Figure 16.1 Processing and posttranscriptional regulation of miRNA. *Source:* Ref [2].

to induce cleavage with near-perfect complementarity or to repress productive translation [3]. MiRNAs function in various cellular processes, such as enzyme regulation, development, differentiation, growth, and death. In particular, they regulate cancer-related processes, so they might themselves function as oncomirs (analogs of oncogenes) or tumor suppressors [4].

Current issues in miRNA studies can be largely classified into the following categories: (1) identification of novel miRNAs, (2) discovering processing mechanisms of miRNAs, (3) finding valid miRNA target genes, (4) functional studies of known miRNAs, (5) design of artificial pri-miRNA or short hairpin (sh) RNA for gene knockdown systems, and (6) systematic modeling of miRNA-target regulation networks. Among these issues, the two problems, that is, the prediction of miRNAs and their target genes, have been mainly studied through computational methods.

16.1.2 miRNA Prediction

Recent efforts to identify miRNA genes have led to the discovery of thousands of miRNAs in animals, plants, and viruses, but many remain to be discovered [5]. Most of such efforts rely on wet-experiment-based methods such as directional cloning of endogenous small RNAs, requiring much time, cost, and labor [6]. A limitation of this experimental approach is that miRNAs expressed at low levels or only in a specific condition or in specific cell types are difficult to detect. Computational approaches can overcome this problem and identify even imprinted miRNAs.

There are three practical issues in the prediction of miRNA: prediction of pri-miRNAs, prediction of miRNA precursors, and prediction of mature miRNAs. These issues correspond to the processing mechanisms: pri-miRNA to pre-miRNA and pre-miRNA to mature miRNA. In fact, because a pri-miRNA, a primary transcript, can contain several pre-miRNAs and its regulation modules still remain unknown, the prediction of pri-miRNAs is one of the most challenging problems. Also, the prediction of mature miRNA requires us to search the cleavage site cropped by two RNase III-type enzymes, Drosha and Dicer, and is the most important problem in determining the efficiency of mature miRNA prediction. However, because the biologically relevant cleavage data contain some errors, a predictive method that is robust against errors is required. Thus, most current work focuses on the prediction of miRNA precursors. The generation of experimental data through biochemical-based studies for the recognition of pri-miRNAs by the Drosha–DGCR8 complex [7] and the introduction of supervised machine learning methods will help address these problems.

16.1.3 miRNA Target Prediction

The behaviors of miRNAs in regulating target genes differ between animals and plants. Those of plants tend to show near-perfect complementarity to their target messenger RNAs (mRNAs), but the miRNAs of animals usually have imperfect characteristics, including mismatches, gaps, G : U wobble pairs, and others. This makes it hard to find animal target genes using only sequence complementarity [8]. Nevertheless, strong sequence conservation observed in target sites and in miRNA sequences makes it possible to develop programs for the prediction of potential targets [9–13]. This evolutionarily meaningful evidence shows the importance of sequence preservation as a requirement for function. Particularly, no specific role has been explained for the $3'$ region of miRNAs even though they tend to be evolutionarily conserved over their entire lengths.

16.1.4 Other miRNA Studies

Recently, based on mass conservation principles and kinetic rate laws, biochemical RNA interference (RNAi) pathway has been modeled with a set of ordinary differential equations that describe the dynamics of siRNA-mediated translational regulation [14]. Similarly, the systematic modeling of miRNA-based posttranscriptional regulation needs to understand the dynamics of the large gene regulation system rewired by miRNAs and thus, in part, will compensate the difference between RNA and protein expression level and more close to the complete gene regulation network. For this, the model consisting of differential equations should be trained properly from given data though various supervised learning methods and optimization algorithms.

The Drosha–DGCR8 complex initiates microRNA maturation by precise cleavage of the stem loops that are embedded in pri-miRNAs. A model for this process, based upon evidence from both computational and biochemical analyses, has been proposed [7]. For insight into the common structural features of pri-miRNAs, the model is based on the thermodynamic profiling, performed to deduce the general structure of these RNA molecules. Systematic mutagenesis and *in vitro* processing assay provided many evidences for significant structural features in pri-miRNA processing. These results will help develop improved miRNA prediction methods as well as perform the rational design of small RNA hairpins for RNA interference systems.

On the other hand, understanding such pri-miRNA processing mechanism helps to design efficient artificial pri-miRNA systems to be applied for RNA interference system. In recent years, the success of RNAi systems has greatly affected medical science and genetics. Nevertheless, many in the biomedical field still expect more stable and tractable interference systems. The rational design of shRNAs mimicking a natural pri-miRNA is promising as an alternative technique of RNAi because hairpin structures is more stable in cells, and their expression is easily controlled through the reengineering of vector systems [15–17].

In the following section, we review the computational approaches proposed for solving these problems.

16.2 COMPUTATIONAL METHODS FOR miRNAs STUDIES

16.2.1 Nonlearning Methods

Several computational approaches have been proposed for the prediction of miRNAs and their target genes. In Table 16.1, we summarize the early approaches where no

Table 16.1   The nonlearning computational methods in miRNA studies

Heuristics and pattern identification	MicroRNA prediction	Target prediction
Comparative method		
Phylogenetic shadowing	[24, 25]	
Cross-species conservation	[19, 26]	[13]
Alignment	[20]	[10, 11]
Rule-based methods	[18, 27]	
Motif search	[28]	
Seed match		[9, 12]

learning methods have been adopted. For the prediction of miRNAs, most of the methods are based on homology searching or rule-based methods [18, 19]. They mainly rely on the context of sequences to predict pre-miRNAs in genomes. These methods search for miRNAs that are closely homologous to known ones but fail to detect any new families that lack clear homologs [20, 21] (Table 16.1). In particular, discovery of miRNAs with genus- or species-specific patterns requires a predictive method to search for unrelated miRNAs [22, 23]. However, a problem in addressing it is that the hairpin structure of pre-miRNA is conserved relatively well, but sequence similarity among known miRNAs is low. Hence, using only a part of the sequence, structural, and thermodynamic features describing miRNAs does not improve the performance of miRNA prediction.

Also, computational methods have been widely used for the prediction of miRNA target genes [9–13, 29–32] (Table 16.1). Different approaches have been used for miRNA target prediction in plants and animals. For plant sequences, similarity-based approaches have shown high performance because complementarity is nearly perfect [30]. However, such approaches are not appropriate for animal genomes because of the imperfect nature of the miRNA:mRNA interaction. Studies for animal sequences have been based on both the complementarity to the $5'$ part of miRNAs and the conserved motifs over species [9, 10, 12, 13, 29]. These can be implemented by a model containing weighted position features and comparative information to detect target mRNA sites and to reduce false positives. Scoring methods using dynamic programming and a complementarity-based strategy are generally preferred to rank the prediction results. They have been quite successful for a few top-ranked results. However, the results are often limited by the conserved nature of the data set used. Lack of experimentally verified target sites makes it difficult to learn rules with the common set of features or an efficient classifier.

16.2.2   Learning Methods for miRNA Studies

Recently, several machine learning-based methods have been used for miRNA research. Most of them belong to the class of supervised learning. In supervised learning, a training data set D consisting of pairs (x_i, y_i) of input and output is given:

$$D = \{(x_i, y_i)|i = 1, \ldots, N\}.$$

The goal is to find a function that $f_\theta(x)$ best describes by the mapping given as the training data, that is, that correctly predicts the output y given any input x (including unobserved ones):

$$f_\theta(x) = y.$$

The input is typically represented as a vector of features $x = (x_1, x_2, x_3, \ldots, x_n)$. The output takes generally continuous values but is usually converted into discrete values or binary values representing classes or categories. The objective of the supervised learning is to search for the optimized model θ^* of a target function, that is, the functional structure and involved parameters. To do this, a loss function $L(f_\theta(x_i), y_i)$ is defined to measure the discrepancy between the predictive output $f_\theta(x_i)$ and the desired output y_i. Then, the model θ is chosen that minimizes the expected value of the loss function for the training data set:

$$\theta^* = \arg\min_{\theta \in \Theta} \left\{ \frac{1}{N} \sum_{i=1}^{N} L(f_\theta(x_i), y_i) \right\},$$

where Θ is the space of all possible models in consideration.

In Table 16.2 we summarize the supervised learning methods found in the literature on miRNA studies. We classified them into three main categories: kernel machines, graphical models, and evolutionary algorithms. In kernel machines, the radial basis function (RBF) kernels have been used for prediction of miRNA precursors [33–35] and target prediction [36]. Linear kernel machines have been used for siRNA efficacy analysis [37]. In graphical models, the hidden Markov models have been applied to find the precursors and mature miRNAs [38, 39]. Also, the naïve Bayes classifiers found their application in this problem [40]. Evolutionary algorithms, such as genetic algorithms, genetic programming, evolution strategies, and evolutionary programming, are an emerging class of methods for computational biology. It is interesting that genetic programming has been already applied to miRNA research in several problems including precursor prediction [42], target prediction [43], and siRNA efficacy analysis [44]. In the table, we also include some methods that seem promising for the specific miRNA research. These include Bayesian networks, which are very popular and found applications in other bioinformatics areas that seem promising for precursor and mature miRNA prediction. These and others methods will be discussed at the end of this chapter when we give outlook an of the future of machine learning for further miRNA studies.

In the following three sections, we give more detailed description of the supervised learning methods with an emphasis on the three categories in Table 16.2. In each section, we start with a description of the learning methods and then describe their

Table 16.2   The supervised learning methods used for miRNA studies

| Learning methods | MicroRNA prediction | | Target prediction | siRNA efficacy | Artificial shRNA design |
	Precursor	Precursor/ mature			
Kernel machines					
RBF kernels	[33–35]		[36]		
Linear kernels				[37]	
Kernel mixture	Not published				
Graphical models					
Hidden Markov models		[38, 39]			
Naïve Bayes classifiers		[40]			
Bayesian networks		Not published	[41]		Not published
Evolutionary algorithms					
Genetic programming	[42]		[43]	[44]	

The methods are divided into three main categories: kernel machines, graphical models, and evolutionary algorithms. In addition to the published works, the table also shows some promising methods, such as Bayesian networks, that have not found their application in miRNA research yet (see text for discussion).

application in the specific problem domain. We shall keep our description of the learning methods minimal to focus more on the general idea of their application to the problem solving in miRNA research.

16.3   KERNEL MACHINES FOR miRNA ANALYSIS

16.3.1   Kernel Methods

Kernel methods have been used for miRNA analysis. The basic idea behind the kernel methods is that the classification problem in the original input space can be made easier if it is transformed into another space, that is, feature space. This nonlinear transformation can be made by various kernel functions.

To implement a kernel method for miRNA prediction, let us denote $S = (x_1, \ldots, x_n)$ as a training set of miRNA data. We suppose that each object x_i is an element of a set X of all possible target data. The data set S is then represented as the set of features, $\Theta(S) = (\Theta(x_1), \ldots, \Theta(x_n))$, where $\Theta(x)$ can be defined as a real-valued vector. The size of the vector is the number of features. This classification method is designed to process a set of pairwise comparisons of data items x_i and x_j. It is represented by an $n \times n$ squared matrix of pairwise comparisons $k_{i,j} = k(x_i, x_j)$. It should satisfy $k_{i,j} = k_{j,i}$ for any i, j between 1 and n, and $c^\mathrm{T} k c \geq 0$ for any $c \in \Re^n$. It defines positive semi-definite kernels as follows:

$$\sum_{i,j=1}^{n} c_i c_j k(x_i, x_j) \geq 0,$$

for any $x_1, \ldots, x_n \in X$, $n \geq 1$, $c_1, \ldots, c_n \in \Re$. For any kernel k on a space X, there is a Hilbert space F, mapped $\Theta : X \rightarrow F$:

$$k(x_i, x_j) = \langle \Theta(x_i), \Theta(x_j) \rangle,$$

where x_1, $x_j \in X$. For example, a Gaussian RBF kernel is described as follows:

$$k(x_i, x_j) = \exp\left(-\frac{d(x_i, x_j)^2}{2\sigma^2}\right)$$

where σ is a standard deviation and d is the distance.

Support vector machines (SVMs) find an optimal hyperplane separating the training data on a Hilbert feature space, represented by a mapping function Θ. In practice, however, a separating hyperplane may not exist when a problem is very noisy or complex. To accommodate this case, slack variables $\xi_t \geq 0$ for all $i = 1, \ldots, n$ are introduced to loosen the constraints as follows [45]:

$$y_i(\langle \mathbf{w}, \mathbf{x}_i \rangle + b) \geq 1 - \xi_i \quad \text{for all } i = 1, \ldots, n.$$

A classifier that generalizes well is then obtained by adjusting both the classifier capacity $\|\mathbf{w}\|$ and the sum of the slacks $\sum_i \xi_i$. The latter can be shown to provide an upper bound on the number of training errors. Such a soft margin classifier can be realized by minimizing the following objective function:

$$\frac{1}{2}\|\mathbf{w}\|^2 + C\sum_{i=1}^{n} \xi_i$$

subject to the constraints on ξ_t, where the constant $C > 0$ determines the trade-off between margin maximization and training error minimization.

Such SVM methods incorporating kernel functions have been successfully applied to predict new miRNA genes and to search for valid miRNA–target interaction sites.

16.3.2 miRNA Prediction Using Kernel Methods

To date, most studies have used RBF kernel function to perform a quadratic classification with specific feature sets uniquely defined in each study (Table 16.2). Also, they mainly focused on only prediction of miRNA precursors. For example, Xue et al. [34] defined 32 local structure units (triplet structures and one base combination) on the stem-loop structure of miRNA precursor (Fig. 16.2) and implemented a SVM classifier with RBF kernel functions to classify pseudo/real miRNA precursors on the feature space consisting the local structure units. The methods showed about 91% accuracy on average over 12 species including human, but the simple feature set seemed to be improved to fully reflect conserved motifs on structure and sequence of miRNA

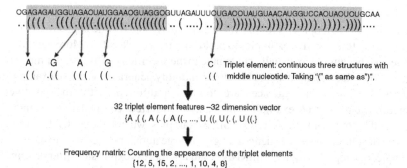

Figure 16.2   The Xue et al.'s triplet elements as a feature set. They represent the local structure–sequence features of the hairpin using the triplet elements. The triplet elements consist of three consecutive substructure (i.e., "(" or ".") and a nucleotide at the middle position. Thus, there are all 32 possible triplet elements whose numbers are counted along a hairpin sequence, forming 32-dimensional vector. The frequency is then normalized to be input vector for kernel function.

precursors. On the contrary, Sewer et al. [33] introduced various types of features (compositional and secondary structure properties) over miRNA precursors and learned a SVM classifier with the RBF kernel functions to classify novel miRNA precursors and clustered miRNAs (Table 16.2). Recently, Hertel and Sladler [35] introduced 12 heterogeneous features consisting of structure, sequence composition, thermodynamic parameters, and conservation level and used RBF kernels to predict miRNA precursors conserved hairpin candidates extracted by the program RNAz (Table 16.2).

Though above RBF kernel functions are suitable to apply a feature set consisting of real values, we are required to prudently choose kernel functions by considering the characteristics of each feature to develop more efficient classifiers. For example, sequence composition and conservation can be suitable for using string kernels rather than RBF kernels for more comprehensive prediction. The features applied to miRNA prediction can be divided to two types, that is, content-based features and position-based features. Content-based features mainly consider compositions or character-istics of sequence and structure such as free energy, GC ratio, entropy, k-spectrum, stem length, loop size, and bulge size. On the contrary, position-based features include base or base pair (bp) preferences of each position and position weighted matrix, and Markov probabilities. However, all previous kernel functions used only content-based features and have not considered position-based features. Thus, they mainly focused on only the prediction of mature miRNAs. The definition of the position-based features will be a shortcut to make the comprehensive results.

16.3.3   Target Prediction Using RBF Kernel Methods with a Biologically Relevant Data Set

Several experiments to confirm miRNA–target interaction have been performed for only a few miRNAs. Moreover, there is no database where such annotated data are organized well. Thus, we need to manually collect the training data set over various species from the literature. We gained the training data set from the literature, containing 235 examples including 152 true positives and 83 true negatives. However, there are too few negative examples to build an effective classifier. More negative data are needed because these usually contribute to the specificity of a classifier much more significantly than positive data. However, randomly generated negative examples are useless because such sequences are too much different with true positives or sometimes can be true examples. Instead, we inferred 163 negative examples as described below. For the inferred negative examples, we noted that deletion of target sites on the target mRNA sequence can give a large number of negative examples. For example, in a report [46], let-7 miRNA could not repress expression after deleting the target sites of let-7 miRNA on lin-41. That is, the remaining region on the lin-41 3' UTR will now not work with let-7 miRNA. Thus, all the other remaining sites with favorable seed pairings can be apposite to negative examples. In practice, we collected examples with more than 4-mer matches at their seed part and discarded the rest to improve the quality of the data set. As a result, we gained 163 inferred negative examples: 50 from lin-41 and 113 from LIN-28. Finally, the final size of the data set was 398 (152 positives, 246 negatives).

For the target prediction using kernel method, features are categorized into three elements: structural features, thermodynamic features, and position-based features. All features are designed on the RNA secondary structure prediction results produced by the RNAfold program in the Vienna RNA Package [47] (Table 16.2). The general scheme of miRNAs and their interactions with target mRNAs are illustrated in Fig. 16.3a [36]. There are 41 different features, as shown in Fig. 16.3b. The RNAfold program requires a single linear RNA sequence as input, so the 3' end of the target mRNA sequence and the 5' end of miRNA sequence are connected by a linker sequence, "LLLLLL." The "L" denotes that it is not an RNA nucleotide, and thus it does not match with any nucleotide and so prevent mRNA and miRNA nucleotides from binding with sequence-specific linker sequences [29]. Thus, the RNAfold program produces an RNA secondary structure alignment with a linker sequence, exemplified in Fig. 16.3a. The positions in the alignment are numbered from the 5'-most position of the seed region. Alignments are extended until the 20th position and the rest positions are discarded.

For structural and thermodynamic features, we divided the secondary alignment into three parts consisting of the 5' part (seed part), the 3' part, and the total alignment as shown in Fig. 16.3. Each count value of matches, mismatches, G:C matches, A:U matches, G:U matches, and other mismatches from the three parts was considered as a structural feature. The free energy values of the 5' part, the 3' part, and the total miRNA:mRNA alignment structure are thermodynamic features that are also calculated by RNAfold. Here, the sequence "AAAGGGLLLLLLLCCCUUU" was used as a linker sequence to ensure that each part of the subsequence was paired. The sequences "AAAGGG" and "CCCUUU" were designed to prevent any unexpected alignment of the short matches. Although such linkers may change the original signal, the thermodynamic effect of the linker sequence will be the same for all short matches.

Position-based features are important because they imitate the shape and mechanism of the seed pairing. Doench et al. [48] and Brennecke et al. [49] focused on the

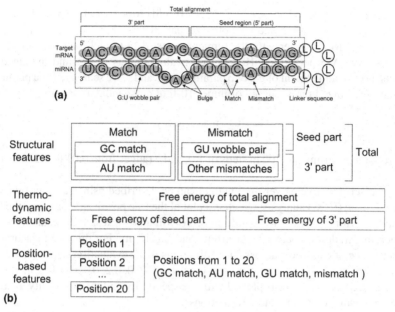

Figure 16.3 **(a)** General scheme of miRNA:mRNA interactions. The interaction between a miRNA and a target gene consists of seed region and 3′ part according to base pairing degree. In the structure of the interaction, we can consider distinct features such as bulge, match/mismatch, and G:U pair. **(b)** Three categories of SVM features. As the miRNA prediction problem, the features can be classified into two types: content-based and position-based features. There are structural features and thermodynamic features in the content-based features that are mostly the number match/mismatch or free energy values. The position-based features consist of the frequencies of specific base pairs at each position.

sequence specificity of miRNA:mRNA interaction. They found that a single point mutation could inhibit the miRNA's function depending on its position. In contrast to the earlier belief, their research revealed that examples with favorable thermodynamic free energy might not regulate expression. Therefore, it needs to investigate the binding mechanism. Position-based features corresponded to point mutations in the above two experiments. Each position had one of the four nominal values consisting of a G:C match, an A:U match, a G:U match, and a mismatch.

With the prepared feature set, the quadratic kernel value $k(x_i, x_j)$ is computed using RBF as follows:

$$k(x_i, x_j) = \exp(-\gamma \|x_i - x_j\|^2),$$

where the parameter γ determines the similarity level of the features so that the classifier becomes optimal.

To select an efficient model, we evaluated the classifier with a completely independent test data set. For this, we repeatedly performed three steps as follows. First, we divided the data equally into training and test sets through random sampling (without replacement). Then we performed 10-fold cross-validation with the training data to train a classifier and to optimize parameters. Finally, we evaluated the optimized SVM classifier with the remaining test data (which must be completely independent). We performed 10 repeated evaluations as above and averaged the results. For the adjustment of the two parameters, C and γ, we searched for a parameter set that maximized the accuracy of upper 10-fold cross-validation using

$$\arg\max_{C,\gamma} A(C, \gamma),$$

where C ranges from 1 to 200 in steps of 1.0, and γ ranges from 0.01 to 2.0 in steps of 0.01.

The discriminative power of a method can be described using receiver operating characteristic (ROC) analysis, which is a plot of the true positive rate against the false positive rate for the different possible cutoffs of a diagnostic test. ROC analysis reveals all possible trade-offs between sensitivity and specificity. For this, we measured the performance of classifiers across 24 cutoff points in the evaluation step ($-4, -3, -2, -1.8, -1.6, -1.4, -1.2, -1, -0.8, -0.6, -0.4, -0.2, 0, 0.2, 0.4, 0.6, 0.8, 1, 1.2, 1.4, 1.6, 1.8, 2, 3$). The ROC was plotted with the specificity and the sensitivity averaged from the results of 10 repeated evaluations.

The ROC result is presented in Fig. 16.4 [36]. Because the previous methods did not have cutoffs, their ROC curves were not represented. Instead, their performances were

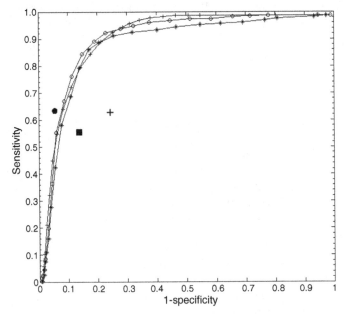

Figure 16.4 The ROC curves of classifiers created on three combinations of features: an entire set (-○-), position-based features only (-✱-), and without position-based features (-+-). The dark circle denotes the performance of TargetScan, the dark rectangle shows the performance of RNAhybrid, and the plus symbol shows the performance of miRanda.

indicated as rectangles on the graph and compared specificities were based on their sensitivity. Overall, the kernel-based method gave a more stable performance than that of miRanda and RNAhybrid, but a slightly lower specificity (0.93) at a sensitivity of 0.63 than TargetScan (0.94). The higher specificity of TargetScan seems to arise from its strong constraint on the seed region. As it requires six continuous pairings on positions 2–7 of the seed region, this constraint made it predict many of the examples in our data set as negative. However, the limitation of TargetScan is that its best sensitivity was 0.63. Indeed, TargetScan seems to be rather conservative and is less flexible than our classifier, which can predict with optional accuracy across a broad range of specificity and sensitivity. Unlike other methods, the kernel-based classifier is trainable and can be improved continuously if we can obtain more biologically relevant data.

16.4 GRAPHICAL MODELS FOR miRNA ANALYSIS

16.4.1 Learning with Graphical Models

Kernel machines are a very efficient method for learning the input–output mapping in a training set to predict the target class of a new input data. However, the resulting models are not easy for the biologists (or humans in general) to elucidate the underlying structure. For computational biologists, it is sometimes very important to understand the underlying relationships between the variables. Graphical models (or probabilistic graphical models) are a class of learning models useful for identifying the structural relationships between various features or variables.

Formally, a graphical model consists of nodes, defined as attributes (or random variables), and edges, representing conditional relations between nodes. Depending on the types of edges, a graphical model can be directed or undirected. Markov networks and Boltzmann machines are typical examples of undirected graphical models and allow the cyclic structure. Bayesian networks are an example of directed graph having a directed acyclic graph (DAG) structure. Graphical models can be applied to both prediction and description tasks. The possibility of visualization inherent in the graphical models is especially interesting for biological applications where comprehensibility and explainability of the discovered results are important. In the rest of this section, we give a very brief description of the Bayesian networks and the hidden Markov model. The following subsections present some case studies for their use in miRNA research.

A Bayesian network represents the joint probability distribution by specifying a set of conditional independence assumption, together with sets of local conditional probabilities. There are two types of information in the Bayesian network. The first one is conditional edges and the second one is conditional probability tables assigned to all nodes. If there is a direct edge from Y to X, we say Y is a parent of X and X is a child of Y. The conditional probability table describes the probability distribution for that variable given the value of its parents. If for each variable $X_i\{X_i | 1 \leq i \leq n\}$, the set of parent variables is denoted by $Pa(X_i)$ then the joint distribution of the variables is the product of the local distributions:

$$P(X_1, \ldots, X_n) = \prod_{i=1}^{n} P(X_i | Pa(X_i)).$$

Here, if X_t has no parents, its local probability distribution is said to be unconditional, otherwise it is conditional. If all X_t have no parents excluding a class parent, the Bayesian network becomes the simple naïve Bayes classifier:

$$P(X_1, \ldots, X_n) = \prod_{i=1}^{n} P(X_i | C),$$

where C is a class node (the top node).

A Bayesian network can be automatically constructed from data by structural learning and parametric learning. Structural learning finds an optimal network structure (determining parents and descendents) given a training data set. For this, exhaustive algorithms like K2 or global search algorithms like Markov chain Monte Carlo are usually used. Parametric learning constructs all local conditional probability tables at each node given the parent nodes. To find the best model parameters, we search for an optimal parameter set maximizing the posterior or likelihood.

Another class of popular probabilistic graphical models is hidden Markov models (HMMs). They are considered to be a simplest dynamic Bayesian network located between Markov networks and Bayesian networks. The HMMs are based on directed graphs and also allow cyclic structure. It usually has been applied to sequential pattern recognition problems such as speech recognition. If we have a series data $D = \{X_i | 0 \leq i \leq L\}$ and specific states $S = \{s_i | 0 \leq i \leq L\}$, the joint probability of an observed data set D and given states S is given as

$$P(D, S) = a_{os_1} \prod_{i=1}^{L} e_{s_i}(x_i) a_{s_i s_{i+1}},$$

where a_{os_1} is the probability of state s_i in the initiation and $a_{s_i s_{i+1}}$ is the transition probability from state s_i to s_{i+1}. $e_{s_i}(x_i)$ is an emission probability where data x_i are emitted at the state s_i. Hidden Markov models are used to predict the sequential data, which is distinguished from other graphical models.

16.4.2 miRNA Prediction Using HMMs

One of the first graphical models ever used for miRNA prediction is the probabilistic colearning model [38] (Table 16.2). This model is based on the paired HMM to implement a general miRNA prediction method identifying close homologs as well as distant homologs. The probabilistic colearning model combines both sequential and

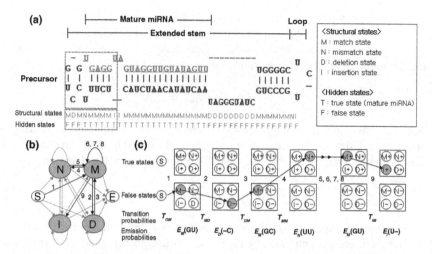

Figure 16.5 Pairwise representation of the stem-loop structures and state sequences of miRNA precursors, where the state of each pair includes structural information and mature miRNA region information (hidden states). **(a)** The structure of the miRNA precursor. **(b)** The transition and emission scheme of the structural states and the hidden states for pairwise sequence in the dotted rectangle shown in **(a)**. TOM, TDM, TMN, and TMI are transition probabilities. EM(GU), ED(-C), EM(GC), EN(UU), EM(GU), and EI(U-) are emission probabilities. **(c)**. The four-state finite state automaton. Finally, the probability of the pairwise sequence is assigned by multiplication of the transition probabilities and the emission probabilities. *Source*: Ref. [38].

structural characteristics of miRNA genes in a probabilistic framework, and simultaneously decides if an miRNA gene and a region of mature miRNA are present by detecting the signals for the site cleaved by Drosha.

An miRNA precursor can be represented as a pairwise sequence. It forms an extended stem–loop structure, and this structure can be formulated as a sequence of matched base pairs (Fig. 16.5a). The pairwise sequence starts from the nonlooped end of the miRNA precursor. Each position of the pairwise sequence has two properties, that is, structural states (match/mismatch/deletion/insertion) and hidden states (information for the mature miRNA region). In the structural states, each match state, M, can emit A–U, U–A, G–C, C–G, U–G, or G–U as an emission symbol. Deletion states, D, can emit ●–A, ●–U, ●–G, or ●–C. Insertion states, I, can emit A–●, U–●, G–●, or C–●. Mismatch states, N, can emit one of the remaining combinations. The possible transitions among the four structural states are shown in Fig. 16.5b. Each emission is represented as a corresponding character in alphabetical order. In the hidden states, T means a true state, namely, a region of mature miRNA, and F means a false state, the precursor region outside mature miRNA sequences (Fig. 16.5). The probabilities of hidden states in this sequence–structure colearning model are estimated from the distribution of all four structural states in pairwise sequences of training data set.

First, we estimate the probability of the pairwise sequence of miRNA precursors. To derive the probability of the pairwise sequence of the structural states and the symbols, we use two parameters: a transition probability and an emission probability as Fig. 16.5b. For the transition probability, let us call the state sequence a path π. The probability of a state depends only on the previous state. If π_i denotes the ith state in the path, the transition probability is defined as

$$T_{kl} = P(\pi_i = l | \pi_{i-1} = k),$$

where the transition is from state $\pi_{i-1} = k$ to state $\pi_i = l$. The probability of starting in state k can be defined as T_{0k}. Let x_i denote the symbol emitted from the ith state. Then, the emission probability of observing symbol b in state k is defined as

$$E_k(b) = P(x_i = b | \pi_i = k).$$

Using the transition and emission probabilities, we can estimate the probability $P(x)$ that sequence x is generated by the probabilistic colearning model. It is easy to define the joint probability of an observed sequence x and a structural state sequence π:

$$P(x, \pi) = T_{0\pi_1} \prod_{i=1}^{L} E_{\pi_i}(x_i) T_{\pi_i \pi_{i+1}},$$

where L is the window size. If we are to choose just a optimal path for our prediction, that one, π^*, with the highest probability should be chosen as follows:

$$\pi^* = \arg \max_{\pi} P(x, \pi).$$

Here, π^* will be an optimal model of miRNA precursor. If we have a pairwise sequence, we search for the maximum $P(x, \pi)$ value by using a sliding window, the size of which is 22 base pairs—the mean length of the mature miRNAs in the pairwise representation. If the value is higher than a threshold selected in advance, then we classify the given candidate as an miRNA precursor. The probabilistic colearning model showed 96% specificity and 73% sensitivity in a generalization test. Next, in order to apply the prediction of mature miRNA to our probabilistic model, we introduced two hidden states indicating whether the position is a mature miRNA region (Fig. 16.5). The probabilities that state that the ith position is true or false are computed as

$$P_t(i) = \max\{P_t(i-1) \cdot T_{\tau_{i-1}\tau_i}, P_f(i-1) \cdot T_{v_{i-1}\tau_i}\} \cdot E_{\tau_i}(x_i),$$
$$P_f(i) = \max\{P_t(i-1) \cdot T_{\tau_{i-1}v_i}, P_f(i-1) \cdot T_{v_{i-1}v_i}\} \cdot E_{v_i}(x_i),$$

where τ_i means that the ith state is true and v_i means that the ith state is false. The initial condition is $P_t(1) = 0$ and $P_f(1) = 1$. Using only the true and the false probabilities, we cannot exactly determine mature miRNA regions, because the transition probability around the cleavage site of miRNA is low. Thus, we focus on the transition probability of false states and compute $S(i)$ as

$$S(i) = \frac{P_t(i-1) \cdot T_{v\tau}}{P_t(i-1) \cdot T_{v\tau} + P_f(i-1)T_{vv}}$$

The mean error of mature miRNA prediction was 2.0 nt. The main reason of the error is that inaccuracy of the cleavage of the miRNA precursor by Dicer bears an error of one nucleotide and from overhanging ends of two nucleotides at the 3′ end. In addition, there are several instances of incompatible data for the locations of mature miRNAs in the miRNA database. It may result in some errors in mature miRNA region prediction. When these limitations are considered, the result indicates that our algorithm gives meaningful results but should be more improved by considering such noises.

16.4.3 miRNA Prediction Using Naïve Bayes and Bayesian Networks

Another probabilistic graphical method used for miRNA prediction is the naive Bayes classifier, which assumes the independence of features. Recently, Yousef et al. [40] used a naïve Bayes classifier with nine kinds of feature, such as sequence composition, structural features, and free energy to predict pre-miRNAs and mature miRNAs (Table 16.2). The naïve Bayes classifier outperformed the HMMs in the prediction of miRNA precursors but not in the prediction of mature miRNAs. In the case of mature prediction, we can expect that positional-based features will be important, and they have conditional relations to each other if we consider the miRNA processing mechanism by Drosha/DGCR8 and Dicer. Therefore, a generative model, for example, Bayesian networks, considering the dependency of positional-based features, such as positional weight matrix and first-order Markov probability, will be more suitable to predict miRNA precursor and mature miRNA.

For the prediction of miRNA precursors using Baysian networks, we used 321 positive data items (precursors) retrieved from the database miRBase release 8.1 [5] and 1459 negative data set extracted from 3′UTR with several structural criteria. For the prediction of mature miRNA, we used 455 mature miRNA sequences retrieved from the miRBase and 2450 negative data set (all positions excluding true mature positions). We next extracted the heterogeneous feature set including positional-based feature set over stem-loop structure of miRNA precursor on the basis of the stem end of mature miRNA (Table 16.3). We constructed 125 and 135 feature sets for the prediction of precursor and mature miRNA, respectively. However, comparing to the number of data set, much number of features makes it intractable to find the optimal structure of Bayesian network because the number of possible structure exponentially increases according to the number of variables. In this case, we usually reduce the dimension of features using several ways to circumvent such a problem. Here, we used a feature selection method, information gain for that, and we finally we select 25 and 28 representative features, respectively.

Table 16.3   The feature set for mature miRNA prediction using Bayesian network

No	Name	Description
1	PWM	Position weighted matrix
2	MCP	Markov chain probability
3	Spectrum (2 bp)	The distribution and frequency of two bases in each miRNA
4	Spectrum (3 bp)	The distribution and frequency of three bases in each miRNA
5	Bulge count	The number of bulges predicted by mfold in miRNA
6	Energy profile	Sequence of energy values in each base of miRNA
7	GC ratio	The ratio of GC/AT

For the structure learning of Bayesian network, we performed well-defined greedy search algorithm, K2 [50], in combination with the Bayesian Dirichlet (BD) scoring metric. K2 algorithm iteratively searches for the best parents of each node (variable) until the BD score does not increase when adding a parent. Thus, we found an optimal Bayesian network for the precursor and mature miRNA using 10-fold cross-validation with the upper data set (Fig. 16.6). The result of precursor prediction was comparable to other methods (Table 16.4). In the results of mature miRNA prediction, the efficiency of the Bayesian method outperformed previous methods (data not shown). Also, the dependent relations of the positional-based features were consistent the results of biochemical study for proving the processing mechanism [7]. It says that the Bayesian network approach can make more comprehensive models, and it is more efficient method than previous methods in the mature miRNA prediction. Moreover, the learned Bayesian model can be used to reconstruct rational miRNA precursors, which will be very useful tool to design new drug or to develop gene knockdown system, more effective and stable than siRNA system.

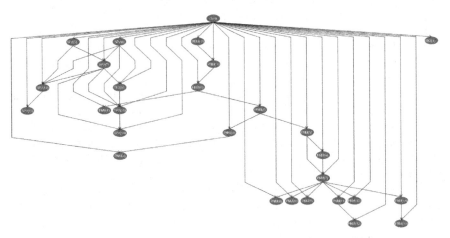

Figure 16.6   The results of prediction of mature miRNAs by using Bayesian networks: the number of total features are 135. After feature selection using information gain, only 25 features are used for construction of the Bayesian network.

Table 16.4 **The 10-fold cross-validation results of miRNA precursor prediction and the comparison with other methods**

	Accuracy	Sensitivity	Specificity
Naïve Bayes	0.9836	0.9840	0.9836
Naïve Bayes[a]	0.9904	0.9904	0.9904
Adaboost	0.9955	0.9744	1.0000
SVM	0.9983	0.9936	0.9993
SVM[a]	0.9955	0.9968	0.9952
Bayesian net[a]	0.9994	0.9968	1.0000

[a]After feature selection.

16.4.4 Target Prediction Using Bayesian Networks

Although many target genes of miRNAs are predicted by various computational methods including a supervised learning method, identification of target genes considering expression profiles of miRNA and target genes is necessary to obtain functional miRNA targets. Huang et al. [41] proposed the Bayesian model and learning algorithm that account for their patterns (Table 16.2). The flow chart of the learning method is described in Fig. 16.7.

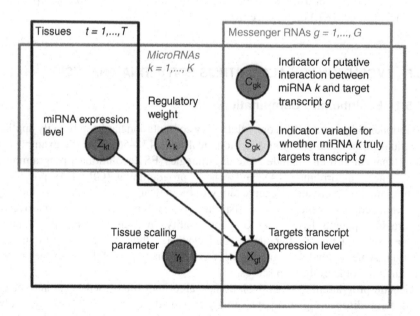

Figure 16.7 The Bayesian network model for the prediction of functional miRNA target. *Source:* Ref. [41].

They modeled the probability of expression pattern X of a set of g genes using target prediction matrix C, miRNA expression pattern Z, and parameter set $\Theta = \{\mu, \Sigma, \pi, \alpha\}$ including prior probabilities.

$$P(X|C, Z, \Theta) = \sum_s \int_\Gamma \int_\Lambda P(X, S, \Gamma, \Lambda | C, Z, \Theta) d\Lambda d\Gamma$$

$$P(X, S, \Gamma, \Lambda | C, Z, \Theta) = P(S|C, \Theta)P(\Gamma|\Theta)P(\Lambda|\Theta) \prod_g P(x_g | Z, S, \Gamma, \Lambda, \Theta),$$

where Γ is the diagonal matrix of the tissue scaling parameters, Λ is a set of regulatory weights that determines the relative amount of down regulation incurred by miRNAs. x_g denotes the expression pattern of gth gene. S is a set of s_{gk} latent variables, a random variable to be assigned to each interaction between kth miRNA and gth target gene. If s_{gk} is equal to 1, it means that the interaction is true. Finally, the posterior probabilities of S given that X, Z, C, Θ should be estimated to predict functional miRNA target as follows:

$$P(S|X, C, Z, \Theta) \propto \int_\Gamma \int_\Lambda P(X, S, \Gamma, \Lambda | C, Z, \Theta) d\Lambda d\Gamma.$$

However, considering the complex combinatorial nature of miRNA–target gene regulation, its computing is nearly intractable. Thus, they approximated the probability of each component using expectation-maximization (EM) method incorporating Kullback–Liebler (KL) divergence measure [41].

16.5 EVOLUTIONARY ALGORITHMS FOR miRNA ANALYSIS

16.5.1 Evolutionary Computation

Evolutionary algorithms (EA) or evolutionary computation has been broadly applied to various optimization problems in biological fields [51–56]. Specific examples of evolutionary algorithms include evolution strategies (ES), evolutionary programming (EP), genetic algorithms (GA), and genetic programming (GP). EAs provide a learning method motivated by an analogy to biological evolution. An evolutionary algorithm iteratively generates new offspring hypotheses by mutation and crossover of the previous parents hypotheses of good quality. In each step, EA maintains a population consisting of hypotheses and evaluates them with a specific fitness function. A new population for the next generation is probabilistically generated according to some evaluation scores.

Genetic programming is an evolutionary algorithm often used as an automated method for discovering computer programs called genetic programs. Typically, genetic programs are tree structured, representing parse trees of computer programs. A genetic tree consists of elements from a function set and a terminal set. The function symbols appear as internal nodes. Terminal symbols are used to denote actions taken by the program. Genetic programming does this by genetically breeding a population of genetic programs using the principles of Darwinian natural selection and biologically inspired genetic operations. Genetic programming uses crossover and mutation as the variation operators, which change candidate solutions into new candidate solutions.

In the miRNA studies, the applications of genetic programming have been introduced to learn the common structure of miRNA precursors [42] and predict target genes [43] (Table 16.2).

16.5.2 Prediction of miRNA Precursors Using Parameterized Genetic Programming

This section describes the parameterized genetic programming (PGP) method to learn an optimal common structure of miRNA precursors [42] (Table 16.2). PGP is a new type of genetic programming where genetic programs are augmented with parameters and thus natural for searching of RNA common-structural descriptors (CSDs) without aligning the sequences. Alignment needs much computational time and causes the unwanted biases. In addition, structural alignment is a computationally intractable problem. PGP evolves CSDs based on a training data set through grammatical tree-structure encoding the parameterized rules for structural description. The definition of the rules made it possible to optimize the CSDs via genetic operators used in genetic programming. More important, the optimized CSDs can be used as classifiers to search for new RNAs in a database using the program RNAmotif [57].

We adopt a structural descriptor [57] for a GP tree using context-free grammar. An example of structural descriptor created by the given grammar is shown in Fig. 16.8. Here, R denotes a root node of a tree-structured program and consecutively generates n functions of $F1$ and $F2$ with the user-defined proportion. The nonterminal $F1$ consecutively generates a parameter set, $P1$, and m functions of $F1$ and $F2$ with the user-defined proportion, and emits structural symbols, "h5" and "h3" to both sides. "h5"–"h3" denote 5' and 3' side strand of a helix structure and should be paired always (Fig. 16.8b). Then, the terminal node $F2$ generates the other parameter set, $P2$, and emits a structural symbol "ss", which means a single strand. Parameters have a numeric value or a 7-mer sequence, an average length of structural symbols such as "h5"–"h3" and "ss." Finally, the structural descriptors are converted to tree-structured programs with parameters at each node. Thus, we refer to this as the parameterized tree-structured program (PTS), describing the structural descriptor of RNAs. Figure 16.8c shows an example of the parameterized tree-structured program. This PTS represents the structural descriptor (Fig. 16.8b) of Gln-tRNA (Fig. 16.8a).

The PGP evolves the PTS for the optimization of CSDs, which are used to search for arbitrary ncRNAs in the program RNAmotif [57]. It consists of two learning steps: the structure learning step and the sequence learning step. In the structure learning step, PGP evolves the structure and structural parameters of PTSs. In the sequence learning step, it optimizes the sequence and sequential parameters of the PTSs.

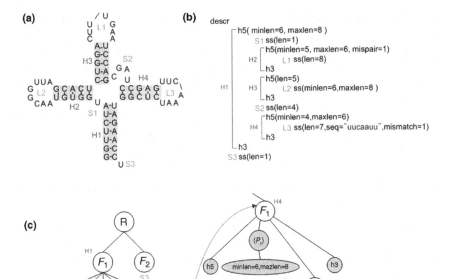

Figure 16.8 **(a)** The secondary structure of Gln-tRNA. The structure consists of three loops (L1, L2, and L3), four stems (helix, H1–H4), two bulges (S1, S2) and a flanking sequence (S3). **(b)** An example of Gln-tRNA structural descriptor. The pair of h5–h3 denotes a helix structure and "ss" (single strand) denotes a single strand, bulge, or loop. Helixes and single strands include structural parameters, such as the length of helix (len), the maximal and minimal length of helix (minlen, maxlen), and the number of pair (mispair), and sequential parameters, such as mismatch of sequence (mismatch) and sequence (seq). **(c)** An example of the PTS with parameterized nodes F1 and F2. All F1 and F2 consist of leaf nodes (h5 h3, or ss), and parameters nodes (P1 or P2) including several parameters. The child nodes of the root R of **(a)** conform to the first indentation of **(b)** and the second depth of **(a)** conforms to the second indentation of **(b)**.

Using the PGP method, we evaluated the efficiency in the prediction of miRNA precursors and compared our method to other prediction methods. For the comparison, we used HMMer, based on profile hidden Markov model of multiple sequences [58]. For the pre-miRNA, PGP outperformed the HMMer as shown in Table 16.5. It is because the efficiency of the HMMer depends on multiple sequence alignment, but the

Table 16.5   Comparison of miRNA precursor prediction methods using *F*-measure

F score	miRNA precursor
HMMer	0.121
SVM	0.594
PGP	0.819

pre-miRNAs have low similar sequences. We next compared with SVM implemented by SMO algorithm and spectrum kernel [59]. We constructed 3-mer spectrum kernel and performed fivefold cross-validation to estimate the accuracy. In the comparison, SVM presented much lower efficiency than PGP method (Table 16.5). These tendency results from the point that HMMer and SVM are sequence-based search methods but our method is simultaneously learning both structure and sequence.

16.5.3 miRNA Target Prediction Using Boosted Genetic Programming

Boosted genetic programming has been used in the studies for miRNA target prediction and RNAi efficacy prediction in small RNA fields [43, 44]. For the target prediction, they performed a supervised learning on a set of validated microRNA targets in lower organisms to create weighted sequence motifs that capture the binding characteristics between microRNAs and their targets [43]. GP evolves the sequence patterns from a population of candidate miRNA–target interaction patterns, and the boosting algorithm guides GP's search by adjusting the importance of each individual in the training set. Then, the boosting algorithm assigns weights to the sequence patterns based on the patterns' performance in the corresponding training set. The final classifier is the average of several such boosted GP classifiers. They showed that the weighted sequence motif approach is favorable to using both the duplex stability and the sequence complementarity steps [43].

16.6 SUMMARY AND OUTLOOK

We reviewed computational methods used for miRNA research with a special emphasis on machine learning algorithms. Most of learning methods belong to the class of supervised learning and the most studied areas include miRNA prediction and target gene prediction in various miRNAs. In particular, we give detailed descriptions for the case studies based on the kernel methods (support vector machines), probabilistic graphical models (Bayesian networks and hidden Markov models), and evolutionary algorithms (genetic programming). The effectiveness of these methods was validated by various approaches including wet experiments, and their contributions were successful in the domain of miRNA. However, there is still much room for the improvement of the algorithms for the prediction of miRNA and its target gene because the discoveries by the prediction methods are still not perfect and because many miRNAs and their targets are still unknown.

There are still unexplored areas in miRNA research. For example, finding the functionally coherent regulatory modules in the regulatory network of miRNAs and target genes is an important issue for the functional study of miRNAs and for analyzing the intrinsic characteristics in regulatory systems. This requires superexponential time since the networks are of combinatorial nature. It is necessary to develop efficient learning methods to avoid the time complexity problem. Unsupervised learning methods, such as clustering algorithms, might help in this regard. Also, the reconstruction of a grand regulatory network consisting of miRNAs and transcription factors will provide a key solution to fill the gap of inconsistency between the RNA level and the protein level. Developing learning methods handling dynamic or perturbed expression data is required in the domain. In fact, large-scale miRNA microarray data are rapidly increasing to study the function and systemic aspects of miRNAs in the field. Learning methods applied to microarray data mining will also be useful in the field in understanding the difference between miRNA and mRNA expression profile and their roles.

Recently, the success of RNAi systems has made great impact on various fields including medical science and genetics. The techniques have been applied for drug design and gene knockdown systems. Nevertheless, many researchers in the biomedical field expect more stable or tractable interference systems. Rational design of artificial pre-miRNA or shRNA is promising as an alternative technique of RNAi because the hairpin structure is more stable in cells, and its expression is easily controlled through the reengineering of vector systems. A well-defined generative model, such as Bayesian networks or hidden Markov models, constructed from a known data set in the prediction of miRNAs can be used for the rational design of artificial pre-/shRNAs.

REFERENCES

1. Mendes Soares, L. M. and Valcarcel, J. The expanding transcriptome: the genome as the 'Book of Sand'. *EMBO Journal*, 25(5): 923–931, 2006.

2. He, L. and Hannon, G. J. MicroRNAs: small RNAs with a big role in gene regulation. *Nature Reviews Genetics*, 5(7): 522–531, 2004.

3. Bartel, D. P. MicroRNAs: genomics, biogenesis, mechanism, and function. *Cell*, 116(2): 281–297, 2004.

4. Esquela-Kerscher, A. and Slack, F. J. Oncomirs—microRNAs with a role in cancer. *Nature Reviews Cancer*, 6(4): 259–269, 2006.

5. Griffiths-Jones, S., Grocock, R. J., van Dongen, S., Bateman, A., and Enright, A. J. miRBase: microRNA sequences, targets and gene nomenclature. *Nucleic Acids Research*, 34(Database issue): D140–D144, 2006.

6. Suh, M. R., Lee, Y., Kim, J. Y., Kim, S. K., Moon, S. H., Lee, J. Y., Cha, K. Y., Chung, H. M., Yoon, H. S., Moon, S. Y., Kim, V. N., and Kim, K. S. Human embryonic stem cells express a unique set of microRNAs. *Devlopmental Biology*, 270(2): 488–498, 2004.

7. Han, J., Lee, Y., Yeom, K. H., Nam, J. W., Heo, I., Rhee, J. K., Sohn, S. Y., Cho, Y., Zhang, B. T., and Kim, V. N. Molecular basis for the recognition of primary microRNAs by the Drosha-DGCR8 complex. *Cell*, 125(5): 887–901, 2006.

8. Rajewsky, N. microRNA target predictions in animals. *Nature Genetics*, 38 Suppl: S8–S13, 2006.

9. Lewis, B. P., Shih, I. H., Jones-Rhoades, M. W., Bartel, D. P., and Burge, C. B. Prediction of mammalian microRNA targets. *Cell*, 115(7): 787–798, 2003.

10. Enright, A. J., John, B., Gaul, U., Tuschl, T., Sander, C., and Marks, D. S. MicroRNA targets in Drosophila. *Genome Biology*, 5(1): R1, 2003.

11. Kiriakidou, M., Nelson, P. T., Kouranov, A., Fitziev, P., Bouyioukos, C., Mourelatos, Z., and Hatzigeorgiou, A. A combined computational–experimental approach predicts human microRNA targets. *Genes& Devlopment*, 18(10): 1165–1178, 2004.

12. Lewis, B. P., Burge, C. B., and Bartel, D. P. Conserved seed pairing, often flanked by adenosines, indicates that thousands of human genes are microRNA targets. *Cell*, 120(1): 15–20, 2005.

13. Krek, A., Grun, D., Poy, M. N., Wolf, R., Rosenberg L, Epstein, E. J., MacMenamin, P., da Piedade, I., Gunsalus, K. C., Stoffel, M., and Rajewsky, N. Combinatorial microRNA target predictions. *Nature Genetics*, 37(5): 495–500, 2005.

14. Malphettes, L. and Fussenegger, M. Impact of RNA interference on gene networks. *Metabolic Engineering*, 8(6): 672–683, 2006.

15. Silva, J. M., Li, M. Z., Chang, K., Ge, W., Golding, M. C., Rickles, R. J., Siolas, D., Hu, G., Paddison, P. J., Schlabach, M. R., Sheth, N., Bradshaw, J., Burchard, J., Kulkarni, A., Cavet, G., Sachidanandam, R., McCombie, W. R., Cleary, M. A., Elledge, S. J., and Hannon, G. J. Second-generation shRNA libraries covering the mouse and human genomes. *Nature Genetics*, 37(11): 1281–1288, 2005.

16. Chang, K., Elledge, S. J., and Hannon, G. J. Lessons from nature: microRNA-based shRNA libraries. *Nature Methods*, 3(9): 707–714, 2006.

17. Shin, K. J., Wall, E. A., Zavzavadjian, J. R., Santat, L. A., Liu, J., Hwang, J. I., Rebres, R., Roach, T., Seaman, W., Simon, M. I., and Fraser, I. D. A single lentiviral vector platform for microRNA-based conditional RNA interference and coordinated transgene expression. *Proceedings of the National Academy of Sciences of the United States of America*, 103(37): 13759–13764, 2006.

18. Lim, L. P., Lau, N. C., Weinstein, E. G., Abdelhakim, A., Yekta, S., Rhoades, M. W., Burge, C. B., and Bartel, D. P. The microRNAs of *Caenorhabditis elegans*. *Genes Devlopment*, 2: 2, 2003.

19. Lai, E. C., Tomancak, P., Williams, R. W., and Rubin, G. M. Computational identification of *Drosophila* microRNA genes. *Genome Biology*, 4(7): R42, 2003.

20. Legendre, M., Lambert, A., and Gautheret, D. Profile-based detection of microRNA precursors in animal genomes. *Bioinformatics*, 21(7): 841–845, 2005.

21. Wang, X., Zhang, J., Li, F., Gu, J., He, T., Zhang, X., and Li, Y. MicroRNA identification based on sequence and structure alignment. *Bioinformatics*, 21(18): 3610–3614, 2005.

22. Kim, V. N. and Nam, J. W. Genomics of microRNA. *Trends in Genetics*, 22(3): 165–173, 2006.

23. Berezikov, E., Cuppen, E., and Plasterk, R. H. Approaches to microRNA discovery. *Nature Genetics*, 38 Suppl 1: S2–S7, 2006.

24. Berezikov, E., Guryev, V., van de Belt, J., Wienholds, E., Plasterk, R. H., and Cuppen, E. Phylogenetic shadowing and computational identification of human microRNA genes. *Cell*, 120(1): 21–24, 2005.

25. Boffelli, D., McAuliffe, J., Ovcharenko, D., Lewis, K. D., Ovcharenko, I., Pachter, L., and Rubin, E. M. Phylogenetic shadowing of primate sequences to find functional regions of the human genome. *Science*, 299(5611): 1391–1394, 2003.

26. Lim, L. P., Glasner, M. E., Yekta, S., Burge, C. B., and Bartel, D. P. Vertebrate microRNA genes. *Science*, 299(5612): 1540, 2003.

27. Bentwich, I., Avniel, A., Karov, Y., Aharonov, R., Gilad, S., Barad, O., Barzilai, A., Einat, P., Einav, U., Meiri, E., Sharon, E., Spector, Y., and Bentwich, Z. Identification of hundreds of conserved and nonconserved human microRNAs. *Nature Genetics*, 37(7): 766–770, 2005.

28. Xie, X., Lu, J., Kulbokas, E. J., Golub, T. R., Mootha, V., Lindblad-Toh, K., Lander, E. S., and Kellis, M. Systematic discovery of regulatory motifs in human promoters and 3′ UTRs by comparison of several mammals. *Nature*, 434(7031): 338–345, 2005.

29. Stark, A., Brennecke, J., Russell, R. B., and Cohen, S. M. Identification of *Drosophila* MicroRNA targets. *PLoS Biology*, 1(3): E60, 2003.

30. Rhoades, M. W., Reinhart, B. J., Lim, L. P., Burge, C. B., Bartel, B., and Bartel, D. P. Prediction of plant microRNA targets. *Cell*, 110(4): 513–520, 2002.

31. Kruger, J. and Rehmsmeier, M. RNAhybrid: microRNA target prediction easy, fast and flexible. *Nucleic Acids Research*, 34(Web Server issue): W451–W454, 2006.

32. John, B., Enright, A. J., Aravin, A., Tuschl, T., Sander, C., and Marks, D. S. Human MicroRNA targets. *PLoS Biology*, 2(11): e363, 2004.

33. Sewer, A., Paul, N., Landgraf, P., Aravin, A., Pfeffer, S., Brownstein, M. J., Tuschl, T., van Nimwegen, E., and Zavolan, M. Identification of clustered microRNAs using an ab initio prediction method. *BMC Bioinformatics*, 6: 267, 2005.

34. Xue, C., Li, F., He, T., Liu, G. P., Li, Y., and Zhang, X. Classification of real and pseudo microRNA precursors using local structure-sequence features and support vector machine. *BMC Bioinformatics*, 6, 310, 2005.

35. Hertel, J. and Stadler, P. F. Hairpins in a Haystack: recognizing microRNA precursors in comparative genomics data. *Bioinformatics*, 22(14): e197–e202, 2006.

36. Kim, S. K., Nam, J. W., Rhee, J. K., Lee, W. J., and Zhang, B. T. miTarget: microRNA target-gene prediction using a support vector machine. *BMC Bioinformatics*, 7(1): 411, 2006.

37. Jia, P., Shi, T., Cai, Y., and Li, Y. Demonstration of two novel methods for predicting functional siRNA efficiency. *BMC Bioinformatics*, 7: 271, 2006.

38. Nam, J. W., Shin, K. R., Han, J., Lee, Y., Kim, V. N., and Zhang, B. T. Human microRNA prediction through a probabilistic co-learning model of sequence and structure. *Nucleic Acids Research*, 33(11): 3570–3581, 2005.

39. Nam, J. W., Kim, J., Kim, S. K., and Zhang, B. T. ProMiR II: a web server for the probabilistic prediction of clustered, nonclustered, conserved and nonconserved micro-RNAs. *Nucleic Acids Research,*, 34(Web Server issue): W455–W458, 2006.

40. Yousef, M., Nebozhyn, M., Shatkay, H., Kanterakis, S., Showe, L. C., and Showe, M. K. Combining multi-species genomic data for microRNA identification using a Naive Bayes classifier. *Bioinformatics*, 22(11): 1325–1334, 2006.

41. Huang, J. C., Morris, Q. D., and Frey, B. J. Detecting microRNA targets by linking sequence, microRNA and gene expression data. *Proceedings of the Tenth Annual International Conference on Research in Computational Molecular Biology (RECOMB)*, Venice, Italy, April 2–5, 2006.

42. Nam, J. W., Joung, J. G., and Zhang, B. T. Two-step genetic programming for optimization of RNA common-structure. *Lecture Notes in Computer Science*, 3005: 73–83, 2004.

43. Saetrom, O., Snove, O., Jr., and Saetrom, P. Weighted sequence motifs as an improved seeding step in microRNA target prediction algorithms. *RNA*, 11: 995–1003, 2005.

44. Saetrom, P. Predicting the efficacy of short oligonucleotides in antisense and RNAi experiments with boosted genetic programming. *Bioinformatics*, 20(17): 3055–3063, 2004.

45. Bennett, K. P. and Mangasarian, O. L. Robust linear programming discrimination Of two linearly inseparable sets. *Optimization Methods and Software*, 1: 23–24, 1992.

46. Vella, M. C., Reinert, K., and Slack, F. J. Architecture of a validated microRNA:target interaction. *Chemistry & Biology*, 11(12): 1619–1623, 2004.

47. Hofacker, I. L. Vienna RNA secondary structure server. *Nucleic Acids Research*, 31(13): 3429–3431, 2003.

48. Doench, J. G. and Sharp, P. A. Specificity of microRNA target selection in translational repression. *Genes & Development*, 18(5): 504–511, 2004.

49. Brennecke, J., Stark, A., Russell, R. B., and Cohen, S. M. Principles of microRNA-target recognition. *PLoS Biology*, 3(3): e85, 2005.

50. Cooper, G. and Herskovits, E. A Bayesian method for the induction of probabilistic networks from data. *Machine Learning*, 9: 309–347, 1992.

51. Notredame, C., O'Brien, E. A. and Higgins, D. G. RAGA: RNA sequence alignment by genetic algorithm. *Nucleic Acids Research*, 25(22): 4570–4580, 1997.

52. Szustakowski, J. D. and Weng, Z. Protein structure alignment using a genetic algorithm. *Proteins*, 38(4): 428–440, 2000.

53. Foster, J. A. Evolutionary computation. *Nature Review Genetics*, 2(6): 428–436, 2001.

54. Cooper, L. R., Corne, D. W., and Crabbe, M. J. Use of a novel Hill-climbing genetic algorithm in protein folding simulations. *Computational Biology and Chemistry*, 27(6): 575–580, 2003.

55. Fogel, G. B. and Corne, D. W. *Evolutionary Computation in Bioinformatics*, Morgan Kaufmann Publishers, San Francisco, 2003.

56. Fogel, G. B., Weekes, D. G., Varga, G., Dow, E. R., Harlow, H. B., Onyia, J. E., and Su, C. Discovery of sequence motifs related to coexpression of genes using evolutionary computation. *Nucleic Acids Research*, 32(13): 3826–3835, 2004.

57. Macke, T. J., Ecker, D. J., Gutell, R. R., Gautheret, D., Case, D. A., and Sampath, R. RNAMotif, an RNA secondary structure definition and search algorithm. *Nucleic Acids Research*, 29(22): 4724–4735, 2001.

58. Eddy, S. R. Profile hidden Markov models. *Bioinformatics*, 14(9): 755–763, 1998.

59. Leslie, C., Eskin, E., and Noble, W. S. The spectrum kernel: a string kernel for SVM protein classification. *Proceedings of the Pacific Symposium on Biocomputing*, 564–575, 2002.

17

MACHINE LEARNING FOR COMPUTATIONAL HAPLOTYPE ANALYSIS

Phil H. Lee and Hagit Shatkay

17.1 INTRODUCTION

Understanding genomic differences in the human population is one of the primary challenges of current genomics research [1]. The human genome can be viewed as a sequence of three billion letters from the nucleotide alphabet {A,G,C,T}. In more than 99% of the positions on the genome, the same nucleotide is shared across the population. However, 1% of the genome includes numerous genetic variations such as different nucleotide occurrences, deletion/insertion of a nucleotide, or variations in the number of multiple nucleotide repetitions. Differences in human traits, as obvious as physical appearance or as subtle as susceptibility to disease, typically originate from these variations in the human DNA.

Early large-scale studies of genetic variations focused on identifying which positions of the human genome are commonly variant and which are typically invariant. In general, when a variation occurs in at least a certain percentage of a population (typically around 1–5%), it is considered a common variation [1]. To date, millions of the common DNA variations have been identified and are accessible in public databases [2, 3]. These identified common variations usually involve the substitution of a single nucleotide and are called single nucleotide polymorphisms (SNPs—pronounced as snips). The nucleotide at a position in which a SNP occurs is called an allele. The nucleotide that most commonly occurs within a population is called the major allele, whereas the others are called the minor alleles. For example, if

Machine Learning in Bioinformatics. Edited by Yan-Qing Zhang and Jagath C. Rajapakse
Copyright © 2009 by John Wiley & Sons, Inc.

80% of a population has the nucleotide A at a certain position of the genome, whereas 20% of the population has the nucleotide T at the same position, A is called the major allele of the SNP, whereas T is the minor allele.

Once variations are identified, the next step is the study of disease–gene association, that is, identifying which DNA variation or set of variations is highly associated with a specific disease. Simple Mendelian diseases (e.g., Huntington disease, Sickle cell anemia) are caused by an abnormal alteration of a single gene. However, most current common diseases (e.g., cancer, heart disease, obesity) are known to be affected by a combination of two or more mutated genes along with certain environmental factors; thus, they are often called complex diseases. To identify the relations among mutations in multiple genes, at a statistically significant level, it is necessary to obtain genetic information from a large-scale population. Thus, traditional family-based analysis methods, which were useful for a simple Mendelian disease [4], do not perform well for complex common disease studies [5].

Recently, haplotype[1] analysis has been successfully applied to the identification of DNA variations relevant to several common and complex diseases [6–10], and is now considered the most promising method for studying disease-gene association for complex diseases [11, 12]. Numerous computational approaches have been proposed for effective haplotype analysis, and among them, machine learning algorithms have proven to be of great practical value. In this chapter, we provide an overview of computational haplotype analysis and introduce machine learning methods in two main computational procedures of haplotype analysis: haplotype phasing and tag SNP selection.

The rest of this chapter is organized as follows: Section 17.2 provides an overview of computational haplotype analysis; Sections 17.3 and 17.4 introduce and discuss machine learning methods in haplotype phasing and tag SNP selection, respectively; and Section 17.5 concludes and outlines future research.

17.2 COMPUTATIONAL HAPLOTYPE ANALYSIS

This section begins with defining the basic genetic concepts in computational genetic analysis. It then provides an overview of computational haplotype analysis, including its general objective, distinguishing features from previous approaches, and essential computational procedures.

17.2.1 Basic Concepts in Computational Genetic Analysis

Population genetics studies genetic changes in populations to understand the evolutionary significance of genetic variations, both within and between species [4]. Thus, it provides the basis for common and complex disease–gene association, that is, identifying a set of DNA variations that is common enough to be prevalent in the

---

[1]A haplotype is a set of SNPs present on one chromosome. All definitions and terms pertaining to computational haplotype analysis will be introduced and defined in the next section.

human population and has a causal connection to the elevated risk of a complex disease [11]. Since the ultimate aim of computational haplotype analysis is disease–gene association, we first define several basic concepts in population genetics, which we use throughout the chapter.

17.2.1.1 Haplotypes, Genotypes, and Phenotypes

Consider an example in which we have chromosome samples from six individuals. Three of them have lung cancer and the others do not. We aim to identify a set of DNA variations associated with lung cancer using these chromosome samples. Due to experimental cost and time, only a limited region of the chromosome, which was previously suggested to be related to lung cancer by other molecular experiments, is examined. The chromosomal location of the target region is referred to as a locus. A locus can be as large as a whole chromosome or as small as a part of a gene.

Let us look at the chromosome samples in detail. All species that reproduce sexually have two sets of chromosomes: one inherited from the father and the other inherited from the mother. Thus, every individual in our sample also has two alleles for each SNP, one on the paternal chromosome and the other on the maternal chromosome. For each SNP, the two alleles can be either identical or different. When they are the same, the SNP is called homozygous. When the alleles are different, the SNP is called heterozygous.

Suppose that our target locus contains six SNPs, and each SNP has only two different alleles (i.e., SNPs are assumed to be bi-allelic). The allele information is as shown in Fig. 17.1a. The major allele of the SNP is colored gray, and the minor is colored black. Each individual has two sets of six SNPs constructed from his/her two chromosomes. A set of SNPs present on one chromosome is referred to as a haplotype [13]. Note that there are 12 haplotypes stemming from the six pairs of chromosomal samples, where each pair is associated with one individual.

Several experimental methods can directly identify the haplotype information from chromosomes, but due to high cost and long operation time, they are mainly used for

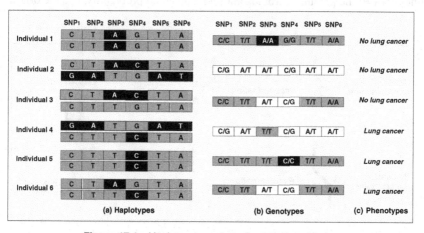

Figure 17.1 Haplotypes, genotypes, and phenotypes.

small- to moderate-size samples (typically from several to tens of individuals) [13]. For large-scale samples (typically hundreds or thousands of individuals), high-throughput biomolecular methods are used to identify the alleles of the target locus for each individual. The main limitation of the high-throughput methods lies in their lack of ability to distinguish the source chromosomes of each allele. Typically, such methods simply associate the two alleles with the SNP position, but do not determine the source chromosome for each. This combined allele information of a target locus is called a genotype, and the experimental procedure obtaining the genotype information is called genotyping.

Figure 17.1b displays the genotype information for our example. When the combined allele information of the SNP consists of two major alleles, it is colored gray. SNPs with two minor alleles are colored black, and SNPs with one major and one minor allele are colored white. The number of genotypes is six, the same as the number of individuals.

While haplotypes and genotypes represent the allele information of a target locus on the chromosome, a phenotype is the physical, observed manifestation of a genetic trait. In this example, the phenotype of an individual is either lung cancer or no lung cancer. In general, the individuals with disease are referred to as cases, whereas the ones with no disease are referred to as controls. Figure 17.1c displays the phenotype information for our example.

17.2.1.2 Linkage Disequilibrium and Block Structure of the Human Genome One interesting feature of a haplotype is the nonrandom association among the SNPs comprising it, called linkage disequilibrium (LD) [5]. As mentioned earlier, humans have two copies of each chromosome: paternal and maternal. Each of these two copies is generated by recombination of the parents' two copies of their own chromosomes, and is passed to the descendant. Figure 17.2 illustrates this process.

Theoretically, recombination can occur at any position along the chromosome any number of times. Thus, a SNP inherited on one chromosome could originate from either copy of the parents' chromosomes with an equal probability, and the origin of

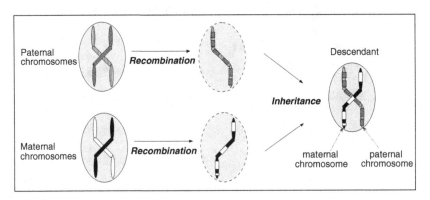

Figure 17.2 Recombination and inheritance.

one SNP is not affected by the origin of the others. This characteristic of independence between SNPs is called linkage equilibrium.

For instance, consider two SNPs s_1 and s_2. Let $|s_1|$ and $|s_2|$ denote the number of alleles that the SNPs s_1 and s_2 have, respectively. Let s_{1i} denote the ith allele of the first SNP s_1, and s_{2j} denote the jth allele of the second SNP s_2, where $i = 1, \ldots, |s_1|$ and $j = 1, \ldots, |s_2|$. Under linkage equilibrium, the joint probability of two alleles s_{1i} and s_{2j} is expected to be equal to the product of the alleles' individual probabilities, since s_1 and s_2 are independent. Thus, under the independence assumption,

$$\forall_{i,j} \Pr(s_{1i}, s_{2j}) = \Pr(s_{1i}) \cdot \Pr(s_{2j}). \tag{17.1}$$

When Equation 17.1 is not satisfied by the two SNPs, that is, when their alleles are not independent, they are said to be in a state of LD. When the allele dependence between the two SNPs is high,[2] the two are considered to be in a state of high LD.

In general, SNPs within physical proximity are assumed to be in a state of high LD. The probability of recombination increases with the distance between two SNPs [13]. Thus, SNPs within proximity tend to be passed together from an ancestor to his/her descendants. As a result, their alleles are often highly correlated with each other, and the number of distinct haplotypes, consisting of the SNPs, within a population is much smaller than expected under linkage equilibrium.

Recently, large-scale LD studies [16–18] have been conducted to understand the comprehensive LD structure of the human genome. The results strongly support the hypothesis that genomic DNA can be partitioned into discrete regions, known as blocks, such that recombination has been very rare (i.e., high LD) within each block, and very common (i.e., low LD) between the blocks. As a result, high LD exists among SNPs within a block, while the number of distinct haplotypes across a population is strikingly small. This observation is referred to as the block structure of the human genome. At this point, there is no agreement upon way to define blocks on the genome [19, 20]. However, there seems to be no disagreement that the human genome indeed has the block structure regardless of our ability to uniquely identify these blocks.

High LD among SNPs within physical proximity and the limited number of haplotypes due to the block structure of the human genome have provided the basis of computational haplotype analysis for disease–gene association. We next introduce computational haplotype analysis in detail.

17.2.2 Computational Haplotype Analysis

Our ultimate goal is to identify a set of DNA variations that is highly associated with a specific disease. Haplotypes, genotypes, or even single-SNP information can be used to examine the association of a genetic variation with a target disease. The use of haplotype information for studying disease–gene association is called haplotype

---

[2] The absolute threshold differs in each LD measure. For details, refer to LD review articles [14, 15].

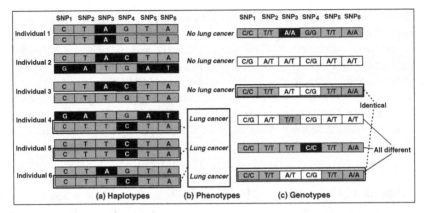

Figure 17.3 Difference between haplotype analysis and genotype analysis.

analysis. Single-SNP analysis and genotype analysis refer to the studies that use single-SNP information and genotype information, respectively.

Haplotype analysis has several advantages compared with single-SNP analysis and genotype analysis. Single-SNP analysis cannot identify the association when a combination of several SNPs on one chromosome (i.e., a haplotype) is required to affect the phenotype of an individual [17, 21, 22]. Figure 17.3 exemplifies this case. All and only the three individuals with lung cancer share the haplotype CTTCTA, marked by a solid box in Fig. 17.3a. Thus, we can conclude that the lung-cancer phenotype is associated with the haplotype CTTCTA. However, if we examine each of the six SNPs individually, no direct association is found between any one of them and the lung-cancer phenotype. For example, both individuals with lung cancer and individuals with no lung cancer have the allele C or the allele G on the first SNP, the allele T or the allele A on the second SNP, and so on.

Because genotypes do not contain the source chromosome information, known as *phase*, they often hide the obvious association existing between a haplotype and a target disease. For example, in Fig. 17.3a, each individual with lung cancer (i.e., case) has two haplotypes: one haplotype is CTTCTA, that is, the one associated with the lung cancer phenotype, whereas the other one is unique for each case. Although all cases share exactly the same haplotype CTTCTA, their genotypes, in Fig. 17.3c, look different due to their other unique haplotype. Worse, the genotype of individual 6, who has lung cancer, is identical to that of individual 3, who has no lung cancer. Thus, we cannot identify a specific genotype that is highly associated with lung cancer, and as a result, miss the real association between the haplotype CTTCTA and lung cancer.

Despite its advantages, the use of haplotype analysis has been limited by the high cost and long operation time of biomolecular methods for obtaining the haplotype information. However, two computational procedures, haplotype phasing and tag SNP selection aim to address this problem and support the use of haplotype analysis for disease–gene association. Haplotype phasing deduces haplotype information from

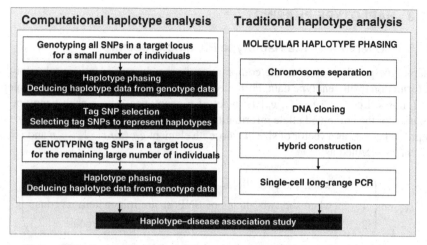

Figure 17.4 Computational haplotype analysis and traditional haplotype analysis.

genotype data. Tag SNP selection selects a subset of SNPs on a haplotype that is sufficiently informative to study disease–gene association but still small enough to reduce the genotyping overhead. When these computational procedures are used for haplotype analysis, we refer to the whole procedure as computational haplotype analysis.

Figure 17.4 summarizes the general procedures of computational haplotype analysis and of traditional haplotype analysis. Biomolecular experiments are displayed in white boxes, and computational and/or statistical procedures are displayed in black boxes. Computational haplotype analysis includes haplotype phasing, tag SNP selection, and haplotype–disease association study along with two genotyping experiments. Initially, a relatively small number of individuals are genotyped from a target population, and their haplotypes are inferred using haplotype phasing methods. Then, tag SNP selection methods select a small subset of SNPs on the haplotypes, which can represent the identified haplotypes with little loss of information. Using the selected small number of SNPs, a second genotyping is done for a large number of individuals. Again, computational haplotype phasing methods are used to infer the haplotypes from these genotype data. Finally, haplotype–disease association study, which is identifying the association of a haplotype or a set of haplotypes with a target disease, is performed on the haplotypes.

In contrast to computational haplotype analysis, traditional haplotype analysis relies on low-throughput biomolecular experiments to directly obtain haplotype information. Thus, it can provide more accurate haplotype information than computational procedures, and, in the near future, the biomolecular methods might become a standard technique for haplotype analysis [23]. However, until biomolecular experiments become less costly and time consuming, the two computational procedures, haplotype phasing and tag SNP selection, are expected to be of much use for large-scale association studies.

17.3 HAPLOTYPE PHASING

17.3.1 Overview of the Haplotype Phasing Problem

Haplotype phasing refers to the computational procedure of identifying haplotype information from genotype data. Formally, we define the haplotype phasing problem as follows: Let $G = \{g_1, \ldots, g_n\}$ be a set of n genotypes, where each genotype g_i consists of the combined allele information of m SNPs, $s_1, \ldots, s_m$. For simplicity, we represent $g_i \in G$ as a vector of size m whose jth element g_{ij}, (for $i = 1, \ldots, n$ and $j = 1, \ldots, m$), is defined as

$$g_{ij} = \begin{cases} 0 & : \text{when the two alleles of SNP } s_j \text{ are major homozygous,} \\ 1 & : \text{when the two alleles of SNP } s_j \text{ are minor homozygous,} \\ 2 & : \text{when the two alleles of SNP } s_j \text{ are heterozygous.} \end{cases}$$

Let H be the set of all[3] haplotypes consisting of the same m SNPs, $s_1, \ldots, s_m$. Like the genotype, each haplotype $h_i \in H$ is also a vector of size m. However, as introduced in Section 17.2.1, haplotypes represent the allele information of SNPs on one chromosome, whereas genotypes represent the combined allele information of SNPs on two chromosomes. Thus, the jth element, h_{ij}, of the haplotype h_i ($i = 1, \ldots, 2^m$ and $j = 1, \ldots, m$) is defined as:

$$h_{ij} = \begin{cases} 0 & : \text{when the allele of SNP } s_j \text{ is major,} \\ 1 & : \text{when the allele of SNP } s_j \text{ is minor.} \end{cases}$$

When the combined allele information of two haplotypes, $h_j, h_k \in H$, comprises the genotype g_i, we say that h_j and h_k resolve g_i and denote the relationship[4] as $h_j \oplus h_k = g_i$. The haplotypes h_j and h_k are referred to as the complementary mates of each other to resolve g_i, and each of them is considered to be compatible with g_i. The haplotype phasing problem can thus be defined as follows:

Problem : Haplotype phasing

Input : A set of genotypes $G = \{g_1, \ldots, g_n\}$

Output : A set of n haplotype pairs
$$O = \{\langle h_{i1}, h_{i2} \rangle | h_{i1} \oplus h_{i2} = g_i, h_{i1}, h_{i2} \in H, 1 \leq i \leq n\}.$$

To solve the haplotype phasing problem, one needs to find a set of haplotype pairs that can resolve all genotypes in G.

However, the solution to the haplotype phasing problem is not straightforward due to resolution ambiguity. Figure 17.5 illustrates the problem. The genotype data of three

[3] Since the allele of a SNP is either major or minor, the total number of haplotypes with m SNPs is 2^m.

[4] The order of the haplotype pairs does not matter, that is, $h_j \oplus h_k = g_i$ is the same as $h_k \oplus h_j = g_i$.

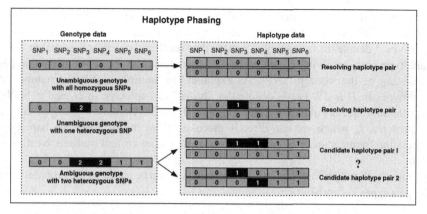

Figure 17.5 Haplotype phasing and ambiguous genotypes.

individuals are displayed on the left. Each genotype consists of six SNPs. When two alleles of a SNP in a genotype are homozygous (i.e., either 0 or 1), the SNP is colored gray. When two alleles of a SNP in a genotype are heterozygous (i.e., 2), the SNP is colored black. The first genotype consists of all homozygous SNPs, whereas the second genotype contains one heterozygous SNP. For both of these genotypes, the resolving haplotype pairs can be identified unambiguously as shown on the top right. However, in the case when there are c heterozygous SNPs in the genotype ($c > 1$) such as the third one in Fig. 17.5, there are 2^{c-1} pairs of haplotypes that can resolve the genotype. Thus, the genotype cannot be uniquely resolved without additional biological insights or constraints. In this case, the genotype is considered to be ambiguous.

Numerous computational and/or statistical algorithms have been developed for addressing this ambiguity in haplotype phasing. The methods are typically based on one of the four main principles: (1) parsimony; (2) phylogeny; (3) maximum likelihood (ML); and (4) Bayesian inference. The first two principles formulate the haplotype phasing problem as a combinatorial problem [24], and try to directly resolve each genotype with a pair of compatible haplotypes within certain constraints. In contrast, the latter two are based on a rather indirect statistical approach, haplotype frequency estimation; they first estimate haplotype frequencies for a population and then resolve each genotype with a most probable pair of haplotypes. Here, we introduce in detail the latter two approaches, namely, ML using an expectation-maximization (EM) algorithm and Bayesian inference using Markov Chain Monte Carlo (MCMC) techniques. Both of these are examples of machine learning methods.

17.3.2 Maximum Likelihood Using Expectation-Maximization Method

The EM method [25] is a widely used machine learning tool that enables learning when some of the information is unobserved. In principle, it searches for the value of certain parameters Θ that maximize the likelihood of the given data D. In haplotype phasing, unobserved haplotype population frequencies are considered to be the hidden parameters Θ, and their ML values, given the genotype data, are estimated.

Let D be the genotype data of n individuals, where each genotype consists of m SNPs, and the number of distinct genotypes in D is n'. Let g_i denote the ith distinct genotype, f_i denote the frequency of g_i in the data set D, and c_i denote the number of the haplotype pairs resolving g_i, where $i = 1, \ldots, n'$. Let H be the set of all haplotypes consisting of the same m SNPs. As explained in Section 17.3.1, the number of haplotypes in H is 2^m. Let h_j denote the jth distinct haplotype in H, and p_j be the population frequency of haplotype h_j, where $j = 1, \ldots, 2^m$. Unlike the genotype sample frequencies, f_i, which we can directly calculate from the data set, the haplotype population frequencies, $\Theta = \{p_1, p_2, \ldots, p_{2^m}\}$, are unknown, and we need to estimate them.

Initially, the likelihood L of the genotype data D can be stated as the probability of the genotypes comprising D:

$$L(D) = \Pr(D|\Theta) \approx \prod_{i=1}^{n'} \Pr(g_i|\Theta)^{f_i} = \prod_{i=1}^{n'} \left(\sum_{j=1}^{c_i} \Pr(h1_{ij}, h2_{ij}|\Theta) \right)^{f_i}, \qquad (17.2)$$

where $h1_{ij}$ and $h2_{ij}$ are the jth haplotype pair resolving the genotype g_i. That is, the likelihood of the data D is the product of the probabilities of all the genotypes in D. Each genotype, g_i, occurs f_i times in D, and its probability, $Pr(g_i | \Theta)$, can be computed by summing the joint probability of each haplotype pair that can resolve the genotype. Under a model known as the Hardy–Weinberg equilibrium (HWE), which assumes random mating, the joint probability $\Pr(h_k, h_l|\Theta)$ of two haplotypes can be computed as the product of the two population haplotype frequencies, p_k and p_l. When $k = l$, $\Pr(h_k, h_l|\Theta) = (p_k)^2$. Otherwise[5], $\Pr(h_k, h_l|\Theta) = 2p_k p_l$. Thus, the joint probability $\Pr(h_k, h_l|\Theta)$ in Equation 17.2 can be substituted by the product of the two population haplotype frequencies, and the population frequencies that maximize Equation 17.2 are computed.

Several groups [26–29] independently proposed an EM method to estimate the ML haplotype frequencies. In this context, the EM procedure is defined as follows: Initially, arbitrary values are assigned to the target haplotype frequencies $p_1, \ldots, p_{2^m}$, which we refer to as $p_1^{(0)}, \ldots, p_{2^m}^{(0)}$. An iterative process involving two steps then follows: In the expectation step, the current haplotype frequencies are used to estimate the expected genotype frequency $\hat{\Pr}(h_k, h_l|\Theta)^{(t)}$ where (t) denotes the tth iteration. In the maximization step, the expected genotype frequency $\hat{\Pr}(h_k, h_l|\Theta)^{(t)}$, computed in the previous step, is used to reestimate the haplotype frequencies $p_1^{(t+1)}, \ldots, p_{2^m}^{(t+1)}$. The expectation and maximization steps are repeated until the change in the haplotype frequency between consecutive iterations falls below a predefined threshold.

In general, the performance of EM-based methods for the haplotype phasing problem was demonstrated to be accurate and robust under a wide range of parameter

---

[5]Since we do not know the phase information of the given genotype, a genotype consisting of two nonidentical haplotypes, h_k and h_l, can have two phases: h_k on the maternal chromosome and h_l on the paternal, or vice versa.

settings [30–32]. However, several shortcomings still exist. First, the EM method strongly depends on the initialization assignment and does not guarantee a global optimum of the likelihood function. To overcome this, EM-based methods are typically run multiple times with different initial settings. Second, the variance of the haplotype frequency estimation is not accurately known [27, 29]. Last, calculating confidence intervals and conducting statistical tests under the EM algorithm typically involve approximations that require a large sample to be accurate [33].

17.3.3 Bayesian Inference Using Markov Chain Monte Carlo

Like the ML methods introduced above, Bayesian inference methods take a statistical approach. However, while the EM algorithm, introduced in the previous section, aims to find a set of exact model parameters Θ that maximize the probability of the genotype data D given the model, that is, argmax $\Pr(D|\Theta)$, Bayesian inference methods aim to find the posterior distribution of the model parameters given the genotype data D, which is $\Pr(D|\Theta)$. Moreover, in EM-based methods, Θ denotes a set of unknown haplotype frequencies in a population, whereas in Bayesian inference methods, Θ denotes a set of each genotype's resolved haplotype pairs. Thus, where H is a set of haplotype pairs resolving the given genotypes, Bayesian inference methods aim to find the posterior probability $\Pr(H|D)$. However, computing $\Pr(H|D)$ exactly is not feasible in the general case [33]. Thus, MCMC techniques are used to obtain approximate samples from $\Pr(H|D)$, and their expectation is presented as the solution.

One popular MCMC technique is Gibbs sampling. This is again an iterative procedure, and it is applied to haplotypes as follows: Let $H^{(t)}$ denote the set of haplotype pairs resolving all genotypes at the tth iteration, $H_{-i}^{(t)}$ denote the set of haplotype pairs resolving all genotypes except g_i at the tth iteration, and $H_i^{(t)}$ denote the set including only the haplotype pair resolving the genotype g_i at the tth iteration. Gibbs sampling starts with an initial guess $H^{(0)}$. An unresolved genotype g_i is then randomly selected. Under the assumption that a current estimation of $H^{(t)}$ is correct for all genotypes except g_i, the new haplotype pair resolving g_i, that is, $H_i^{(t+1)}$, is sampled from the distribution $\Pr(H_i|D, H_{-i}^{(t)})$. This random selection and update is iterated, until the transition distribution, $\Pr(H_i|D, H_{-i}^{(t)})$ converges to its target posterior distribution $\Pr(H|D)$. We can therefore use the sampled haplotype pairs to approximate the haplotype distribution for the genotypes D, that is, $\Pr(H|D)$.

The first Bayesian inference method in the context of haplotype phasing was proposed by Stephens et al. [34]. Their algorithm exploits ideas from the coalescent theory to guide the Gibbs sampling procedure. The coalescent is a popular genetic model that denotes a rooted tree describing the evolutionary history of a set of DNA sequences [4]. Along the coalescent, haplotypes evolve one SNP at a time. Thus, whenever a genotype cannot be resolved using existing haplotypes, new haplotypes that are most similar to the existing common[6] ones are generated to resolve the genotype.

---

[6]Due to its preference to common haplotypes, some [35] have interpreted this approach as a kind of a parsimony approach.

Other Bayesian inference methods [36–37] use similar MCMC sampling techniques, but employ priors of different forms: simple Dirichlet [35, 36] and Dirichlet process [37]. In addition, Lin et al. [35] use neighboring information of heterozygous SNPs to resolve each genotype.

It is undetermined whether either one of the two methods, namely, Bayesian inference using the MCMC techniques or ML using the EM algorithm is superior to the other. The performance of the two varies under different genotype compositions and different measures of accuracy [38–42]. Most importantly, both approaches have their merits and shortcomings. Unlike ML methods, Bayesian inference can be applied to samples consisting of a large number of SNPs or to samples in which a substantial portion of haplotypes occur only once [33, 43]. In addition, given the sufficient running time, MCMC techniques can potentially explore the whole state space and avoid local maxima [44]. Finally, Bayesian inference methods can incorporate prior knowledge to guide their estimation procedure. In contrast, ML methods require less computing time [33], and are easier to check for convergence than Bayesian inference methods [45]. Furthermore, the performance of ML methods is robust even under the violation of their basic assumption, the Hardy–Weinberg equilibrium, whereas the performance of Bayesian inference methods is reported to be affected by the deviation of the data from their basic assumption, namely, the coalescent theory [42].

17.3.4 Discussion

Many empirical studies have shown that two machine learning-based approaches, namely, ML using the EM algorithm and Bayesian inference using MCMC techniques, are quite promising [31, 39, 40]. However, there are still several difficulties that none of the current haplotype phasing methods address well. First, the phasing accuracy of all methods decreases as LD drops [30, 40]. This poor accuracy occurs more often when a large number of SNPs are examined, since LD tends to decrease as the distance between SNPs increases.

Second, phasing algorithms work well for data sets with few or no genotyping errors or missing alleles [31, 46]. However, very often, allele information is incorrect or missing due to imperfection of current genotyping technology. Missing allele information increases the combinatorial complexity of the haplotype phasing problem. The genotyping error problem is even more difficult to solve, since, in general, we do not know which alleles are incorrect.

Finally, all haplotype phasing algorithms show a poor phasing accuracy for rare haplotypes (i.e., ones with a population frequency of at most 1–5%) [24, 31, 40, 47]. This problem occurs since most haplotype phasing algorithms are based on population genetics assumptions that prefer common haplotypes (i.e., occurring in more than 5–20% of the population). However, it is not clear yet whether rare haplotypes or common ones are important in the etiology of complex diseases [13].

In conclusion, future research of haplotype phasing should focus on improving the performance of algorithms for data sets with low LD, genotyping errors, missing alleles, and rare haplotypes.

17.4 TAG SNP SELECTION

In most large-scale disease studies, genotyping all the SNPs in a candidate region for a large number of individuals is still costly and time consuming. Thus, selecting a subset of SNPs that is sufficiently informative to conduct disease–gene association but small enough to reduce the genotyping overhead, a process known as tag SNP selection, is a critical problem to solve. In general, the selected SNPs on a haplotype are referred to as haplotype tag SNPs (htSNPs) and the unselected SNPs are referred to as tagged SNPs.

17.4.1 Overview of the Tag SNP Selection Problem

Formally, we define the tag SNP selection problem as follows: Let $S = \{s_1, \ldots, s_m\}$ be a set of m SNPs in a candidate region, and $D = \{h_1, \ldots, h_n\}$ be a data set of n haplotypes consisting of the m SNPs. As defined in Section 17.3.1, $h_i \in D$ is a vector of size m whose vector element is 0 when the allele of a SNP is major and 1 when it is minor. Suppose that the maximum number of htSNPs is k, and a function $f(T',D)$ evaluates how well the subset $T' \subset S$ represents the original data D. The tag SNP selection problem can then be stated as follows:

Problem	:	Tag SNP Selection
Input	:	A set of SNPs S; a set of haplotypes D; a maximum number of htSNPs k;
Output	:	A set of htSNPs $T = \underset{T' \; s.t. \; T' \subset S \text{ and } |T'| \leq k}{\operatorname{argmax}} f(T',D)$.

To solve the tag SNP selection problem, one needs to find an optimal subset of SNPs, T, of size at most k based on the given evaluation function, f, among all possible subsets of the original SNPs.

Initially, tag SNP selection was motivated by LD, as introduced in Section 17.2.1. When high LD exists among SNPs, their allele information may be almost the same. Thus, although only one SNP is selected from the ones with redundant allele information, most genetic variation in the haplotype can still be retained. However, what comprises the best tag SNP selection strategy is still an open problem.

Researchers proposed a variety of measures to represent the information of haplotypes, and tried to identify the subset of SNPs that optimizes these measures. The relations among these measures and their effect on the selection of htSNPs are still the subject of ongoing research. Most important, unlike haplotype phasing, there is no gold standard to evaluate the performance of different approaches [20]. Thus, the performance of tag SNP selection algorithms is often evaluated based on their own information measure, which makes comparison among different approaches difficult.

Here we introduce two representative tag SNP selection methods using machine learning techniques: (1) pairwise association using clustering; and (2) tagged SNP prediction using Bayesian networks (BNs). Further methods in tag SNP selection are surveyed in a more extensive report [48].

17.4.2 Pairwise Association Using Clustering

Pairwise association-based approaches rely on the idea that a set of htSNPs should be the smallest subset of available SNPs that are capable of predicting a disease locus on a haplotype. However, the disease locus is generally the one we are looking for, and thus not known ahead of time. Instead, pairwise association between SNPs is used as an estimate of the predictive power with respect to the disease locus.

In principle, a set of htSNPs is selected such that all SNPs on the haplotype are highly associated with at least one of the htSNPs. This way, although the SNP that is directly relevant to the disease may not be selected as an htSNP, the relationship between the disease and that SNP can be indirectly inferred from the htSNP that is highly associated with the actual disease SNP. In most studies, nonrandom association of SNPs (i.e., LD, introduced in Section 17.2.1) is used to estimate the pairwise association.

Byng et al. [49] first proposed to use cluster analysis for pairwise association-based tag SNP selection. The original set of SNPs is partitioned into hierarchical clusters, where each SNP has at least a prespecified level, σ, of pairwise LD with at least one of the other SNPs within the cluster (typically $\sigma > 0.6$). After clustering is performed, one SNP from each cluster is selected based on practical feasibility such as ease of genotyping, importance of physical location, or significance of the SNP mutation.

Others [50–52] proposed new clustering methods. These methods ensure that within each cluster, the pairwise LD between an htSNP and all the other SNPs is greater than the threshold level, σ. Examples for such methods are minimax clustering [52] and greedy binning algorithm [50, 51].

In minimax clustering, the minimax distance between the two clusters C_i and C_j is defined as $D_{\text{minimax}}(C_i, C_j) = \min_{\forall s \in (C_i \cup C_j)} (D_{\max}(s))$, where $D_{\max}(s)$ is the maximum distance between the SNP s and all the other SNPs in the two clusters. Initially, every SNP constitutes its own cluster. The two closest clusters, based on their minimax distance, are iteratively merged, and the SNP defining the minimax distance is denoted as a representative SNP for the cluster. The merging stops when the smallest distance between the two clusters is larger than the prespecified level, $1 - \sigma$, and the representative SNPs for the final clusters are selected as the set of htSNPs.

The greedy binning algorithm works as follows: First, it examines all pairwise LD relationship between SNPs, and for each SNP, counts the number of other SNPs whose pairwise LD with the SNP is greater than the prespecified level, σ. The SNP that has the largest count is then clustered together with its associated SNPs, and becomes the htSNP for the cluster. This procedure is iterated with the remaining SNPs until all the SNPs are clustered. The SNPs whose pairwise LD is not greater than σ with respect to any other SNPs are individually placed in singleton clusters.

Pairwise association-based methods are straightforward, and the selected htSNPs can be directly used in disease–gene association studies. In addition, the complexity of clustering ranges between $O(nm^2 \log m)$ [49] and $O(cnm^2)$ [50–52], where the number of clusters is c, the number of haplotypes is n, and the number of SNPs is m. Thus, pairwise association-based methods are typically faster than other selection approaches that examine multiple relationships among SNPs. The major shortcoming

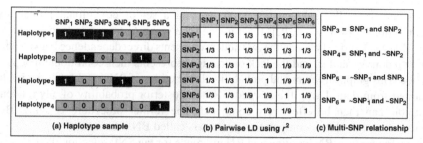

Figure 17.6   Pairwise LD among SNPs and multi-SNP dependencies.

of pairwise association-based methods lies in their lack of ability to capture multi-SNP dependencies [53] and in their tendency to select more htSNPs than other methods [19, 54, 55].

Figure 17.6 illustrates this weakness of pairwise association-based methods. Suppose that our sample consists of four haplotypes with six SNPs each, as shown in Fig. 17.6a. If we measure pairwise LD between the SNPs using the most common LD measure, the correlation coefficient, r^2 [56], no two SNPs have pairwise LD greater than 0.5, as shown in Fig. 17.6b. Thus, pairwise association-based methods will select all the six SNPs as htSNPs. However, as shown in Fig. 17.6c, the allele of SNPs 3–6 can be perfectly represented by the alleles of SNPs 1 and 2. Thus, if we consider multi-SNP dependencies, only two SNPs, namely, SNP_1 and SNP_2, are sufficient to represent all of the six SNPs.

17.4.3   Tagged SNP Prediction Using Bayesian Networks

Tagged SNP prediction-based approaches consider tag SNP selection as a reconstruction problem of the original haplotype data using only a subset of SNPs. Thus, they aim to select a set of SNPs that can predict the unselected (i.e., tagged) SNPs with little error. In general, after the selected htSNPs are genotyped, the alleles of the tagged SNPs are predicted using the alleles of the htSNPs, and disease–gene association is conducted based on the reconstructed full haplotype data. Therefore, these methods present a prediction rule for each tagged SNP along with the selected set of htSNPs.

Bafna et al. [53] first proposed to select htSNPs based on tagged SNP prediction accuracy, and presented a dynamic programming algorithm. Halperin et al. [57] also proposed a polynomial time dynamic programming algorithm. However, the prediction performance of these methods was somewhat limited because they fix the number or the location of predictive htSNPs for each tagged SNP. Recently, we have introduced a new prediction-based tag SNP selection system, called BNTagger [58], which is not limited by these previous restrictions and, as a result, improves prediction performance.

BNTagger aims to select a set of independent and highly predictive htSNPs using the formalism of BNs. A BN is a graphical model of joint probability distributions that captures conditional independence and dependence relations among its variables [59]. Given a finite set of random variables, $V = \{X_1, \ldots, X_p\}$, a BN consists of two components: a directed acyclic graph, G, and a set of conditional probability distributions, $\Theta = \{\theta_1, \ldots, \theta_p\}$. Each node in the graph G corresponds to a random variable X_j. An edge between the two nodes represents direct dependence between the two random variables, and the lack of an edge represents their conditional independence. This graph structure and probability distributions of BNs can be automatically learned from data. Using the learned BN, the posterior probability of any random variable can then be effectively computed. The former procedure is called BN learning, and the latter computation process is called BN inference.

BNTagger uses BN learning and BN inference for tag SNP selection. It considers SNPs as random variables, and learns their conditional independence and dependence relations using BN learning. Following the topological order (from parents to children) in the BN, BNTagger examines the prediction accuracy for each SNP, and decides whether it should be selected as an htSNP or not. This sequential evaluation in a topological order ensures that all previously selected htSNPs are always the ancestors of the currently examined SNP, which has at least two advantages. First, because the parent–child relation in the BN encodes the direct dependence between the SNPs, the state of a child node depends primarily on the information of its parents. Thus, parents htSNPs are typically good predictors for a child SNP.

Second, BNTagger calculates the prediction accuracy for each SNP X_j using the posterior probability of X_j, given the allele information of the htSNPs. When parents are htSNPs, the computation of this posterior probability can be omitted because it is already provided in the conditional probability distribution table associated with X_j. As a result, the topological structure and the conditional probability parameters of the BN enable BNTagger to find a set of highly predictive htSNPs as well as to expedite the evaluation procedure.

Unlike pairwise association-based methods, tagged SNP prediction-based methods use multi-SNP dependencies to select a set of htSNPs. As a result, the number of selected htSNPs is often smaller than that resulting from pairwise association-based methods [60]. In addition, tagged SNP prediction-based methods neither assume the limited diversity of haplotypes nor the high LD among SNPs. Accordingly, they work well for low LD regions of the genome [58]. However, as the tagged SNP prediction-based methods aim to reconstruct the allele information of all the SNPs, the number of SNPs to analyze in disease association studies is much larger than that of other tag SNP selection approaches. The efficient handling of this large number of variables in association studies requires further research.

17.4.4 Discussion

The feasibility of tag SNP selection has been empirically demonstrated by simulation studies [21, 54–56, 61]. The results suggest that tag SNP selection can yield about two to fivefold savings in the genotyping efforts. Most important, Zhang et al. [21] demonstrate that tag SNP selection shows little loss of power[7] in subsequent association studies. Based on 1000 simulated data sets, the average difference in

---

[7]The power of association tests is the probability that the test rejects the false null hypotheses [62].

power between a whole set of SNPs and a set of htSNPs whose size is one-fourth of the original SNP set is only 4%.

However, several problems still need to be addressed:

(1) Most tag SNP selection algorithms focus on covering common haplotypes or common SNPs rather than rare ones [11]. Common variations are of interest because many common diseases have been explained by common DNA variations rather than by rare ones [63]. Furthermore, practically, a much larger sample size is needed to identify rare haplotypes [1]. However, as discussed in Section 17.3.4, it is still an open question whether common variations or rare ones influence the susceptibility to common and complex disease.

(2) Many of the algorithms presented above require haplotype data rather than genotype data. When only genotype data are available, haplotype phasing is performed on the genotype data and the identified haplotype information is used. However, haplotype phasing may lead to incorrect resolution. To address this, several statistical methods produce multiple solutions along with their respective uncertainty measures [64] or the distribution of haplotype pairs for each genotype rather than a single resolved pair [11]. No current tag SNP selection method considers this uncertainty in the inferred haplotype data.

(3) All the algorithms described above assume that the set of htSNPs selected from a given sample will work well for another sample from the same population. However, to ensure such generalization, a sufficient number of individuals should be sampled for tag SNP selection. Therefore, tag SNP selection should be applied only when a sufficient number of individuals can indeed be sampled. In addition, it is important to ensure that the model for SNP selection does not overfit the given data set when the sample data are insufficient.

Along with the ways to address these limitations, more comparative studies are needed to understand the merits and shortcomings of different tag SNP selection approaches. To date, little comparative study has been done, and existing studies arrive at conflicting conclusions. Burkett et al. [55] concluded that different approaches produce considerably different sets of htSNPs. In contrast, Duggal et al. [65] and Ke et al. [54, 61] reported that the proportion of commonly selected htSNPs among different selection methods is, in general, consistently high (about 50–95%). It is difficult to generalize any of these conclusions since they are based on different data sets and on different evaluation measures. Further research should clarify these results as well as establish a common test bed to evaluate the performance of different selection approaches.

17.5 CONCLUSION

Computational haplotype analysis and, specifically its two procedures, haplotype phasing and tag SNP selection, provide a practical framework for conducting large-scale association studies. They enable an inexpensive and fast haplotype analysis, and numerous empirical studies support their feasibility and practical performance [21, 31, 39, 40, 54–56, 61]. Thus, until the overhead of low-throughput biomolecular experiments becomes less formidable, computational haplotype analysis is expected to be in demand.

In this chapter, we introduced several approaches to haplotype phasing and tag SNP selection along with their biological motivation. Genetic variations in the human genome are the result of a long and nondeterministic genetic evolution process. Thus, robust probabilistic machine learning methods and inference algorithms are well suited for the identification of the latent structure in DNA variations, and can help to generate plausible hypotheses for further analysis. As specific examples, we discussed the EM and the MCMC techniques for haplotype phasing and cluster analysis and Bayesian networks for tag SNP selection. Of course, certain limitations still exist in current approaches, and future research should focus on addressing them.

Missing alleles, genotyping errors, and low LD among SNPs are the common difficulties with which all haplotype phasing algorithms are confronted. Further improvement of tag SNP selection requires the ability to handle rare haplotypes, uncertainty in haplotype data, and small sample size. In addition, thorough evaluation of different approaches and development of a common test bed are important current goals.

REFERENCES

1. Kruglyak, L. and Nickerson, D. A. Variation is the spice of life. *Nature Genetics*, 27: 234–236, 2001.

2. Sherry, S. T., Ward, M. H., Kholodov, M., Baker, J., Phan, L., Smigielski, E. M., and Sirotkin, K. dbSNP: the NCBI database of genetic variation. *Nucleic Acids Research*, 29: 308–311, 2001.

3. Haga, H., Yamada, R., Ohnishi, Y., Nakamura, Y., and Tanaka, T. Gene-based SNP discovery as part of the Japanese millennium genome project: identification of 190,562 genetic variations in the human genome. Single-nucleotide polymorphism. *Journal of Human Genetics*, 47(11): 605–610, 2002.

4. Hedrick, P. W. *Genetics of Population*, 3rd edition, Jones and Bartlett Publishers, Sudbury, 2004.

5. Goldstein, D. B. Islands of linkage disequilibrium. *Nature Genetics*, 29: 109–211, 2001.

6. Reif, A., Herterich, S., Strobel, A., Ehlis, A. C., Saur, D., Jacob, C. P., Wienker, T., Töpner, T., Fritzen, S., Walter, U., Schmitt, A., Fallgatter, A. J., and Lesch, K. P. A neuronal nitric oxide synthase (NOS-I) haplotype associated with schizophrenia modifies prefrontal cortex function. *Molecular Psychiatry*, 11: 286–300, 2006.

7. Mas, A., Blanco, E., Moñux, G., Urcelay, E., Serrano, F. J., de la Concha, E. G., and Martínez, A. DRB1-TNF-alpha-TNF-beta haplotype is strongly associated with severe aortoiliac occlusive disease, a clinical form of atherosclerosis. *Human Immunology*, 66 (10): 1062–1067, 2005.

8. Bonnen, P. E., Wang, P. J., Kimmel, M., Chakraborty, R., and Nelson, D. L. Haplotype and linkage disequilibrium architecture for human cancer-associated genes. *Genome Research*, 12, 1846–1853, 2002.

9. Fallin, D., Cohen, A., Essioux, L., Chumakov, I., Blumenfeld, M., Cohen, D., and Schork, N. J. Genetic analysis of case/control data using estimated haplotype frequencies: application to APOE locus variation and Alzheimer's disease. *Genome Research*, 11: 143–151, 2001.

10. Hugot, J. P., Chamaillard, M., Zouali, H., Lesage, S., Cézard, J. P., Belaiche, J., Almer, S., Tysk, C., O'Morain, C. A., Gassull, M., Binder, V., Finkel, Y., Cortot, A., Modigliani, R., Laurent-Puig, P., Gower-Rousseau, C., Macry, J., Colombel, J. F., Sahbatou M., and Thomas, G. Association of NOD2 leucine-rich repeat variants with susceptibility to Crohn's disease. *Nature*, 411(6837): 599–603, 2001.

11. Zhao, H., Pfeiffer, R., and Gail, M. H. Haplotype analysis in population genetics and association studies. *Pharmacogenomics*, 4(2): 171–178, 2003.

12. Shastry, B. S. SNPs and haplotypes: genetic markers for disease and drug response (review). *International Journal of Molecular Medicine*, 11: 379–382, 2003.

13. Crawford, D. C. and Nickerson, D. A. Definition and clinical importance of haplotypes. *Annual Reviews of Medicine*, 56: 303–320, 2005.

14. Jorde, L. B. Linkage disequilibrium and the search for complex disease genes. *Genome Research*, 10: 1435–1444, 2000.

15. Devlin, B. and Risch, N. A comparison of linkage disequilibrium measures for fine scale mapping. *Genomics*, 29: 311–322, 1995.

16. Patil, N., Berno, A. J., Hinds, D. A., Barrett, W. A., Doshi, J. M., Hacker, C. R., Kautzer, C. R., Lee, D. H., Marjoribanks, C., McDonough, D. P., Nguyen, B. T., Norris, M. C., Sheehan, J. B., Shen, N., Stern, D., Stokowski, R. P., Thomas, D. J., Trulson, M. O., Vyas, K. R., Frazer, K. A., Fodor, S. P., and Cox, D. R. Blocks of limited haplotype diversity revealed by high resolution scanning of human chromosome 21. *Science*, 294: 1719–1722, 2001.

17. Daly, M. J., Rioux, J. D., Schaffner, S. F., Hudson, T. J., Lander, E. S. High-resolution haplotype structure in the human genome. *Nature Genetics*, 29(2): 229–232, 2001.

18. Gabriel, S. B., Schaffner, S. F., Nguyen, H., Moore, J. M., Roy, J., Blumenstiel, B., Higgins, J., DeFelice, M., Lochner, A., Faggart, M., Liu-Cordero, S. N., Rotimi, C., Adeyemo, A., Cooper, R., Ward, R., Lander, E. S., Daly, M. J., and Altshuler, D. The structure of haplotype blocks in the human genome. *Science*, 296: 2225–2229, 2002.

19. Schulze, T. G., Zhang, K., Chen, Y., Akula, N., Sun, F., and McMahon, F. J. Defining haplotype blocks and tag single-nucleotide polymorphisms in the human genome. *Human Molecular Genetics*, 13(3): 335–342, 2004.

20. Ding, K., Zhou, K., Zhang, J., Knight, J., Zhang, X., and Shen, Y. The effect of haplotype-block definitions on inference of haplotype-block structure and htSNPs selection. *Molecular Biology and Evolution*, 22(1): 148–159, 2005.

21. Zhang, K., Calabrese, P., Nordborg, M., and Sun, F. Haplotype block structure and its application to association studies: power and study designs. *American Journal of Human Genetics*, 71: 1386–1394, 2002.

22. Akey, J., Jin, L., and Xiong, M. Haplotypes vs single marker linkage disequilibrium tests: what do we gain? *European Journal of Human Genetics*, 9: 291–300, 2001.

23. Nothnagel, M. The definition of multilocus haplotype blocks and common diseases. PhD thesis, University of Berlin, 2004.

24. Gusfield, D. and Orzack, S. H. Haplotype inference. *CRC Handbook in Bioinformatics*, CRC Press, Boca Raton, FL, pp. 1–25, 2005.

25. Dempster, A. P., Laird, N. M., and Rubin, D. B. Maximum likelihood from incomplete data via the EM algorithm. *Journal of the Royal Statistical Society*, 39(1): 1–38, 1977.

26. Hawley, M. E., Pakstis, A. J., and Kidd, K. K. A computer program implementing the EM algorithm for haplotype frequency estimation. *American Journal of physiological Anthropololy*, 18: 104, 1994.

27. Hawley, M. and Kidd, K. HAPLO: a program using the EM algorithm to estimate the frequencies of multi-site haplotypes. *Journal of Heredity*, 86: 409–411, 1995.

28. Long, J. C., Williams, R. C., and Urbanek, M. An EM algorithm and testing strategy for multiple locus haplotypes. *American Journal of Human Genetics*, 56(3): 799–810, 1995.

29. Excoffier, L. and Slatkin, M. Maximum-likelihood estimation of molecular haplotype frequencies in a diploid population. *Molecular Biology of Evolution*, 12: 921–927, 1995.

30. Fallin, D. and Schork, N. J. Accuracy of haplotype frequency estimation for biallelic loci, via the expectation-maximization algorithm for unphased diploid genotype data. *American Journal of Human Genetics*, 67: 947–959, 2000.

31. Tishkoff, S. A., Pakstis, A. J., Ruano, G., and Kidd, K. K. The accuarcy of statistical methods for estimation of haplotype frequencies: an example from the CD4 locus. *American Journal of Human Genetics*, 67: 518–522, 2000.

32. Kelly, E. D., Sievers, F., and McManus, R. Haplotype frequency estimation error analysis in the presence of missing genotype data. *BMC Bioinformatics*, 5(188): 1–13, 2004.

33. Beaumont, M. A. and Rannala, B. The Bayesian revolution in genetics. *Nature Reviews: Genetics*, 5: 251–260, 2004.

34. Stephens, M., Smith, N. J., and Donnelly, P. A new statistical method for haplotype reconstruction from population data. *American Journal of Human Genetics*, 68(4): 978–989, 2001.

35. Lin, S., Culter, D., Zwick, M., and Chakravarti, A. Haplotype inference in random population samples. *American Journal of Human Genetics*, 71: 1129–1137, 2002.

36. Niu, T., Qin, Z. S., Xu, X., and Liu, J. S. Bayesian haplotype inference for multiple linked single-nucleotide polymorphisms. *American Journal of Human Genetics*, 70: 157–169, 2002.

37. Xing, E. P., Sharan, R., and Jordan, M. I. Bayesian haplotype inference via the Dirichlet process. *Proceedings of the 21st International Conference on Machine Learning*, pp. 879–886, 2004,

38. Zhang, S., Pakstis, A. J., Kidd, K. K., and Zhao, H. Comparison of two methods for haplotype reconstruction and haplotype frequency estimation from population data. *American Journal of Human Genetics*, 69: 906–914, 2001.

39. Xu, C. F., Lewis, K., Cantone, K. L., Khan, P., Donnelly, C., White, N., Crocker, N., Boyed, P. R., Zaykin, D. V., and Purvis, I. J. Effectiveness of computational methods in haplotype prediction. *Human Genetics*, 110: 148–156, 2002.

40. Adkins, R. M. Comparison of the accuracy of methods of computational haplotype inference using a large empirical dataset. *BMC Genetics*, 5(22): 1–7, 2004.

41. Lee, H., Cho, H., and Song, H. Comparison of EM algorithm and Phase for haplotype frequency estimation with diverse accuracy measures. *Proceedings of the Spring Conference*, Korea Statistical Society, pp. 229–234, 2004.

42. Stephens, M. and Donnelly, P. A comparison of Bayesian methods for haplotype reconstruction. *American Journal of Human Genetics*, 73(5): 1162–1169, 2003.

43. Halldörsson, B. V., Bafna, V., Edwards, N., and Lippert, R. SNPs and Haplotype Inference. *A Survey of Computational Methods for Determining Haplotypes*, Springer-Verlag, Berlin, Heidelberg, pp. 26–47, 2004.

44. Olund, G. Analysis and implementation of statistical algorithms capable of estimating haplotypes in phase-unknown genotype data. Master's thesis, Kungliga Tekniska Hogskolan, 2004.

45. Qin, Z. S., Niu, T., and Liu, J. S. Partition-ligation-expectation-maximization algorithm for haplotype inference with single-nucleotide polymorphisms. *American Journal of Human Genetics*, 71: 1242–1247, 2002.

46. Kirk, K. M. and Cardon, L. R. The impact of genotyping error on haplotype reconstruction and frequency estimation. *European Journal of Human Genetics*, 10: 616–622, 2002.

47. Gusfield, D. Inference of haplotypes from samples of diploid populations: complexity and algorithms. *Journal of Computational Biology*, 8(3): 305–323, 2001.

48. Lee, P. H. Computational haplotype analysis: an overview of computational methods in genetic variation study. Technical Report 2006-512, Queen's University, 2006.

49. Byng, M. C., Whittaker, J. C., Cuthbert, A. P., Mathew, C. G., and Lewis, C. M. SNP subset selection for genetic association studies. *Annals of Human Genetics*, 67: 543–556, 2003.

50. Wu, X., Luke, A., Rieder, M., Lee, K., Toth, E. J., Nickerson, D., Zhu, X., Kan, D., and Cooper, R. S. An association study of angiotensiongen polymorphisms with serum level and hypertension in an African-American population. *Journal of Hypertension*, 21(10): 1847–1852, 2003.

51. Carlson, C. S., Eberle, M. A., Rieder, M. J., Yi, Q., Kruglyak, L., and Nickerson, D. A. Selecting a maximally informative set of single-nucleotide polymorphisms for association analyses using linkage disequilibrium. *American Journal of Human Genetics*, 74: 106–120, 2004.

52. Ao, S. I., Yip, K., Ng, M., Cheung, D., Fong, P., Melhado, I., and Sham, P. C. CLUSTAG: Hierarchical clustering and graph methods for selecting tag SNPs. *Bioinformatics*, 21: 1735–1736, 2005.

53. Bafna, V., Halldörsson, B. V., Schwartz, R., Clark, A. G., and Istrail, S. Haplotypes and informative SNP selection algorithms: don't block out information. *Proceedings of the Seventh International Conference on Computational Molecular Biology*, pp. 19–26, 2003.

54. Ke, X., Miretti, M. M., Broxholme, J., Hunt, S., Beck, S., Bentley, D. R., Deloukas, P., and Cardon, L. R. A comparison of tagging methods and their tagging space. *Human Molecular Genetics*, 14(18): 2757–2767, 2005.

55. Burkett, K. M., Ghadessi, M., McNeney, B., Graham, J., and Daley, D. A comparison of five methods for selecting tagging single-nucleotide polymorphisms. *BMC Genetics*, 6 (Suppl1): S71, 2005.

56. Goldstein, D. B., Ahmadi, K. R., Weale, M. E., and Wood, N. W. Genome scans and candidate gene approaches in the study of common diseases and variable drug responses. *Trends in Genetics*, 19(11): 615–622, 2003.

57. Halperin, E., Kimmel, G., and Sharmir, R. Tag SNP selection in genotype data for maximizing SNP prediction accuracy. *Bioinformatics*, 21(Suppl 1): i195–i203, 2005.

58. Lee, P. H. and Shatkay, H. BNTagger: improved tagging SNP selection using Bayesian networks. *The 14th Annual International Conference on Intelligent Systems for Molecular Biology (ISMB)*, 2006.

59. Jensen, F. *Bayesian Networks and Decision Graphs*, Springer-Verlag, New York, 1997.

60. De Bakker, P. I. W., Graham, R. R., Altshuler, D., Henderson, B. E., and Haiman, C. A. Transferability of tag SNPs to capture common genetic variation in DNA repair genes across multiple population. *Proceedings of the Pacific Symposium on Biocomputing*, 2006.

61. Ke, X., Durrant, C., Morris, A. P., Hunt, S., Bentley, D. R., Deloukas, P., and Cardon, L. R. Efficiency and consistency of haplotype tagging of dense SNP maps in multiple samples. *Human Molecular Genetics*, 13(21): 2557–2565, 2004.

62. Pagano, M. and Gauvreau, K. *Principles of Biostatistics*, 2nd edition, Duxbury Thomson Learning, Boston, 2000.

63. Pritchard, J. K. and Cox, N. J. The allelic architecture of human disease genes: common disease-common variant. . . or not? *Human Molecular Genetics*, 11(20): 2417–2423, 2002.

64. Lu, X., Niu, T., and Liu, J. S. Haplotype information and linkage disequilibrium mapping for single nucleotide polymorphisms. *Genome Research*, 13: 2112–2117, 2003.

65. Duggal, P., Gillanders, E. M., Mathias, R. A., Ibay, G. P., Klein, A. P., Baffoe-Bonnie, A. B., Ou, L., Dusenberry, I. P., Tsai, Y. Y., Chines, P. S., Doan, B. Q., and Bailey-Wilson, J. E. Identification of tag single-nucleotide polymorphisms in regions with varying linkage disequilibrium. *BMC Genetics*, 6(Suppl 1): S73, 2005.

18

MACHINE LEARNING APPLICATIONS IN SNP–DISEASE ASSOCIATION STUDY

Pritam Chanda, Aidong Zhang, and Murali Ramanathan

There are several biological domains where machine learning techniques are applied for knowledge extraction from data. These can be classified into six different domains: genomics, proteomics, microarrays, systems biology, evolution, and text mining [1]. In this chapter, we start with the description of the most common form of human genetic variation called single nucleotide polymorphism (abbreviated as SNP and pronounced as snip) and present several machine learning methods to explore the association of human genetic variations and complex diseases.

Genomic DNA consists of two complementary strands, each being a chain of subunits called nucleotides. Each nucleotide consists three parts: deoxyribose (a sugar), a phosphate group, and a base that is made of chemical rings containing nitrogen compounds. There are four types of nucleotides based on four kinds of bases: adenine, thymine, guanine and cytosine (commonly abbreviated A, T, G, and C). Human DNA from two different individuals is almost similar except small variations in these nucleotides, 90% of which is single nucleotide differences between the pairs of homologous chromosomes. Each such variation in the nucleotide sequence of the template strand of the DNA is called SNP. Within a population, the proportion of the chromosomes in the population carrying the lesser common variant is termed minor allele frequency (abbreviated MAF) and each such variation or SNP is assigned an

Machine Learning in Bioinformatics. Edited by Yan-Qing Zhang and Jagath C. Rajapakse
Copyright © 2009 by John Wiley & Sons, Inc.

MAF. For the purpose of this chapter, SNPs are single base pair positions in genomic DNA at which different sequence alternatives (alleles) exist among individuals in some population(s) [2], wherein the least frequent allele has an abundance of 1% or greater. Although theoretically a nucleotide can take four different values, practically most SNPs are biallelic, that is, only two of the four possible DNA bases (A, T, G, C) are seen; multiple base variations at a single SNP site are usually rare.

SNPs represent the most frequent form of polymorphism in the human genome. The amount of nucleotide diversity in the human genome is estimated to fall in between 3.7×10^{-4} and 8.3×10^{-4} differences per base pair [3–6]. From these and other studies, it has been estimated that SNPs with MAF $> 10\%$ occur once every ~600 base pairs [7]. SNPs may occur within coding sequences of genes, noncoding regions of genes, or in the regions between genes. SNPs within a coding sequence will not necessarily change the amino acid sequence of the protein that is produced. Due to redundancy in the genetic code, some SNPs do not alter the amino acid sequence of the protein produced. An SNP in which both alleles lead to the same protein sequence is termed synonymous—if different proteins are produced they are nonsynonymous. SNPs that are not in protein coding regions may still have consequences for gene splicing, transcription factor binding, or the sequence of noncoding RNA. SNPs in humans can affect how humans develop diseases and respond to pathogens, chemicals, drugs, and so on. Because SNPs are inherited and do not change much from generation to generation in an individual with time, SNPs are of great value to biomedical research and in developing diagnostic and pharmaceutical products.

After human genome was sequenced, there has been a rapid increase in the amount of SNP data and the number of SNP databases, which has been driven in part by the International HapMap project that began in October 2002. The Phase 1 data have been published and analysis of Phase 2 data is underway as of October 2006. The aim of the project is to record the significant SNPs. These efforts provide the scientific community access to a vast amount of human genome sequence and SNP data, which can provide valuable insight into the root cause behind several genetic and hereditary diseases and can have significant impact upon population genetics, forensics, and genetic disease research and drug development. In the dBSNP database Build 126, there are over two million human reference SNPs documented.

In a disease association study, the objective is to locate genetic factors that are responsible for or associated with the disease. Intensive research to map the genes involved in a disease to the heritable quantitative traits (biological traits that exhibit continuous variations, e.g., phenotypic variations like height) that identify the risk factors for diseases is underway. The mapping approaches can be broadly categorized into genome-wide association studies and candidate-gene-based approaches that use either association or resequencing methods [8]. Genome-wide linkage analysis has been traditionally applied to identify genetic causes behind the monogenic Mendelian diseases since the genetic markers that flank a diseased gene segregate with the disease in families and are also highly penetrant in a population showing the disease phenotype. The correlation between the genetic markers and diseased markers is measured using linkage disequilibrium (LD). It also refers to association between tightly linked SNPs. Alleles in different loci are sometimes found together more or less often than expected based on their frequency of occurrence. Two SNPs are said to be in LD if their alleles are in statistical association. Several popular statistics exist for describing LD; the two most commonly used are D' and r^2 (also referred to as δ^2) [9]. The maximum possible value of LD is 1, which would indicate strong allelic association with no recombination between the sites and no mutation in each site.

In this chapter, we describe the application of some machine learning methods to the disease–SNP association study. We focus on two aspects, namely, tag SNP selection, or selectively choosing some SNPs from a given set of possibly thousands of markers as representatives of the remaining markers (that are not chosen), and machine learning models for detecting markers that have potentially high association with the given disease phenotype(s). Often tag SNP selection is the first step in narrowing down a set of markers that are then used for further analysis.

18.1 TAG SNP SELECTION

Diploid organisms, such as human beings, possess two identical copies of each chromosome, called homologous chromosomes. The set of SNPs on a single chromosome of a pair of homologous chromosomes is referred to as a haplotype. Thus, for a given set of SNPs, an individual possesses two haplotypes, one from each chromosome copy. An SNP where both haplotypes have the same variant (allele) is called a homozygous site (e.g., AA); an SNP site where the haplotypes have different variants (alleles) is called a heterozygous site (e.g., Aa). The data from the two haplotypes taken together are referred to as a genotype. SNPs are inherited from one generation to another in blocks such that each block contains a few common haplotypes and the SNPs in the block are in LD. Because of LD, each block contains a minimal informative set of SNPs that can represent the rest of the SNPs with high accuracy and also can identify all the haplotypes of the block. Discrepancies in the haplotype structure of two different individuals can be identified using various statistical tests, which often point to the culprit genomic regions. Linkage analysis has been successfully used by statisticians to locate genes responsible for simpler monogenic Mendelian diseases. But to study the genetic factor for complex diseases where several genes contribute together to a disease phenotype, one needs to study a relatively large number of SNPs (as well as other genetic variations like microsatellites) and also a bigger sample size of individuals. This is clear because the statistical significance of the study is directly affected by the number of SNPs examined and the number of individuals typed. But genotyping a large number of SNPs is also cost-prohibitive; so, it is essential to choose a set of SNPs to be genotyped such that this set predicts the rest of the SNPs (not typed) with high accuracy. This set of SNPs is called the tag SNPs and this is possible because of the phenomena of LD. Thus, while performing a disease association study, the geneticist would experimentally test for association by only considering the tag SNPs and not the entire set, thereby considerably saving resources (alternatively, increasing the power of the statistical tests by increasing the number of individuals) as well as making the problem computationally tractable. Tag SNP selection deals with finding a set of tag SNPs of minimum size that would have very good prediction ability.

The generic tagging problem can be formulated as follows: Given a sample of a population P of individuals (either haplotypes or genotypes) on m SNPs, select k tag SNPs ($k < m$) such that one can predict (or statistically cover) the remaining $m - k$ SNPs of an entire individual (either haplotype or genotype) using the values of only these k tag SNPs. This thus also identifies the haplotypes of the individual completely when selection of tag SNPs is done with haplotype data.

Several algorithms have been proposed for selecting tagging SNPs using combinatorial and machine learning and data mining approaches. A general approach is to define a criterion for the goodness of a selected tag SNP set and try to find the minimum set that maximizes the goodness measure [10, 11]. From a set theoretic point of view, it is computationally intractable to examine all possible subsets of the given set of SNPs to select a set of tag markers, except for very small data sets. To overcome this difficulty, many of the current tag SNP selection methods utilize haplotype features and partition the given chromosomal region into blocks of limited haplotype diversity [12–14], and then search for tag SNPs within each block that predicts the common haplotypes of each block. Such methods are known as block-based methods. But these have various disadvantages, such as lack of cross-block information and the dependency of the tag SNPs chosen on the block definition [15]. Also, there is little consensus on how the block boundaries are defined and inter-block correlations are ignored in favor of intra-block local correlations [16].

Several researchers have applied entropy-based approaches to select tag SNPs and quantify haplotype diversity [10, 17]. These methods aim to select the minimum set of tag SNPs to retain the maximal information content of the given set. Tag SNP selection problem also has been analyzed from the computational complexity point of view. Bafna et al. [18] have proposed to select tag SNPs based on how well one SNP predicts another and how well a set of SNPs predicts another set of SNPs and have showed that the tag SNP selection problem using this measure is NP-complete (nondeterministic polynomial-time problem).

Researchers also have applied a number of machine learning and data mining techniques to this problem domain. These include clustering of SNPs based on some similarity measures (like LD) and then selecting one SNP per cluster [6, 16, 19], application of principal component analysis (PCA) [20], support vector machines (SVM) [21], and regression-based methods [22]. We discuss some of these approaches in the following sections.

18.1.1 FSFS—A Feature Selection Method

We start with formulating the problem of identifying tag SNPs as a simple feature selection problem as in feature selection using feature similarity (FSFS) algorithm [16]. Assume we are given N haploid sequences consisting of m biallelic SNPs, which can be represented as $N \times m$ matrix M with the sequences as rows and SNPs as columns. Element $M(i, j)$ of the matrix represents the jth allele of the ith sequence and takes values 0, 1, and 2 with 1 and 2 representing the two alleles and 0 representing the missing data. Each row is a learning instance belonging to a class and the SNPs represent the attributes that are to be used to identify the class to which the sequence belongs. As a machine learning problem, the idea is to select a subset of SNPs that can classify all the haplotypes with accuracy close to that of when all SNPs are used. Thus, this is a case of unsupervised learning method. The LD measure r^2 is used as a measure of similarity between pairs of features in the FSFS algorithm

$$r^2 = \frac{(p_{AB} \cdot p_{ab} - p_{Ab} \cdot p_{aB})^2}{p_{AB} \cdot p_{ab} \cdot p_{Ab} \cdot p_{aB}}, \quad 0 \le r \le 1, \tag{18.1}$$

where A, a are the two alleles at a particular locus, p_{xy} denotes the frequency of observing alleles x and y together in the same haplotype and p_x denotes the frequency of allele x alone.

FSFS selects the most informative set of SNPs by first grouping them into homogenous subsets and then choosing a representative SNP from each group. Let the set of all SNPs be given by $S = \{F_1, F_2, \ldots, F_N\}$. $D(F_i, F_j)$ represents the dissimilarity between the two SNPs (F_i and F_j) and is calculated using the LD measure (Eq. 18.1). Let R represent the final set of SNPs chosen as the tag SNPs. FSFS takes as input S and another parameter K (the number of nearest neighbors of an SNP to consider), and produces R as the output. First, the algorithm initializes R to S (step 1). During each iteration, FSFS calculates the distance d_i^k between each SNP F_i in R and its kth nearest neighboring SNP (step 2). The algorithm then finds SNP F_0 for which d_0^k is minimum, retains this SNP in R, and removes its K nearest SNPs from R (step 3) since this SNP serves as the representative SNP for the its K-nearest neighbors. Thus, the algorithm always discards SNPs from the most compact cluster causing the minimum information loss. Step 4 compares the cardinality of R with K and adjusts K if necessary. In step 6, FSFS gradually decreases K and recomputes d_0^k until d_0^k is less than or equal to the threshold θ (θ is an error threshold that is set to d_0^k in the first iteration of the algorithm). The algorithm ends when no SNP in R can be removed with error less than or equal to θ. To predict the value of an SNP using the tag SNPs, the most similar tagging SNP is used which is the seed of the corresponding cluster. The parameter K is chosen such that the desired prediction accuracy is achieved by cross-validation. Experimental results using first 1000 SNPs of chromosome 21 data set with leave-one-out cross-validation yields a good prediction accuracy of about 88% with only 100 tag SNPs and more than 90% with about 200 SNPs. A 10-fold cross-validation using the IBD 5q31 data set [23] also achieves a prediction accuracy of about 96% with only 30 tag SNPs. The detailed algorithm [16] is given below:

```
FSFS algorithm for tag SNP selection
  Input
  S ── original set of SNP.
  K ── parameter K of the algorithm.
  Output
  R ── selected tag SNPs.
  1. R⇐S
  2. for each Fᵢ∈R do
        dᵢᴷ ⇐ D(Fᵢ,Fᵢᴷ) where Fᵢᴷ is the Kth nearest SNP of Fᵢ in R
     end for
```

```
3. find F_0 such that d_0^K ⇐ argmin F_{i∈R}(d_i^K)
   let F_0^1, F_0^2, , F_0^K be the K nearest SNPs of F_0
   R ⇐ R - {F_0^1, F_0^2, ... F_0^K}
   if first iteration then set θ = d_0^K
4. if K>|R|-1 then K⇐|R|-1
5. if K=1 goto 8
6. while d_0^K>θ do
      K⇐K-1
      if K=1 goto 8
      recompute d_0^K
   end while
7. goto 2
8. return R
```

The FSFS clustering algorithm does not involve subset search, rather SNPs are removed individually to form tagging sets based on pairwise similarity. Also, global similarities/correlations between SNPs across chromosomes are used to find redundant markers. Both reduces computational complexity and enables FSFS to handle large amount of SNP data. But the set of tag SNP chosen is not optimal although it gives good prediction accuracy with reasonable number of tag SNPs by using both local and long range LD across chromosomes.

18.1.2 MLR—A Regression-Based Method

Multivariate linear regression (MLR) has been successfully applied in determining tag SNPs with good prediction accuracy in Ref. [22]. The advantage of this method is that, unlike the feature selection problem described earlier, this method can work with both haplotype and genotype data. The algorithm represents a genotype numerically by the values 0, 1, or 2, where 0 represents a homozygous site with major allele, 1 represents a homozygous site with minor allele, and 2 represents a heterozygous site. Each SNP in a haplotype is represented with 0 (major allele) and 1 (minor allele). As an SNP prediction problem, consider a $(n + 1) \times (k + 1)$ matrix M corresponding to n sample individuals and the individual x and k tag SNPs (assume these are already known for prediction purpose) and a single non-tag SNP s (whose value the tag SNPs will predict). All SNP values in M are known except the value of s in x. In the case of haplotypes, there are only two possible resolutions of s, s_0 (for SNP value 0), and s_1 (for SNP value 1). For genotypes, there are three possible resolutions s_0 (SNP value 0), s_1 (SNP value 1), and s_2 (SNP value 2). The SNP prediction method should predict correct resolution of s.

In the SNP prediction scheme (see Fig. 18.1), the multivariate regression method considers set of tag SNPs T as vectors in the $(n + 1)$-dimensional Euclidean space and considers the projections of the vectors s_0, s_1, and s_2 onto the span of the set of tag SNPs. It assumes that the most probable resolution of s should be closest to the span of T. The distance of s_i from T is measured as $D(T, s_i) = |T(T^tT)^{-1} s_i - s_i|$. The prediction scheme is then used in two different selection algorithms: Stepwise tagging algorithm (STA) and local minimization tag (LMT) selection algorithm. STA starts with selecting the best tag t_0 that alone predicts all other tags with minimum prediction

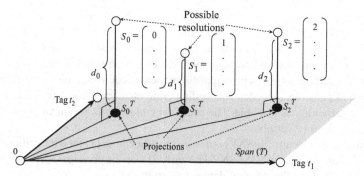

Figure 18.1   MLR SNP prediction algorithm [22]. Three possible resolutions s_0, s_1, and s_2 of s are projected on the span of tag SNPs (the dark plane). The unknown SNP value is predicted as 1 since the distance between s_1 and its projection s_1^T is shorter than the distances between s_0 and s_2 and their respective projections. Reproduced with permission from Springer © 2006 from He et al. 2006.

error, and then in each iteration, continues to add tags to the set T such that T best predicts the remaining tags. Thus, STA follows a greedy approach. LMT improves upon STA by starting with the k tags produced by STA and then iteratively replacing each single tag with the best possible choice while not changing other tags. Such replacements will be continued until no significant improvement occurs in the prediction quality.

Several data sets were used to evaluate the performance of the MLR algorithm: three regions (ENm013, ENr112, and ENr113) from 30 CEPH family trios obtained from HapMap ENCODE Project [24], two gene regions STEAP and TRPM8 from 30 CEPH family trios obtained from HapMap [24], and 616 kb region of human chromosome 5q31 from 129 family trios [23]. Leave-one-out cross-validation is used to evaluate the quality of the solution given by the tag SNP selection and prediction methods. One by one, each individual is removed from the sample. Tag SNPs are selected using only the remaining individuals. The left out individual is reconstructed based on its tag SNPs and the remaining individuals in the sample. The prediction accuracy is the percent of correctly predicted SNPs in all individuals. Both STA and LMT achieves high prediction accuracy, reaching up to 97% accuracy with only 22 tag SNPs out of 361 SNPs in the ENm013 data set. For the remaining data sets, a prediction accuracy of 96% is achieved using on an average 20% of the available SNPs as tag SNPs.

MLR tagging method gives good prediction accuracy with few chosen tag SNPs and it can handle both haplotype and genotype data as input, but it does not perform well with missing data in the input genotype/haplotype and does not scale well with large amount of SNP data when the tagging matrix becomes too large.

18.1.3   Using Principal Component Analysis

Many of the current tag SNP selection methods also aim to identify all the haplotypes in each block by partitioning a chromosomal region into blocks of limited diversity and find a set of tag SNPs that predicts the common haplotypes of each block. This set is called the haplotype tagging SNPs (htSNPs). Lin et al. [20] have used PCA to obtain a set of htSNPs. PCA is a dimensionality reduction method that projects high-dimensional data to lower dimensions for analysis. The set of correlated variables in the data is transformed into a smaller set of uncorrelated variables (principal components) such that the first principal component accounts for the greatest variation in the data as possible, and each succeeding component accounts for as much of the remaining variability as possible. The principal components are linear combinations of the original set of variables.

The problem of locating htSNPs is a two-step procedure: (1) Find the principal SNP components using PCA from the SNP data; and (2) map the principal SNP components to the closest SNP in the original data set as an htSNP.

Let X be the SNP binary matrix with each row corresponding to a chromosome and each column representing an SNP from the chromosome. Each X_{ij} contains values 0 (common allele) and 1 (rare allele). Let μ be the p-dimensional mean vector, R be the $p \times p$ data correlation matrix, $E = \{e_1, \ldots, e_p\}$ be the eigenvectors and $\Lambda = \{\lambda_1, \ldots, \lambda_p\}$ be the set of eigenvalues. E and Λ can be obtained by solving the equation

$$Re_j = \lambda_j e_j, \quad j = 1, \ldots, p. \tag{18.2}$$

The eigenvectors are sorted according to the decreasing value of the eigenvalues and most informative top k eigenvectors are selected. Correspondingly, k eigenSNPs are defined as a linear combination of the p eigenvectors

$$s_i = \sum_{j=1}^{p} e_{ij} x_j, i = 1, \ldots, k, \tag{18.3}$$

where x_j are the weights (or coefficients) of the eigenvectors.

The eigenSNPs are mathematical abstractions and do not represent any measurable quantities. The top k eigenvectors are chosen so that the ratio

$$\frac{\sum_{i=1}^{k} \lambda_i}{\sum_{i=1}^{p} \lambda_i} \geq 0.9,$$

that is, just enough components are included to explain 90% of the total variance.

In the second step of the process, the above eigenSNPs are mapped to the closest SNP in the original data set using two methods:

(1) *Greedy-Discard Method.* Eigenvectors with small eigenvalues are of lesser importance since they do not capture most of the data variance. So, the idea is to discard SNPs correlated with eigenvectors with smaller eigenvalues. Starting from the eigenvector with the smallest eigenvalue to the one with $(p - k)$th smallest eigenvalue, discard SNP that has the largest coefficient in the $(p - k)$th eigenSNP and the remaining k eigenSNPs are mapped in reverse order to the remaining data SNPs as the htSNPs.

(2) *Varimax-Rotation Method.* Pairs of axes defined by PCA are rotated iteratively such that the rotated solution spans the same geometric space and explains the same amount of variance as before, but simplifies the SNP–eigenSNP relationship, and in the optimal case, each SNP has a high coefficient with only one rotated eigenSNP.

Let T be an orthogonal transformation that generates the rotated eigenvector $E^\gamma = \{e_1^\gamma, \ldots, e_p^\gamma\}$. For each SNP i, compute the average weight for all the k rotated eigenSNPs Γ_i and the average weight for the remaining $(p - k)$ eigenSNPs γ_i as follows:

$$\Gamma_i = \frac{1}{k}\sum_{j=1}^{k}|e_{ij}^\gamma|, \quad i = 1, \ldots, n,$$

$$\gamma_i = \frac{1}{p-k}\sum_{j=k+1}^{p}|e_{ij}^\gamma|, \quad i = 1, \ldots, n. \tag{18.4}$$

Select SNP i as an htSNP if $\Gamma_i > \gamma_i$ as then this SNP contributes most significantly to the k eigenSNPs.

The chosen htSNPs should be able to capture haplotype diversity of the genomic region analyzed and therefore should be able to recover the genotypes of the remaining non-htSNP. The efficiency of the above algorithm depends on how accurately the genotypes of the non-htSNPs are reconstructed using the htSNPs. The data set is partitioned into training and test sets, the training data set is used to select the htSNPs and they are used to predict the genotypes of the remaining SNPs from the test data set. The prediction of a non-htSNP genotype depends on how well it correlates with each htSNP genotype in the training set. The htSNP genotype that has the greatest correlation with the non-htSNP genotype determines the value of the non-htSNP genotype. When multiple htSNPs have the same correlation coefficient with a non-htSNP but their predicted genotypes are in conflict, the genotype of the non-htSNP is assigned its common allele as the predicted value. The prediction accuracy is evaluated as

$$\frac{\text{number of correctly predicted}}{\text{all predictions}}. \tag{18.5}$$

Three published experimental SNP data sets are used for the evaluation, and unphased data are preprocessed using PHASE haplotype inference program [25] to obtain the haplotypes. The three data sets are Angiotensin I Converting Enzyme (ACE) data set from a study of 78 SNPs typed on DCP1 [26] in 11 individuals, ABCB1 data set [27] with 48 SNPs typed in 247 individuals, IBD 5q31 data set [23] with 103 SNPs typed in 387 subjects. For each method, the htSNPs explains 90% of the data variance and individually, the greedy-discard method identified 5 htSNPs in the ACE data set, 18 in the ABCB1 data set, and 15 in the IBD data set. The performance of both greedy-discard and varimax-rotation algorithms when compared with previously designed htStep method [28] and PCA with sliding window approach [29] selected the smallest set of htSNPs beyond 80% variance-explained cutoff. The greedy-discard and varimax-rotation methods chose the smallest set of htSNPs to achieve a prediction precision of 90% when compared with Ref. [28] and sliding-window-based PCA.

18.1.4 An Entropy-Based Approach

Finally, we describe an entropy-based approach [17] in choosing tag SNPs to maximize the chance of detecting disease association with the lowest possible genotyping effort. The entropy-based SNP selection method integrates physical distance, haplotype structure, and marker information content in a single parameter (named, mapping utility). A disease causing SNP can be located anywhere in a given genomic interval so that an optimal SNP selection strategy should aim at uniformly maximizing the mapping information over the whole interval.

Before we start with the tag SNP selection method, we introduce the common information theory concepts. Entropy is defined as the amount of uncertainty about an event that follows a given probability distribution and was introduced by Shanon [30]. In terms of genetic diversity and association, consider a marker locus X with k alleles each with frequency of occurrence p_i ($i = 1, \ldots, k$). X can be thought of as a random variable taking on different values (alleles) and its entropy $H(X)$ is given by

$$H(X) = -\sum_{i=1}^{k} p_i \log_2(p_i). \tag{18.6}$$

The joint entropy of two such loci X and Y is given by $H(X, Y) = H(X|Y) + H(Y) = H(Y|X) + H(X)$. The mutual information between two random variables I measures the mutual dependence of the two variables. The information between X and Y is given by $I(X; Y) = H(X) + H(Y) - H(X, Y) = H(X) - H(X|Y) = H(Y) - H(Y|X)$.

In the application to tag SNP selection for disease association mapping, assume that the set of markers $\bar{X} = \{X_1, X_2, \ldots, X_n\}$ are already known markers covering a genetic region. The problem is now to choose an additional set of markers $\bar{Y} = \{Y_1, Y_2, \ldots, Y_m\}$ so that $\bar{X}$ and $\bar{Y}$ together maximize the mapping capability.

Consider a marker $Y \in \bar{Y}$. Let Z be the disease locus (yet unknown). In terms of entropy, choosing the best Y would be to maximize the increase in mutual information $I(Z|Y, X_1, \ldots, X_n) - I(Z|X_1, \ldots, X_n)$. But the computation of this requires frequency estimation for haplotype frequencies and its practicality is limited to small marker sets. The gain in mutual information about Z when Y is considered depends upon the association between Y and the other markers, which, in turn, mainly depends upon the pairwise association between Y and X_i. So, Y is chosen by maximizing the minimum information gained using pairwise $Y - X_j$ haplotypes,

$$Y_{\max} = \max_{j=1,\ldots,m} \min_{i=1,\ldots,n} I(Z|Y_j, X_i) - I(Z|X_i). \tag{18.7}$$

But the above quantity cannot be calculated explicitly since Z is unknown. So, a mapping utility score $\kappa(Y: X, z)$ is constructed as follows:

Information gain about Z when Y is considered fulfils

$$I(Z|Y, X) = I(Z, X|Y) - I(Y|X) \geq I(Z|Y) - I(Y|X). \qquad (18.8)$$

The choice of Y should maximize $I(Z|Y)$ and minimize $I(Y|X)$ to maximize the information gain. The first quantity is a simple function of association probability ρ [31] given by

$$I(Z|Y) = \frac{(1+\rho)\log_2(1+\rho) + (1-\rho)\log_2(1-\rho)}{2}, \qquad (18.9)$$

which can be approximated under a Malecot model for decay of LD [31] as

$$I(Z|Y) \approx \rho_e = e^{-\epsilon\epsilon|y-z|}, \qquad (18.10)$$

where ϵ is a region-specific constant and is estimated in the Malecot model of LD decay by means of log-linear regression between LD (ρ) and genetic distance d (in cM) as ln $(\rho) = -\epsilon d + \text{constant}$

And, $I(Y|X)$ is inversely proportional to $H(X|Y)$. Thus, κ is proposed as

$$\kappa(Y : X, z) = e^{-\epsilon\epsilon|y-z|} \cdot H(Y|X). \qquad (18.11)$$

The maximum of κ is unity when $y = z$ and $H(Y|X) = H(Y) = 1$ (marker Y is located at locus Z, has two equally frequent alleles, and is in perfect equilibrium with X). The minimum is 0, which occurs when either $|y - z| = \infty$ (i.e., Y and Z are unlinked) or $H(X|Y) = 0$ (Y is in complete disequilibrium with X).

Using Equation 18.6 we have,

$$Y_{\max} = \max_{j=1,\ldots,m} \min_{i=1,\ldots,n} \kappa(Y_j : X_i, z). \qquad (18.12)$$

To decide whether the selected marker $Y_{\max}$ is in the final marker set to be analyzed in the laboratory, a threshold α is proposed such that $Y_{\max}$ is tested if

$$\kappa_{\min}(z) = \min_{i=1,\ldots,n} \kappa(Y_{\max} : X_i, z) \geq \alpha, \qquad (18.13)$$

where $\kappa_{\min}(z)$ can be estimated as

$$\kappa_{\min} = \min_{i=1,\ldots,n}[1 - e^{-\epsilon\epsilon|x_i - z|} \cdot H(X_i)] \qquad (18.14)$$

$$= 1 - \max_{i=1,\ldots,n} e^{-\epsilon\epsilon|x_i - z|} \cdot H(X_i). \qquad (18.15)$$

For evaluation purposes, a global mapping utility κ_G is defined as

$$\kappa_G = H(X_1, \ldots, X_n) \cdot \sum_{n+1}^{i=1} \left(1 - e^{-\epsilon\frac{x_i - x_{i-1}}{2}}\right), \qquad (18.16)$$

which is the global entropy of a marker set $\{X_1, \ldots, X_n\}$ scaled by a factor proportional to the average utility of adjacent markers to map a disease marker halfway between them. Four genomic regions (6p21.31, 6p21.33-6p21.21, 16p11.2, and 16q22) containing 136 SNPs were evaluated resulting in 72,000 genotypes. The mapping utility method is compared with three other methods: global utility method, distance method that uses a uniform distribution of the markers across a genomic region, and rho method that uses simple LD criterion to choose a marker SNP. The mapping utility method consistently performed at par with the global utility method, while rho performed the worst.

Traditional combinatorial approaches are block-based and the selected tag SNPs may not achieve fine mapping within haplotype blocks and may be too far away from the recombination breakpoints within the blocks to resolve the disease marker and other markers in the vicinity. In such cases, the entropy-based utility score κ allows efficient marker selection.

18.2 SNP-BASED DISEASE ASSOCIATION STUDY

Recently, there has been increasing research to find genetic markers, haplotypes, and potentially other variables that together contribute to a disease and serve as good predictors of the observable disease phenotypes. Complex diseases are typically associated with multiple genetic loci and several external (e.g., environmental) factors. Mendelian diseases are most commonly monogenic and simpler linkage analysis suffices to locate genetic marker associated with such diseases. But for more complex polygenic diseases, it is essential to investigate all polymorphisms located in the functional regions of candidate genes [32–34], and integrate the information about the network of genes involved in biological systems of major physiological importance, such as lipid metabolism, cellular adhesion, inflammation, and others [35]. Also, external factors such as environmental factors, smoking habits, and alcohol consumption are important parameters of such models. In this section, we shall see several different approaches using a variety of machine learning methods for detecting the association between a disease phenotype and the causative genetic markers.

18.2.1 DICE Algorithm

The DICE algorithm [35] provides an automated machine learning method for detection of informative combined effects among several genetic polymorphisms and nongenetic factors that serve to predict disease phenotypes (either individually or in combination). The relationship between the disease phenotype and the variables (genetic and nongenetic) is modeled using a logistic (binary outcome), linear (quantitative trait), or Cox (censored response) regression model. The algorithm explores by a forward procedure a set of competing models for which an information criterion (IC) is derived and selects the most parsimonious and informative approximating models(s) that minimizes IC within each step and various steps explored. DICE is limited to detecting at most three-locus marker combination association with a phenotype, but extension to considering combinations of an increasing number of markers is possible. In the first step (step 0), the DICE algorithm calculates the IC value associated with the model, including the intercept and possibly variables forced into the model. In the next step (step 1), the competing models are obtained by adding individual markers to the step-0 model and the IC values for these competing models are calculated. If a competing model satisfies a certain composite condition (given subsequently), DICE remembers the particular model and continues to step 2. If the composite condition is not satisfied for any of the step-2 models, DICE repeats step 1 by adding pairs of markers to the variables of model 0. If the composite condition is still not satisfied, DICE continues to explore in the same way all the three marker combinations. DICE stops exploring after considering the three marker combinations. In this case, the algorithm has detected no one-, two- or three-locus combinations associated with the phenotype. If the composite condition has been satisfied at step 1, the algorithm goes to step 2 and replaces model 0 with the best model retained at step 1. The procedure continues iteratively until there is no more improvement of the IC value (see Fig. 18.2).

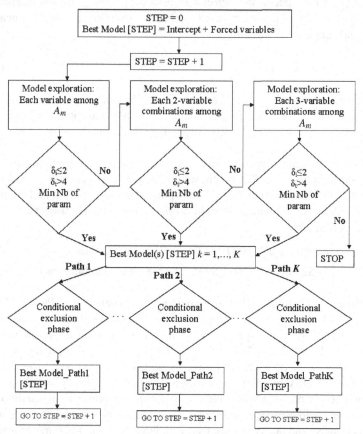

Figure 18.2 Diagram summarizing the steps of the DICE algorithm. A_m denotes the set of remaining variables not included in the model [STEP-1]. When more than one model is chosen at any step, the algorithm proceeds in parallel paths 1, ..., K [35]. Reproduced with permission from Cold Spring Harbor Laboratory Press and Genome Research © 2003, from Tahri-Daizadeh et al. 2003.

The composite condition to be satisfied is implemented using the information criterion AIC_C [36] that uses Akaike's information criterion AIC [37],

$$AIC_C = AIC + \frac{2K(K+1)}{n-K-1}, \qquad (18.17)$$

where $AIC = -2 \, ln[L(\theta|data)] + 2\,K$ represents the expected relative Kullback–Leibler (K–L) divergence, $ln[L(\theta|data)]$ is the value of the maximized log-likelihood over the unknown parameter θ, K is the number of parameters estimated in the approximating model, and n is the total sample size. Since AIC_C is on a relative scale, for each candidate model, AIC_C is rescaled to

$$\delta_i = AIC_C - minAIC_C, \qquad (18.18)$$

where $minAIC_C$ is the minimum AIC_C in that step among all the candidate models. At each step, DICE calculates δ_i for each candidate model i, and selects those that have both $\delta_i \leq 2$ and a "substantial decrease" in AIC_C relative to the previous step. The measure of "substantial decrease" is given by

$$\delta_t = AIC_{C(t-1)} - AIC_{C(t)} > 4, \quad t = 1, \ldots, S, \qquad (18.19)$$

where S is the total number of steps explored so far. In case several competing models fulfill $\delta_i \leq 2$ and $\delta_t > 4$, DICE uses principle of parsimony to choose the models with least number of parameters and evolves in parallel when more than one model is chosen at any step.

The method is applied to several real data samples. Association of several known polymorphisms on SELP gene and myocardial infarction (MI) was investigated with DICE to consider five polymorphisms in the 5′ region and eight in the coding region, as well as the country of origin as one of the parameters. The output of DICE is a model:

$$country + T751P*T741T \quad \text{(two-locus combination)}$$
$$+ S290N*N562D*V599L \quad \text{(three-locus combination)}.$$

Similarly, an application of DICE to association study between SNPs on CETP gene and HDL-cholesterol concentration, including alcohol consumption as an environmental variable, explored 10 previously identified polymorphisms and identified the combinations [alcohol, C-629A] and [alcohol, C+8/in7T] as critical models.

DICE being based on IC avoids problems related to null hypothesis testing theory [38], and can identify complex relationships between variables by alternating exploration and conditional exclusion phases, and uses the principle of parsimony and avoids complicated overparameterized models. On the downside, DICE is computationally intensive and does not scale well with large number of polymorphisms as all possible multilocus combinations (and potential environmental factors) cannot be searched exhaustively.

18.2.2 An Application of Neural Network

Next, we describe an interesting application of artificial neural networks to the SNP and multilocus disease phenotype association study. The advantage of using neural networks is that haplotype calculations need not be done as an intermediate step and genotype data can be directly used instead. Neural networks are computer programs, which are so called because they are designed to resemble animal nervous systems. The study of neural networks has been inspired by the observation that biological learning systems are built of complex web of interconnected neurons. In a rough analogy, they are built of interconnected simple processing units where each unit takes a number of real-valued inputs (possibly the outputs of other units) and produces a single output (which may be the input to another unit). Each unit uses an internal activation function to aggregate its inputs and produce the output. There is a weight associated with each interconnection that multiply the output from one unit by some factor to provide input to another unit and determines how all the inputs to a particular unit are combined before being aggregated by the activation function. A simple neural network will consist of a single layer of input units, a layer of hidden units, and a layer of one or more output units. Typically, each unit of one layer connects with all the units of the next layer, although more complex patterns of connections are possible. The complex structure of neural networks makes it possible to implement high-order interactions between variables. Also, neural networks are known to recognize complex patterns in data and are relatively error tolerant in that they are good at performing discrimination based only on partial matches and minute changes in some aspects of the input data will not necessarily lead to misclassification.

Neural networks can be designed to implement discriminant functions, which aim to classify sets of input values (independent variables) according to their associated output values (dependent variables) so that a given set of input values will produce a set of outputs close to the observed values [39]. The weights in the interconnections and the parameters of the activation function in each unit comprise the parameters of the neural network. The network parameters can be adjusted using standard methods to produce outputs that closely follow the target function response without having to model the discriminant function explicitly.

A neural network is thus a supervised learning method, which implies that the above network parameters are learnt using training samples before the network is used for classification of unknown data. During the training procedure, the network learns to associate particular sets of inputs with particular outputs. Sets of input data and target outputs are repeatedly presented to the network and it adjusts the parameters (connection weights and activation function parameters) to optimize the prediction accuracy.

In the current application [40], a simple feed-forward neural network with four layers is used to predict disease–SNP association. For each individual, four markers are simulated, so the input layer has four inputs each corresponding to a marker. There are two hidden layers each having three units. The output layer has a single unit. The input to each unit can be values 0, 1, or 2, corresponding to genotypes AA, Aa, or aa (where A and a represent the two alleles). The output from the output unit is the affectation status (0 for control, 1 for diseased cases) of the individual as predicted by the neural network.

The output from unit k of the input layer is 0, 1, or 2, according to the genotype of the corresponding marker. The activation function in each neuron of the hidden and output layers is given by the logistic function

$$y_k = \frac{1}{1 + \exp\left(t_k + \sum_{j,k} w_{jk} x_j\right)}, \tag{18.20}$$

where x_j is the output from each unit j of the previous layer, w_{jk} being the weight for the connection between unit k of the current layer and unit j of the previous layer, and t_k being a threshold value applying bias to unit k of the current layer. The entire data set is divided into two equal portions randomly to generate a training data set and a validation data set. Initially the network is trained using the marker values of both cases and controls from the training data set. The neural network adjusts the weights w_{jk} and thresholds t_k on the connections between the units across the hidden and the output layers using backpropagation procedure. Once the network is trained, the validation data set is presented and the output for each subject is recorded, based on their marker genotype. Then the process is reversed: The second half of the data set serves as the training data set and the first half as the validation data set. The process is repeated for over 200 cycles. The extent to which the outputs produced from the network matched the target output (control 0, case 1) is used as a test for association.

Simple chi-square test is applied to test for association of each individual marker with the disease using a 2×2 contingency table, and then Bonferroni [41] correction is applied for the number of markers tested to produce an overall test for association from considering only one marker at a time. When association is tested using t-test based on the neural network outputs obtained from cases and controls, the test is found to be unacceptably anticonservative. So, two overall tests for association are designed based on either single marker data or a combined analysis including the neural network. The most significant statistic from either a single marker or from the network was taken, and then a Bonferroni correction was applied for the total number of tests considered (one more than the number of markers). Ten thousand chromosome data sets are simulated, using five different disease models: Mendelian dominant, Mendelian recessive, dominant with phenocopies, recessive with phenocopies, and codominant case. For each disease model, sample sizes were chosen such that reasonable comparisons of power can be done when other parameters were altered. Both the above-mentioned tests are applied individually as well as with Bonferroni correction under a null hypothesis of no effect from the susceptibility locus. Taking only the most significant of the single locus tests is found to be anticonservative, but applying a Bonferroni correction to take account of four markers being tested produces acceptable performance. When neural network is used in the association test, taking the most significant of all the single locus tests and of the neural network test, and then correcting for the total number of tests produces an overall test for association, which conforms very well to the desired null hypothesis distribution. The tests are repeated under various conditions of LD between a susceptibility locus and the markers. Using a neural network is found to cause no appreciable loss of power of test. The overall power to detect association increases by up to 10% when number of mutations are increased.

18.2.3 Using Support Vector Machine

Recently, SVMs developed by Vapnik and Cortes [42] have attracted lot of attention as supervised machine learning methods for classification and regression. Often we would like to classify data across multiple dimensions and an SVM is utilized to find an optimal maximal margin hyperplane separating the two classes of data. The maximum distance of the hyperplane to the closest data point from both classes is called the margin and the vectors that are closest to this hyperplane are called the support vectors. The goal is to maximize the margin and at the same time minimize classification error. Maximizing the margin can be viewed as an optimization task solvable using linear or quadratic programming methods (see Fig. 18.3). SVM is especially useful where the data classes are not linearly separable. In this case, the data points are projected onto a higher dimensional space and a separating hyperplane is obtained using nonlinear kernel functions. An application of SVM to predict cancer susceptibility using SNPs can be found in Ref. [21]. Although well over 1.8 million SNPs are known, SNP data are error prone and missing values are quite common. SVM is chosen for the association study since it is very robust to missing data and performs well in high dimensions. An added advantage is that unphased genotype data are used instead of

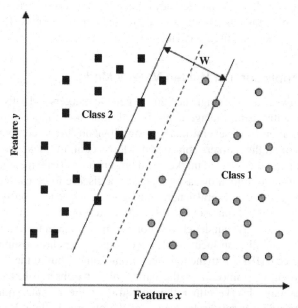

Figure 18.3 The figure shows an SVM for binary classification. The margin to be maximized is *W* that separates the hyperplane (shown with dotted line) from the two classes of data. In this example, although there are only two features, *x* and *y*, number of classifying features would be much higher.

haplotype data to avoid inherent inaccuracies introduced while identifying the haplotypes from the genotypes in the preprocessing stage.

Waddell et al. [21] studied the specific case of multiple myeloma, a cancer of antibody secreting plasma cells that grow and expand in the bone marrow. Since it occurs with relatively high frequency in aged adults (0.035% in people over 70 in the United States) and less in young adults (0.002% aged 30–54 in the United States), it is possible to detect differences in SNP patterns in diseased patients in the above two age groups if people who are diagnosed with the disease at a young age (under 40) have a genetic predisposition to the disease. The data set consists of genotypes from 3000 SNPs for 80 patients who are selected so that they are evenly spaced at about 1 Mb apart to give a good overall coverage of the human genome. SVM with a linear kernel function is used since the output is more comprehensible and it can be easily extended to using nonlinear kernel functions. Each heterozygous SNP data are coded as 0, one homozygous is arbitrarily coded as +1 and the other as −1. An entropy-based feature selection is applied as an initial step to select the most informative top 10% SNPs from the set of 3000 SNPs. The entropy of a data set is given by $-p \cdot \log_2(p) - (1-p) \cdot \log_2(1-p)$ where p is the fraction of examples that belong to class predisposed. The information gain of the split is given by the entropy of the original data set minus the weighted sum of entropies of the two data sets resulting from the split, where these entropies are weighted by the fraction of data points in each set. The SNP features are ranked by information gain, and the top-scoring 10% of the features are selected. Classification of the diseased and control cases using a leave-one-out cross-validation approach yields an overall classification accuracy of 71%, which is significantly better than chance (50%). Although the accuracy is not that high, it is significant considering that only relatively sparse SNP data are used for the purpose of classification, which does not change once the disease occurs.

18.2.4 An Application of Hidden Markov Model

Association between a set of single nucleotide genetic markers and a disease has been studied traditionally using linkage analysis with multilocus models that takes into account information from several markers simultaneously and compute composite log-likelihood of all the markers considered. Morris et al. [43] take into account the correlation between the selected markers and present a fine scale method of mapping genes contributing to human diseases. Consider a disease to be caused by a single mutation at a disease locus several generations ago. At each generation, recombination takes place between the chromosomes and all affected people in the current generation have descended from this founder chromosome. An affected individual will have the disease allele at the disease locus and the corresponding chromosome segment is identical by descent (IBD) with the founder chromosome. Thus, if the chromosome is IBD with the founder chromosome at the marker locus, we observe the ancestral allele; otherwise, we may observe either the ancestral or the nonancestral allele with probability depending upon the frequency of the observed allele. For the markers close to the disease locus, the probability of being IBD with the founder will be higher than with more distant markers.

Let F denote the event that a chromosomal locus is IBD with the founder and N denote the event that it is not. The occurrence of different ancestral states along the chromosome at each locus is the result of the recombination events in the previous generations. Consider two different loci, locus 1 and locus 2, on a chromosome. Then,

$$
\begin{aligned}
P(\text{locus } 2 = F|\text{locus } 1 = F) &= P(\text{NR}) + (1 - P(\text{NR}))P(\text{MRR} = F), \\
P(\text{locus } 2 = F|\text{locus } 1 = N) &= (1 - P(\text{NR}))P(\text{MRR} = F)
\end{aligned}
\tag{18.21}
$$

where NR denotes no recombination and MRR denotes most recent recombination event at locus 2.

This can be generalized to multiple loci on the chromosome. Then, given the ancestral state at one chromosome locus, we can calculate the ancestral states at other loci using two independent Markov chains, one starting from each side of the given locus.

Let S_0 be the disease locus. The markers to the left S_0 are denoted as $S_{-1}, S_{-2}, \ldots, S_{-L}$, and markers to the right are denoted by $S_1, S_2, \ldots, S_R$. Thus, the chromosome can be considered as two independent paths of ancestral states, conditional on the ancestral state at the given disease locus S_0. For a marker locus S_{-L} to the left of S_0, the path is $\{S_0, S_{-1}, \ldots, S_{-L}\}$ and for a marker locus S_R to the right of S_0, the path is $\{S_0, S_1, \ldots, S_R\}$.

Now we show how we can calculate the probability that the chromosome at a given marker locus is IBD with the founder, given the chromosome's ancestral state at the disease locus S_0. Let $\tau^{i+1}_{S_i S_{i+1}}$ denote the transition probability of ancestral state S_{i+1} at locus $i+1$, given the chromosome's ancestral state S_i. Then, we have the following transition probabilities over the two state values F and N:

$$
\begin{aligned}
\tau^{i+1}_{FF} &= \exp(-\gamma d_{i+1}) + [1 - \exp(-\gamma d_{i+1})]\alpha, \\
\tau^{i+1}_{FN} &= [1 - \exp(-\gamma d_{i+1})][1 - \alpha], \\
\tau^{i+1}_{NF} &= [1 - \exp(-\gamma d_{i+1})]\alpha, \\
\tau^{i+1}_{NN} &= \exp(-\gamma d_{i+1}) + [1 - \exp(-\gamma d_{i+1})][1 - \alpha].
\end{aligned}
\tag{18.22}
$$

Here $\exp(-\gamma d_{i+1})$ is the probability of no recombination between the marker loci S_i and S_{i+1} from the founder generation to the current one. The parameter γ is the expected frequency of recombination events over a region of 1 Mb of a chromosome since the founder generation; and α is set to be $P(\text{MRR} = F)$, that is, the probability that a chromosome locus is IBD with the founder. Following a hidden Markov model (the IBD status at a given locus on the chromosome can be thought of as the hidden state), we can calculate the probability $\rho[S_i|S_0]$, that a chromosome is of ancestral state S_i at marker locus i, conditional on the chromosome's ancestral state at the disease locus S_0 as

$$
\begin{aligned}
\rho[S_1|S_0] &= \tau^i_{S_0 S_1}, \quad i = 1, \\
\rho[S_i|S_0] &= \sum_{S_i = F, N} \rho[S_{i-1}|S_0]\tau^i_{S_{i-1} S_i}, \quad i > 1.
\end{aligned}
\tag{18.23}
$$

Consider locus i with two alleles M_{i1} and M_{i2}. Let p_i denote the frequency of M_{i1} (and correspondingly $1 - p_i$ for M_{i2}). Also, let n_{ij}^{Ph} denote the observed frequency of allele M_{ij} in the sample of chromosomes obtained from people having disease phenotype Ph. Assuming a multiplicative model for the disease with parameters β_F and β_N for the ancestral states $S_0 = F$ and $S_0 = N$, respectively, at the disease locus, the penetrance of the genotype $S_0 S'_0$ is given as

$$P(\text{Ph} = A | S_0 S'_0) = \beta_{S_0} \beta_{S'_0}. \tag{18.24}$$

Thus,

$$
\begin{aligned}
P(\text{Ph} = A | S_0 = N) &= \beta_F \beta_N \alpha + \beta_N^2 (1 - \alpha) = f_N, \\
P(\text{Ph} = A | S_0 = F) &= \beta_F^2 \alpha + \beta_F \beta_N (1 - \alpha) = f_F, \\
P(\text{Ph} = U | S_0 = N) &= (1 - \beta_F \beta_N) \alpha + (1 - \beta_N^2)(1 - \alpha) = 1 - f_N, \\
P(\text{Ph} = U | S_0 = F) &= (1 - \beta_F^2) \alpha + (1 - \beta_F \beta_N)(1 - \alpha) = 1 - f_F
\end{aligned}
\tag{18.25}
$$

Ph denotes the disease phenotype with values A (affected) and U (unaffected controls). We do not observe the ancestral state at the disease locus, rather, we can only see the phenotype Ph of the affected individual. S'_0 is the ancestral state at the disease locus in the corresponding homologous chromosome.

The overall log-likelihood of a sample data for a fixed location of disease gene x is determined by two independent Markov processes as

$$L(\text{data}|x, \Gamma, \vec{p}, \vec{w})_{\text{Total}} = L(\text{data}|x, \Gamma, \vec{p}, \vec{w})_{\text{Left}} + L(\text{data}|x, \Gamma, \vec{p}, \vec{w})_{\text{Right}}, \tag{18.26}$$

where $\Gamma = \alpha, \beta_F, \beta_N, \gamma$ is the vector of model parameters, $\vec{p}$ is the vector of allele frequencies, and $\vec{w}$ is the vector of ancestral state indicators. The individual log-likelihood to the right and left of the disease locus are given by

$$
\begin{aligned}
L(\text{data}|x, \Gamma, \vec{p}, \vec{w})_{\text{Right}} &= \sum_{i=1}^{R} \sum_{j=1}^{2} \sum_{\text{Ph} \in \{A,U\}} n_{ij}^{\text{Ph}} \log[\phi_{ij}^{\text{Ph}}] + C_R, \\
L(\text{data}|x, \Gamma, \vec{p}, \vec{w})_{\text{Left}} &= \sum_{i=1}^{L} \sum_{j=1}^{2} \sum_{\text{Ph} \in \{A,U\}} n_{ij}^{\text{Ph}} \log[\phi_{ij}^{\text{Ph}}] + C_L,
\end{aligned}
\tag{18.27}
$$

where $C_R(C_L)$ is a constant for a known population disease frequency. n_{ij}^{Ph} denotes the observed frequency of allele M_{ij} in the sample of chromosomes obtained from people having disease phenotype Ph. ϕ_{ij}^{Ph} denotes the probability that a chromosome is associated with observed disease phenotype Ph and also bears allele M_{ij} at locus i and is given by

$$\phi_{ij}^{\text{Ph}} = P(\text{Ph} \cap M_{ij}) = P(\text{Ph} \cap M_{ij} | S_0 = F) \alpha + P(\text{Ph} \cap M_{ij} | S_0 = N)(1 - \alpha). \tag{18.28}$$

TABLE 18.1   **List of model parameters**

Parameters	Description
x	Location of the disease gene.
α	Frequency of mutation that is, probability that a chromosome locus is IBD with the founder.
β	Disease model parameter.
γ	Expected frequency of recombination events over a region of 1 Mb of a chromosome since the founder generation. Assuming that a physical distance of 1 Mb corresponds to a genetic distance of 1 cM, 100 γ gives the number of generations since the founding mutation.

See Table 18.1 for the parameters of this model.

The model performance is evaluated using an application to the cystic fibrosis (CF) disease and the model parameters are estimated using Metropolis–Hastings rejection-sampling [44] scheme with substantial burn-in period to allow convergence. CF is one of the most common autosomal recessive disorders affecting whites, occurring with an incidence of 1 case/2000 births. CF has been attributed to the CF gene in chromosomal location 7q31, a 3-bp deletion sequence (ΔF508, which accounts for more than 68% of all chromosomes in affected people), and several other rarer mutations in the same gene [45]. Applying the hidden Markov model, the mean estimate of the location of the mutation x is 0.784 Mb from the MET locus, with a 99% credibility interval of 0.731–0.838 Mb. α is estimated as 0.223, which is in agreement with a mutation frequency estimate of 0.223 [45]. Estimated value of β approaches 0, which is expected for a fully penetrant recessive disease. The estimated age of the mutation γ is 2.05, (i.e., 205 generations). All these estimates concur well with already established studies on CF.

18.3 CONCLUSION

We have thus presented several works on tag SNP selection and mapping disease locus based on association study using SNPs. Tag SNP selection uses redundancy in the genotype/haplotype data to select the most informative SNPs that predict the remaining markers as accurately as possible. This not only reduces the immense cost of genotyping a large number of SNPs, but also narrows down the set of candidate SNPs to be analyzed in the subsequent association studies. In general, machine learning methods tend to do better than purely combinatorial methods and also are applicable to bigger data sets with hundreds of SNPs. Identifying SNPs in disease association study is more difficult, largely depends on the population under study, and often faces the problem of replication. Both DICE and neural networks suffer from the lack of scalability with the increase in number of markers. SVM works well even when large number of marker are considered, but the accuracy of disease prediction is not that high, indicating that simply using the markers as attributes for classification purpose cannot reliably conclude the chances of being affected with a disease, and this again varies from one sample population to another. Even when the sophisticated hidden Markov model is considered, the prior distributions on many model parameters need to be assumed. Nevertheless, it is a firm step in identifying the complex interplay between various genes and their products (proteins) (and therefore the mutations, deletions, insertions lying on the corresponding genes) in causing a complex disease.

REFERENCES

1. Larranaga, P., Calvo, B., Santana, R., Bielza, C., Galdiano, J., Inza, I., Lozano, J. A., Armananzas, R., Santafe, G., Perez, A., and Robles, V. Machine learning in bioinformatics. *Briefings in Bioinformatics*, 7(1): 86–112, 2006.

2. Brookes, A. J. The essence of SNPs. *Genetics*, 234(2): 177–186, 1999.

3. Cambien, F., Poirier, O., Nicaud, V., Herrmann, S. M., Mallet, C., Ricard, S., Behague, I., Hallet, V., Blanc, H., Loukaci, V., Thillet, J., Evans, A., Ruidavets, J. B., Arveiler, D., Luc, G., and Tiret, L. Sequence diversity in 36 candidate genes for cardiovascular disorders. *American Journal of Human Genetics*, 65(1): 183–191, 1999.

4. Halushka, M. K., Fan, J. B., Bentley, K., Hsie, L., Shen, N., Weder, A., Cooper, R., Lipshutz, R., and Chakravarti, A. Patterns of single-nucleotide polymorphisms in candidate genes for blood-pressure homeostasis. *Natural Genetics*, 22(3): 239–247, 1999.

5. Sachidanandam R., Weissman D., Schmidt S. C., Kakol J. M., Stein, L. D., Marth, G., Sherry, S., Mullikin, J. C., Mortimore, B. J., Willey, D. L., Hunt, S. E., Cole, C. G., Coggill, P. C., Rice, C. M., Ning, Z., Rogers, J., Bentley, D. R., Kwok, P. Y., Mardis, E. R., Yeh, R. T., Schultz, B., Cook, L., Davenport, R., Dante, M., Fulton, L., Hillier, L., Waterston, R. H., McPherson, J. D, Gilman, B., Schaffner, S., Van Etten, W. J., Reich, D., Higgins, J., Daly, M. J., Blumenstiel, B., Baldwin, J., Stange-Thomann, N., Zody, M. C., Linton, L., Lander, E. S., and Altshuler, D. International SNP Map Working Group. A map of human genome sequence variation containing 1.42 million single nucleotide polymorphisms. *Nature*, 409(6822): 928–933, 2001.

6. Carlson, C. Selecting a maximally informative set of single-nucleotide polymorphisms for association analyses using linkage disequilibrium. *American Journal of Human Genetics*, 2004.

7. Kruglyak, L. and Nickerson, D. A. Variation is the spice of life. *Natural Genetics*, 27: 234–236, 2001.

8. Hirschhorn, J. N. and Daly, M. J. Genome wide association studies for common diseases and complex traits. *Natural Review of Genetics*, 6(2): 95–108, 2005.

9. Devlin, B. and Risch, N. A comparison of linkage disequilibrium measures for fine-scale mapping. *Genomics*, 29: 311–322, 1995.

10. Avi-Itzhak, H. I., Su, X., and De La Vega, F. M. Selection of minimum subsets of single nucleotide polymorphisms to capture haplotype block diversity. *Pacific Symposium on Biocomputing*, 466–477, 2003.

11. Stram, D. O., Pearce, C. L., Bretsky, P., Freedman, M., Hirschhorn, J. N., Altshuler, D., Kolonel, L. N., Henderson, B. E., and Thomas, D. C. Modeling and EM estimation of haplotype-specific relative risks from genotype data for a case-control study of unrelated individuals. *Human Heredity*, 55: 179–190, 2003.

12. Zhang, K., Qin, Z., Liu, J., Chen, T., Waterman, M. S., and Sun, F. Haplotype block partitioning and tag SNP selection using genotype data and their applications to association studies. *Genome Research*, 14(5): 908–916, 2004.

13. Zhang, K., Sun, F., Waterman, M. S., and Chen, T. Dynamic programming algorithms for haplotype block partitioning: applications to human chromosome 21 haplotype data. *Proceedings of the Seventh Annual International Conference on Research in Computational Molecular Biology (RECOMB 03)*, The Association for Computing Machinery, pp. 332–340, 2003.

14. Patil, N., Berno, A., Hinds, D., Barrett, W., Doshi, J., Hacker, C., Kautzer, C., Lee, D., Marjoribanks, C., and McDonough, D. Blocks of limited haplotype diversity revealed by high resolution scanning of human chromosome 21. *Science*, 294: 1719–1723, 2001.

15. Halperin, E., Kimmel, G., and Shamir, R. Tag SNP selection in genotype data for maximizing SNP prediction accuracy. *Bioinformatics*, 21(1): pp. 195–203, 2005.

16. Phuong, T. M., Lin, Z., and Altman, R. B. Choosing SNPs using feature selection. *Proceedings of the IEEE Computational Systems Bioinformatics Conference*, 2005, pp. 301–309.

17. Hampe, J., Schreiber, S., and Krawczak, M. Entropy-based SNP selection for genetic association studies. *Human Genetics*, 114: 36–43, 2003.

18. Bafna, V., Halldorsson, B. V., Schwartz, R., Clark, A. G., and Istrail, S. Haplotypes and informative SNP selection algorithms: don't block out information. *Proceedings of the Seventh Annual International Conference on Research in Computational Molecular Biology*, Berlin, Germany, pp. 19–27, 2003.

19. Wu, X., Luke, A., Rieder, M., Lee, K., Toth, E. J., Nickerson, D., Zhu, X., Kan, D., and Cooper, R. S. An association study of angiotensinogen polymorphisms with serum level and hypertension in an African-American population. *Journal of Hypertension*, 21: 1847–1852, 2003.

20. Lin, Z. and Altman, R. B. Finding haplotype tagging SNPs by use of principal components analysis. *American Journal of Human Genetics*, 75(5): 850–861, 2004.

21. Waddell, M., Page, D., Zhan, F., Barlogie, B., and Shaughnessy, J., Jr. Predicting cancer susceptibility from single-nucleotide polymorphism data: a case study in multiple myeloma. *Proceedings of BIOKDD '05*, Chicago, Illinois, August 2005.

22. He, J. and Zelikovsky, A. Tag SNP selection based on multivariate linear regression. *Proceedings of the International Conference on Computational Science (ICCS 2006)*, Vol. 3992, 2006, pp. 750–757.

23. Daly, M., Rioux, J., Schaffner, S., Hudson, T., and Lander, E. High-resolution haplotype structure in the human genome. *Nature Genetics*, 29: 229–232, 2001.

24. www.hapmap.org.

25. Stephens, M., Smith, N., and Donnelly, P. A new statistical method for haplotype reconstruction from population data. *American Journal of Human Genetics*, 68: 978–989, 2001.

26. Rieder, M. J., Taylor, S. L., Clark, A. G., and Nickerson, D. A. Sequence variation in the human angiotensin converting enzyme. *Nature Genetics*, 22: 59–62, 1999.

27. Kroetz, D. L., Pauli-Magnus, C., Hodges, L. M., Huang, C. C., Kawamoto, M., Johns, S. J., Stryke, D., Ferrin, T. E., DeYoung, J., Taylor, T., Carlson, E. J., Herskowitz, I., Giacomini, K. M., and Clark, A. G. Sequence diversity and haplotype structure in the human ABCB1 (MDR1, multidrug resistance transporter) gene. *Pharmacogenetics*, 13: 481–494, 2003.

28. Johnson, G. C., Esposito, L., Barratt, B. J., Smith, A. N., Heward, J., Di Genova, G., Ueda, H., Cordell, H. J., Eaves, I. A., Dudbridge, F., Twells, R. C., Payne, F., Hughes, W., Nutland,

S., Stevens, H., Carr, P., Tuomilehto-Wolf, E., Tuomilehto, J., Gough, S. C., Clayton, D. G., and Todd, J. A. Haplotype tagging for the identification of common disease genes. *Nature Genetics*, 29: 233–237, 2001.

29. Meng, Z., Zaykin, D. V., Xu, C. F., Wagner, M., and Ehm, M. G. Selection of genetic markers for association analyses, using linkage disequilibrium and haplotypes. *American Journal of Human Genetics*, 73: 115–130, 2003.

30. Shannon, C. E. A mathematical theory of communication. *Bell Systems Technical Journal*, 27: 379–423, 1948.

31. Morton, N. E., Zhang, W., Taillon-Miller, P., Ennis, S., Kwok, P. Y., and Collins, A. The optimal measure of allelic association. *Proceedings of the National Academy of Sciences of the United States of America*, 98: 5217–5221, 2001.

32. Corbex, M., Poirier, O., Fumeron, F., Betoulle, D., Evans, A., Ruidavets, J. B., Arveiler, D., Luc, G., Tiret, L., and Cambien, F. Extensive association analysis between the CETP gene and coronary heart disease phenotypes reveals several putative functional polymorphisms and gene–environment interaction. *Genetic Epidemiology*, 19: 64–80, 2000.

33. Stengard, J. H., Clark, A. G., Weiss, K. M., Kardia, S., Nickerson, D. A., Salomaa, V., Ehnholm, C., Boerwinkle, E., and Sing, C. F. Contributions of 18 additional DNA sequence variations in the gene encoding apolipoprotein E to explaining variation in quantitative measures of lipid metabolism. *American Journal of Human Genetics*, 71: 501–517, 2002.

34. Tregouet, D. A., Barbaux, S., Escolano, S., Tahri, N., Goldmard, J. L., Tiret, L., and Cambien, F. Specific haplotypes of the P-selectin gene are associated with myocardial infarction. *Human Molecular Genetics*, 11: 2015–2023, 2002.

35. Tahri-Daizadeh, N., Tregouet, D. A., Nicaud, V., Manuel, N., Cambien, F., and Tiret, L. Automated detection of informative combined effects in genetic association studies of complex traits. *Genome Research*, 13: 1952–1960, 2003.

36. Hurvich, C. M. and Tsai, C. L. Regression and time series model selection in small samples. *Biometrika*, 76: 297–307, 1989.

37. Akaike, H. A new look at the statistical model identification. *IEEE Transactions on Automated Control*, 19: 716–723, 1974.

38. Johnson, D. H. The insignificance of statistical significance testing. *Journal of Wildlife Management*, 63: 763–772, 1999.

39. Bishop, C. M. *Neural Networks for Pattern Recognition*. Oxford University Press, Oxford, UK, 1995.

40. Curtis, D., North, B. V., and Sham, P. C. Use of an artificial neural network to detect association between a disease and multiple marker genotypes. *Annals of Human Genetics*, 65: 95–107, 2001.

41. Shaffer, J. P. Multiple hypothesis testing. *Annual Review of Psychology*, 46: 561–584, 1995.

42. Vapnik, V. and Cortes, C. Support Vector Networks. *Machine Learning*, 20: 273–293, 1995.

43. Morris, A. P., Whittaker, J. C., and Balding, D. J. Bayesian fine-scale mapping of disease loci, by hidden markov models. *American Journal of Human Genetics*, 67: 155–169, 2000.

44. Berg, B. A. *Markov Chain Monte Carlo Simulations and Their Statistical Analysis*, World Scientific, Singapore, 2004.

45. Kerem, B., Rommens, J. M., Buchanan, J. A., Markiewicz, D., Cox, T. K., Chakravarti, A., and Buchwald, M. Identification of the cystic fibrosis gene: genetic analysis. *Science*, 245: 1073–1080, 1989.

19

NANOPORE CHEMINFORMATICS-BASED STUDIES OF INDIVIDUAL MOLECULAR INTERACTIONS

Stephen Winters-Hilt

19.1 INTRODUCTION

Channel current-based nanopore cheminformatics provides an incredibly versatile method for transducing single-molecule events into discernable channel current blockade levels. Single biomolecules, and the ends of biopolymers such as DNA, have been examined in solution with nanometer-scale precision [1–6]. In early studies [1], it was found that complete base-pair dissociations of dsDNA to ssDNA, "melting," could be observed for sufficiently short DNA hairpins. In later work [3, 5], the nanopore detector attained angstrom resolution and was used to "read" the ends of dsDNA molecules. This operated as a chemical biosensor. In Refs. [1, 2, 4] the nanopore detector was used to observe the conformational kinetics at the termini of single DNA molecules. In Refs. [7, 8], preliminary evidence of single-molecule binding and conformational kinetics was obtained by observation of single-molecule channel blockade currents. The DNA–DNA, DNA–protein, and protein–protein binding experiments that were described were novel in that they made critical use of indirect sensing (described below), where one of the molecules in the binding

Machine Learning in Bioinformatics. Edited by Yan-Qing Zhang and Jagath C. Rajapakse

experiment is either a natural channel blockade modulator or is attached to a blockade modulator.

19.2   NANOPORE DETECTOR BACKGROUND AND METHODS

19.2.1   The Highly Stable, Nanometer-Scale, α-Hemolysin Protein Channel

The nanopore detector is based on the α-hemolysin transmembrane channel, formed by seven identical 33 kDa protein molecules secreted by *Staphylococcus aureus*. The total channel length is 10 nm and is comprised of a 5-nm transmembrane domain and a 5-nm vestibule that protrudes into the aqueous *cis* compartment [9]. The narrowest segment of the pore is a 1.5-nm diameter aperture [9]. By comparison, a single strand of DNA is about 1.3-nm in diameter and able to translocate. Although dsDNA is too large to translocate, about 10 base pairs at one end can still be drawn into the large *cis*-side vestibule (see Fig. 19.1). This actually permits the most sensitive experiments to date, as the ends of "captured" dsDNA molecules can be observed for extensive periods of time to resolve features [1–5].

19.2.2   The Coulter Counter

The notion of using channels as detection devices dates back to the Coulter counter [10], where pulses in channel flow were measured in order to count bacterial cells. Cell transport through the Coulter counter is driven by hydrostatic pressure, and interactions between the cells and the walls of the channel are ignored. Since its original formulation, channel sizes have reduced from millimeter scale to nanometer scale, and the detection mechanism has shifted from measurements of hydrostatically driven fluid flow to measurements of electrophoretically driven ion flow. Analytes observed via channel measurements are likewise reduced in scale and are now at the scale of single biomolecules such as DNA and polypeptides [1–6, 11–16]. In certain situations, intramolecular, angstrom-level features are beginning to be resolved as well [1–5].

For nanoscopic channels, interactions between channel wall and translocating biomolecules cannot, usually, be ignored. On the one hand, this complicates analysis of channel blockade signals and, on the other hand, tell-tale on-off kinetics are revealed for binding between analyte and channel, and this is what has allowed the probing of intramolecular structure on single DNA molecules [1–5].

19.2.3   Coulter Data—Blockades Typically Static

Biophysicists and medical researchers have performed measurements of ion flow through single nanopores since the 1970s [17, 18]. The use of very large (biological) pores as polymer sensors is a relatively new possibility that dates from the pioneering experiments of Bezrukov et al. [16]. Their work proved that resistive

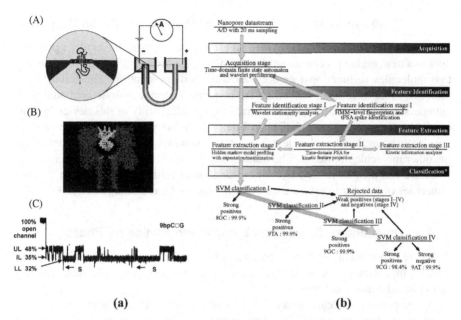

(a) **(b)**

Figure 19.1 The α–hemolysin nanopore detector and cheminformatics architecture. **(a)** (A). A nanopore device based on the α–hemolysin channel (from Ref. 3). It has been used for analysis of single DNA molecules, such as ssDNA, shown, and dsDNA, a nine-base-pair DNA hairpin is shown in (B) superimposed on the channel geometry. The channel current blockade trace for the nine base-pair DNA hairpin blockade from (B) is shown in (C). **(b)** The signal processing architecture that was used to classify DNA hairpins with this approach: Signal acquisition was performed using a time-domain, thresholding finite state automaton, followed by adaptive prefiltering using a wavelet-domain finite state automaton. Hidden Markov Model processing with Expectation–Maximization was used for feature extraction on acquired channel blockades. Classification was then done by support vector machine on five DNA molecules: four DNA hairpin molecules with nine base-pair stem lengths that only differed in their blunt-ended DNA termini and an eight-base-pair DNA hairpin. The accuracy shown is obtained upon completing the 15th single-molecule sampling/classification (in approximately 6 s), where SVM-based rejection on noisy signals was employed.

pulse measurements, familiar from cell counting with the Coulter counter [10], could be reduced to the molecular scale and applied to polymers in solution. A seminal study, by Kasianowicz et al. [11], then showed that individual DNA and RNA polymers could be detected via their translocation blockade of a nanoscale pore formed by α-hemolysin toxin. In such prior nanopore detection work, the data analysis problems were also of a familiar "Coulter event" form, where the event was associated with a current blockade at a certain fixed level. A more informative setting is possible with nanometer-scale channels, however, due to nonnegligible interaction between analyte and channel. In this situation, the blockading molecule will not provide a single, fixed current reduction in the channel but will modulate the ion flow through the channel by imprinting its binding interactions and conformational kinetics on the confined channel flow environment.

19.2.4 Nanopore Detector Augmentation Using Bifunctional Molecules

The improved detector sensitivity with toggling-type auxiliary molecules opens the door to a new, highly precise means for examining the binding affinities between any two molecules (bifunctional or not), all while still in solution. The bifunctional molecules that have been studied on the nanopore detector include antibodies and aptamers that were chosen to demonstrate the specific utility of this device in drug candidate screening (see Ref. [7]). In brief, an auxiliary molecule can be rigidly/ covalently bound to the molecule of interest, and then exposed to a solution containing the other molecule of interest. The transitions between different stationary phases of blockade can then be related to the bound/unbound configuration between the two molecules of interest to reveal their binding kinetics (and binding strength).

19.2.5 Detection of Short-Term Binding and Stationary Phase

There are important distinctions in how a nanopore detector can function: direct versus indirect measurement of static, stationary, dynamic (possibly modulated), or non-stationary channel blockades.

A nanopore-based detector can directly measure molecular characteristics in terms of the blockade properties of individual molecules. This is possible due to the kinetic information that is embedded in the blockade measurements, where the adsorption–desorption history of the molecule to the surrounding channel, and the configurational changes in the molecule itself directly, imprint on the ionic flow through the channel [1–6]. (*Note*: The hypothesis that the current blockade patterns are caused by adsorption–desorption, and conformational flexing, is not conclusively proven, although preliminary work on mechanism [5] and the success of the experimental approaches [1–6] add growing credence to this hypothesis.) This approach offers prospects for DNA sequencing and single nucleotide polymorphism (SNP) analysis.

The nanopore-based detector works indirectly if it uses a reporter molecule that binds to certain molecules, with subsequent distinctive blockade by the bound-molecule complex. One example of this, with the established DNA experimental protocols, is exploration of transcription factor binding sites via the different dsDNA blockade signals that occur with and without DNA binding by a hypothesized transcription factor. Similarly, a channel-captured dsDNA "gauge" that is already bound to an antibody could provide a similar blockade shift upon antigen binding to its exposed antibody. The latter description provides the general mechanism for directly observing the single–molecule antigen-binding affinities of any antibody.

19.2.6 Nanopore Observation of Conformational Kinetics

Two conformational kinetic studies have been done, one on DNA hairpins with HIV-like termini [8] and the other on antibodies (preliminary results shown in Ref. [7]). The objective of the DNA HIV-hairpin conformational study was to systematically test how DNA dinucleotide flexibility (and reactivity) could be discerned using channel current blockade information (see Ref. [8]).

19.3   CHANNEL CURRENT CHEMINFORMATICS METHODS

19.3.1   The Channel Current Cheminformatics Architecture

The signal processing architecture (Fig. 19.1b) is designed to rapidly extract useful information from noisy blockade signals using feature extraction protocols, wavelet analysis, HMMs and SVMs. For blockade signal acquisition and simple, time-domain feature extraction, a finite state automaton (FSA) approach is used [19] that is based on tuning a variety of threshold parameters. A generic HMM can be used to characterize current blockades by identifying a sequence of subblockades as a sequence of state emissions Refs. [20–22]. The parameters of the generic-HMM can then be estimated using a method called EM [23] to effect denoising. The HMM method with EM, denoted HMM/EM, is used in what follows (further background on these methods can be found in Refs. [1–6]). Classification of feature vectors obtained by the HMM for each individual blockade event is then done using SVMs, an approach that automatically provides a confidence measure on each classification.

19.3.2   The Feature Vectors for SVM Analysis

The nanopore detector is operated such that a stream of 100-ms samplings are obtained (throughput was approximately one sampling per 300 ms in Ref. [3]). Each 100-ms signal acquired by the time-domain FSA consists of a sequence of 5000 subblockade levels (with the 20 µs analog-to-digital sampling). Signal preprocessing is then used for adaptive low-pass filtering. For the data sets examined, the preprocessing is expected to permit compression on the sample sequence from 5000 to 625 samples (later HMM processing then only required construction of a dynamic programming table with 625 columns). The signal preprocessing makes use of an off-line wavelet stationarity analysis (off-line wavelet stationarity analysis, Fig. 19.1b).

With completion of preprocessing, an HMM is used to remove noise from the acquired signals and to extract features from them (feature extraction stage, Fig. 19.1b). The HMM is, initially, implemented with 50 states, corresponding to current blockades in 1% increments ranging from 20% residual current to 69% residual current. The HMM states, numbered 0–49, corresponded to the 50 different current blockade levels in the sequences that are processed. The state emission parameters of the HMM are initially set so that the state j, $0 \leq j \leq 49$ corresponding to level $L = j + 20$, can emit all possible levels, with the probability distribution over emitted levels set to a discretized Gaussian with mean L and unit variance. All transitions between states are possible and initially are equally likely. Each blockade signature is denoised by five rounds of EM training on the parameters of the HMM. After the EM iterations, 150 parameters are extracted from the HMM. The 150 feature vectors obtained from the 50-state HMM-EM/Viterbi implementation in Refs. [1–8] are the 50 dwell percentage in the different blockade levels (from the Viterbi trace-back states), the 50 variances of the emission probability distributions associated with the different states, and the 50 merged transition probabilities from the primary and secondary blockade occupation levels (fits to two-state dominant modulatory blockade signals).

19.3.3 τ-FSA Blockade Acquisition and Feature Extraction

A channel current spike detector algorithm was developed in Ref. [8] to characterize the brief, very strong blockade "spike" behavior observed for molecules that occasionally break in the region exposed to the limiting aperture's strong electrophoretic force region. (In Refs. [1–6], where nine base-pair hairpins were studied, the spike events were attributed to a fray/extension event on the terminal base pair.) Together, the formulation of HMM-EM, FSAs, and spike detector provides a robust method for analysis of channel current data. Application of these methods is described in Ref. [8] for radiation-damaged DNA signals. The spike detector software is designed to count "anomalous" spikes, that is, spike noises not attributable to the Gaussian fluctuations about the mean of the dominant blockade level. Spike count plots are generated to show increasing counts as cutoff thresholds are relaxed (to where eventually any downward deflection will be counted as a spike). The plots are automatically generated and automatically fit with extrapolations of their linear phases (exponential phases occur when cutoffs begin to probe the noise band of a blockade state—typically Gaussian noise "tails"). The extrapolations provide an estimate of "true" anomalous spike counts.

19.3.4 Markov Chains

Key "short-term memory" property of a Markov chain $P(x_i|x_{i-1},\ldots,x_1) = P(x_i|x_{i-1}) = a_{x_{i-1}x_i}$, where $a_{x_{i-1}x_i}$ are sometimes referred to as "transition probabilities," and we have $P(x) = P(x_L, x_{L-1}, \ldots, x_1) = P(x_1)\,\Pi_{i=2..L} a_{x_{i-1}x_i}$. If we denote C_y for the count of events y, C_{xy} for the count of simultaneous events x and y, T_y for the count of strings of length 1, and T_{xy} for the count of strings of length 2, $a_{x_{i-1}x_i} = P(x|y) = P(x, y)/P(y) = [C_{xy}/T_{xy}]/[C_y/T_y]$. Note that since $T_{xy} + 1 = T_y \rightarrow T_{xy}$ (T_y (sequential data sample property if one long training block), $a_{x_{i-1}x_i} \cong C_{xy}/C_y = C_{xy}/\sum_x C_{xy}$, so C_{xy} is complete information for determining transition probabilities.

19.3.5 Viterbi Path

The recursive algorithm for the most likely state path given an observed sequence (the Viterbi algorithm) is expressed in terms of v_{ki}, the probability of the most probable path that ends with observation $Z_i = z_i$ and state $S_i = k$. The recursive relation is $v_{ki} = \max_n\{e_{ki} a_{nk} v_{n(i-1)}\}$, where the $\max_n\{\ldots\}$ operation returns the maximum value of the argument over different values of index n, and the boundary condition on the recursion is $v_{k0} = e_{k0} p_k$. The a_{kl} are the transition probabilities $P(S_i = l|S_{i-1} = k)$ to go from state k to state l. The e_{kb} are the emission probabilities $P(Z_i = b|S_i = k)$ while in state k. The emission probabilities are the main place where the data is brought into the HMM–EM algorithm (An inversion on the emission probability is possible when the states and emissions share the same alphabet of states/quantized emissions, and it is described in the results). The Viterbi path labelings are then recursively defined by $p\,(S_i|S_{(i+1)} = n) = \mathrm{argmax}_k\{v_{ki} a_{kn}\}$, where the $\mathrm{argmax}_n\{\ldots\}$ operation returns the index n with maximum value of the argument. The evaluation of sequence probability (and its Viterbi labeling) takes the emission and transition probabilities as a given. Estimates on these emissions and transition probabilities are usually obtained by the EM algorithm.

19.3.6 Forward and Backward Probabilities

The forward/backward probabilities are used in the HMM/EM algorithm. The probabilities occur when evaluating $p(Z_{0...L-1})$ by breaking the sequence probability $p(Z_{0...L-1})$ into two pieces via use of a single hidden variable treated as a Bayesian parameter: $p(Z_{0...L-1}) = \Sigma_k p(Z_{0...i}, s_{i...} = k) p(Z_{i+1...L-1}, s_i = k) = \Sigma_k f_{ki} b_{ki}$, where $f_{ki} = p(Z_{0...i}, s_i = k)$ and $b_{ki} = p(Z_{i+1...L-1}, s_i = k)$. Given stationarity, the state transition probabilities and the state probabilities at the ith observation satisfy the trivial relation $p_{qi} = \Sigma_k a_{kq} p_{k(i-1)}$, where $p_{qi} = p(S_i = q)$, and $p_{q0} = p(S = q)$, and the latter probabilities are the state priors. The trivial recursion relation that is implied can be thought of as an operator equation, with operation the product by a_{kq} followed by summation (contraction) on the k index. The operator equation can be rewritten using an implied summation convention on repeated Greek-font indices (Einstein summation convention): $p_q = a_{\beta q} p_\beta$. Transition probabilities in a similar operator role, but now taking into consideration local sequence information via the emission probabilities, are found in recursively defined expressions for the forward variables, $f_{ki} = e_{ki}(a_{\beta k} f_{\beta(i-1)})$, and backward variables, $b_{ki} = a_{k\beta} e_{\beta(i+1)} b_{\beta(i+1)}$. The recursive definitions on forward and backward variables permit efficient computation of observed sequence probabilities using dynamic programming tables. It is at this critical juncture that side information must mesh well with the states (column components in the table), that is, in a manner like the emission or transition probabilities. Length information, for example, can be incorporated via length-distribution-biased transition probabilities (as described in a new method in Ref. [24]).

19.3.7 HMM-with-Duration Channel Current Signal Analysis

The HMM-with-duration implementation, described in Ref. [24], has been tested in terms of its performance at parsing synthetic blockade signals. The synthetic data range over an exhaustive set of possibilities for thorough testing of the HMM-with-duration. The synthetic data used in Ref. [24] were designed to have two levels, with lifetime in each level determined by a governing distribution (Poisson and Gaussian distributions with a range of mean values were considered). The results clearly demonstrate the superior performance of the HMM-with-duration over its simpler, HMM-without-Duration, formulation. With the use of the EVA-projection method, this affords a robust means to obtain kinetic feature extraction. The HMM with duration is critical for accurate kinetic feature extraction, and the results in Ref. [24] suggest that this problem can be elegantly solved with a pairing of the HMM-with-duration stabilization with EVA projection.

19.3.8 HMM-with-Duration via Cumulant Transition Probabilities

The transition probabilities for state "s" to remain in state "s," a "ss" transition can be computed as $\text{Prob}(ss|s_{\text{length}} = L) = \text{Prob}(s_{\text{length}} \geq L + 1)/\text{Prob}(s_{\text{length}} \geq L)$. The transition probabilities out of state 's' can have some subtleties, as shown in the following, where the states are exon (e), intron (i), and junk (j). In this case, the transition probabilities governing the following transitions, $(jj) \rightarrow (je)$, $(ee) \rightarrow (ej)$, $(ee) \rightarrow (ei)$, $(ii) \rightarrow (ie)$, are computed as $\text{Prob}(ei|e_{\text{length}} = L) = \text{Prob}(e_{\text{length}} = L)/\text{Prob}(e_{\text{length}} \geq L) = 40/(40 + 60)$ and $\text{Prob}(ej|e_{\text{length}} = L) = \text{Prob}(e_{\text{length}} = L)/\text{Prob}(e_{\text{length}} \geq L) \times 60/(40 + 60)$, where the total number of (ej) transitions is 60 and the total number of (ei) transitions is 40. The pseudocode to track the critical length information, on a cellular basis in the dynamic programming table, goes as follows:

(1) Maintain separate counters for the junk, exon, and intron regions.
(2) The counters are updated as
    (a) The exon counter is set to 2 for a $(je) \rightarrow (ee)$ transition
    (b) The exon counter gets incremented by 1 for every $(ee) \rightarrow (ee)$ transition.
(3) $\text{Prob}(e_{\text{length}} \geq L + 1)$ is computed as $\text{Prob}(e_{\text{length}} \geq L + 1) = 1 - \sum_{i=1\ldots L} \text{Prob}(e_{\text{length}} = i)$. Hence, we generate a list such that for each index "$k > 0$," the value $1 - \sum_{i=1,\ldots,k} \text{Prob}(e_{\text{length}} = i)$ is stored.

19.3.9 EVA Projection

The HMM method is based on a stationary set of emission and transition probabilities. Emission broadening, via amplification of the emission state variances, is a filtering heuristic that leads to level projection that strongly preserves transition times between major levels (see Ref. [24] for further details). This approach does not require the user to define the number of levels (classes). This is a major advantage compared to existing tools that require the user to determine the levels (classes) and perform a state projection. This allows kinetic features to be extracted with a "simple" FSA that requires minimal tuning. One important application of the HMM-with-duration method used in Ref. [24] includes kinetic feature extraction from EVA-projected channel current data (the HMM-with-duration is shown to offer a critical stabilizing capability in an example in Ref. [24]). The EVA-projected/HMMwDur processing offers a handsoff (minimal tuning) method for extracting the mean dwell times for various blockade states (the core kinetic information).

19.3.10 Support Vector Machines

SVMs are fast, easily trained discriminators [25, 26], for which strong discrimination is possible without the over-fitting complications common to neural net discriminators [16]. The SVM approach also encapsulates a significant amount of model fitting and discriminatory information in the choice of kernel in the SVM, and a number of novel kernels have been developed. In application to channel current signal analysis, there is generally an abundance of experimental data available, and if not, the experimenter can usually just take more samples and make it so. In this situation, it is appropriate to seek a method good at both classifying data and evaluating a confidence in the classifications given. In this way, data that are low confidence can simply be dropped. The structural risk minimization at the heart of the SVM method's robustness also provides a strong confidence measure. For this reason, SVMs are the classification method of choice for channel current analysis as they have excellent performance at 0% data drop and as weak data are allowed to be dropped, the SVM-based approaches far exceed the performance of most other methods known.

In Ref. [3], novel information-theoretic kernels were introduced for notably better performance over standard kernels, with discrete probability distributions as part of feature vector data. The use of probability vectors, and L_1-norm feature vectors in general, turns out to be a very general formulation, wherein feature extraction makes use of signal decomposition into a complete set of separable states that can be interpreted or represented as a probability vector (or normalized collection of such, etc.). A probability vector formulation also provides a straightforward handoff to the SVM classifiers since all feature vectors have the same length with such an approach. What this means for the SVM, however, is that geometric notions of distance are no longer the best measure for comparing feature vectors. For probability vectors (i.e., discrete distributions), the best measures of similarity are the various information-theoretic divergences: Kullback–Leibler, Renyi, and so on. By symmetrizing over the arguments of those divergences, a rich source of kernels is obtained that works well with the types of probabilistic data obtained, as shown in Ref. [3, 7, 27].

The SVM discriminators are trained by solving their Karush–Kuhn–Tucker (KKT) relations using the sequential minimal optimization (SMO) procedure [28]. A chunking [29, 30] variant of SMO is also employed to manage the large training task at each SVM node. The multiclass SVM training generally involves thousands of blockade signatures for each signal class.

19.3.11   Binary Support Vector Machines

Binary SVMs are based on a decision-hyperplane heuristic that incorporates structural risk management by attempting to impose a training-instance void, or "margin," around the decision hyperplane [25].

Feature vectors are denoted by x_{ik}, where index i labels the M feature vectors ($1 \leq i \leq M$) and index k labels the N feature vector components ($1 \leq i \leq N$). For the binary SVM, labeling of training data is done using label variable $y_i = \pm 1$ (with sign according to whether the training instance was from the positive or negative class). For hyperplane separability, elements of the training set must satisfy the following conditions: $w_\beta x_{i\beta} - b \geq +1$ for i such that $y_i = +1$, and $w_\beta x_{i\beta} - b \leq -1$ for $y_i = -1$, for some values of the coefficients $w_1, \ldots, w_N$, and b (using the convention of implied sum on repeated Greek indices). This can be written more concisely as: $y_i(w_\beta x_{i\beta} - b) - 1 \geq 0$. Data points that satisfy the equality in the above are known as "support vectors" (or "active constraints").

Once training is complete, discrimination is based solely on position relative to the discriminating hyperplane: $w_\beta x_{i\beta} - b = 0$. The boundary hyperplanes on the two classes of data are separated by a distance $2/w$, known as the "margin," where

$w^2 = w_\beta w_\beta$. By increasing the margin between the separated data as much as possible, the optimal separating hyperplane is obtained. In the usual SVM formulation, the goal to maximize w^{-1} is restated as the goal to minimize w^2. The Lagrangian variational formulation then selects an optimum defined at a saddle point of $L(w, b; \alpha) = (w_\beta w_\beta)/2 - (\alpha_\gamma y_\gamma(w_\beta x_{\gamma\beta} - b) - \alpha_0$, where $\alpha_0 = \Sigma_\gamma \alpha_\gamma$, $\alpha_\gamma \geq 0$ $(1 \leq \gamma \leq M)$. The saddle point is obtained by minimizing with respect to $\{w_1, \ldots, w_N, b\}$ and maximizing with respect to $\{\alpha_1, \ldots, \alpha_M\}$. If $y_i(w_\beta x_{i\beta} - b) - 1 \geq 0$, then maximization on α_i is achieved for $\alpha_i = 0$. If $y_i(w_\beta x_{i\beta} - b) - 1 = 0$, then there is no constraint on α_i. If $y_i(w_\beta x_{i\beta} - b) - 1 < 0$, there is a constraint violation, and $\alpha_i \to \infty$. If absolute separability is possible, the last case will eventually be eliminated for all α_i, otherwise it is natural to limit the size of α_i by some constant upper bound, that is, $\max(\alpha_i) = C$, for all i. This is equivalent to another set of inequality constraints with $\alpha_i \leq C$. Introducing sets of Lagrange multipliers, ξ_γ and μ_γ $(1 \leq \gamma \leq M)$, to achieve this, the Lagrangian becomes

$$L(w, b; \alpha, \xi, \mu) = (w_\beta w_\beta)/2 - \alpha_\gamma[y_\gamma(w_\beta x_{\gamma\beta} - b) + \xi_\gamma] + \alpha_0 + \xi_0 C - \mu_\gamma \xi_\gamma, \text{ where } \xi_0$$
$$= \Sigma_\gamma \xi_\gamma, \alpha_0 = \Sigma_\gamma \alpha_\gamma, \text{ and } \alpha_\gamma \geq 0 \text{ and } \xi_\gamma \geq 0 (1 \leq \gamma \leq M).$$

At the variational minimum on the $\{w_1, \ldots, w_N, b\}$ variables, $w_\beta = \alpha_\gamma y_\gamma x_{\gamma\beta}$, and the Lagrangian simplifies to $L(\alpha) = \alpha_0 - (\alpha_\delta y_\delta x_{\delta\beta} \alpha_\gamma y_\gamma x_{\gamma\beta})/2$, with $0 \leq \alpha_\gamma \leq C (1 \leq \gamma \leq M)$ and $\alpha_\gamma y_\gamma = 0$, where only the variations that maximize in terms of the α_γ remain (known as the Wolfe Transformation). In this form, the computational task can be greatly simplified. By introducing an expression for the discriminating hyperplane $f_i = w_\beta x_{i\beta} - b = \alpha_\gamma y_\gamma x_{\gamma\beta} x_{i\beta} - b$, the variational solution for $L(\alpha)$ reduces to the following set of relations (known as the Karush–Kuhn–Tucker, or KKT, relations): (1) $\alpha_i = 0 \Leftrightarrow y_i f_i \geq 1$, (2) $0 < \alpha_i < C \Leftrightarrow y_i f_i = 1$, and (3) $\alpha_i = C \Leftrightarrow y_i f_i \leq 1$. When the KKT relations are satisfied for all of the α_γ (with $\alpha_\gamma y_\gamma = 0$ maintained), the solution is achieved. (The constraint $\alpha_\gamma y_\gamma = 0$ is satisfied for the initial choice of multipliers by setting the α's associated with the positive training instances to $1/N^{(+)}$ and the α's associated with the negatives to $1/N^{(-)}$, where $N^{(+)}$ is the number of positives and $N^{(-)}$ is the number of negatives.) Once the Wolfe transformation is performed, it is apparent that the training data (support vectors, in particular, KKT class (2) above) enter into the Lagrangian solely via the inner product $x_{i\beta} x_{j\beta}$. Likewise, the discriminator f_i, and KKT relations are also dependent on the data solely via the $x_{i\beta} x_{j\beta}$ inner product.

Generalizations of the SVM formulation to data-dependent inner products other than $x_{i\beta} x_{j\beta}$ are possible and are usually formulated in terms of the family of symmetric positive definite functions (reproducing kernels) satisfying Mercer's conditions [25].

19.3.12 Binary SVM Discriminator Implementation

The SVM discriminators are trained by solving their KKT relations using the SMO procedure [28]. The method described here follows the description of Ref. [28] and begins by selecting a pair of Lagrange multipliers, $\{\alpha_1, \alpha_2\}$, where at least one of the multipliers has a violation of its associated KKT relations (for simplicity, it is assumed in what follows that the multipliers selected are those associated with the first and second feature vectors: $\{x_1, x_2\}$). The SMO procedure then "freezes" variations in all but the two selected Lagrange multipliers, permitting much of the computation to be circumvented by use of analytical reductions:

$$L(\alpha_1, \alpha_2; \alpha_{\beta' \geq 3}) = \alpha_1 + \alpha_2 - (\alpha_1^2 K_{11} + \alpha_2^2 K_{22} + 2\alpha_1 \alpha_2 y_1 y_2 K_{12})/2$$
$$- \alpha_1 y_1 v_1 - \alpha_2 y_2 v_2 + \alpha_{\beta'} U_{\beta'} - (\alpha_{\beta'} \alpha_{\gamma'} y_{\beta'} y_{\gamma'} K_{\beta' \gamma'})/2,$$

with $\beta', \gamma' \geq 3$, and where $K_{ij} \equiv K(x_i, x_j)$ and $v_i \equiv \alpha_{\beta'} y_{\beta'} K_{i\beta'}$ with $\beta' \geq 3$. Due to the constraint $\alpha_\beta y_\beta = 0$, we have the relation $\alpha_1 + s\alpha_2 = -\gamma$, where $\gamma \equiv y_1 \alpha_{\beta'} y_{\beta'}$ with $\beta' \geq 3$ and $s \equiv y_1 y_2$. Substituting the constraint to eliminate Ref. to α_1, and performing the variation on α_2: $\partial L(\alpha_2; \alpha_{\beta' \geq 3}/\partial \alpha_2 = (1 - s) + \eta \alpha_2 + s\gamma(K_{11} - K_{22}) + sy_1 v_1 - y_2 v_2$, where $\eta \equiv (2K_{12} - K_{11} + K_{22})$. Since v_i can be rewritten as $v_i = w_\beta x_{i\beta} - \alpha_1 y_1 K_{i1} - \alpha_2 y_2 K_{i2}$, the variational maximum $\partial L(\alpha_2; \alpha_{\beta' \geq 3})/\partial \alpha_2 = 0$ leads to the following update rule:

$$\alpha_2^{\text{new}} = \alpha_2^{\text{old}} - y_2((w_\beta x_{1\beta} - y_1) - (w_\beta x_{2\beta} - y_2))/\eta.$$

Once α_2^{new} obtained, the constraint $\alpha_2^{\text{new}} \leq C$ must be reverified in conjunction with the $\alpha_\beta y_\beta = 0$ constraint. If the $L(\alpha_2; \alpha_{\beta' \geq 3})$ maximization leads to a α_2^{new} that grows too large, the new α_2 must be "clipped" to the maximum value satisfying the constraints. For example, if $y_1 \neq y_2$, then increases in α_2 are matched by increases in α_1. So, depending on whether α_2 or α_1 is nearer its maximum of C, we have max $(\alpha_2) = \text{argmin}\{\alpha_2 + (C - \alpha_2); \alpha_2 + (C - \alpha_1)\}$. Similar arguments provide the following boundary conditions: (1) if $s = -1$, $\max(\alpha_2) = \text{argmin}\{\alpha_2; C + \alpha_2 - \alpha_1\}$ and $\min(\alpha_2) = \text{argmax}\{0; \alpha_2 - \alpha_1\}$, and (2) if $s = +1$, $\max(\alpha_2) = \text{argmin}\{C; \alpha_2 + \alpha_1\}$ and $\min(\alpha_2) = \text{argmax}\{0; \alpha_2 + \alpha_1 - C\}$. In terms of the new $\alpha_2^{\text{new,clipped}}$, clipped as indicated above if necessary, the new α_1 becomes

$$\alpha_1^{\text{new}} = \alpha_1^{\text{old}} + s(\alpha_2^{\text{old}} - \alpha_2^{\text{new,clipped}}),$$

where $s \equiv y_1 y_2$ as before. After the new α_1 and α_2 values are obtained, there still remains the task of obtaining the new b value. If the new α_1 is not "clipped," then the update must satisfy the nonboundary KKT relation: $y_1 f(x_1) = 1$, that is, $f^{\text{new}}(x_1) - y_1 = 0$. By relating f^{new} to f^{old}, the following update on b is obtained:

$$b^{\text{new1}} = b - (f^{\text{new}}(x_1) - y_1) - y_1(\alpha_1^{\text{new}} - \alpha_1^{\text{old}})K_{11} - y_2(\alpha_2^{\text{new,clipped}} - \alpha_2^{\text{old}})K_{12}.$$

If α_1 is clipped but α_2 is not, the above argument holds for the α_2 multiplier, and the new b is

$$b^{\text{new2}} = b - (f^{\text{new}}(x_2) - y_2) - y_2(\alpha_2^{\text{new}} - \alpha_2^{\text{old}})K_{22} - y_1(\alpha_1^{\text{new,clipped}} - \alpha_1^{\text{old}})K_{12}.$$

If both α_1 and α_2 values are clipped, then any of the b values between $b^{\text{new}1}$ and $b^{\text{new}2}$ is acceptable and following the SMO convention, the new b is chosen to be

$$b^{\text{new}} = (b^{\text{new}1} + b^{\text{new}2})/2.$$

19.3.13 SVM Kernel/Algorithm Variants

The SVM kernels that are used are based on "regularized" distances or divergences as those used in Refs. [3, 7, 27], where regularization is achieved by exponentiating the negative of a distance-measure squared ($d^2(x, y)$) or a symmetrized divergence measure ($D(x, y)$), the former if using a geometric heuristic for comparison of feature vectors and the latter if using a distributional heuristic. For the Gaussian Kernel, $d^2(x, y) = \Sigma_k(x_k - y_k)^2$; for the Absdiff Kernel $d^2(x, y) = [\Sigma_k|x_k - y_k|]^{1/2}$; and for the symmetrized relative entropy kernel $D(x, y) = D(x||y) + D(y||x)$, where $D(x||y)$ is the standard relative entropy.

19.3.14 SVM—External Clustering

As with the multiclass SVM discriminator implementations, the strong performance of the binary SVM enables SVM-external as well as SVM-internal approaches to clustering. The external-SVM clustering algorithm introduced in Ref. [27] clusters data vectors with no *a priori* knowledge of each vector's class. The algorithm works by first running a binary SVM against a data set, with each vector in the set randomly labeled, until the SVM converges. To obtain convergence, an acceptable number of KKT violators must be found. This is done through running the SVM on the randomly labeled data with different numbers of allowed violators until the number of violators allowed is near the lower bound of violators needed for the SVM to converge on the particular data set. Choice of an appropriate kernel and an acceptable sigma value will also affect convergence. After the initial convergence is achieved, the sensitivity plus specificity will be low, likely near 1. The algorithm now improves this result by iteratively relabeling the worst misclassified vectors that have confidence factor values beyond some threshold, followed by rerunning the SVM on the newly relabeled data set. This continues until no more progress can be made. Progress is determined by an increasing value of sensitivity plus specificity, hopefully nearly reaching 2. This method provides a way to cluster data sets without prior knowledge of the data's clustering characteristics, or the number of clusters.

19.4 RECENT ARCHITECTURAL REFINEMENTS

19.4.1 Data Inversion

A new form of "inverted" data injection is possible when the states and quantized emission values share the same alphabet. This new form may not have any clear

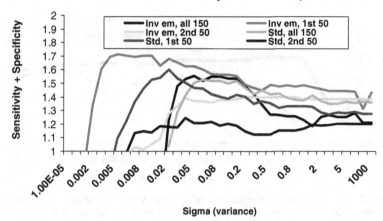

Figure 19.2 The binary classification performance using features extracted with HMM data inversion versus HMM standard. Blockade data were extracted from channel measurements of 9AT and 9CG hairpins (both hairpins with nine base-pair stems), and the data extraction involved either standard (std) emission data representations or inverted (inv) emission data, and was based on feature sets of the full 150 features, or the first 50, with the Viterbi-path level dwell time percentages, or the second 50, the emission variances (much weaker features as expected). The inverted data offer consistently better discriminatory performance by the SVM classifier.

probabilistic interpretation (use of "time-reversed" conditional probabilities or "absorption" instead of emission perhaps) but can be clearly defined in terms of the core data injection that occurs via the forward/backward variables, with emissions conditional probabilities taken with reversed conditional probabilities. Results shown in Fig. 19.2 are part of an extensive study that consistently shows approximately 5% improvement in accuracy (sensitivity + specificity) with the aforementioned data inversion (upon SVM classification), and this holds true over wide ranges of kernel parameters and collections of feature sets in all cases.

SVM performance on the same train/test data splits, but with 2600 uncompressed component feature vectors instead of 150 component feature vectors, offered similar performance after drop optimization. SVM performance with Adaboost on the 2600 components (taken as naive Bayes stubs), with selection for the top 150 "experts," demonstrates a significant robustness to what the SVM can "learn" in the presence of noise (some of the 2600 components have richer information, but even more are noise contributors). This also validates the effectiveness with which the 150-parameter compression was able to describe the two-state dominant blockade data found for the nine-base-pair hairpin and other types of "toggler" blockades.

19.4.2 Automated Feature Selection Using AdaBoost

Two new methods are being pursued for automated feature selection/feature compression. This is particularly important for handling the transition probabilities

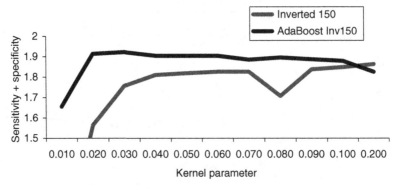

9GC versus 9TA classification results for AdaBoost versus Inv150

Figure 19.3   Adaboost feature selection strengthens the SVM performance of the Inverted HMM feature extraction set. Classification improvement with Adaboost taking the best 50 from the inverted emission 150-feature set. 95% accuracy is possible for discriminating 9GC from 9TA hairpins with no data dropped with use of Adaboost, and without Adaboosting, the accuracy is approximately 91%.

obtained by the HMM (if it has 50 states, it has 50^2 transition probabilities). The first builds on the transition probability compression in other ways optimized for the signals observed, and the second uses boosting (AdaBoost) over the individual emission and transition probabilities (which are used to provide a pool of weak, naïve Bayes, classifiers) to select the best features, and then use those features when passing feature vectors to the SVM classifiers, among other things. It is found, however, that boosting from the set of 150 features worked better than from the 2600 naive Bayes, and boosting from the 50 features in the first group worked best (see Fig. 19.3). (This result is also consistent with the PCA filtering in Ref. [31], mostly reducing the 150-feature set to the first 50 features.)

19.4.3 The Machine Learning Software Interface Project

Web-accessible machine-learning tools have been developed for general pattern recognition tasks, with specific application to channel current analysis, kinetic analysis, and computational genomics. The core machine learning tools are primarily based on SVM algorithms, HMM algorithms, and FSAs. The group Web site at http://logos.cs.uno.edu/~nano/ provides interfaces to (1) several binary SVM variants (with novel kernel selections and heuristics), (2) a multiclass (internal) SVM, (3) an SVM-based clustering tool, (4) an FSA-based nanopore spike detector, (5) an HMM channel current feature extraction tool, and (6) a kinetic feature extraction tool. The Web site is designed using HTML and CGI scripts that are executed to process the data sent when a form filled in by the user is received at the web server—results are then e-mailed to the address indicated by the user.

19.5 DISCUSSION

19.5.1 Individual Reaction Histories—Single Molecule Kinetics

Channel current-based kinetic feature extraction not only appears to be practical but are also the next key step in the study of individual reaction histories. In essence, binding strength (K_d) between molecules in solution, and conformational state transitions, can be determined via channel blockade observations corresponding to lifetimes on the different states. The ergodic hypothesis, that time averages can replace ensemble averages, can now be explored in this context as well. Nanopore detection promises to be a very precise method for evaluating binding strengths and observing single-molecule conformational changes. Adaptive software techniques to manage the complex data analysis are needed, and they are in growing demand. Recent advances have been made in channel current cheminformatics to address these issues, including new developments in distributed and unsupervised learning processes.

19.5.2 Deciphering the Transcriptome and Transcription Factor-based Drug Discovery

The examination of transcription factor binding to target transcription factor binding site (TF/TFBS interactions) affords the possibility to understand, quantitatively, much of the transcriptome. This same information, coupled with new interaction information upon introduction of synthetic TFs (possible medicines), provides a very powerful, directed approach to drug discovery.

19.5.3 A New Window into Understanding Antibody Function

Upon binding to antigen, a series of events are initiated by the interaction of the antibody carboxy-terminal region with serum proteins and cellular receptors. Biological effects resulting from the carboxy-terminal interactions include activation of the complement cascade, binding of immune complexes by carboxy-terminal receptors on various cells, and the induction of inflammation. Nanopore detection provides a new way to study the binding/conformational histories of individual antibodies. Many critical questions regarding antibody function are still unresolved, questions that can be approached in a new way with the nanopore detector. The different antibody binding strengths to target antigen, for example, can be ranked according to the observed lifetimes of their bound states. Questions of great interest include: Are allosteric changes transmitted through the molecule upon antigen binding? Can effector function activation be observed and used to accelerate drug discovery efforts?

19.5.4 Hybrid Clustering and Scan Clustering for Indirect-Interaction Kinetic Information

An exciting area of machine learning research is being brought to bear on the kinetic signal decomposition of channel currents. The external-SVM approach described in the background and Ref. [27] offers to provide one of the most powerful, unsupervised methods for clustering. Part of the strength of the method is that it is nonparametric. Part of the weakness is in obtaining an initial clustering. To improve on this, efforts are underway to graft the external SVM onto an initial clustering using bisect-K-means (that is seeded by principle direction divisive partitioning [32–34], or principle component analysis [35], when random seeding does poorly. External-SVM clustering, along the lines of Ref. [27], may allow precise cluster regrowth by its ability to operate on a shifting support vector structure as direct label operations (binary "flipping") are performed.

REFERENCES

1. Winters-Hilt, S. Single-molecule Biochemical Analysis Using Channel Current Cheminformatics. *UPoN 2005: Fourth International Conference on Unsolved Problems of Noise and Fluctuations in Physics, Biology, and High Technology, June 6–10, 2005. AIP Conference Proceeding*, Vol. 800, pp. 337–342, 2005.
2. Winters-Hilt, S. and Akeson, M. Nanopore cheminformatics. *DNA and Cell Biology*, 23 (10):675–683, 2004.
3. Winters-Hilt, S., Vercoutere, W., DeGuzman, V. S., Deamer, D. W., Akeson, M., and Haussler, D. Highly accurate classification of Watson–Crick basepairs on termini of single DNA molecules. *Biophysical Journal*, 84: 967–976, 2003.
4. Winters-Hilt, S. Highly accurate real-time classification of channel-captured DNA termini. *Third International Conference on Unsolved Problems of Noise and Fluctuations in Physics, Biology, and High Technology*, pp. 355–368, 2003.
5. Vercoutere, W., Winters-Hilt, S., DeGuzman, V. S., Deamer, D., Ridino, S., Rogers, J. T., Olsen, H. E., Marziali, A., and Akeson, M. Discrimination among individual Watson–Crick base-pairs at the termini of single DNA hairpin molecules. *Nucleic Acids Research* 31: 1311–1318, 2003.
6. Vercoutere, W., Winters-Hilt, S., Olsen, H., Deamer, D. W., Haussler, D., and Akeson, M. Rapid discrimination among individual DNA hairpin molecules at single-nucleotide resolution using an ion channel. *Nature Biotechnology*, 19(3): 248–252, 2001.
7. Winters-Hilt, S. Nanopore detector based analysis of single-molecule conformational kinetics and binding interactions. *BMC Bioinformatics*, 7 (Suppl 2): S21, 2006.
8. Winters-Hilt, S., Landry, M., Akeson, M., Tanase, M., Amin, I., Coombs, A., Morales, E., Millet, J., Baribault, C., and Sendamangalam, S. Cheminformatics methods for novel nanopore analysis of HIV DNA termini. *BMC Bioinformatics*, 7 (Suppl 2): S22, 2006.
9. Song, L., Hobaugh, M. R., Shustak, C., Cheley, S., Bayley, H., and Gouaux, J. E. Structure of staphylococcal alpha-hemolysin, a heptameric transmembrane pore. *Science*, 274 (5294):1859–1866, 1996.
10. Coulter, W. H. High speed automatic blood cell counter and cell size analyzer. *Proceedings of the National Electronics Conference*, 12: 1034–1042, 1957.
11. Kasianowicz, J. J., Brandin, E., Branton, D., and Deamer, D. W. Characterization of individual polynucleotide molecules using a membrane channel. *Proceedings of the National Academy of Sciences of the United States of America*, 93(24): 13770–13773, 1996.

12. Akeson, M., Branton, D., Kasianowicz, J. J., Brandin, E., and Deamer, D. W. Microsecond time-scale discrimination among polycytidylic acid, polyadenylic acid, and polyuridylic acid as homopolymers or as segments within single RNA molecules. *Biophysical Journal*, 77(6): 3227–3233, 1999.

13. Meller, A., Nivon, L., Brandin, E., Golovchenko, J., and Branton, D. Rapid nanopore discrimination between single polynucleotide molecules. *Proceedings of the National Academy of Sciences of the United States of America*, 97(3): 1079–1084, 2000.

14. Meller, A., Nivon, L., and Branton, D. Voltage-driven DNA translocations through a nanopore. *Physical Review Letters*, 86(15): 3435–3438, 2001.

15. Bezrukov, S. M. Ion channels as molecular coulter counters to probe metabolite transport. *The Journal of Membrane Biology*, 174: 1–13, 2000.

16. Bezrukov, S. M., Vodyanoy, I., and Parsegian, V. A. Counting polymers moving through a single ion channel. *Nature*, 370(6457): 279–281, 1994.

17. Sakmann, B., and Neher, E. *Single-Channel Recording*. Plenum Press, New York 1995.

18. Ashcroft, F. *Ion Channels and Disease*. Academic Press, 2000.

19. Cormen, T. H., Leiserson, C. E., and Rivest, R. L. *Introduction to Algorithms*. MIT-Press, Cambridge, MA, 1989.

20. Chung, S.-H., Moore, J. B., Xia, L., Premkumar, L. S., and Gage, P. W. Characterization of single channel currents using digital signal processing techniques based on Hidden Markov models. *Philosophical Transactions of the Royal Society of London, Series B, Biological Sciences*, 329: 265–285, 1990.

21. Chung, S.-H., and Gage, P. W. Signal processing techniques for channel current analysis based on hidden Markov models. In: P. M. Conn,ed., *Methods in Enzymology; Ion Channels, Part B*, Academic Press, Inc., San Diego, pp. 420–437, 1998.

22. Colquhoun, D., and Sigworth, F. J. Fitting and statistical analysis of single-channel products. In: B. Sakmann and E. Neher, eds, *Single-Channel Recording*, 2nd edition, Plenum Publishing Corp., New York, pp. 483–587, 1995.

23. Durbin, R. *Biological Sequence Analysis: Probabilistic Models of Proteins and Nucleic Acids*. Cambridge University Press, Cambridge, UK, 1998.

24. Winters-Hilt, S. Hidden Markov model variants and their application. *BMC Bioinformatics*, 7 (Suppl 2): S14, 2006.

25. Vapnik, V. N. *The Nature of Statistical Learning Theory*, 2nd edition, Springer-Verlag, New York, 1998.

26. Burges, C. J. C. A tutorial on support vector machines for pattern recognition. *Data Mining Knowledge Discovery*, 2: 121–167, 1998.

27. Winters-Hilt, S., Yelundur, A., McChesney, C., and Landry, M. Support vector machine implementations for classification & clustering. *BMC Bioinformatics*, 7 (Suppl 2): S4, 2006.

28. Platt, J. C. Fast training of support vector machines using sequential minimal optimization. In: B. Scholkopf, C. J. C. Burges, and A. J. Smola, eds, *Advances in Kernel Methods— Support Vector Learning*, MIT Press, Cambridge, MA, 1998.

29. Osuna, E., Freund, R., and Girosi, F. An improved training algorithm for support vector machines. In: J. Principe, L. Gile, N. Morgan, and E. Wilson,eds, *Neural Networks for Signal Processing VII*, IEEE, New York, pp. 276–285, 1997.

30. Joachims, T. Making large-scale SVM learning practical. In: B. Scholkopf, C. J.C. Burges, and A. J. Smola,eds, *Advances in Kernel Methods—Support Vector Learning*, MIT Press, Cambridge, MA, 1998.

31. Iqbal, R., Landry, M., and Winters-Hilt, S. DNA molecule classification using feature primitives. *BMC Bioinformatics*, 7 (Suppl 2): S15, 2006.

32. Hand, D., Mannila, H., and Smyth P. *Principles of Data Mining*, The MIT Press, Cambridge, MA, 2001.

33. O'Connel, M. J. Search program for significant variables. *Computer Physics Communications*, 8: 49–55, 1974.

34. Wall, M. E., Rechtsteiner, A., and Rocha, L. M. *Singular value decomposition and principle component analysis*. In: D. P. Berrar, W. Dubitsky, and M. A. Granzow,eds, *A Practical Approach to Microarray Data Analysis*, Kluwer, Norwell, MA, pp. 91–109, 2003.

35. Boley, D. L. Principle direction divisive partitioning. *Data Mining and Knowledge Discovery*, 2(4): 325–344, 1998.

20

AN INFORMATION FUSION FRAMEWORK FOR BIOMEDICAL INFORMATICS

Srivatsava R. Ganta, Anand Narasimhamurthy,
Jyotsna Kasturi, and Raj Acharya

20.1 INTRODUCTION

Biomedical informatics deals with the study and analysis of various clinical and genomic data. This involves highly heterogeneous data sets such as patient demographics, tissue and pathology data, treatment history, and patient outcomes, as well as genomic data such as DNA sequences, and gene expression data. The goal is to discover hidden trends and patterns in these information sources that could be used to verify more complex hypotheses. Such studies could lead to better disease diagnosis approaches, treatment procedures, and drug development.

The data sets used in biomedical informatics research are usually distributed among various hospitals, research centers, and government agencies and are controlled independently. Hence, researchers have so far been limited to islands of data and informatics tools. This scenario poses a serious challenge to the study of the disease from a global point of view. The heavy distribution of data and tools imposes restrictions on the extent of data analysis and exploration possible. However, with the advent of nationwide grid systems such as CaBIG (Cancer Bioinformatics Grid) [1], sharing data and tools across organizational and even international boundaries has been made feasible. Such infrastructures facilitate a comprehensive collaborative

Machine Learning in Bioinformatics. Edited by Yan-Qing Zhang and Jagath C. Rajapakse
Copyright © 2009 by John Wiley & Sons, Inc.

research platform for global disease-related studies. The availability of various data sets and tools provides an opportunity for integrated data exploration and simultaneous analysis of multiple data sets.

"Data exploration" involves the application of various technologies to examine large collections of data for structure, patterns, faults, and other characteristics. While related to the large universe of data mining, data exploration involves a more human-centered approach to this pattern discovery. This involves tools and techniques that are used by analysts and researchers to retrieve and examine data through what is often an interactive- and intuitive-based process of trial and error. Data exploration generally involves four broad technologies: statistical descriptions of the data, structured queries against databases, multidimensional visualization, and the automatic clustering and organization of data around common features. In the context of clinical data, there has been some work [2, 3] on data exploration using visualization techniques. These techniques lay emphasis on visualization of a single data set, whereas a query-based exploration of a set of related information sources is often more critically important process for the analyst.

Data analysis in biomedical informatics is usually carried out using techniques such as clustering [4], association rule mining [5], and so on. However, these studies mainly involve only either clinical or biological data sets. For example, application of clustering techniques has by far been limited to gene expression data. It has been observed in recent studies that the richness of the knowledge extracted from information sources separately is limited. "Information fusion" involves the use of multiple, heterogeneous information sources to extract "richer" knowledge. Several techniques [6, 7] have been proposed to signify the use of multiple data sources to perform information fusion-based analysis of data sets. Hence, there is a need for a platform to facilitate integrated data exploration and simultaneous analysis of multiple biomedical informatics data.

In this chapter, we present an online data warehouse platform to perform and verify data exploration and analysis across heterogeneous biomedical informatics data sets using information fusion-based techniques. Figure 20.1 gives an overall view of the goal. The platform is aimed at providing two main functionalities: (1) A data warehouse that serves as a one-point access for biomedical informatics data sets related to various diseases and (2) an environment that provides a suite of tools to perform information fusion on biomedical informatics data. We demonstrate the functionalities of the system using prostate cancer-related data sets. However, the platform can be used with any disease-related data sets. We also present example scenarios in which the results obtained are used to verify complex hypotheses. The data were collected from the leading cancer research centers in the state of Pennsylvania as part of The Pennsylvania Cancer Alliance for Bioinformatics Consortium (PCABC) in addition to some publicly available data sets. The rest of the chapter is organized as follows: Section 20.2 motivates the need for an information fusion-based data exploration and analysis tools for biomedical informatics. Section 20.3 provides a brief background on data warehousing and the system architecture followed by the suite of tools available on the platform. Section 20.4 provides the summary and conclusion.

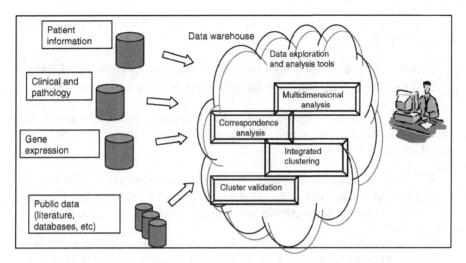

Figure 20.1 Information fusion-based data exploration and analysis.

20.2 MOTIVATION

Prostate cancer is the most commonly diagnosed nonskin cancer in the United States with an estimated 198,100 new cases and 31,500 deaths in 2001 [8]. One in six American men develops prostate cancer in the course of his or her lifetime. Disease data can be broadly categorized into two types: Clinical and Genomic. Clinical data consist of patient demographics, diagnosis information, treatment records, and follow-up data while genomic data consist of more complex data such as gene expression, DNA sequences, and so on. Patient demographics are collected before the start of the diagnosis and consist of the patient's age, sex, family history of cancer, and so on. Exploration and analysis of this information set gives a global view of the trends in various disease-related attributes and insights into spread of the disease in different races and geographic locations. Examples of some generic queries are as follows

1. What is the average age at diagnosis for African-American patients?
2. What percentage of the patients recorded have a family history of prostate cancer?

Similarly, such queries can be posed on other types of clinical data such as treatment and follow-up. For example, pathologic studies are done during the diagnostic process to categorize the type and aggressiveness of the tumor (histo-logical type and grade) as well as the extent of tumor spread (stage). In most cases, the disease is cured by the initial therapy, however, in some cases the tumor recurs. Documentation of the time and extent of this recurrence is important in better understanding the biologic nature of aggressive tumors. Example queries on these data sets are as follow

1. How many patients have been categorized with different pathologic T-stages?
2. What is the average tumor size for cases with Gleason grades of greater than 3?

While the above-mentioned scenarios deal with only clinical information, the studies can be extended to multiple data sets including gene expression and sequence data. Example queries are as follows

1. What is the average gene expression vector for patients with Gleason score of 5?
2. Based on certain ontology for cellular functions, what is the average expression vector of all the genes corresponding to a particular cellular function?

These queries help the researcher in exploring the data, understanding the relationships between initial grade and stage, and therapy and recurrence (and outcome), and how this correlates with gene expression and sequences, which is a major area of biomedical informatics research. Thus, there is a need for a tool that helps the user pose such queries across multiple data sets.

The various data components of biomedical informatics research such as patient records, clinical and pathology data, treatment history, patient outcomes, as well as gene expression data and other genomic experiments are associated to each other in different ways. These associations may lead to trends that correlate parameters corresponding to disparate sources. For example, clinical data of certain diseases indicate that patients from certain races have lesser age at diagnosis when compared to others. In this sense, the race of patient is related to average age of diagnosis of the patient. This may possibly indicate that people of certain races may be prone to a disease much earlier than people of other races. Although associations such as the above can be detected by simple queries (finding average age at diagnosis for the different races), more complex associations are usually hidden and cannot be detected easily from the data. Thus, exploratory analysis of these data sets is needed to detect these associations.

Another key facet of the biomedical informatics studies involves mining gene expression data to identify disease biomarkers [8, 9]. This is usually carried out using clustering techniques such as hierarchical, K-means, SOM, and so on [4]. However, in the context of biomedical informatics, application of clustering techniques so far has been limited to gene expression data. Recent work on this problem [6, 7] emphasizes the use of multiple data sets through "information fusion" to obtain better clustering results. This suggests the inclusion of additional information sources such as sequences and clinical data for clustering genes. Thus, there is a need for tools that provide the user with clustering capabilities on multiple data sets.

Clustering is an "unsupervised" exploratory technique, and hence most of the clustering algorithms need the user to specify the number of clusters prior to the analysis. Cluster validation studies [10–12] deal with the identification of the "optimal" number of clusters in a given data set. This is achieved by evaluating the clustering result for a given number of clusters. However, these studies offer only a generic quantification mechanism to evaluate the results of a clustering algorithm on a specific data set. In the context of information fusion, there is a need for specialized extension of these techniques to account for multiple data sets and different distance measures.

The current platform aims to provide the user with a suite of tools to perform fusion-based exploration, analysis, and validation as noted above. The "multidimensional analysis" tool helps the user to formulate exploratory queries across multiple data sets in a generic manner. This is discussed in Section 20.3.1 in more detail. The "correspondence analysis" tool aims to help the user detect associations among various biomedical informatics data sets. This is based on a novel application of a statistical technique [13] and is described in Section 20.3.2. The "integrated clustering" tool provides the user a means to cluster gene expression data along with other data sources such as sequence data and is presented in Section 20.3.3. Finally, the "cluster validation" tool provides means to verify the quality of the results obtained and justify the use of information fusion.

20.3 PLATFORM

In this section, we present a brief background on data warehousing and the system architecture followed by the suite of tools available on the platform. As mentioned in the previous sections, the goal of the platform is to facilitate two main functionalities: (1) A data warehouse that serves as a one-point access to biomedical informatics data sets related to various diseases and (2) an environment that provides a suite of tools to perform information fusion-based data exploration and analysis on biomedical informatics data clustering capabilities on multiple data sets.

"Data warehousing" encompasses the architectures, algorithms, and tools for bringing together selected data from multiple databases and information sources. Traditionally, warehousing is done by the lazy or on-demand approach that involves a two-step process: (1) accept a query, determine the appropriate set of information sources to answer the query, and then fire the subqueries to corresponding data sources, and (2) retrieve the results back from each repository and compute the final answer for the user. The disadvantage of this approach is that data are not retrieved until a query is fired and involves some delay. Data warehousing involves the alternative approach of prefetching the data so that specific analysis queries can be answered in an optimized way. This approach yields better results than the *on-demand* approach when the system is targeted at specific data analysis and exploration operations.

Our system architecture is based on a three-tier design and is presented in Fig. 20.2. It consists of three layers: the presentation layer, the application layer and the data warehouse layer. The presentation layer essentially deals with the presentation of results from various information fusion-based techniques. It is currently implemented using Java applets along with JSPs to provide the required visualization functionalities.

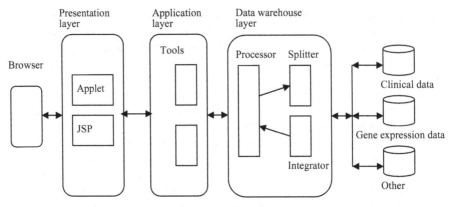

Figure 20.2 System architecture.

The application logic layer deals with the implementation of the various tools offered through the system. The data warehouse layer takes care of the necessary-query processing capabilities required for integration from heterogeneous data sets stored in the data warehouse. The processor module accomplishes this by sending in preprocessed queries to the splitter, which directs the data requests to appropriate section of the data warehouse. The data integrator collects the data and processes it to cross-link each of the data items obtained and submits it to the processing engine. The final results are submitted to the presentation layer for visualization. This design offers the flexibility and robustness required for extending the functionalities offered on the system.

Information fusion involves analysis of multiple heterogeneous data sets through specialized techniques. The idea is to extract knowledge from these data sets simultaneously with the aim that the intrinsic relationship between the data sets yields better results. Different data sets are stored in different formats and models based on their individual properties. A major hurdle in conducting fusion-based studies is the lack of common data model for the various data sets involved. A data model that can support fusion-based studies should be able to capture the data sets involved in the most generic way possible and at the same time preserve their relationship. We can identify the requirements of such a data model as follows:

(1) The data model needs to capture multiple interrelated data sets.
(2) The model should capture each of the data sets without adding any additional information.
(3) The model should capture each of the data sets in a lossless manner.
(4) The model should highlight the relationship between two data sets.
(5) The semantics of the model should be easily comprehensible.

In our system, we use an adapted version of the data cube [14] proposed by Gray et al. to serve as the data model for information fusion. The basic idea is to model information as "facts" and "dimensions." The "facts" are the "data of interest," that is, the data to be analyzed with respect to the "dimensions." A data cube can have multiple attributes as facts for the same set of dimensions. Each of the dimensions and facts belongs to specific domains. We define the cube as a set of tuples:

$$\mathbf{CUBE}() = \{(d_1, d_2, \ldots, d_n, f_1, f_2, \ldots, f_m), \ldots\}, \qquad (20.1)$$

where d_i is the dimension value and f_i is the fact value, and there are n dimensions and m fact values in each tuple. Each dimension d_i belongs to the domain D_i. Figure 20.3 shows a conceptual three-dimensional data cube over patient demographic data. It consists of three dimensions: "age of the patient," "Tissue sample type," and "race to which the patient belongs to" and a single fact "number of patients." In the data warehousing literature, there has been some previous work [15] that formalizes data modeling using a cube in a similar manner. Tao et al. [16] applied this model for simultaneous visualization of genotypic and phenotypic data sets.

The rest of the section presents the information fusion tools offered through the platform. We use prostate cancer-related data sets collected from the leading cancer research centers in the state of Pennsylvania as part of PCABC and some publicly available data to demonstrate the functionalities of the system. However, the platform can be used with any disease-related data sets.

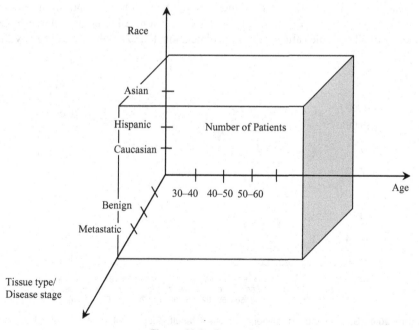

Figure 20.3 Data cube.

20.3.1 Multidimensional Analysis

The "multidimensional analysis" tool facilitates the exploration and visualization of global statistics and associations among various biomedical informatics data sets. The understanding and interpretation of these statistics and associations helps identify disease hot spots, geographical-spread patterns, and other global insights. The tool achieves this by providing a means to formulate queries to explore the data available on the warehouse.

The tool consists of a query formulation module that helps the user formulate the global statistic query and a visualization module that helps in the visual interpretation of the result. The queries are formulated by first selecting the subset of the information sources to be considered for the analysis. Once this done, the user selects a certain attribute (fact) from one of the data sets as the focus of analysis, and some attribute, (dimensions) along which the fact needs to be analyzed. For example, consider the query "What percentage of the prostate cancer patients belong to the African-American race and fall in the age range of 50–60?" In this case, the fact is "percentage of patients" and the dimensions are "race" and "age at diagnosis." Figure 20.4 presents the result obtained for a similar query depicting the number of prostate cancer cases registered with respect to the age of the patient. Visual interpretation of the result leads to the conclusion that the graph peaks for the age values 63–67. Looking at the same piece of information at different granularity levels is usually very important in exploring the data set. The tool facilitates this by allowing the user to perform some operations on the result obtained to further focus on the knowledge of interest. The operations are as follows:

Summary: This operation takes in the current information view and summarizes it based on a certain hierarchy defined over the dimension. A hierarchy is a tree-based grouping of all possible values for a given dimension. For example, gene ontology can

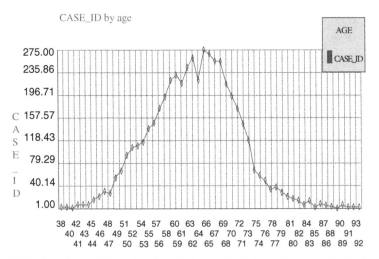

Figure 20.4 Sample multidimensional analysis result: Fact–number of patients diagnosed against the dimension–age of the patient.

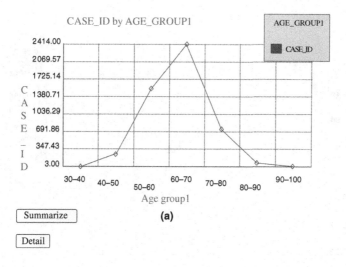

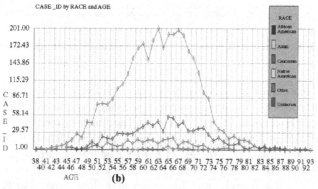

Figure 20.5 Sample multidimensional analysis result (**a**) using summerization and (**b**) using multidimensional analysis on clinical and gene expression data.

be looked as a hierarchy defined on genes based on the cellular functions they belong to. Using this operation, the user can view the fact values for various levels of dimension values. Figure 20.5a shows the result obtained by summarizing the age of the patient into "age group" categories for the one shown in Fig. 20.4. The platform supports some of the most commonly used aggregation operations such as average, maximum, minimum, and so on.

Detail: This operation is the complement of summarization. It presents the user with a detailed view of the cube. In effect, it is used to go down the hierarchy defined of the dimension. The tool allows the user to select hierarchies on each of the dimensions and also add dimensions to the resulting cube. Figure 20.5b depicts the result obtained by adding the dimension "race" to the result obtained earlier in Fig. 20.4.

The sample queries dealt with so far involved only patient demographics data. Association queries on multiple data sets can be posed in similar manner by choosing one data set as the fact and other data set(s) as the dimensions Fig. 20.6. shows a

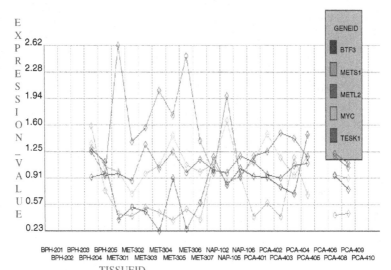

EXPRESSION_VALUE by GENEID and TISSUEID

Figure 20.6   Sample result: using multidimensional analysis on clinical and gene expression data.

sample result that depicts the gene expression values of a set of genes along the clinical stages of the tissues from which the samples are taken in a study [9] involving the identification of biomarkers for prostate cancer. The clinical stages are coded as BPH, benign; NAP, normal adjacent; PCA, localized; and MET, metastatic. In the actual literature [9], it is found that the genes MYBL2 and MYC are overexpressed in malignant tumors. Visual interpretation of the result leads to a similar conclusion and quantifies the amount of overexpression observed in these genes. Further, it can be observed that MYBL2 is much more dominant in metastatic tumors than when compared to MYC. This could be justified by using the summarize operation on this result along the clinical stages as shown in Fig. 20.7.

20.3.2   Correspondence Analysis

The correspondence analysis tool helps the user further explore the data cube presented by multidimensional analysis tool by providing visualization and quantification of the association between the facts and the dimensions of the cube. The tool is based on an exploratory analysis technique proposed in Ref. [13]. The primary goal here is to facilitate the discovery and analysis of associations present in the various data sets involved in biomedical informatics. For example, consider a simple two-dimensional cube with "number of patients" as the fact and "age group of the patient" and "race of the patient" as dimensions. Given this, the user is interested to know "for patients belonging to race A, which is the most likely age group at which they are diagnosed" from a global point of view of the data set. Although associations such as the above can be detected by simple queries (finding average age-at-diagnosis for the

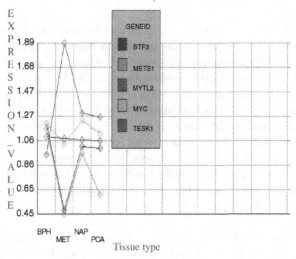

EXPRESSION_VALUE by GENEID and TISSUEID

Figure 20.7   Sample result: using multidimensional analysis on clinical and gene expression data.

different races), more complex associations are usually hidden and cannot be detected easily from the data.

The central idea behind correspondence analysis is a profile. The profile of a set of entries (either row or column of a data cube) is simply the set of entries normalized by the corresponding total. If the input data cube (for simplicity, consider 2D data cube) is denoted by $N(i, j)$ $1 \leq i \leq I;$ $1 \leq j \leq J$, the row and column profile vectors are given by

where

$$r_i = \frac{N(i,:)}{\sum\limits_{j=1}^{J} N(i,j)} \qquad c_j = \frac{N(:,j)}{\sum\limits_{i=1}^{I} N(i,j)}$$

r_i is the profile of ith row

and

c_j is the profile of the jth column.

The output of the tool is a graphical display known as a *map*, a plot in which the row and column profiles are depicted as points. We briefly summarize the symmetric map as below:

(1) The origin corresponds to the centroid of the data profile, that is, the average profile.

(2) It is comprised of the "optimal displays" of the row and column profiles, although strictly speaking these two sets of points occupy different spaces.

(3) The map is scaled such that row and column points are equally spread out along each principal axis (for a 2D plot, along the horizontal and vertical directions).

(4) Although there is no direct interpretation of the distance between a row and a column point, there is certainly a joint interpretation of the row and column points with respect to the principal axes of the map.

However, unlike some of the other methods such as principal component analysis (PCA), which depict a low-dimensional projection, correspondence analysis displays both the row and column profiles simultaneously on the same plot. Another difference between PCA and correspondence analysis is that whereas in PCA the Euclidean distance is used, in correspondence analysis the ξ^2 distance is used. We skip further details here for the sake of simplicity, and the interested reader is referred to Ref. 13 for a thorough exposition and computational details of correspondence analysis.

We now demonstrate the functionalities of the tool using patient demographic data and gene expression data. Consider a 2D data cube along the dimensions "race" and "age" of the patient with "total number of patients" as the fact. We take the result obtained using the "multidimensional analysis" tool by adding "race" dimension to the result presented in Fig. 20.5. Running the correspondence analysis tool on such a data set would result in a map shown in Fig. 20.8. The following are some observations that can be drawn from this result.

1. The proximity of the points age group "(50–60)" and race group "Caucasian" to the origin indicates a relatively strong association between them and the average profile. We may hypothesize that middle aged and older Caucasians are more prone to prostate cancer as compared to other age groups.

2. The age groups (<40) and (80–90), which are far away from the origin, are atypical profiles.

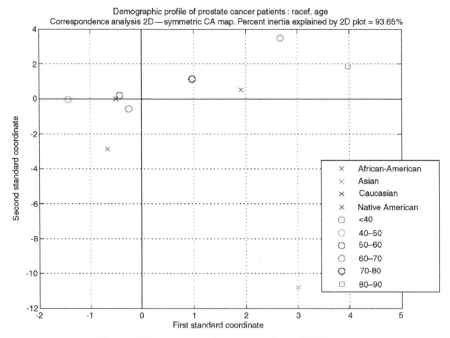

Figure 20.8   Correspondence analysis on clinical data.

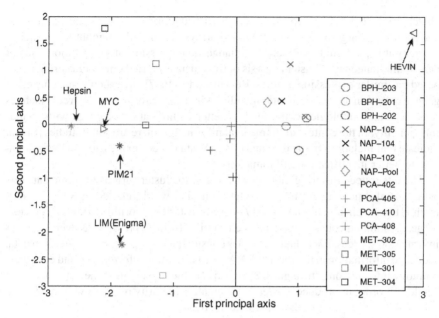

Figure 20.9 Correspondence analysis on gene expression data.

Similar studies can be made on other data sets such as gene expression data. Figure 20.9 displays the result obtained for gene expression data. For consistency we use the same data set that we used earlier [9] for demonstrating the multidimensional analysis tool. The input data set contained gene expression profiles of 9984 human cDNA to analyze gene expression profiles in benign and malignant prostate tissues. The tissues are categorized as BPH, benign prostate hyperlysia; NAP, normal adjacent prostate; PCA, localized; and MET, metastatic.

It can be observed that the benign and normal samples (BPH and NAP) lie to the right of the origin (with respect to the first principal axis), while the malignant states (PCA and MET samples) lie on the other side. Thus, for this data set, we see a separation between the benign and the malignant states along one of the principal axes that was observed by the original paper through clustering. The genes that lie to the left of the origin tend to have a "positive association" with the malignant states. Such genes may be candidates for biomarkers especially if they are close to any of the clinical sample points in the graph. Ex. Hepsin, LIM (Enigma), PIM1, MYC were identified in the original paper. These conclusions were actually hypothesized and verified in the original study [9], and the tool helps the user visualize and interpret them in a comprehensible way.

20.3.3 Integrated Clustering

Clustering of gene expression data is one of the most widely used analysis techniques in biomedical informatics. Cluster analysis aims at identifying groups of similar objects and, therefore, helps to discover distribution of patterns and interesting correlations in large data sets. DNA microarrays are high-throughput experiments, whereby it is possible to measure the changes in expression of several thousands of genes simultaneously. Cluster analysis of these data is carried out by clustering genes based on similar expression profiles. Recent studies [6, 7] have shown that clustering gene expression data by including other kinds of information sources such as sequences and ontologies could lead to better identification of genes with similar functionality. In the context of biomedical informatics, there are some studies [8] that indicate that including tissue data and clinical data in gene expression analysis might lead to better identification of biomarkers.

The integrated clustering tool allows the user to cluster various genomic data sets using an information fusion, based algorithm that we developed in Ref. [7]. This method uses a self-organizing map [17]-based clustering algorithm to identify clusters of genes simultaneously based on their similarity in expression as well as other information sources. Each data set used is considered a *category*, like expression data, sequence information in the form of DNA motifs, location information, and even gene ontology information. The algorithm is also capable of weighting the data sources, if needed, in order to produce clusters with greater similarity of genes within one data source when compared to the other data sources.

The algorithm uses an iterative procedure by which the probability distribution of the data is reproduced as closely as possible. At each iteration step, a category is randomly selected based on the weighting scheme P. The chosen category r and its associated distance function dr are used to train the network of neurons and the weights for the entire input tuple (of dimension $N_1 + N_2 + \cdots + N_m$) are updated using the Kohonen learning rule [17], although distances are calculated on each segment of the input vector independently using the appropriate distance. Information-theoretic similarity measures such as Kullback–Liebler are preferred for clustering of intensity values [18]. To measure similarity between genes based on frequency of motif occurrence, we use a measure based on the extended Jaccard similarity coefficient. Interested readers are referred to Ref. [7] for further details on the algorithm.

We now present the usage of the tool and some sample results obtained by running the tool on gene expression data collected from Spellman et al. [19]. The data set consists of gene expression data for yeast cell cycle data. The data set was chosen since it was well studied in the literature, and several interesting observations have been made based on cluster analysis. The usage of the tool involves the selection of data sets to be clustered and the weights to be assigned to each data set. The tool outputs the result in a visualizable format depicting the resulting clusters. Figures 20.10 and 20.11 present the results obtained by running this tool by considering only gene expression data DNA sequence data (motif frequency data), respectively. Figure 20.12 presents the clustering results obtained by using the integrated clustering algorithm on both the gene expression data and the sequence data. The results obtained through the integrated clustering provide tighter clusters and correlate well with both sequence data and gene expression data [7]. Further analysis of one of the clusters obtained using integrated clustering showed that four out of five genes in the cluster, namely, CTF4, POL30,

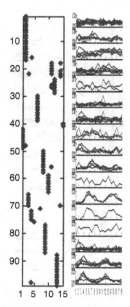

Figure 20.10   Clustering of sequence data.

HYS2, and POL32 were mentioned in the original paper to be DNA Syn-related genes. These genes also share a common transcription factory MCBa. These results suggest that genes that have a similar function might share a common expression profile and also a common motif. A quantified approach to signify the integrated clustering results is provided in the next section using the cluster validation tool.

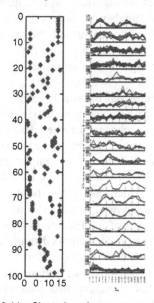

Figure 20.11   Clustering of gene expression data.

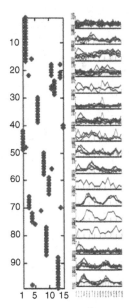

Figure 20.12 Clustering of gene expression and sequence data.

20.3.4 Cluster Validation

Clustering techniques offer an "unsupervised" means of mining gene expression data. Most of the clustering algorithms need to have the "initial" number of clusters as input. The performance of the algorithm and the quality of results usually vary with the quality of the input number of clusters. Furthermore, one would need to evaluate the quality of results obtained through the particular clustering algorithm. Cluster validity deals with the problem of identifying the "optimal" number of clusters in a given data set using a measurable index value to compare the quality of clustering.

Cluster validation procedures may be broadly categorized as internal and external validation. Internal validation procedures look at the data alone and typically employ criteria such as inter- and intracluster distances. External validation on the other hand involves comparing the given clustering with an external clustering. A number of cluster validation indices (CVIs) and techniques have been proposed in the clustering literature. Examples of internal CVIs include the Dunn [11], Hartigan [20], Calinski-Harabasz [10], and Davies and Bouldin [12] indices. Meila [21] classifies methods for comparing two clusterings as pair counting (e.g., Rand index [22]), set matching and variation-of-information (VI) (e.g., Ref. [21]). More recent work such as Zhou et al. [23] propose a Mallows distance-based metric for comparing clustering results obtained from different algorithms.

The cluster validation tool helps the user perform cluster validation on standard clustering techniques as well as information fusion-based algorithms. To achieve this, we propose a two-staged cluster validation protocol. The first stage predominantly employs internal CVIs (although, in general, external CVIs could be used for this stage too) to aid the user in estimating the "true" number of clusters in the data that reflects the structure in the data as best as possible. In the second stage, the optimal clustering thus determined may be compared against an appropriately chosen "reference" clustering. Considering that our system deals with heterogeneous data sets, we provide a limited number of indices. This is because most cluster validity indices implicitly make assumptions about the structure of the clusters, the distribution of the data, and so on, and thus only a few are general enough to be applicable for a variety of data types.

Figures 20.13 and 20.14 present some sample results obtained using this tool. To maintain consistency, we use the same data set [19] that we used to demonstrate the functionalities of the integrated clustering tool. The user starts by selecting the data set that needs to be analyzed and then a cluster validity index along with the clustering algorithm and the distance measure to be used. Figure 20.13 is the result obtained for Davies–Bouldin (DB) [12] index value on the clustering of above-mentioned data set using K-means algorithm and the Euclidean distance measure. The tool automatically runs the clustering algorithm 200 times for each number of clusters. Thus the index value plotted is the median index value observed for the corresponding number of clusters. Since lower values of DB index indicate better partitions, the user can conclude that optimal number of clusters is three.

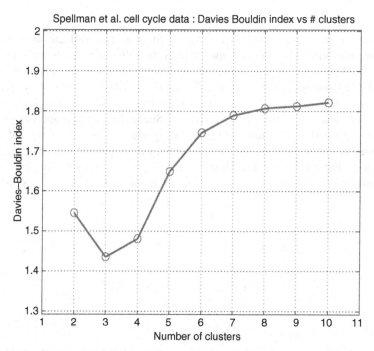

Figure 20.13 Sample cluster validation result: Stage 1—index value versus number of clusters.

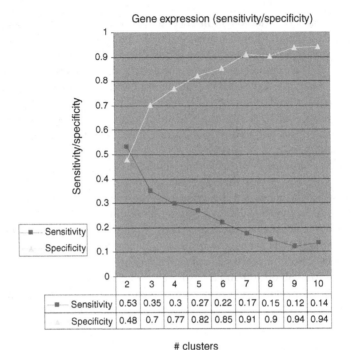

	2	3	4	5	6	7	8	9	10
Sensitivity	0.53	0.35	0.3	0.27	0.22	0.17	0.15	0.12	0.14
Specificity	0.48	0.7	0.77	0.82	0.85	0.91	0.9	0.94	0.94

clusters

Figure 20.14   Sample cluster validation result: Stage 2—comparison with reference clustering.

Based on the results obtained in stage 1, the user can evaluate them using a reference clustering. For this, the user selects the reference clustering to be used. Figure 20.14 presents the results obtained by "comparing" the previous results using the MIPS [24] database as the standard. The result depicts the sensitivity–specificity curves with respect to all pairs of genes. Sensitivity is defined as the proportion of "true pairs" detected by the cluster algorithm. Specificity is the proportion of "true nonpairs" correctly identified by the cluster algorithm. The sensitivity and specificity may be defined in terms of the contingency matrix as shown in Fig. 20.15. In most cases, there is a trade-off between high sensitivity and specificity as can be observed in Fig. 20.14, which shows the sensitivity–specificity curves for gene expression data.

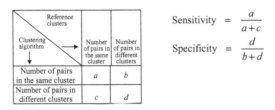

$$\text{Sensitivity} = \frac{a}{a+c}$$

$$\text{Specificity} = \frac{d}{b+d}$$

Figure 20.15   Contingency matrix.

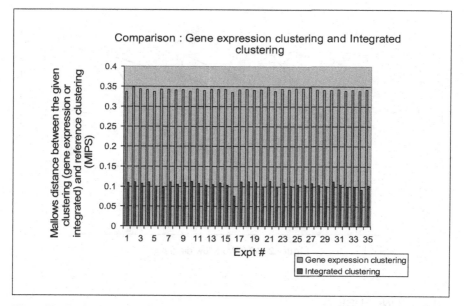

Figure 20.16   Comparative analysis of integrated clustering and clustering gene expression data.

Finally, we justify the use of information fusion-based algorithms to cluster gene expression data through analysis as shown in Fig. 20.16. The figure plots the Mallows index proposed in Ref. [23] for clustering based on 1. Gene expression data only 2. Gene expression data and sequence data. Multiple sets of 500 genes were obtained by sampling with replacement, and the procedure was repeated to assess the statistical significance of the result. It was observed that in every sample the integrated clusters had a better score (lower distance between the integrated clusters and the MIPS clusters as compared to that between expression clusters and MIPS clusters) as shown in Figure 20.16. This suggests that including multiple data types could indeed produce more functionally relevant clusters.

20.3.5 WorkFlow

The suite of tools presented so far provide an environment that helps the user perform information fusion on various biomedical informatics data sets. The results obtained from each tool could be pipelined to others to perform a comprehensive fusion study. Figure 20.17 provides a workflow diagram for using the platform on patient demographic, clinical, gene expression and other data sets. The multidimensional analysis tool can be used to explore the data sets, and the resulting "cube" information can be fed as input to the correspondence analysis and integrated clustering tools. The results obtained from the integrated clustering tool and possibly other clustering mechanisms can be verified using the cluster validation tool.

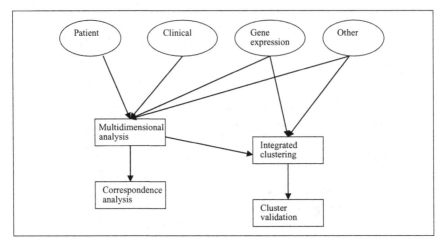

Figure 20.17   Workflow diagram.

20.4   CONCLUSION

In this chapter, we present an online platform for performing exploratory analysis of clinical and biological data sets. The goal is to demonstrate the significance of information fusion-based tools for biomedical informatics research that could take advantage of the nationwide data sharing infrastructures such as grids. The current platform serves two purposes: (1) as a data warehouse for various data sets involved in biomedical informatics studies and (2) to provide and demonstrate a set of information fusion tools for disease research. Our future work involves extending the system to a distributed environment and automated discovery of interesting patterns. This project is funded, in part, under a grant with the Pennsylvania Department of Health using tobacco settlement funds.

REFERENCES

1. Fenstermacher, D., Street, C., McSherry, T., Nayak, V., Overby, C., and Feldman, M. The cancer biomedical informatics grid. *Proceedings of IEEE Engineering in Medicine and Biology*, September 1–4, 2005, Shanghai, China.
2. Falkman, G. Information visualisation in clinical Odontology: multidimensional analysis and interactive data exploration. *Artificial Intelligence in Medicine*, 22: 133–158, 2001.
3. Shahar et al. Interactive visualization and exploration of time-oriented clinical data using a distributed temporal-abstraction architecture. *Proceedings of the AMIA Annual Symposiun*, 1004, 2003.
4. Granzow, M., Berrar, D., Dubitzky, W., Schuster, A., Azuaje, F., and Eils, R. Tumor identification by gene expression profiles: a comparison of five different clustering methods. *ACM-SIGBIO Newsletter*, 21, 16–22, 2001.

5. Mullins, I. M., Siadaty, M. S., Lyman, J., Scully, K., Garrett, C.T., et al. Data mining and clinical data repositories: Insights from a 667,000 patient data set. *Computer in Biology and Medicine*, 36(12): 1351–1377, 2005.

6. Holmes, I. and Bruno, W. J. Finding Regulatory Elements Using Joint Likelihoods for Sequence and Expression Profile Data, *ISMB*, 8: 202–210, 2000.

7. Kasturi, J. Acharya, R. Clustering Diverse Genomic Data using Information Fusion. *Proceedings of ACM Symposium on Applied Computing*, pp. 116–120, March 2004.

8. Singh, D., et al. Gene Expression correlates of clinical prostate cancer behaviour, *Cancer Cell*, 1(2): 203–209, 2002.

9. Dhanasekaran, S. M., Barrette, T. R., Ghosh, D., Shah, R., Varambally, S., Kurachi, K., Pienta, K. J., Rubin, M.A., and Chinnaiyan, A. M. Delineation of prognostic biomarkers in prostate cancer. *Nature*, 412, 822–826, 2001.

10. Calinski, R. and Harabasz, J. A dendrite method for cluster analysis. *Communications in Statistics*, 3: 1–27, 1974.

11. Dunn, J. C. Well separated clusters and optimal fuzzy partitions. *Journal of Cybernetics*, 4: 95–104, 1974.

12. Davies, D. L. and Bouldin, D. W. A cluster separation measure. *IEEE Transactions on Pattern Analysis and Machine Intelligence*, 1(2): 224–227, 1979.

13. Greenacre, M. J. *Correspondence Analysis in Practice*, Academic Press, London, 1993.

14. Gray, J., et al. Data cube: a relational aggregation operator generalizing group-by, cross-tab, and sub totals. *Data Mining and Knowledge Discovery*, 1(1): 29–53, 1997.

15. Agrawal, R., Gupta, A., and Sarawagi, S. Modeling multidimensional databases, *Proceedings of 13th International Conference on Data Engineering*, 1995.

16. Tao, Y., Friedman, C., and Lussier, Y. A. Visualizing information across multidimensional post-genomic structured and textual databases, *Bioinformatics*, 21(8): 1659–1667, 2004.

17. Kohonen, T. *Self-Organizing Maps, Springer Series in Information Sciences*, Springer, Berlin, 1995.

18. Kasturi, J., Acharya, R., and Ramanathan, M. An information theoretic approach for analyzing temporal patterns of gene expression. *Bioinformatics*, 19, 449–458, 2003.

19. Spellman, P. T., Sherlock, G., Zhang, M. Q., Iyer, V. R., Eisen, M. B., Brown, P. O., Botstein, D., and Futcher, B. Comprehensive identification of cell cycle-regulated genes of the yeast *Saccharomyces cerevisiae* by microarray hybridization. *Molecular Biology of the Cell*, 9: 3273–3297, 1998.

20. Hartigan, J. A. Statistical theory in clustering. *Journal of Classification*, 2: 63–76, 1985.

21. Meila, M. *Comparing Clusterings*, Department of Statistics, University of Washington, 2002.

22. Rand, W. M. Objective criteria for the evaluation of clustering methods. *Journal of the American Statistical Association*, 66: 846–850, 1971.

23. Zhou, D., Li, J., and Zha, H. A new Mallows distance based metric for comparing clusterings. *Proceedings of the International Conference on Machine Learning (ICML)*, Bonn, Germany, 2005.

24. MIPS Comprehensive Yeast Genome Database. Available at http://mips.gsf.de/genre/proj/yeast.

INDEX

Acharya Raj, 431
Adaptive classifiers, 82
Adaptive-network-based fuzzy inference
    system (ANFIS), 125
Adaptive nonlinear kernel discriminant
    analysis, 70
Ahmad Shandar, 157
Araúzo-Bravo Marcos J., 157
Artificial neural networks, 90

Backpropagation algorithm, 125
Backward elimination, 12–14
Bagging, 135
Bayesian inference, 377–378
Bayesian model, 90, 357
Bayesian network, 351–352, 381
Bayes method, 176
Bidirectional promoters, 322
Binding sites, 166
Biological networks, 3
Biological pathways, 329
Biomarkers, 3, 443
Biomedical informatics, 431
Black-box methods, 112
Boosted genetic programming, 361
Boosting, 135, 309
Bootstrapping approach, 89–90
Bootstrapping consistency gene selection, 95
Bootstrapping consistency method, 91
Breast cancer data, 103–106
Bu Dongbo, 189

CaBIG (Cancer Bioinformatics Grid), 431
Cancer, 47–49, 90, 96, 98, 113, 368, 433, 438

Chanda Pritam, 389
Cheminformatics, 413
Chiang Hsiao-Dong, 263
Cholesky decomposition, 70
Classification, 8, 90
Classification consistency, 92–94
Clinical data, 433
Clustering algorithms, 434
Cluster kernels, 222
Coexpressed genes, 3
Colon cancer data set, 116
Common genes, 93
Complex diseases, 368
Composite kernels, 218
Computational genetic analysis, 368
Computational haplotype
    analysis, 373
Consensus model, 265
Consensus prediction, 191–194
Conserved motifs, 287
Consistency modeling, 89
Correspondence analysis, 440
Coulter counter, 414
Cross-validation, 5

Data-adaptive method (DA), 93
Data exploration, 432
Data mining, 2, 283, 292, 432
Data overfitting, 3
Data warehousing, 435
Decision forests, 135
Decision trees (DTs), 112, 323–324
Dimension reduction, 3
Disease association, 390

Machine Learning in Bioinformatics. Edited by Yan-Qing Zhang and Jagath C. Rajapakse
Copyright © 2009 by John Wiley & Sons, Inc.

Disease-gene association, 368–369
Disease–SNP association, 390
Disulfide bridge prediction, 224
DNA-binding sites, 166
DNA samples, 113
DNA sequences, 23, 243, 264, 284, 303, 322
Domain-specific knowledge, 241
Dynamic programming, 221

Eigenvectors, 271
Elastic net, 55
Elnitski Laura L., 321
EM algorithm, 265–280
Energy minimization method, 176
Ensemble learning, 126, 135
Entropy-based approach, 398
Eukaryotic protein, 5
Evolutionary algorithm, 90, 339
Evolutionary computations, 297, 358
Evolutionary granular kernel trees, 230
Expectation-Maximization Method, 375

Feature dimension, 3–4
Feature extraction, 306, 418
Feature granules, 230
Feature reduction, 3
Feature representation, 1–3
Feature selection, 1–2, 135, 250, 309, 392
Feature spaces, 3
Filter method, 14
Fisher-based method, 5
Fisher discriminant analysis, 8, 69
Fisher discriminant ratio, 16
Fisher kernels, 219
Fisher's linear discriminate, 90
F-ratio, 49–50,
Forward/backward probabilities, 419
Forward selection, 11
Fuzzy association rules, 131
Fuzzy-based classifiers, 115
Fuzzy-based gene selection approach, 119
Fuzzy clustering methods, 119
Fuzzy C-mean clustering (FCC)
    method, 120–122
Fuzzy decision surfaces, 130
Fuzzy k-nearest neighbors (k-NN)
    algorithm, 230
Fuzzy rule-based models, 112
Fuzzy rules, 112, 125
Fuzzy systems, 112, 125

Gao Xin, 189
Ganta Srivatsava R., 431
Gaussian radial basis function, 74
Genbank, 313–314
Gene expression data, 241, 443–444
Gene profiling, 2
Gene selection, 2, 89, 91
Generic tagging problem, 392
Genetic algorithm, 136, 232, 277, 344
Genetic algorithm-based feature
    selection, 146
Genetic information, 242
Genetic programming, 344
Genomic data, 1, 2, 5, 284, 431
Genomic data mining, 2
Genomic features, 3
Genomics, 389
Gibbs sampling, 265, 377
Granular computing (GrC), 231
Granular feature transformation, 230–232
Granular kernels, 232
Graphical models, 351
Grid systems, 431
Groupwise information, 11
Gubbi Jayavardhana, 209

Haplotype phasing methods, 378
Haplotype phasing problem, 374
Havukkala Ilkka, 89
Hessian matrix, 271
Heterogeneous data sets, 431
Heuristic multitask learning, 146
Heuristic search methods, 145
Hidden Markov models (HMM), 219,
    245, 352, 406, 417
High dimensionality, 3
Ho L. S., 283
Hu Yingjie, 89
Human Genome Project (HGP), 284
Human genome sequence, 301

Information fusion, 432
Informative genes, 90
Integrated clustering tool, 444
Interfeature relationship, 9
Irrelevant genes, 3

Jin Bo, 229

Kasabov Nikola, 89
Kasturi Jyotsna, 431

Kernel discriminant analysis, 79
Kernel function, 29, 72, 231
Kernel matrix, 75–78
Kernel method, 72, 209, 339, 345
Kim Hyunsoo, 69
King David C., 321
K-means, 434
K-nearest neighbor (KNN) method, 112
Knowledge discovery, 283
Kullback–Leibler divergence, 303–317, 358
Kung Sun-Yuan, 1

Lee Phil H., 367
Leukemia cancer data set, 116
Leukemia data set, 5
Leukemia gene expression data set, 82
Li Guo-Zheng, 135
Li Ming, 189
Li Shuai Cheng, 189
Liao Li, 241
Linear combination features, 22
Linear kernels, 216
Linguistic terms, 112
Linguistic values, 119
Long-range features, 252
Lung cancer data, 103–106
Lymphoma cancer data set, 116

Machine learning, 2
Majority-voting method, 193
Mak Man-Wai, 1
Markov chain models, 287
Markov chain monte carlo, 377
Markov models, 283
Markov/neural hybrid approach, 286
MATLAB software, 66
Matrix decompositions, 70
Membrane proteins, 246
Menjoge Rajiv S., 47
Mercer's kernel, 70
Microarray data, 2, 107, 122
Microarray experiments, 111–113
MicroRNAs (miRNAs), 339
MiRNA precursors, 344
Mismatch kernels, 220
Motif discovery problem, 263
Motif finding algorithms, 263–264
Motif refinement, 264
Multiclassification problem, 233
Multidimensional analysis, 438

Multiple linear regression (MLR), 171
Multitask learning, 145

Naive Bayes classifier, 355
Nam Jin-Wu, 339
Narasimhamurthy Anand, 431
Nearest neighbor method, 26, 49
Neural fuzzy ensemble method, 90
Neural networks, 136, 172, 194, 283, 302, 403
Neuro-fuzzy ensemble (NFE) model, 113
Noise sampling method, 90
Nonpromoters, 306
Nonrandom association, 370
Nuclear magnetic resonance spectroscopy (NMR), 189

Oncogenes, 339
One-versus-one voting approach, 234
Optimal feature selection, 3
Orthogonal encoding, 289
Overfitting, 16
Overtraining, 4

Pairwise association-based methods, 380
Pairwise feature information, 9
Pairwise interaction energy, 192
Pairwise scoring kernels, 29
Palade Vasile, 111
Palaniswami Marimuthu, 209
Pang Shaoning, 89
Parameterized genetic programming (PGP) method, 359
Park Haesun, 69
Partial least squares algorithm, 136
Polynomial kernels, 216
Position-specific scoring matrix (PSSM), 267
Principal component analysis (PCA), 90, 395, 442
Probabilistic graphical models, 339
Probabilistic models, 285
Profile alignment support vector machines, 5
Profile-based kernels, 222
Promoter, 284, 301, 322
Promoter prediction algorithm, 306
Protein-coding regions prediction, 301–303
Protein data bank (PDB), 246
Protein–DNA interactions, 166, 284
Protein folding, 205
Protein–protein interactions, 168
Protein sequence, 2, 209

456 INDEX

Protein structure, 157, 189, 210
Protein structure prediction, 217
Proteomics, 389

QR decomposition, 70
Quaternary structure, 212

Radial basis function (RBF), 80–81, 344
Rajapakse Jagath C., 283
Ramanathan Murali, 389
Random features, 138
Random forests, 50
Random number, 122
Random variable, 120
Receiver operating characteristics
   (ROC), 294–296, 333–335, 350
Recursive feature elimination (RFE), 26
Recursive SVM, 26
Reddy Chandan K., 263
Redundancy, 8, 9
Regression-based methods, 394
RNA secondary structure, 253
Rule-based systems, 125

Sarai Akinori, 157
Secondary structure of proteins,
   159–162, 211
Secondary structure prediction,
   159–170, 197
Sequence analysis, 244
Shatkay Hagit, 367
Shilton Alistair, 209
Shrunken centroids, 54
Signal processing architecture, 417
Signal-to-noise ratio, 2
Signed-FDR, 15
Significance analysis of microarrays, 90
Singh Yumlembam Hemjit, 157
Single nucleotide polymorphisms (SNPs),
   367, 389–410
Stochastic context-free grammars (SCFG),
   252
Subcellular localization, 5, 36
Subcellular locations of proteins, 229
Suboptimal search, 4
Supervised feature selection, 14
Supervised learning algorithm, 309
Supervised learning methods, 344

Supervised selection criteria, 14
   closed-loop approach, 14
   open-loop approach, 14
Support vectors, 31, 70
Support Vector Machines (SVM), 23, 25, 28,
   52, 70, 173, 195, 210, 231, 361, 405, 420
   based wrapper methods, 25
   RFE, 26
Symmetric divergence, 17
Systems biology, 389

Tag SNP selection problem, 379
T-Statistics, 16, 49, 90,
Three-dimensional structure, 166, 303
Three-dimensional (3D) structure
   of a protein, 189
Three-party problem, 12
Time-delay neural network (TDNN), 287
Top-ranked genes, 92
Traditional haplotype analysis, 373
Tumor suppressors, 339

Unsupervised methods, 111

Variable order markov Model (VOMM), 331
Visualization, 59, 61
Viterbi algorithm, 418

Wang Zhenyu, 111
Wavelet analysis, 417
Weighted Voting, 22
Welsch Roy E., 47
Weng Yao-Chung, 263
Winters-Hilt Stephen, 413
Wrapper methods, 117
Wu Shuanhu, 301

Xie Xudong, 301
Xu Jinbo, 189

Yan Hong, 301
Yang Jack Y., 135
Yang Mary Q., 321
Yu Libo, 189

Zhang Aidong, 389
Zhang Byoung-Tak, 339
Zhang Yan-Qing, 229